TABLE A.2 Cumulative normal distribution (continued)

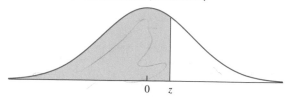

z	0.00	0.01	0.02	0.03	0.04	0.05	0.06	0.07	0.08	0.09
0.0	.5000	.5040	.5080	.5120	.5160	.5199	.5239	.5279	.5319	.5359
0.1	.5398	.5438	.5478	.5517	.5557	.5596	.5636	.5675	.5714	.5753
0.2	.5793	.5832	.5871	.5910	.5948	.5987	.6026	.6064	.6103	.6141
0.3	.6179	.6217	.6255	.6293	.6331	.6368	.6406	.6443	.6480	.6517
0.4	.6554	.6591	.6628	.6664	.6700	.6736	.6772	.6808	.6844	.6879
0.5	.6915	.6950	.6985	.7019	.7054	.7088	.7123	.7157	.7190	.7224
0.6	.7257	.7291	.7324	.7357	.7389	.7422	.7454	.7486	.7517	.7549
0.7	.7580	.7611	.7642	.7673	.7704	.7734	.7764	.7794	.7823	.7852
0.8	.7881	.7910	.7939	.7967	.7995	.8023	.8051	.8078	.8106	.8133
0.9	.8159	.8186	.8212	.8238	.8264	.8289	.8315	.8340	.8365	.8389
1.0	.8413	.8438	.8461	.8485	.8508	.8531	.8554	.8577	.8599	.8621
1.1	.8643	.8665	.8686	.8708	.8729	.8749	.8770	.8790	.8810	.8830
1.2	.8849	.8869	.8888	.8907	.8925	.8944	.8962	.8980	.8997	.9015
1.3	.9032	.9049	.9066	.9082	.9099	.9115	.9131	.9147	.9162	.9177
1.4	.9192	.9207	.9222	.9236	.9251	.9265	.9279	.9292	.9306	.9319
1.5	.9332	.9345	.9357	.9370	.9382	.9394	.9406	.9418	.9429	.9441
1.6	.9452	.9463	.9474	.9484	.9495	.9505	.9515	.9525	.9535	.9545
1.7	.9554	.9564	.9573	.9582	.9591	.9599	.9608	.9616	.9625	.9633
1.8	.9641	.9649	.9656	.9664	.9671	.9678	.9686	.9693	.9699	.9706
1.9	.9713	.9719	.9726	.9732	.9738	.9744	.9750	.9756	.9761	.9767
2.0	.9772	.9778	.9783	.9788	.9793	.9798	.9803	.9808	.9812	.9817
2.1	.9821	.9826	.9830	.9834	.9838	.9842	.9846	.9850	.9854	.9857
2.2	.9861	.9864	.9868	.9871	.9875	.9878	.9881	.9884	.9887	.9890
2.3	.9893	.9896	.9898	.9901	.9904	.9906	.9909	.9911	.9913	.9916
2.4	.9918	.9920	.9922	.9925	.9927	.9929	.9931	.9932	.9934	.9936
2.5	.9938	.9940	.9941	.9943	.9945	.9946	.9948	.9949	.9951	.9952
2.6	.9953	.9955	.9956	.9957	.9959	.9960	.9961	.9962	.9963	.9964
2.7	.9965	.9966	.9967	.9968	.9969	.9970	.9971	.9972	.9973	.9974
2.8	.9974	.9975	.9976	.9977	.9977	.9978	.9979	.9979	.9980	.9981
2.9	.9981	.9982	.9982	.9983	.9984	.9984	.9985	.9985	.9986	.9986
3.0	.9987	.9987	.9987	.9988	.9988	.9989	.9989	.9989	.9990	.9990
3.1	.9990	.9991	.9991	.9991	.9992	.9992	.9992	.9992	.9993	.9993
3.2	.9993	.9993	.9994	.9994	.9994	.9994	.9994	.9995	.9995	.9995
3.3	.9995	.9995	.9995	.9996	.9996	.9996	.9996	.9996	.9996	.9997
3.4	.9997	.9997	.9997	.9997	.9997	.9997	.9997	.9997	.9997	.9998
3.5	.9998	.9998	.9998	.9998	.9998	.9998	.9998	.9998	.9998	.9998
3.6	.9998	.9998	.9999	.9999	.9999	.9999	.9999	.9999	.9999	.9999

Principles of Statistics for Engineers and Scientists

William Navidi

 **Higher Education**

Boston   Burr Ridge, IL   Dubuque, IA   New York   San Francisco   St. Louis
Bangkok   Bogotá   Caracas   Kuala Lumpur   Lisbon   London   Madrid   Mexico City
Milan   Montreal   New Delhi   Santiago   Seoul   Singapore   Sydney   Taipei   Toronto

PRINCIPLES OF STATISTICS FOR ENGINEERS AND SCIENTISTS

Published by McGraw-Hill, a business unit of The McGraw-Hill Companies, Inc., 1221 Avenue of the Americas, New York, NY 10020. Copyright © 2010 by The McGraw-Hill Companies, Inc. All rights reserved. No part of this publication may be reproduced or distributed in any form or by any means, or stored in a database or retrieval system, without the prior written consent of The McGraw-Hill Companies, Inc., including, but not limited to, in any network or other electronic storage or transmission, or broadcast for distance learning.

Some ancillaries, including electronic and print components, may not be available to customers outside the United States.

This book is printed on acid-free paper.

1 2 3 4 5 6 7 8 9 0 DOC/DOC 0 9

ISBN 978–0–07–337634–9
MHID 0–07–337634–5

Global Publisher: *Raghothaman Srinivasan*
Sponsoring Editor: *Debra B. Hash*
Director of Development: *Kristine Tibbetts*
Developmental Editor: *Lora Neyens*
Senior Marketing Manager: *Curt Reynolds*
Senior Project Manager: *April R. Southwood*
Lead Production Supervisor: *Sandy Ludovissy*
Associate Design Coordinator: *Brenda A. Rolwes*
Cover Designer: *Studio Montage, St. Louis, Missouri*
Compositor: *Macmillan Publishing Solutions*
Typeface: *10.5/12 Times Roman*
Printer: *R. R. Donnelley Crawfordsville, IN*

Library of Congress Cataloging-in-Publication Data

Navidi, William Cyrus.
 Principles of statistics for engineers and scientists / William Navidi. -- 1st ed.
   p. cm.
 Includes bibliographical references and index.
  ISBN 978–0–07–337634–9 — ISBN 0–07–337634–5 (hard copy : alk. paper)   1. Mathematical statistics –
Textbooks. 2. Science – Statistical methods – Textbooks. 3. Engineering – Statistical methods – Textbooks. I. Title.
 QA276.4.N379 2010
 519.5–dc22

                                                                                    2008049028

www.mhhe.com

To Catherine, Sarah, and Thomas

ABOUT THE AUTHOR

William Navidi is Professor of Mathematical and Computer Sciences at the Colorado School of Mines. He received the B.A. degree in mathematics from New College, the M.A. in mathematics from Michigan State University, and the Ph.D. in statistics from the University of California at Berkeley. Professor Navidi has authored more than 50 research papers both in statistical theory and in a wide variety of applications including computer networks, epidemiology, molecular biology, chemical engineering, and geophysics.

CONTENTS

Preface   vii

Chapter 1
Summarizing Univariate Data   1

Introduction   1

1.1   Sampling   3

1.2   Summary Statistics   11

1.3   Graphical Summaries   20

Chapter 2
Summarizing Bivariate Data   37

Introduction   37

2.1   The Correlation Coefficient   37

2.2   The Least-Squares Line   49

2.3   Features and Limitations of the
Least-Squares Line   56

Chapter 3
Probability   66

Introduction   66

3.1   Basic Ideas   66

3.2   Conditional Probability and
Independence   74

3.3   Random Variables   84

3.4   Functions of Random Variables   104

Chapter 4
Commonly Used Distributions   119

Introduction   119

4.1   The Binomial Distribution   119

4.2   The Poisson Distribution   127

4.3   The Normal Distribution   134

4.4   The Lognormal Distribution   143

4.5   The Exponential Distribution   147

4.6   Some Other Continuous
Distributions   151

4.7   Probability Plots   157

4.8   The Central Limit Theorem   161

Chapter 5
**Point and Interval Estimation for
a Single Sample   172**

Introduction   172

5.1   Point Estimation   173

5.2   Large-Sample Confidence Intervals for
a Population Mean   176

5.3   Confidence Intervals for
Proportions   189

5.4   Small-Sample Confidence Intervals for
a Population Mean   195

5.5   Prediction Intervals and Tolerance
Intervals   204

Chapter 6
Hypothesis Tests for a Single Sample   212

Introduction   212

6.1   Large-Sample Tests for a Population
Mean   212

6.2   Drawing Conclusions from the Results
of Hypothesis Tests   221

6.3   Tests for a Population Proportion   229

6.4   Small-Sample Tests for a Population
Mean   233

6.5   The Chi-Square Test   239

6.6   Fixed-Level Testing   248

6.7 Power   253

6.8 Multiple Tests   262

Chapter 7
Inferences for Two Samples   268

   Introduction   268

7.1 Large-Sample Inferences on the Difference Between Two Population Means   268

7.2 Inferences on the Difference Between Two Proportions   276

7.3 Small-Sample Inferences on the Difference Between Two Means   284

7.4 Inferences Using Paired Data   294

7.5 The F Test for Equality of Variance   303

Chapter 8
Inference in Linear Models   312

   Introduction   312

8.1 Inferences Using the Least-Squares Coefficients   313

8.2 Checking Assumptions   335

8.3 Multiple Regression   345

8.4 Model Selection   362

Chapter 9
Factorial Experiments   396

   Introduction   396

9.1 One-Factor Experiments   396

9.2 Pairwise Comparisons in One-Factor Experiments   415

9.3 Two-Factor Experiments   421

9.4 Randomized Complete Block Designs   441

9.5 2^p Factorial Experiments   448

Chapter 10
Statistical Quality Control   477

   Introduction   477

10.1 Basic Ideas   477

10.2 Control Charts for Variables   480

10.3 Control Charts for Attributes   499

10.4 The CUSUM Chart   504

10.5 Process Capability   507

Appendix A: Tables   514

Appendix B: Bibliography   537

Answers to Selected Exercises   540

Index   574

PREFACE

MOTIVATION

This book is based on the author's more comprehensive text *Statistics for Engineers and Scientists*, 2nd edition (McGraw-Hill, 2008), which is used for both one- and two-semester courses. The key concepts from that book form the basis for this text, which is designed for a one-semester course. The emphasis is on statistical methods and how they can be applied to problems in science and engineering, rather than on theory. While the fundamental principles of statistics are common to all disciplines, students in science and engineering learn best from examples that present important ideas in realistic settings. Accordingly, the book contains many examples that feature real, contemporary data sets, both to motivate students and to show connections to industry and scientific research. As the text emphasizes applications rather than theory, the mathematical level is appropriately modest. Most of the book will be mathematically accessible to those whose background includes one semester of calculus.

COMPUTER USE

Over the past 30 years, the development of fast and cheap computing has revolutionized statistical practice; indeed, this is one of the main reasons that statistical methods have been penetrating ever more deeply into scientific work. Scientists and engineers today must not only be adept with computer software packages; they must also have the skill to draw conclusions from computer output and to state those conclusions in words. Accordingly, the book contains exercises and examples that involve interpreting, as well as generating, computer output, especially in the chapters on linear models and factorial experiments. Many instructors integrate the use of statistical software into their courses; this book may be used effectively with any package.

CONTENT

Chapter 1 covers sampling and descriptive statistics. The reason that statistical methods work is that samples, when properly drawn, are likely to resemble their populations. Therefore Chapter 1 begins by describing some ways to draw valid samples. The second part of the chapter discusses descriptive statistics for univariate data.

Chapter 2 presents descriptive statistics for bivariate data. The correlation coefficient and least-squares line are discussed. The discussion emphasizes that linear models are appropriate only when the relationship between the variables is linear, and it describes the effects of outliers and influential points. Placing this chapter early enables instructors to present some coverage of these topics in courses where there is not enough time for a full treatment from an inferential point of view. Alternatively, this chapter may be postponed and covered just before the inferential procedures for linear models in Chapter 8.

Chapter 3 is about probability. The goal here is to present the essential ideas without a lot of mathematical derivations. I have attempted to illustrate each result with an example or two, in a scientific context where possible, to present the intuition behind the result.

Chapter 4 presents many of the probability distribution functions commonly used in practice. Probability plots and the Central Limit Theorem are also covered. Only the normal and binomial distribution are used extensively in the remainder of the text, instructors may choose which of the other distributions to cover.

Chapters 5 and 6 cover one-sample methods for confidence intervals and hypothesis testing, respectively. Point estimation is covered as well, in Chapter 5. The *P*-value approach to hypothesis testing is emphasized, but fixed-level testing and power calculations are also covered. A discussion of the multiple testing problem is also presented.

Chapter 7 presents two-sample methods for confidence intervals and hypothesis testing. There is often not enough time to cover as many of these methods as one would like; instructors who are pressed for time may choose which of the methods they wish to cover.

Chapter 8 covers inferential methods in linear regression. In practice, scatterplots often exhibit curvature or contain influential points. Therefore this chapter includes material on checking model assumptions and transforming variables. In the coverage of multiple regression, model selection methods are given particular emphasis, because choosing the variables to include in a model is an essential step in many real-life analyses.

Chapter 9 discusses some commonly used experimental designs and the methods by which their data are analyzed. One-way and two-way analysis of variance methods, along with randomized complete block designs and 2^p factorial designs, are covered fairly extensively.

Chapter 10 presents the topic of statistical quality control, covering control charts, CUSUM charts, and process capability, and concluding with a brief discussion of six-sigma quality.

RECOMMENDED COVERAGE

The book contains enough material for a one-semester course meeting four hours per week. For a three-hour course, it will probably be necessary to make some choices about coverage. One option is to cover the first three chapters, going lightly over the last two sections of Chapter 3, then cover the binomial, Poisson, and normal distributions in Chapter 4, along with the Central Limit Theorem. One can then cover the confidence intervals and hypothesis tests in Chapters 5 and 6, and finish either with the two-sample procedures in Chapter 7 or by covering as much of the material on inferential methods in regression in Chapter 8 as time permits.

For a course that puts more emphasis on regression and factorial experiments, one can go quickly over the power calculations and multiple testing procedures, and cover Chapters 8 and 9 immediately following Chapter 6. Alternatively, one could substitute Chapter 10 on statistical quality control for Chapter 9.

THE ARIS COURSE MANAGEMENT SYSTEM

The ARIS (Assessment, Review, and Instruction System) online course management system is available to instructors who adopt this text. With ARIS, instructors can assign and grade text-based homework within a versatile homework management system. In addition, ARIS contains algorithmic problems for student practice, along with Java applets that allow students to interactively explore ideas in the text. Customizable PowerPoint lecture notes for each chapter are available as well, along with additional tools and resources such as a guide to simulation in MINITAB, suggested syllabi, and other features. More information can be found at www.mhhe.com/navidi.

ELECTRONIC TEXTBOOK OPTION

This text may be purchased in electronic form through an online resource known as CourseSmart. Students can access the complete text online through their browsers at approximately one-half the cost of a traditional text. In addition, purchasing the eTextbook allows students to use CourseSmart's web tools, which include full text search, notes and highlighting, and email tools for sharing notes between classmates. More information can be found at www.CourseSmart.com.

ACKNOWLEDGMENTS

I am indebted to many people for contributions at every stage of development. I received many valuable suggestions from my colleagues Barbara Moskal, Gus Greivel, Ashlyn Hutchinson, and Melissa Laeser at the Colorado School of Mines. Mike Colagrosso of the School of Mines developed some excellent applets. I am particularly grateful to Jackie Miller of The Ohio State University, who read the entire manuscript, found many errors, and made many valuable suggestions for improvement.

The staff at McGraw-Hill has been extremely capable and supportive. In particular, I would like to express thanks to my sponsoring editor Michael Hackett and developmental editor Lora Neyens for their patience and guidance in the preparation of this book.

William Navidi

Summarizing Univariate Data

Introduction

Advances in science and engineering occur in large part through the collection and analysis of data. Proper analysis of data is challenging, because scientific data are subject to random variation. That is, when scientific measurements are repeated, they come out somewhat differently each time. This poses a problem: How can one draw conclusions from the results of an experiment when those results could have come out differently? To address this question, a knowledge of statistics is essential. The methods of statistics allow scientists and engineers to design valid experiments and to draw reliable conclusions from the data they produce.

While our emphasis in this book is on the applications of statistics to science and engineering, it is worth mentioning that the analysis and interpretation of data are playing an ever-increasing role in all aspects of modern life. For better or worse, huge amounts of data are collected about our opinions and our lifestyles, for purposes ranging from the creation of more effective marketing campaigns to the development of social policies designed to improve our way of life. On almost any given day, newspaper articles are published that purport to explain social or economic trends through the analysis of data. A basic knowledge of statistics is therefore necessary not only to be an effective scientist or engineer, but also to be a well-informed member of society.

The Basic Idea

The basic idea behind all statistical methods of data analysis is to make inferences about a population by studying a relatively small sample chosen from it. As an illustration, consider a machine that makes steel balls for ball bearings used in clutch systems. The specification for the diameter of the balls is 0.65 ± 0.03 cm. During the last hour, the machine has made 2000 balls. The quality engineer wants to know approximately how many of these balls meet the specification. He does not have time to measure all 2000 balls. So he draws a random sample of 80 balls, measures them, and finds that 72 of them (90%) meet the diameter specification. Now, it is unlikely that the sample of 80 balls represents the population of 2000 perfectly. The proportion of good balls in the population is likely to differ somewhat from the sample proportion of 90%. What the engineer needs to know is just how large that difference is likely to be. For example, is it plausible that the population percentage could be as high as 95%? 98%? As low as 85%? 80%?

Here are some specific questions that the engineer might need to answer on the basis of these sample data:

1. The engineer needs to compute a rough estimate of the likely size of the difference between the sample proportion and the population proportion. How large is a typical difference for this kind of sample?

2. The quality engineer needs to note in a logbook the percentage of acceptable balls manufactured in the last hour. Having observed that 90% of the sample balls were good, he will indicate the percentage of acceptable balls in the population as an interval of the form $90\% \pm x\%$, where x is a number calculated to provide reasonable certainty that the true population percentage is in the interval. How should x be calculated?

3. The engineer wants to be fairly certain that the percentage of good balls is at least 85%; otherwise, he will shut down the process for recalibration. How certain can he be that at least 85% of the 1000 balls are good?

Much of this book is devoted to addressing questions like these. The first of these questions requires the computation of a **standard deviation**, which we will discuss in Chapter 3. The second question requires the construction of a **confidence interval**, which we will learn about in Chapter 5. The third calls for a **hypothesis test**, which we will study in Chapter 6.

The remaining chapters in the book cover other important topics. For example, the engineer in our example may want to know how the amount of carbon in the steel balls is related to their compressive strength. Issues like this can be addressed with the methods of **correlation** and **regression**, which are covered in Chapters 2 and 8. It may also be important to determine how to adjust the manufacturing process with regard to several factors, in order to produce optimal results. This requires the design of **factorial experiments**, which are discussed in Chapter 9. Finally, the engineer will need to develop a plan for monitoring the quality of the product manufactured by the process. Chapter 10 covers the topic of **statistical quality control**, in which statistical methods are used to maintain quality in an industrial setting.

The topics listed here concern methods of drawing conclusions from data. These methods form the field of **inferential statistics**. Before we discuss these topics, we must first learn more about methods of collecting data and of summarizing clearly the basic information they contain. These are the topics of **sampling** and **descriptive statistics**, and they are covered in the rest of this chapter.

1.1 Sampling

As mentioned, statistical methods are based on the idea of analyzing a **sample** drawn from a **population**. For this idea to work, the sample must be chosen in an appropriate way. For example, let us say that we wished to study the heights of students at the Colorado School of Mines by measuring a sample of 100 students. How should we choose the 100 students to measure? Some methods are obviously bad. For example, choosing the students from the rosters of the football and basketball teams would undoubtedly result in a sample that would fail to represent the height distribution of the population of students. You might think that it would be reasonable to use some conveniently obtained sample, for example, all students living in a certain dorm or all students enrolled in engineering statistics. After all, there is no reason to think that the heights of these students would tend to differ from the heights of students in general. Samples like this are not ideal, however, because they can turn out to be misleading in ways that are not anticipated. The best sampling methods involve **random sampling**. There are many different random sampling methods, the most basic of which is **simple random sampling**.

Simple Random Samples

To understand the nature of a simple random sample, think of a lottery. Imagine that 10,000 lottery tickets have been sold and that 5 winners are to be chosen. What is the fairest way to choose the winners? The fairest way is to put the 10,000 tickets in a drum, mix them thoroughly, and then reach in and one by one draw 5 tickets out. These 5 winning tickets are a simple random sample from the population of 10,000 lottery tickets. Each ticket is equally likely to be one of the 5 tickets drawn. More important, each collection of 5 tickets that can be formed from the 10,000 is equally likely to comprise the group of 5 that is drawn. It is this idea that forms the basis for the definition of a simple random sample.

Summary

- A **population** is the entire collection of objects or outcomes about which information is sought.
- A **sample** is a subset of a population, containing the objects or outcomes that are actually observed.
- A **simple random sample** of size n is a sample chosen by a method in which each collection of n population items is equally likely to comprise the sample, just as in a lottery.

Since a simple random sample is analogous to a lottery, it can often be drawn by the same method now used in many lotteries: with a computer random number generator. Suppose there are N items in the population. One assigns to each item in the population an integer between 1 and N. Then one generates a list of random integers between

1 and N and chooses the corresponding population items to comprise the simple random sample.

A utility company wants to conduct a survey to measure the satisfaction level of its customers in a certain town. There are 10,000 customers in the town, and utility employees want to draw a sample of size 200 to interview over the telephone. They obtain a list of all 10,000 customers, and number them from 1 to 10,000. They use a computer random number generator to generate 200 random integers between 1 and 10,000 and then telephone the customers who correspond to those numbers. Is this a simple random sample?

Solution

Yes, this is a simple random sample. Note that it is analogous to a lottery in which each customer has a ticket and 200 tickets are drawn.

A quality engineer wants to inspect electronic microcircuits in order to obtain information on the proportion that are defective. She decides to draw a sample of 100 circuits from a day's production. Each hour for 5 hours, she takes the 20 most recently produced circuits and tests them. Is this a simple random sample?

Solution

No. Not every subset of 100 circuits is equally likely to comprise the sample. To construct a simple random sample, the engineer would need to assign a number to each circuit produced during the day and then generate random numbers to determine which circuits comprise the sample.

Samples of Convenience

In some cases, it is difficult or impossible to draw a sample in a truly random way. In these cases, the best one can do is to sample items by some convenient method. For example, imagine that a construction engineer has just received a shipment of 1000 concrete blocks, each weighing approximately 50 pounds. The blocks have been delivered in a large pile. The engineer wishes to investigate the crushing strength of the blocks by measuring the strengths in a sample of 10 blocks. To draw a simple random sample would require removing blocks from the center and bottom of the pile, which might be quite difficult. For this reason, the engineer might construct a sample simply by taking 10 blocks off the top of the pile. A sample like this is called a **sample of convenience**.

Definition

A **sample of convenience** is a sample that is not drawn by a well-defined random method.

The big problem with samples of convenience is that they may differ systematically in some way from the population. For this reason samples of convenience should only be used in situations where it is not feasible to draw a random sample. When it is necessary to take a sample of convenience, it is important to think carefully about all the ways in which the sample might differ systematically from the population. If it is reasonable to believe that no important systematic difference exists, then it may be acceptable to treat the sample of convenience as if it were a simple random sample. With regard to the concrete blocks, if the engineer is confident that the blocks on the top of the pile do not differ systematically in any important way from the rest, then he may treat the sample of convenience as a simple random sample. If, however, it is possible that blocks in different parts of the pile may have been made from different batches of mix or may have different curing times or temperatures, a sample of convenience could give misleading results.

Some people think that a simple random sample is guaranteed to reflect its population perfectly. This is not true. Simple random samples always differ from their populations in some ways, and occasionally they may be substantially different. Two different samples from the same population will differ from each other as well. This phenomenon is known as **sampling variation**. Sampling variation is one of the reasons that scientific experiments produce somewhat different results when repeated, even when the conditions appear to be identical. For example, suppose that a quality inspector draws a simple random sample of 40 bolts from a large shipment, measures the length of each, and finds that 32 of them, or 80%, meet a length specification. Another inspector draws a different sample of 40 bolts and finds that 36 of them, or 90%, meet the specification. By chance, the second inspector got a few more good bolts in her sample. It is likely that neither sample reflects the population perfectly. The proportion of good bolts in the population is likely to be close to 80% or 90%, but it is not likely that it is exactly equal to either value.

Since simple random samples don't reflect their populations perfectly, why is it important that sampling be done at random? The benefit of a simple random sample is that there is no systematic mechanism tending to make the sample unrepresentative. The differences between the sample and its population are due entirely to random variation. Since the mathematical theory of random variation is well understood, we can use mathematical models to study the relationship between simple random samples and their populations. For a sample not chosen at random, there is generally no theory available to describe the mechanisms that caused the sample to differ from its population. Therefore, nonrandom samples are often difficult to analyze reliably.

Tangible and Conceptual Populations

The populations discussed so far have consisted of actual physical objects—the customers of a utility company, the concrete blocks in a pile, the bolts in a shipment. Such populations are called **tangible populations**. Tangible populations are always finite. After an item is sampled, the population size decreases by 1. In principle, one could in some cases return the sampled item to the population, with a chance to sample it again, but this is rarely done in practice.

Engineering data are often produced by measurements made in the course of a scientific experiment, rather than by sampling from a tangible population. To take a simple example, imagine that an engineer measures the length of a rod five times, being as careful as possible to take the measurements under identical conditions. No matter how carefully the measurements are made, they will differ somewhat from one another, because of variation in the measurement process that cannot be controlled or predicted. It turns out that it is often appropriate to consider data like these to be a simple random sample from a population. The population, in these cases, consists of all the values that might possibly have been observed. Such a population is called a **conceptual population**, since it does not consist of actual objects.

Definition

A simple random sample may consist of values obtained from a process under identical experimental conditions. In this case, the sample comes from a population that consists of all the values that might possibly have been observed. Such a population is called a **conceptual population**.

Example 1.3 involves a conceptual population.

A geologist weighs a rock several times on a sensitive scale. Each time, the scale gives a slightly different reading. Under what conditions can these readings be thought of as a simple random sample? What is the population?

Solution

If the physical characteristics of the scale remain the same for each weighing, so that the measurements are made under identical conditions, then the readings may be considered to be a simple random sample. The population is conceptual. It consists of all the readings that the scale could in principle produce.

Determining Whether a Sample Is a Simple Random Sample

We saw in Example 1.3 that it is the physical characteristics of the measurement process that determine whether the data are a simple random sample. In general, when deciding whether a set of data may be considered to be a simple random sample, it is necessary to have some understanding of the process that generated the data. Statistical methods can sometimes help, especially when the sample is large, but knowledge of the mechanism that produced the data is more important.

A new chemical process has been designed that is supposed to produce a higher yield of a certain chemical than does an old process. To study the yield of this process, we run it 50 times and record the 50 yields. Under what conditions might it be reasonable to treat this as a simple random sample? Describe some conditions under which it might not be appropriate to treat this as a simple random sample.

Solution

To answer this, we must first specify the population. The population is conceptual and consists of the set of all yields that will result from this process as many times as it will ever be run. What we have done is to sample the first 50 yields of the process. *If, and only if,* we are confident that the first 50 yields are generated under identical conditions and that they do not differ in any systematic way from the yields of future runs, then we may treat them as a simple random sample.

Be cautious, however. There are many conditions under which the 50 yields could fail to be a simple random sample. For example, with chemical processes, it is sometimes the case that runs with higher yields tend to be followed by runs with lower yields, and vice versa. Sometimes yields tend to increase over time, as process engineers learn from experience how to run the process more efficiently. In these cases, the yields are not being generated under identical conditions and would not comprise a simple random sample.

Example 1.4 shows once again that a good knowledge of the nature of the process under consideration is important in deciding whether data may be considered to be a simple random sample. Statistical methods can sometimes be used to show that a given data set is *not* a simple random sample. For example, sometimes experimental conditions gradually change over time. A simple but effective method to detect this condition is to plot the observations in the order they were taken. A simple random sample should show no obvious pattern or trend.

Figure 1.1 presents plots of three samples in the order they were taken. The plot in Figure 1.1a shows an oscillatory pattern. The plot in Figure 1.1b shows an increasing trend. Neither of these samples should be treated as a simple random sample. The plot in Figure 1.1c does not appear to show any obvious pattern or trend. It might be appropriate to treat these data as a simple random sample. However, before making that decision, it

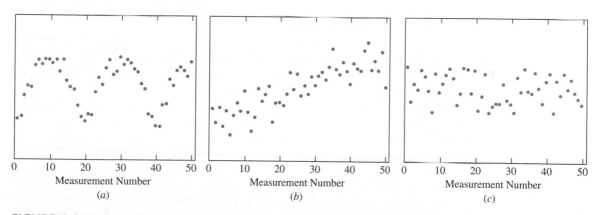

FIGURE 1.1 Three plots of observed values versus the order in which they were made. *(a)* The values show a definite pattern over time. This is not a simple random sample. *(b)* The values show a trend over time. This is not a simple random sample. *(c)* The values do not show a pattern or trend. It may be appropriate to treat these data as a simple random sample.

is still important to think about the process that produced the data, since there may be concerns that don't show up in the plot.

Independence

The items in a sample are said to be **independent** if knowing the values of some of them does not help to predict the values of the others. With a finite, tangible population, the items in a simple random sample are not strictly independent, because as each item is drawn, the population changes. This change can be substantial when the population is small. However, when the population is very large, this change is negligible and the items can be treated as if they were independent.

To illustrate this idea, imagine that we draw a simple random sample of 2 items from the population

For the first draw, the numbers 0 and 1 are equally likely. But the value of the second item is clearly influenced by the first; if the first is 0, the second is more likely to be 1, and vice versa. Thus, the sampled items are dependent. Now assume we draw a sample of size 2 from this population:

Again on the first draw, the numbers 0 and 1 are equally likely. But unlike the previous example, these two values remain almost equally likely the second draw as well, no matter what happens on the first draw. With the large population, the sample items are for all practical purposes independent.

It is reasonable to wonder how large a population must be in order that the items in a simple random sample may be treated as independent. A rule of thumb is that when sampling from a finite population, the items may be treated as independent so long as the sample comprises 5% or less of the population.

Interestingly, it is possible to make a population behave as though it were infinitely large, by replacing each item after it is sampled. This method is called **sampling with replacement**. With this method, the population is exactly the same on every draw, and the sampled items are truly independent.

With a conceptual population, we require that the sample items be produced under identical experimental conditions. In particular, then, no sample value may influence the conditions under which the others are produced. Therefore, the items in a simple random sample from a conceptual population may be treated as independent. We may think of a conceptual population as being infinite or, equivalently, that the items are sampled with replacement.

Summary

- The items in a sample are **independent** if knowing the values of some of the items does not help to predict the values of the others.
- Items in a simple random sample may be treated as independent in many cases encountered in practice. The exception occurs when the population is finite and the sample comprises a substantial fraction (more than 5%) of the population.

Other Sampling Methods

In addition to simple random sampling there are other sampling methods that are useful in various situations. In **weighted sampling**, some items are given a greater chance of being selected than others, like a lottery in which some people have more tickets than others. In **stratified random sampling**, the population is divided up into subpopulations, called **strata**, and a simple random sample is drawn from each stratum. In **cluster sampling**, items are drawn from the population in groups, or clusters. Cluster sampling is useful when the population is too large and spread out for simple random sampling to be feasible. For example, many U.S. government agencies use cluster sampling to sample the U.S. population to measure sociological factors such as income and unemployment. A good source of information on sampling methods is Cochran (1977).

  Simple random sampling is not the only valid method of sampling. But it is the most fundamental, and we will focus most of our attention on this method. From now on, unless otherwise stated, the terms "sample" and "random sample" will be taken to mean "simple random sample."

Types of Data

When a numerical quantity designating how much or how many is assigned to each item in a sample, the resulting set of values is called **numerical** or **quantitative**. In some cases, sample items are placed into categories, and category names are assigned to the sample items. Then the data are **categorical** or **qualitative**. Sometimes both quantitative and categorical data are collected in the same experiment. For example, in a loading test of column-to-beam welded connections, data may be collected both on the torque applied at failure and on the location of the failure (weld or beam). The torque is a quantitative variable, and the location is a categorical variable.

Controlled Experiments and Observational Studies

Many scientific experiments are designed to determine the effect of changing one or more factors on the value of a response. For example, suppose that a chemical engineer wants to determine how the concentrations of reagent and catalyst affect the yield of a process. The engineer can run the process several times, changing the concentrations each time, and compare the yields that result. This sort of experiment is called a **controlled experiment**, because the values of the factors—in this case, the concentrations of reagent and catalyst—are under the control of the experimenter. When designed and conducted properly, controlled experiments can produce reliable information about cause-and-effect relationships between factors and response. In the yield example just mentioned, a well-done experiment would allow the experimenter to conclude that the differences in yield were caused by differences in the concentrations of reagent and catalyst.

There are many situations in which scientists cannot control the levels of the factors. For example, many studies have been conducted to determine the effect of cigarette smoking on the risk of lung cancer. In these studies, rates of cancer among smokers are compared with rates among nonsmokers. The experimenters cannot control who smokes and who doesn't; people cannot be required to smoke just to make a statistician's job easier. This kind of study is called an **observational study**, because the experimenter simply observes the levels of the factor as they are, without having any control over them. Observational studies are not nearly as good as controlled experiments for obtaining reliable conclusions regarding cause and effect. In the case of smoking and lung cancer, for example, people who choose to smoke may not be representative of the population as a whole, and may be more likely to get cancer for other reasons. For this reason, although it has been known for a long time that smokers have higher rates of lung cancer than nonsmokers, it took many years of carefully done observational studies before scientists could be sure that smoking was actually the cause of the higher rate.

Exercises for Section 1.1

1. Each of the following processes involves sampling from a population. Define the population, and state whether it is tangible or conceptual.

   a. A shipment of bolts is received from a vendor. To check whether the shipment is acceptable with regard to shear strength, an engineer reaches into the container and selects 10 bolts, one by one, to test.

   b. The resistance of a certain resistor is measured five times with the same ohmmeter.

   c. A graduate student majoring in environmental science is part of a study team that is assessing the risk posed to human health of a certain contaminant present in the tap water in their town. Part of the assessment process involves estimating the amount of time that people who live in that town are in contact with tap water. The student recruits residents of the town to keep diaries for a month, detailing day by day the amount of time they were in contact with tap water.

   d. Eight welds are made with the same process, and the strength of each is measured.

   e. A quality engineer needs to estimate the percentage of parts manufactured on a certain day that are defective. At 2:30 in the afternoon he samples the last 100 parts to be manufactured.

2. If you wanted to estimate the mean height of all the students at a university, which one of the following sampling strategies would be best? Why? Note that none of the methods are true simple random samples.

   i. Measure the heights of 50 students found in the gym during basketball intramurals.

   ii. Measure the heights of all engineering majors.

   iii. Measure the heights of the students selected by choosing the first name on each page of the campus phone book.

3. True or false:

   a. A simple random sample is guaranteed to reflect exactly the population from which it was drawn.

   b. A simple random sample is free from any systematic tendency to differ from the population from which it was drawn.

4. A quality engineer draws a simple random sample of 50 O-rings from a lot of several thousand. She measures the thickness of each and finds that 45 of them, or 90%, meet a certain specification. Which of the following statements is correct?

   i. The proportion of O-rings in the entire lot that meet the specification is likely to be equal to 90%.

ii. The proportion of O-rings in the entire lot that meet the specification is likely to be close to 90% but is not likely to equal 90%.

5. A certain process for manufacturing integrated circuits has been in use for a period of time, and it is known that 12% of the circuits it produces are defective. A new process that is supposed to reduce the proportion of defectives is being tested. In a simple random sample of 100 circuits produced by the new process, 12 were defective.

   a. One of the engineers suggests that the test proves that the new process is no better than the old process, since the proportion of defectives in the sample is the same. Is this conclusion justified? Explain.

   b. Assume that there had been only 11 defective circuits in the sample of 100. Would this have proven that the new process is better? Explain.

   c. Which outcome represents stronger evidence that the new process is better: finding 11 defective cir-cuits in the sample, or finding 2 defective circuits in the sample?

6. Refer to Exercise 5. True or false:

   a. If the proportion of defectives in the sample is less than 12%, it is reasonable to conclude that the new process is better.

   b. If the proportion of defectives in the sample is only slightly less than 12%, the difference could well be due entirely to sampling variation, and it is not rea-sonable to conclude that the new process is better.

   c. If the proportion of defectives in the sample is a lot less than 12%, it is very unlikely that the difference is due entirely to sampling variation, so it is reason-able to conclude that the new process is better.

7. To determine whether a sample should be treated as a simple random sample, which is more important: a good knowledge of statistics, or a good knowledge of the process that produced the data?

1.2 Summary Statistics

A sample is often a long list of numbers. To help make the important features of a sample stand out, we compute summary statistics. The two most commonly used summary statistics are the **sample mean** and the **sample standard deviation**. The mean gives an indication of the center of the data, and the standard deviation gives an indication of how spread out the data are.

The Sample Mean

The sample mean is also called the "arithmetic mean," or, more simply, the "average." It is the sum of the numbers in the sample, divided by how many there are.

Definition

Let $X_1, \ldots, X_n$ be a sample. The **sample mean** is

$$\overline{X} = \frac{1}{n} \sum_{i=1}^{n} X_i \tag{1.1}$$

It is customary to use a letter with a bar over it (e.g., $\overline{X}$) to denote a sample mean. Note also that the sample mean has the same units as the sample values $X_1, \ldots, X_n$.

xample

1.5

A simple random sample of five men is chosen from a large population of men, and their heights are measured. The five heights (in cm) are 166.4, 183.6, 173.5, 170.3, and 179.5. Find the sample mean.

Solution

We use Equation (1.1). The sample mean is

$$\overline{X} = \frac{1}{5}(166.4 + 183.6 + 173.5 + 170.3 + 179.5) = 174.66 \text{ cm}$$

The Standard Deviation

Here are two lists of numbers: 28, 29, 30, 31, 32 and 10, 20, 30, 40, 50. Both lists have the same mean of 30. But the second list is much more spread out than the first. The **standard deviation** is a quantity that measures the degree of spread in a sample.

Let $X_1, \ldots, X_n$ be a sample. The idea behind the standard deviation is that when the spread is large, the sample values will tend to be far from their mean, but when the spread is small, the values will tend to be close to their mean. So the first step in calculating the standard deviation is to compute the differences (also called deviations) between each sample value and the sample mean. The deviations are $(X_1 - \overline{X}), \ldots, (X_n - \overline{X})$. Now, some of these deviations are positive and some are negative. Large negative deviations are just as indicative of spread as large positive deviations are. To make all the deviations positive, we square them, obtaining the squared deviations $(X_1 - \overline{X})^2, \ldots, (X_n - \overline{X})^2$. From the squared deviations, we can compute a measure of spread called the **sample variance**. The sample variance is the average of the squared deviations, except that we divide by $n - 1$ instead of n. It is customary to denote the sample variance by s^2.

Definition

Let $X_1, \ldots, X_n$ be a sample. The **sample variance** is the quantity

$$s^2 = \frac{1}{n-1} \sum_{i=1}^{n} (X_i - \overline{X})^2 \tag{1.2}$$

An equivalent formula, which can be easier to compute, is

$$s^2 = \frac{1}{n-1} \left(\sum_{i=1}^{n} X_i^2 - n\overline{X}^2 \right) \tag{1.3}$$

While the sample variance is an important quantity, it has a serious drawback as a measure of spread. Its units are not the same as the units of the sample values; instead, they are the squared units. To obtain a measure of spread whose units are the same as those of the sample values, we simply take the square root of the variance. This quantity is known as the **sample standard deviation**. It is customary to denote the sample standard deviation by the letter s (the square root of s^2).

> ### Definition
>
> Let $X_1, \ldots, X_n$ be a sample. The **sample standard deviation** is the quantity
>
> $$s = \sqrt{\frac{1}{n-1} \sum_{i=1}^{n} (X_i - \overline{X})^2} \qquad (1.4)$$
>
> An equivalent formula, which can be easier to compute, is
>
> $$s = \sqrt{\frac{1}{n-1} \left(\sum_{i=1}^{n} X_i^2 - n\overline{X}^2 \right)} \qquad (1.5)$$
>
> The sample standard deviation is the square root of the sample variance.

It is natural to wonder why the sum of the squared deviations is divided by $n - 1$ rather than n. The purpose in computing the sample standard deviation is to estimate the amount of spread in the population from which the sample was drawn. Ideally, therefore, we would compute deviations from the mean of all the items in the population, rather than the deviations from the sample mean. However, the population mean is in general unknown, so the sample mean is used in its place. It is a mathematical fact that the deviations around the sample mean tend to be a bit smaller than the deviations around the population mean and that dividing by $n - 1$ rather than n provides exactly the right correction.

Example
1.6

Find the sample variance and the sample standard deviation for the height data in Example 1.5.

Solution

We'll first compute the sample variance by using Equation (1.2). The sample mean is $\overline{X} = 174.66$ (see Example 1.5). The sample variance is therefore

$$s^2 = \frac{1}{4}[(166.4 - 174.66)^2 + (183.6 - 174.66)^2 + (173.5 - 174.66)^2$$
$$+ (170.3 - 174.66)^2 + (179.5 - 174.66)^2] = 47.983$$

Alternatively, we can use Equation (1.3):

$$s^2 = \frac{1}{4}[166.4^2 + 183.6^2 + 173.5^2 + 170.3^2 + 179.5^2 - 5(174.66^2)] = 47.983$$

The sample standard deviation is the square root of the sample variance:

$$s = \sqrt{47.983} = 6.93$$

What would happen to the sample mean, variance, and standard deviation if the heights in Example 1.5 were measured in inches rather than centimeters? Let's denote the heights in centimeters by X_1, X_2, X_3, X_4, X_5, and the heights in inches by

Y_1, Y_2, Y_3, Y_4, Y_5. The relationship between X_i and Y_i is then given by $Y_i = 0.3937X_i$. If you go back to Example 1.5, convert to inches, and compute the sample mean, you will find that the sample means in inches and in centimeters are related by the equation $\overline{Y} = 0.3937\overline{X}$. Thus, if we multiply each sample item by a constant, the sample mean is multiplied by the same constant. As for the sample variance, you will find that the deviations are related by the equation $(Y_i - \overline{Y}) = 0.3937(X_i - \overline{X})$. It follows that $s_Y^2 = 0.3937^2 s_X^2$ and that $s_Y = 0.3937 s_X$.

What if each man in the sample stood on a 2-centimeter platform? Then each sample height would increase by 2 cm, and the sample mean would increase by 2 cm as well. In general, if a constant is added to each sample item, the sample mean increases (or decreases) by the same constant. The deviations, however, do not change, so the sample variance and standard deviation are unaffected.

Summary

- If $X_1, \ldots, X_n$ is a sample and $Y_i = a + bX_i$, where a and b are constants, then $\overline{Y} = a + b\overline{X}$.
- If $X_1, \ldots, X_n$ is a sample and $Y_i = a + bX_i$, where a and b are constants, then $s_Y^2 = b^2 s_X^2$, and $s_Y = |b|s_X$.

Outliers

Sometimes a sample may contain a few points that are much larger or smaller than the rest. Such points are called **outliers**. See Figure 1.2 for an example. Sometimes outliers result from data entry errors; for example, a misplaced decimal point can result in a value that is an order of magnitude different from the rest. Outliers should always be scrutinized, and any outlier that is found to result from an error should be corrected or deleted. Not all outliers are errors. Sometimes a population may contain a few values that are much different from the rest, and the outliers in the sample reflect this fact.

FIGURE 1.2 A data set that contains an outlier.

Outliers are a real problem for data analysts. For this reason, when people see outliers in their data, they sometimes try to find a reason, or an excuse, to delete them. An outlier should not be deleted, however, unless it is reasonably certain that it results from an error. If a population truly contains outliers, but they are deleted from the sample, the sample will not characterize the population correctly.

The Sample Median

The **median**, like the mean, is a measure of center. To compute the median of a sample, order the values from smallest to largest. The sample median is the middle number. If the sample size is an even number, it is customary to take the sample median to be the average of the two middle numbers.

Definition

If n numbers are ordered from smallest to largest:

- If n is odd, the sample median is the number in position $\dfrac{n+1}{2}$.

- If n is even, the sample median is the average of the numbers in positions $\dfrac{n}{2}$ and $\dfrac{n}{2}+1$.

Example **1.7**

Find the sample median for the height data in Example 1.5.

Solution

The five heights, arranged in increasing order, are 166.4, 170.3, 173.5, 179.5, 183.6. The sample median is the middle number, which is 173.5.

The median is often used as a measure of center for samples that contain outliers. To see why, consider the sample consisting of the values 1, 2, 3, 4, and 20. The mean is 6, and the median is 3. It is reasonable to think that the median is more representative of the sample than the mean is. See Figure 1.3.

Median   Mean

FIGURE 1.3 When a sample contains outliers, the median may be more representative of the sample than the mean is.

Quartiles

The median divides the sample in half. **Quartiles** divide it as nearly as possible into quarters. A sample has three quartiles. There are several different ways to compute quartiles, and all of them give approximately the same result. The simplest method when computing by hand is as follows: Let n represent the sample size. Order the sample values from smallest to largest. To find the first quartile, compute the value $0.25(n+1)$. If this is an integer, then the sample value in that position is the first quartile. If not, then take the average of the sample values on either side of this value. The third quartile is computed in the same way, except that the value $0.75(n+1)$ is used. The second quartile uses the value $0.5(n+1)$. The second quartile is identical to the median. We note that some computer packages use slightly different methods to compute quartiles, so their results may not be quite the same as the ones obtained by the method described here.

Example **1.8**

In the article "Evaluation of Low-Temperature Properties of HMA Mixtures" (P. Sebaaly, A. Lake, and J. Epps, *Journal of Transportation Engineering*, 2002:578–583), the following values of fracture stress (in megapascals) were measured for a sample of 24 mixtures of hot-mixed asphalt (HMA).

$$30 \quad 75 \quad 79 \quad 80 \quad 80 \quad 105 \quad 126 \quad 138 \quad 149 \quad 179 \quad 179 \quad 191$$
$$223 \quad 232 \quad 232 \quad 236 \quad 240 \quad 242 \quad 245 \quad 247 \quad 254 \quad 274 \quad 384 \quad 470$$

Find the first and third quartiles.

Solution
The sample size is $n = 24$. To find the first quartile, compute $(0.25)(25) = 6.25$. The first quartile is therefore found by averaging the 6th and 7th data points, when the sample is arranged in increasing order. This yields $(105 + 126)/2 = 115.5$. To find the third quartile, compute $(0.75)(25) = 18.75$. We average the 18th and 19th data points to obtain $(242 + 245)/2 = 243.5$.

Percentiles

The pth percentile of a sample, for a number p between 0 and 100, divides the sample so that as nearly as possible $p\%$ of the sample values are less than the pth percentile, and $(100 - p)\%$ are greater. There are many ways to compute percentiles, all of which produce similar results. We describe here a method analogous to the method described for computing quartiles. Order the sample values from smallest to largest, and then compute the quantity $(p/100)(n + 1)$, where n is the sample size. If this quantity is an integer, the sample value in this position is the pth percentile. Otherwise, average the two sample values on either side. Note that the first quartile is the 25th percentile, the median is the 50th percentile, and the third quartile is the 75th percentile. Some computer packages use slightly different methods to compute percentiles, so their results may differ slightly from the ones obtained by this method.

Percentiles are often used to interpret scores on standardized tests. For example, if a student is informed that her score on a college entrance exam is on the 64th percentile, this means that 64% of the students who took the exam got lower scores.

Example **1.9**

Find the 65th percentile of the asphalt data in Example 1.8.

Solution
The sample size is $n = 24$. To find the 65th percentile, compute $(0.65)(25) = 16.25$. The 65th percentile is therefore found by averaging the 16th and 17th data points, when the sample is arranged in increasing order. This yields $(236 + 240)/2 = 238$.

In practice, the summary statistics we have discussed are often calculated on a computer, using a statistical software package. The summary statistics are sometimes called **descriptive statistics**, because they describe the data. We present an example of the calculation of summary statistics from the software package MINITAB. Then we will show how these statistics can be used to discover some important features of the data.

For a Ph.D. thesis that investigated factors affecting diesel vehicle emissions, J. Yanowitz of the Colorado School of Mines obtained data on emissions of particulate matter (PM) for a sample of 138 vehicles driven at low altitude (near sea level) and for a sample of 62 vehicles driven at high altitude (approximately one mile above sea level). All the vehicles were manufactured between 1991 and 1996. The samples contained roughly equal proportions of high- and low-mileage vehicles. The data, in units of grams of particulates per gallon of fuel consumed, are presented in Tables 1.1 and 1.2. At high altitude, the barometric pressure is lower, so the effective air/fuel ratio is lower as well. For this reason, it was thought that PM emissions might be greater at higher altitude. We would like to compare the samples to determine whether the data support this assumption. It is difficult to do this simply by examining the raw data in the tables. Computing summary statistics makes the job much easier. Figure 1.4 (page 18) presents summary statistics for both samples, as computed by MINITAB.

In Figure 1.4, the quantity labeled "N" is the sample size. Following that is the sample mean. The next quantity (SE Mean) is the **standard error of the mean**. The standard error of the mean is equal to the standard deviation divided by the square root of the sample size. This quantity is not used much as a descriptive statistic, although it is important for applications such as constructing confidence intervals and hypothesis tests, which we will cover in Chapters 5, 6, and 7. Following the standard error of the mean is the standard deviation. Finally, the second line of the output provides the minimum, median, and maximum, as well as the first and third quartiles (Q1 and Q3). We note that the values of the quartiles produced by the computer package differ slightly from the values that would be computed by the methods we describe. This is not surprising, since

TABLE 1.1 Particulate matter (PM) emissions (in g/gal) for 138 vehicles driven at low altitude

1.50	0.87	1.12	1.25	3.46	1.11	1.12	0.88	1.29	0.94	0.64	1.31	2.49
1.48	1.06	1.11	2.15	0.86	1.81	1.47	1.24	1.63	2.14	6.64	4.04	2.48
2.98	7.39	2.66	11.00	4.57	4.38	0.87	1.10	1.11	0.61	1.46	0.97	0.90
1.40	1.37	1.81	1.14	1.63	3.67	0.55	2.67	2.63	3.03	1.23	1.04	1.63
3.12	2.37	2.12	2.68	1.17	3.34	3.79	1.28	2.10	6.55	1.18	3.06	0.48
0.25	0.53	3.36	3.47	2.74	1.88	5.94	4.24	3.52	3.59	3.10	3.33	4.58
6.73	7.82	4.59	5.12	5.67	4.07	4.01	2.72	3.24	5.79	3.59	3.48	2.96
5.30	3.93	3.52	2.96	3.12	1.07	5.30	5.16	7.74	5.41	3.40	4.97	11.23
9.30	6.50	4.62	5.45	4.93	6.05	5.82	10.19	3.62	2.67	2.75	8.92	9.93
6.96	5.78	9.14	10.63	8.23	6.83	5.60	5.41	6.70	5.93	4.51	9.04	7.71
7.21	4.67	4.49	4.63	2.80	2.16	2.97	3.90					

TABLE 1.2 Particulate matter (PM) emissions (in g/gal) for 62 vehicles driven at high altitude

7.59	6.28	6.07	5.23	5.54	3.46	2.44	3.01	13.63	13.02	23.38	9.24	3.22
2.06	4.04	17.11	12.26	19.91	8.50	7.81	7.18	6.95	18.64	7.10	6.04	5.66
8.86	4.40	3.57	4.35	3.84	2.37	3.81	5.32	5.84	2.89	4.68	1.85	9.14
8.67	9.52	2.68	10.14	9.20	7.31	2.09	6.32	6.53	6.32	2.01	5.91	5.60
5.61	1.50	6.46	5.29	5.64	2.07	1.11	3.32	1.83	7.56			

```
Descriptive Statistics: LowAltitude, HiAltitude

Variable             N      Mean   SE Mean     StDev
LoAltitude         138     3.715     0.218     2.558
HiAltitude          62     6.596     0.574     4.519

Variable       Minimum        Q1    Median        Q3    Maximum
LoAltitude       0.250     1.468     3.180     5.300     11.230
HiAltitude       1.110     3.425     5.750     7.983     23.380
```

FIGURE 1.4 MINITAB output presenting descriptive statistics for the PM data in Tables 1.1 and 1.2.

there are several ways to compute these values. The differences are not large enough to have any practical importance.

The summary statistics tell a lot about the differences in PM emissions between high- and low-altitude vehicles. First, note that the mean is indeed larger for the high-altitude vehicles than for the low-altitude vehicles (6.596 vs. 3.715), which supports the hypothesis that emissions tend to be greater at high altitudes. Now note that the maximum value for the high-altitude vehicles (23.38) is much higher than the maximum for the low-altitude vehicles (11.23). This shows that there are one or more high-altitude vehicles whose emissions are much higher than the highest of the low-altitude vehicles. Could the difference in mean emissions be due entirely to these vehicles? To answer this, compare the medians and the first and third quartiles. These statistics are not affected much by a few large values, yet all of them are noticeably larger for the high-altitude vehicles. Therefore, we can conclude that the high-altitude vehicles not only contain a few very high emitters, they also have higher emissions than the low-altitude vehicles in general. Finally, note that the standard deviation is larger for the high-altitude vehicles, which indicates that the values for the high-altitude vehicles are more spread out than those for the low-altitude vehicles. At least some of this difference in spread must be due to the one or more high-altitude vehicles with very high emissions.

Exercises for Section 1.2

1. A vendor converts the weights on the packages she sends out from pounds to kilograms (1 kg ≈ 2.2 lb).
   a. How does this affect the mean weight of the packages?
   b. How does this affect the standard deviation of the weights?

2. Refer to Exercise 1. The vendor begins using heavier packaging, which increases the weight of each package by 50 g.
   a. How does this affect the mean weight of the packages?
   b. How does this affect the standard deviation of the weights?

3. True or false: For any list of numbers, half of them will be below the mean.

4. Is the sample mean always the most frequently occurring value? If so, explain why. If not, give an example.

5. Is the sample mean always equal to one of the values in the sample? If so, explain why. If not, give an example.

6. Is the sample median always equal to one of the values in the sample? If so, explain why. If not, give an example.

7. Find a sample size for which the median will always equal one of the values in the sample.

8. For a list of positive numbers, is it possible for the standard deviation to be greater than the mean? If so, give an example. If not, explain why not.

9. Is it possible for the standard deviation of a list of numbers to equal 0? If so, give an example. If not, explain why not.

10. A sample of 100 adult women was taken, and each was asked how many children she had. The results were as follows:

Children	0	1	2	3	4	5
Number of Women	27	22	30	12	7	2

   a. Find the sample mean number of children.
   b. Find the sample standard deviation of the number of children.
   c. Find the sample median of the number of children.
   d. What is the first quartile of the number of children?
   e. What proportion of the women had more than the mean number of children?
   f. For what proportion of the women was the number of children more than one standard deviation greater than the mean?
   g. For what proportion of the women was the number of children within one standard deviation of the mean?

11. In a sample of 20 men, the mean height was 178 cm. In a sample of 30 women, the mean height was 164 cm. What was the mean height for both groups put together?

12. In a study of visual perception, five people were asked to estimate the length of a line by eye and then to measure it with a ruler. The results, in cm, were:

   By eye: 8.0, 9.0, 7.5, 9.5, 8.5
   With a ruler: 8.1, 8.2, 8.1, 8.1, 8.3

   a. Compute the mean measurement for each method.
   b. Compute the median measurement for each method.
   c. Compute the standard deviation of the measurements for each method.
   d. For which method is the standard deviation the largest? Why should one expect this method to have the largest standard deviation?

e. Other things being equal, is it better for a measurement method to have a smaller standard deviation or a larger standard deviation? Or doesn't it matter? Explain.

13. Refer to Exercise 12.

   a. If the measurements for one of the methods were converted to inches (1 inch = 2.54 cm), how would this affect the mean? The median? The standard deviation?
   b. If each person remeasured the line, using a ruler marked in inches, would the effects on the mean, median, and standard deviation be the same as in part (a)? Explain.

14. There are 10 employees in a particular division of a company. Their salaries have a mean of $70,000, a median of $55,000, and a standard deviation of $60,000. The largest number on the list is $100,000. By accident, this number is changed to $1,000,000.

   a. What is the value of the mean after the change?
   b. What is the value of the median after the change?
   c. What is the value of the standard deviation after the change?

15. Quartiles divide a sample into four nearly equal pieces. In general, a sample of size n can be broken into k nearly equal pieces by using the cutpoints $(i/k)(n+1)$ for $i = 1, \ldots, k - 1$. Consider the following ordered sample:

   $$2 \ 18 \ 23 \ 41 \ 44 \ 46 \ 49 \ 61$$
   $$62 \ 74 \ 76 \ 79 \ 82 \ 89 \ 92 \ 95$$

   a. Tertiles divide a sample into thirds. Find the tertiles of this sample.
   b. Quintiles divide a sample into fifths. Find the quintiles of this sample.

16. In each of the following data sets, tell whether the outlier seems certain to be due to an error or whether it could conceivably be correct.

   a. A rock is weighed five times. The readings in grams are 48.5, 47.2, 4.91, 49.5, 46.3.
   b. A sociologist samples five families in a certain town and records their annual income. The incomes are $34,000, $57,000, $13,000, $1,200,000, $62,000.

1.3 Graphical Summaries

Stem-and-Leaf Plots

In addition to numerical summaries such as the mean, median, and standard deviation, graphical summaries can be used to help visualize a list of numbers. We will begin by discussing a simple graphical summary known as the **stem-and-leaf plot**.

As an example, the data in Table 1.3 concern a study of the bioactivity of a certain antifungal drug. The drug was applied to the skin of 48 subjects. After three hours, the amount of drug remaining in the skin was measured in units of ng/cm^2. The list has been sorted into numerical order.

TABLE 1.3 Amount of drug in skin

3	4	4	7	7	8	9	9	12	12
15	16	16	17	17	18	20	20	21	21
22	22	22	23	24	25	26	26	26	26
27	33	34	34	35	36	36	37	38	40
40	41	41	51	53	55	55	74		

Figure 1.5 presents a stem-and-leaf plot of the data in Table 1.3. Each item in the sample is divided into two parts: a **stem**, consisting of the leftmost one or two digits, and the **leaf**, which consists of the next digit. In Figure 1.5, the stem consists of the tens digit, and the leaf consists of the ones digit. Each line of the stem-and-leaf plot contains all of the sample items with a given stem. The stem-and-leaf plot is a compact way to represent the data. It also gives some indication of its shape. For these data, we can see that there are equal numbers of subjects in the intervals 0–9, 10–19, and 30–39, and somewhat more subjects in the interval 20–29. In addition, the largest value (74) appears to be an outlier.

```
Stem  Leaf
   0  34477899
   1  22566778
   2  001122234566667
   3  34456678
   4  0011
   5  1355
   6
   7  4
```

FIGURE 1.5 Stem-and-leaf plot for the data in Table 1.3

When there are a great many sample items with the same stem, it is often necessary to assign more than one row to that stem. As an example, Figure 1.6 presents a computer-generated stem-and-leaf plot, produced by MINITAB, for the PM data in Table 1.2 in Section 1.2. The middle column, consisting of 0s, 1s, and 2s, contains the stems, which

```
Stem-and-leaf of HiAltitude    N = 62
Leaf Unit = 1.0

    4   0   1111
   19   0   222222223333333
  (14)  0   44445555555555
   29   0   66666666777777
   15   0   8889999
    8   1   0
    7   1   233
    4   1
    4   1   7
    3   1   89
    1   2
    1   2   3
```

FIGURE 1.6 Stem-and-leaf plot of the PM data in Table 1.2 on page 17, as produced by MINITAB.

are the tens digits. To the right of the stems are the leaves, consisting of the ones digits for each of the sample items. Since many numbers are less than 10, the 0 stem must be assigned several lines, five in this case. Specifically, the first line contains the sample items whose ones digits are either 0 or 1, the next line contains the items whose ones digits are either 2 or 3, and so on. For consistency, all the stems are assigned several lines in the same way, even though there are few enough values for the 1 and 2 stems that they could have fit on fewer lines.

The output in Figure 1.6 contains a cumulative frequency column to the left of the stem-and-leaf plot. The upper part of this column provides a count of the number of items at or above the current line, and the lower part of the column provides a count of the number of items at or below the current line. Next to the line that contains the median is the count of items in that line, shown in parentheses.

A good feature of stem-and-leaf plots is that they display all the sample values. One can reconstruct the sample in its entirety from a stem-and-leaf plot—with one important exception: The order in which the items were sampled cannot be determined.

Dotplots

A **dotplot** is a graph that can be used to give a rough impression of the shape of a sample. It is useful when the sample size is not too large and when the sample contains some repeated values. Figure 1.7 (page 22) presents a dotplot for the data in Table 1.3. For each value in the sample, a vertical column of dots is drawn, with the number of dots in the column equal to the number of times the value appears in the sample. The dotplot gives a good indication of where the sample values are concentrated and where the gaps are. For example, it is easy to see from Figure 1.7 that the sample contains no subjects with values between periods between 42 and 50. In addition, the outlier is clearly visible as the rightmost point on the plot.

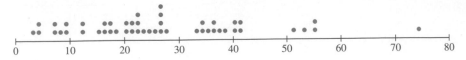

FIGURE 1.7 Dotplot for the data in Table 1.3.

Stem-and-leaf plots and dotplots are good methods for informally examining a sample, and they can be drawn fairly quickly with pencil and paper. They are rarely used in formal presentations, however. Graphics more commonly used in formal presentations include the histogram and the boxplot, which we will now discuss.

Histograms

A **histogram** is a graphic that gives an idea of the "shape" of a sample, indicating regions where sample points are concentrated and regions where they are sparse. We will construct a histogram for the PM emissions of 62 vehicles driven at high altitude, as presented in Table 1.2 (Section 1.2). The sample values range from a low of 1.11 to a high of 23.38, in units of grams of emissions per gallon of fuel. The first step is to construct a **frequency table**, shown as Table 1.4.

TABLE 1.4 Frequency table for PM emissions of 62 vehicles driven at high altitude

Class Interval (g/gal)	Frequency	Relative Frequency
$1 \leq x < 3$	12	0.1935
$3 \leq x < 5$	11	0.1774
$5 \leq x < 7$	18	0.2903
$7 \leq x < 9$	9	0.1452
$9 \leq x < 11$	5	0.0806
$11 \leq x < 13$	1	0.0161
$13 \leq x < 15$	2	0.0323
$15 \leq x < 17$	0	0.0000
$17 \leq x < 19$	2	0.0323
$19 \leq x < 21$	1	0.0161
$21 \leq x < 23$	0	0.0000
$23 \leq x < 25$	1	0.0161

The intervals in the left-hand column are called **class intervals**. They divide the sample into groups. For the histograms that we will consider, the class intervals will all have the same width. In Table 1.4, all classes have width 2. There is no hard and fast rule as to how to decide how many class intervals to use. In general, it is good to have more intervals rather than fewer, but it is also good to have large numbers of sample points in the intervals. Striking the proper balance is a matter of judgment and of trial and error. When the number of observations n is large (several hundred or more), some have suggested that reasonable starting points for the number of classes may be $\log_2 n$ or $2n^{1/3}$. When the number of observations is smaller, more classes than these are often needed.

The column labeled "Frequency" in Table 1.4 presents the numbers of data points that fall into each of the class intervals. The column labeled "Relative Frequency" presents the frequencies divided by the total number of data points, which for these data is 62. The relative frequency of a class interval is the proportion of data points that fall into the interval. Note that since every data point is in exactly one class interval, the relative frequencies must sum to 1 (allowing for round-off error).

Figure 1.8 presents a histogram for Table 1.4. The units on the horizontal axis are the units of the data—in this case, grams per gallon. Each class interval is represented by a rectangle. The heights of the rectangles may be set equal to the frequencies or to the relative frequencies. Since these quantities are proportional, the shape of the histogram will be the same in each case. For the histogram in Figure 1.8, the heights of the rectangles are the relative frequencies.

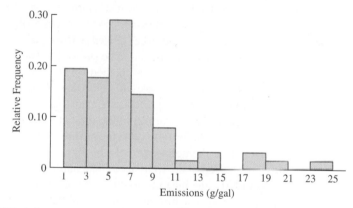

FIGURE 1.8 Histogram for the data in Table 1.4. In this histogram the heights of the rectangles are the relative frequencies. The frequencies and relative frequencies are proportional to each other, so it would have been equally appropriate to set the heights equal to the frequencies.

Summary

To construct a histogram:

- Determine the number of classes to use, and construct class intervals of equal width.
- Compute the frequency and relative frequency for each class.
- Draw a rectangle for each class. The heights of the rectangles may be set equal to the frequencies or to the relative frequencies.

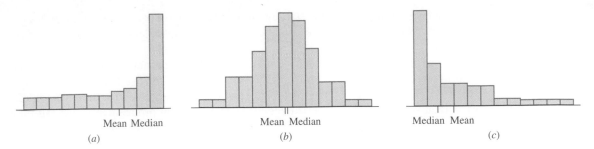

FIGURE 1.9 *(a)* A histogram skewed to the left. The mean is less than the median. *(b)* A nearly symmetric histogram. The mean and median are approximately equal. *(c)* A histogram skewed to the right. The mean is greater than the median.

Symmetry and Skewness

A histogram is perfectly **symmetric** if its right half is a mirror image of its left half. Histograms that are not symmetric are referred to as **skewed**. In practice, virtually no sample has a perfectly symmetric histogram; almost all exhibit some degree of skewness. In a skewed histogram, one side, or tail, is longer than the other. A histogram with a long right-hand tail is said to be **skewed to the right**, or **positively skewed**. A histogram with a long left-hand tail is said to be **skewed to the left**, or **negatively skewed**. While there is a formal mathematical method for measuring the skewness of a histogram, it is rarely used; instead, people judge the degree of skewness informally by looking at the histogram. Figure 1.9 presents some histograms for hypothetical samples. Note that for a histogram that is skewed to the right (Figure 1.9c), the mean is greater than the median. The reason for this is that the mean is near the center of mass of the histogram; that is, it is near the point where the histogram would balance if supported there. For a histogram skewed to the right, more than half the data will be to the left of the center of mass. Similarly, the mean is less than the median for a histogram that is skewed to the left (Figure 1.9a). The histogram for the PM data (Figure 1.8) is skewed to the right. The sample mean is 6.596, which is greater than the sample median of 5.75.

Unimodal and Bimodal Histograms

We have used the term "mode" to refer to the most frequently occurring value in a sample. This term is also used in regard to histograms and other curves to refer to a peak, or local maximum. A histogram is **unimodal** if it has only one peak, or mode, and **bimodal** if it has two clearly distinct modes. In principle, a histogram can have more than two modes, but this does not happen often in practice. The histograms in Figure 1.9 are all unimodal. Figure 1.10 presents a bimodal histogram for a hypothetical sample.

In some cases, a bimodal histogram indicates that the sample can be divided into two subsamples that differ from each other in some scientifically important way. Each sample corresponds to one of the modes. As an example, the data in Table 1.5 concern the geyser Old Faithful in Yellowstone National Park. This geyser alternates periods of

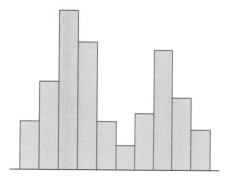

FIGURE 1.10 A bimodal histogram.

eruption, which typically last from 1.5 to 4 minutes, with periods of dormancy, which are considerably longer. Table 1.5 presents the durations, in minutes, of 60 dormant periods. Along with the durations of the dormant period, the duration of the eruption immediately preceding the dormant period is classified either as short (less than 3 minutes) or long (more than 3 minutes).

Figure 1.11a (page 26) presents a histogram for all 60 durations. Figures 1.11b and 1.11c present histograms for the durations following short and long eruptions, respectively. The histogram for all the durations is clearly bimodal. The histograms for the durations following short or long eruptions are both unimodal, and their modes form the two modes of the histogram for the full sample.

TABLE 1.5 Durations of dormant periods (in minutes) and of the previous eruptions of the geyser Old Faithful

Dormant	Eruption	Dormant	Eruption	Dormant	Eruption	Dormant	Eruption
76	Long	90	Long	45	Short	84	Long
80	Long	42	Short	88	Long	70	Long
84	Long	91	Long	51	Short	79	Long
50	Short	51	Short	80	Long	60	Long
93	Long	79	Long	49	Short	86	Long
55	Short	53	Short	82	Long	71	Long
76	Long	82	Long	75	Long	67	Short
58	Short	51	Short	73	Long	81	Long
74	Long	76	Long	67	Long	76	Long
75	Long	82	Long	68	Long	83	Long
80	Long	84	Long	86	Long	76	Long
56	Short	53	Short	72	Long	55	Short
80	Long	86	Long	75	Long	73	Long
69	Long	51	Short	75	Long	56	Short
57	Long	85	Long	66	Short	83	Long

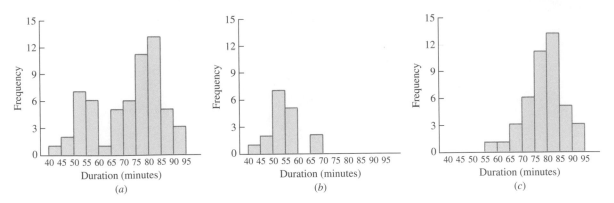

FIGURE 1.11 *(a)* Histogram for all 60 durations in Table 1.5. This histogram is bimodal. *(b)* Histogram for the durations in Table 1.5 that follow short eruptions. *(c)* Histogram for the durations in Table 1.5 that follow long eruptions. The histograms for the durations following short eruptions and for those following long eruptions are both unimodal, but the modes are in different places. When the two samples are combined, the histogram is bimodal.

Boxplots

A **boxplot** is a graphic that presents the median, the first and third quartiles, and any outliers that are present in a sample. Boxplots are easy to understand, but there is a bit of terminology that goes with them. The **interquartile range** is the difference between the third quartile and the first quartile. Note that since 75% of the data is less than the third quartile, and 25% of the data is less than the first quartile, it follows that 50%, or half, of the data is between the first and third quartiles. The interquartile range is therefore the distance needed to span the middle half of the data.

We have defined outliers as points that are unusually large or small. If IQR represents the interquartile range, then for the purpose of drawing boxplots, any point that is more than 1.5 IQR above the third quartile, or more than 1.5 IQR below the first quartile, is considered an outlier. Some texts define a point that is more than 3 IQR from the first or third quartile as an **extreme outlier**. These definitions of outliers are just conventions for drawing boxplots and need not be used in other situations.

Figure 1.12 presents a boxplot for some hypothetical data. The plot consists of a box whose bottom side is the first quartile and whose top side is the third quartile. A horizontal line is drawn at the median. The "outliers" are plotted individually and are indicated by crosses in the figure. Extending from the top and bottom of the box are vertical lines called "whiskers." The whiskers end at the most extreme data point that is not an outlier.

Apart from any outliers, a boxplot can be thought of as having four pieces: the two parts of the box separated by the median line, and the two whiskers. Again, apart from outliers, each of these four parts represents one-quarter of the data. The boxplot therefore indicates how large an interval is spanned by each quarter of the data, and in this way it can be used to determine the regions in which the sample values are more densely crowded and the regions in which they are more sparse.

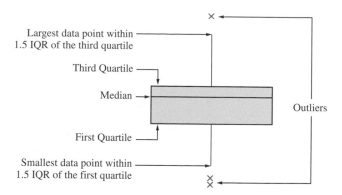

FIGURE 1.12 Anatomy of a boxplot.

Steps in the Construction of a Boxplot

- Compute the median and the first and third quartiles of the sample. Indicate these with horizontal lines. Draw vertical lines to complete the box.

- Find the largest sample value that is no more than 1.5 IQR above the third quartile, and the smallest sample value that is no more than 1.5 IQR below the first quartile. Extend vertical lines (whiskers) from the quartile lines to these points.

- Points more than 1.5 IQR above the third quartile, or more than 1.5 IQR below the first quartile, are designated as outliers. Plot each outlier individually.

Figure 1.13 (page 28) presents a boxplot for the geyser data presented in Table 1.5. First note that there are no outliers in these data. Comparing the four pieces of the boxplot, we can tell that the sample values are comparatively densely packed between the median and the third quartile, and more sparse between the median and the first quartile. The lower whisker is a bit longer than the upper one, indicating that the data have a slightly longer lower tail than an upper tail. Since the distance between the median and the first quartile is greater than the distance between the median and the third quartile, and since the lower quarter of the data produces a longer whisker than the upper quarter, this boxplot suggests that the data are skewed to the left.

A histogram for these data was presented in Figure 1.11a. The histogram presents a more general impression of the spread of the data. Importantly, the histogram indicates that the data are bimodal, which a boxplot cannot do.

Comparative Boxplots

A useful feature of boxplots is that several of them may be placed side by side, allowing for easy visual comparison of the features of several samples. Tables 1.1 and 1.2 (in Section 1.2) presented PM emissions data for vehicles driven at high and low altitudes. Figure 1.14 (page 28) presents a side-by-side comparison of the boxplots for these two samples.

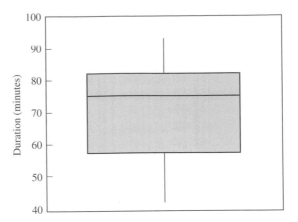

FIGURE 1.13 Boxplot for the Old Faithful dormant period data in Table 1.5.

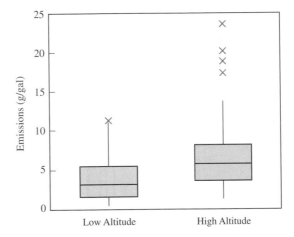

FIGURE 1.14 Comparative boxplots for PM emissions for vehicles driven at high versus low altitudes.

The comparative boxplots in Figure 1.14 show that vehicles driven at low altitude tend to have lower emissions. In addition, there are several outliers among the data for high-altitude vehicles whose values are much higher than any of the values for the low-altitude vehicles (there is also one low-altitude value that barely qualifies as an outlier). We conclude that at high altitudes, vehicles have somewhat higher emissions in general and that a few vehicles have much higher emissions. The box for the high-altitude vehicles is a bit taller, and the lower whisker a bit longer, than that for the low-altitude vehicles. We conclude that apart from the outliers, the spread in values is slightly larger for the high-altitude vehicles and is much larger when the outliers are considered.

In Figure 1.4 (in Section 1.2), we compared the values of some numerical descriptive statistics for these two samples and reached some conclusions similar to the previous ones. The visual nature of the comparative boxplots in Figure 1.14 makes comparing the features of samples much easier.

We have mentioned that it is important to scrutinize outliers to determine whether they have resulted from errors, in which case they may be deleted. By identifying outliers, boxplots can be useful in this regard. The following example provides an illustration.

The article "Virgin Versus Recycled Wafers for Furnace Qualification: Is the Expense Justified?" (V. Czitrom and J. Reece, in *Statistical Case Studies for Industrial Process Improvement*, ASA and SIAM, 1997:87–104) describes a process for growing a thin silicon dioxide layer onto silicon wafers that are to be used in semiconductor manufacture. Table 1.6 presents thickness measurements, in angstroms (Å), of the oxide layer for 24 wafers. Nine measurements were made on each wafer. The wafers were produced in two separate runs, with 12 wafers in each run.

TABLE 1.6 Oxide layer thicknesses for silicon wafers

	Wafer				Thicknesses (Å)					
Run 1	1	90.0	92.2	94.9	92.7	91.6	88.2	92.0	98.2	96.0
	2	91.8	94.5	93.9	77.3	92.0	89.9	87.9	92.8	93.3
	3	90.3	91.1	93.3	93.5	87.2	88.1	90.1	91.9	94.5
	4	92.6	90.3	92.8	91.6	92.7	91.7	89.3	95.5	93.6
	5	91.1	89.8	91.5	91.5	90.6	93.1	88.9	92.5	92.4
	6	76.1	90.2	96.8	84.6	93.3	95.7	90.9	100.3	95.2
	7	92.4	91.7	91.6	91.1	88.0	92.4	88.7	92.9	92.6
	8	91.3	90.1	95.4	89.6	90.7	95.8	91.7	97.9	95.7
	9	96.7	93.7	93.9	87.9	90.4	92.0	90.5	95.2	94.3
	10	92.0	94.6	93.7	94.0	89.3	90.1	91.3	92.7	94.5
	11	94.1	91.5	95.3	92.8	93.4	92.2	89.4	94.5	95.4
	12	91.7	97.4	95.1	96.7	77.5	91.4	90.5	95.2	93.1
Run 2	1	93.0	89.9	93.6	89.0	93.6	90.9	89.8	92.4	93.0
	2	91.4	90.6	92.2	91.9	92.4	87.6	88.9	90.9	92.8
	3	91.9	91.8	92.8	96.4	93.8	86.5	92.7	90.9	92.8
	4	90.6	91.3	94.9	88.3	87.9	92.2	90.7	91.3	93.6
	5	93.1	91.8	94.6	88.9	90.0	97.9	92.1	91.6	98.4
	6	90.8	91.5	91.5	91.5	94.0	91.0	92.1	91.8	94.0
	7	88.0	91.8	90.5	90.4	90.3	91.5	89.4	93.2	93.9
	8	88.3	96.0	92.8	93.7	89.6	89.6	90.2	95.3	93.0
	9	94.2	92.2	95.8	92.5	91.0	91.4	92.8	93.6	91.0
	10	101.5	103.1	103.2	103.5	96.1	102.5	102.0	106.7	105.4
	11	92.8	90.8	92.2	91.7	89.0	88.5	87.5	93.8	91.4
	12	92.1	93.4	94.0	94.7	90.8	92.1	91.2	92.3	91.1

The 12 wafers in each run were of several different types and were processed in several different furnace locations. The purpose in collecting the data was to determine whether the thickness of the oxide layer was affected by either the type of wafer or the furnace location. This was therefore a factorial experiment, with wafer type and furnace location as the factors and oxide layer thickness as the outcome. The experiment was intended to produce no systematic difference in the thicknesses between one run and another. The first step in the analysis was to construct a boxplot for the data in each

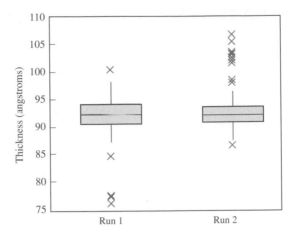

FIGURE 1.15 Comparative boxplots for oxide layer thickness.

run to help determine whether this condition was in fact met and whether any of the observations should be deleted. The results are presented in Figure 1.15.

The boxplots show that there were several outliers in each run. Note that apart from these outliers, there are no striking differences between the samples and therefore no evidence of any systematic difference between the runs. The next task is to inspect the outliers, to determine which, if any, should be deleted. By examining the data in Table 1.6, we can see that the eight largest measurements in run 2 occurred on a single wafer: number 10.

It was then determined that this wafer had been contaminated with a film residue, which caused the large thickness measurements. It would therefore be appropriate to delete these measurements. In the actual experiment, the engineers had data from several other runs available, and for technical reasons, they decided to delete the entire run, rather than to analyze a run that was missing one wafer. In run 1, the three smallest measurements were found to have been caused by a malfunctioning gauge and were therefore appropriately deleted. No cause could be determined for the remaining two outliers in run 1, so they were included in the analysis.

Exercises for Section 1.3

1. As part of a quality-control study aimed at improving a production line, the weights (in ounces) of 50 bars of soap are measured. The results are as follows, sorted from smallest to largest.

11.6 12.6 12.7 12.8 13.1 13.3 13.6 13.7 13.8 14.1
14.3 14.3 14.6 14.8 15.1 15.2 15.6 15.6 15.7 15.8
15.8 15.9 15.9 16.1 16.2 16.2 16.3 16.4 16.5 16.5
16.5 16.6 17.0 17.1 17.3 17.3 17.4 17.4 17.4 17.6
17.7 18.1 18.3 18.3 18.3 18.5 18.5 18.8 19.2 20.3

a. Construct a stem-and-leaf plot for these data.
b. Construct a histogram for these data.
c. Construct a dotplot for these data.
d. Construct a boxplot for these data. Does the boxplot show any outliers?

2. Forty-five specimens of a certain type of powder were analyzed for sulfur trioxide content. Following are the

results, in percent. The list has been sorted into numerical order.

14.1	14.4	14.7	14.8	15.3	15.6	16.1	16.6	17.3
14.2	14.4	14.7	14.9	15.3	15.7	16.2	17.2	17.3
14.3	14.4	14.8	15.0	15.4	15.7	16.4	17.2	17.8
14.3	14.4	14.8	15.0	15.4	15.9	16.4	17.2	21.9
14.3	14.6	14.8	15.2	15.5	15.9	16.5	17.2	22.4

a. Construct a stem-and-leaf plot for these data.

b. Construct a histogram for these data.

c. Construct a dotplot for these data.

d. Construct a boxplot for these data. Does the boxplot show any outliers?

3. Refer to Table 1.2 (page 17). Construct a stem-and-leaf plot with the ones digit as the stem (for values greater than or equal to 10 the stem will have two digits) and the tenths digit as the leaf. How many stems are there (be sure to include leafless stems)? What are some advantages and disadvantages of this plot, compared to the one in Figure 1.6 (page 21)?

4. Two methods were studied for the recovery of protein. Thirteen runs were made using each method, and the fraction of protein recovered was recorded for each run. The results are as follows:

Method 1	Method 2
0.32	0.25
0.35	0.40
0.37	0.48
0.39	0.55
0.42	0.56
0.47	0.58
0.51	0.60
0.58	0.65
0.60	0.70
0.62	0.76
0.65	0.80
0.68	0.91
0.75	0.99

a. Construct a histogram for the results of each method.

b. Construct comparative boxplots for the two methods.

c. Using the boxplots, what differences can be seen between the results of the two methods?

5. A certain reaction was run several times using each of two catalysts, A and B. The catalysts were supposed to control the yield of an undesirable side product. Results, in units of percentage yield, for 24 runs of catalyst A and 20 runs of catalyst B are as follows:

Catalyst A			
4.4	3.4	2.6	3.8
4.9	4.6	5.2	4.7
4.1	2.6	6.7	4.1
3.6	2.9	2.6	4.0
4.3	3.9	4.8	4.5
4.4	3.1	5.7	4.5

Catalyst B			
3.4	1.1	2.9	5.5
6.4	5.0	5.8	2.5
3.7	3.8	3.1	1.6
3.5	5.9	6.7	5.2
6.3	2.6	4.3	3.8

a. Construct a histogram for the yields of each catalyst.

b. Construct comparative boxplots for the yields of the two catalysts.

c. Using the boxplots, what differences can be seen between the results of the yields of the two catalysts?

6. Sketch a histogram for which

a. The mean is greater than the median.

b. The mean is less than the median.

c. The mean is approximately equal to the median.

7. The following histogram presents the distribution of systolic blood pressure for a sample of women. Use it to answer the following questions.

a. Is the percentage of women with blood pressures above 130 mm closest to 25%, 50%, or 75%?

b. In which interval are there more women: 130–135 or 140–150 mm?

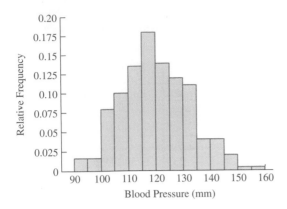

8. The following histogram presents the amounts of silver (in parts per million [ppm]) found in a sample of rocks. One rectangle from the histogram is missing. What is its height?

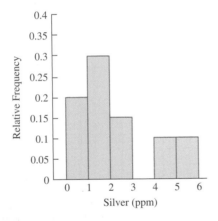

9. An engineer wants to draw a boxplot for the following sample:

37 82 20 25 31 10 41 44 4 36 68

Which of these values, if any, will be labeled as outliers?

10. Which of the following statistics *cannot* be determined from a boxplot?

   i. The median
   ii. The mean
   iii. The first quartile
   iv. The third quartile
   v. The interquartile range

11. A sample of 100 men has average height 70 in. and standard deviation 2.5 in. A sample of 100 women has average height 64 in. and standard deviation 2.5 in. If both samples are combined, the standard deviation of all 200 heights will be _____.

   i. less than 2.5 in.
   ii. greater than 2.5 in.
   iii. equal to 2.5 in.
   iv. can't tell from the information given

(*Hint:* Don't do any calculations. Just try to sketch, very roughly, histograms for each sample separately, and then one for the combined sample.)

12. Following are boxplots comparing the charge (in coulombs per mole [C/mol] $\times$ 10^{-25}) at pH 4.0 and pH 4.5 for a collection of proteins (from the article "Optimal Synthesis of Protein Purification Processes," E. Vasquez-Alvarez, M. Leinqueo, and J. Pinto, *Biotechnology Progress* 2001:685–695). True or false:

   a. The median charge for the pH of 4.0 is greater than the 75th percentile of charge for the pH of 4.5.
   b. Approximately 25% of the charges for pH 4.5 are less than the smallest charge at pH 4.0.
   c. About half the sample values for pH 4.0 are between 2 and 4.
   d. There is a greater proportion of values outside the box for pH 4.0 than for pH 4.5.
   e. Both samples are skewed to the right.
   f. Both samples contain outliers.

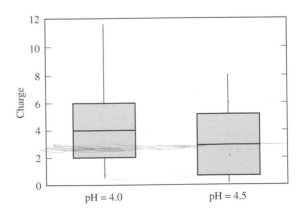

13. Following are summary statistics for two data sets, A and B.

	A	B
Minimum	0.066	−2.235
1st Quartile	1.42	5.27
Median	2.60	8.03
3rd Quartile	6.02	9.13
Maximum	10.08	10.51

a. Compute the interquartile ranges for both A and B.

b. Do the summary statistics for A provide enough information to construct a boxplot? If so, construct the boxplot. If not, explain why.

c. Do the summary statistics for B provide enough information to construct a boxplot? If so, construct the boxplot. If not, explain why.

14. Match each histogram to the boxplot that represents the same data set.

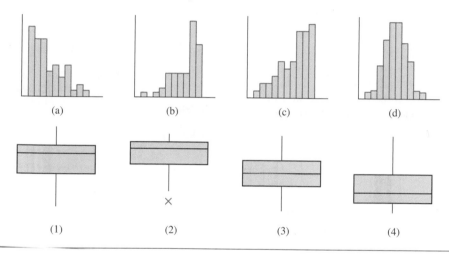

15. Refer to the asphalt data in Example 1.8 (page 16).

   a. Construct a boxplot for the asphalt data.

   b. Which values, if any, are outliers?

   c. Construct a dotplot for the asphalt data.

   d. For purposes of constructing boxplots, an outlier is defined to be a point whose distance from the nearest quartile is more than 1.5 IQR. A more general, and less precise, definition is that an outlier is any point that is detached from the bulk of the data. Are any points in the asphalt data set outliers under this more general definition but not under the boxplot definition? If so, which are they?

Supplementary Exercises for Chapter 1

1. In a certain company, every worker received a 5% raise. How does this affect the mean salary? The standard deviation of the salaries?

2. Suppose that every worker in a company received a $2000-per-year raise. How does this affect the mean salary? The standard deviation of the salaries?

3. Integrated circuits consist of electric channels that are etched onto silicon wafers. A certain proportion of circuits are defective because of "undercutting," which occurs when too much material is etched away so that the channels, which consist of the unetched portions of the wafers, are too narrow. A redesigned process, involving lower pressure in the etching chamber, is being investigated. The goal is to reduce the rate of undercutting to less than 5%. Out of the first 100 circuits manufactured by the new process, only 4 show evidence of undercutting. True or false:

   a. Since only 4% of the 100 circuits had undercutting, we can conclude that the goal has been reached.

b. Although the sample percentage is under 5%, this may represent sampling variation, so the goal may not yet be reached.

c. There is no use in testing the new process, because no matter what the result is, it could just be due to sampling variation.

d. If we sample a large enough number of circuits, and if the percentage of defective circuits is far enough below 5%, then it is reasonable to conclude that the goal has been reached.

4. A coin is tossed twice and comes up heads both times. Someone says, "There's something wrong with this coin. A coin is supposed to come up heads only half the time, not every time."

a. Is it reasonable to conclude that something is wrong with the coin? Explain.

b. If the coin came up heads 100 times in a row, would it be reasonable to conclude that something is wrong with the coin? Explain.

5. The smallest number on a list is changed from 12.9 to 1.29.

a. Is it possible to determine by how much the mean changes? If so, by how much does it change?

b. Is it possible to determine by how much the median changes? If so, by how much does it change? What if the list consists of only two numbers?

c. Is it possible to determine by how much the standard deviation changes? If so, by how much does it change?

6. There are 15 numbers on a list, and the smallest number is changed from 12.9 to 1.29.

a. Is it possible to determine by how much the mean changes? If so, by how much does it change?

b. Is it possible to determine the value of the mean after the change? If so, what is the value?

c. Is it possible to determine by how much the median changes? If so, by how much does it change?

d. Is it possible to determine by how much the standard deviation changes? If so, by how much does it change?

7. There are 15 numbers on a list, and the mean is 25. The smallest number on the list is changed from 12.9 to 1.29.

a. Is it possible to determine by how much the mean changes? If so, by how much does it change?

b. Is it possible to determine the value of the mean after the change? If so, what is the value?

c. Is it possible to determine by how much the median changes? If so, by how much does it change?

d. Is it possible to determine by how much the standard deviation changes? If so, by how much does it change?

8. The article "The Selection of Yeast Strains for the Production of Premium Quality South African Brandy Base Products" (C. Steger and M. Lambrechts, *Journal of Industrial Microbiology and Biotechnology*, 2000:431–440) presents detailed information on the volatile compound composition of base wines made from each of 16 selected yeast strains. Following are the concentrations of total esters (in mg/L) in each of the wines.

```
284.34   173.01   229.55   312.95   215.34   188.72
144.39   172.79   139.38   197.81   303.28   256.02
658.38   105.14   295.24   170.41
```

a. Compute the mean concentration.

b. Compute the median concentration.

c. Compute the first quartile of the concentrations.

d. Compute the third quartile of the concentrations.

e. Construct a boxplot for the concentrations. What features does it reveal?

9. Concerning the data represented in the following boxplot, which one of the following statements is true?

i. The mean is greater than the median.

ii. The mean is less than the median.

iii. The mean is approximately equal to the median.

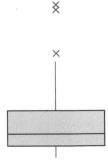

10. In the article "Occurrence and Distribution of Ammonium in Iowa Groundwater" (K. Schilling, *Water Environment Research*, 2002:177–186), ammonium concentrations (in mg/L) were measured at a total of 349 alluvial wells in the state of Iowa. The mean concentration was 0.27, the median was 0.10, and the standard deviation was 0.40. If a histogram of these 349 measurements were drawn, _____

   i. it would be skewed to the right.

   ii. it would be skewed to the left.

   iii. it would be approximately symmetric.

   v. its shape could not be determined without knowing the relative frequencies.

11. The article "Vehicle-Arrival Characteristics at Urban Uncontrolled Intersections" (V. Rengaraju and V. Rao, *Journal of Transportation Engineering*, 1995: 317–323) presents data on traffic characteristics at 10 intersections in Madras, India. One characteristic measured was the speeds of the vehicles traveling through the intersections. The accompanying table gives the 15th, 50th, and 85th percentiles of speed (in km/h) for two intersections.

	Percentile		
Intersection	15th	50th	85th
A	27.5	37.5	40.0
B	24.5	26.5	36.0

  a. If a histogram for speeds of vehicles through intersection A were drawn, do you think it would be skewed to the left, skewed to the right, or approximately symmetric? Explain.

  b. If a histogram for speeds of vehicles through intersection B were drawn, do you think it would be skewed to the left, skewed to the right, or approximately symmetric? Explain.

12. The *cumulative frequency* and the *cumulative relative frequency* for a given class interval are the sums of the frequencies and relative frequencies, respectively, over all classes up to and including the given class. For example, if there are five classes, with frequencies 11, 7, 3, 14, and 5, the cumulative frequencies would be 11, 18, 21, 35, and 40, and the cumulative relative frequencies would be 0.275, 0.450,

0.525, 0.875, and 1.000. Construct a table presenting frequencies, relative frequencies, cumulative frequencies, and cumulative relative frequencies, for the data in Exercise 1 of Section 1.3, using the class intervals $11 \leq x < 12, 12 \leq x < 13, \ldots, 20 \leq x < 21$.

13. The article "The Ball-on-Three-Ball Test for Tensile Strength: Refined Methodology and Results for Three Hohokam Ceramic Types" (M. Beck, *American Antiquity*, 2002:558–569) discusses the strength of ancient ceramics. Several specimens of each of three types of ceramic were tested. The loads (in kg) required to crack the specimens are as follows:

Ceramic Type	Loads (kg)
Sacaton	15 30 51 20 17 19 20 32 17 15 23 19 15 18 16 22 29 15 13 15
Gila Plain	27 18 28 25 55 21 18 34 23 30 20 30 31 25 28 26 17 19 16 24 19 9 31 19 27 20 43 15
Casa Grande	20 16 20 36 27 35 66 15 18 24 21 30 20 24 23 21 13 21

  a. Construct comparative boxplots for the three samples.

  b. How many outliers does each sample contain?

  c. Comment on the features of the three samples.

14. The article "Hydrogeochemical Characteristics of Groundwater in a Mid-Western Coastal Aquifer System" (S. Jeen, J. Kim, et al., *Geosciences Journal*, 2001:339–348) presents measurements of various properties of shallow groundwater in a certain aquifer system in Korea. Following are measurements of electrical conductivity (in microsiemens per centimeter) for 23 water samples.

2099	528	2030	1350	1018	384	1499
1265	375	424	789	810	522	513
488	200	215	486	257	557	260
461	500					

  a. Find the mean.

  b. Find the standard deviation.

c. Find the median.

d. Construct a dotplot.

e. Find the first quartile.

f. Find the third quartile.

g. Find the interquartile range.

h. Construct a boxplot.

i. Which of the points, if any, are outliers?

j. If a histogram were constructed, would it be skewed to the left, skewed to the right, or approximately symmetric?

Chapter 2

Summarizing Bivariate Data

Introduction

Scientists and engineers often collect data in order to determine the nature of a relationship between two quantities. For example, a chemical engineer may run a chemical process several times in order to study the relationship between the concentration of a certain catalyst and the yield of the process. Each time the process is run, the concentration x and the yield y are recorded. The experiment thus generates a collection of ordered pairs $(x_1, y_1), \ldots, (x_n, y_n)$, where n is the number of runs. Data that consist of ordered pairs are called **bivariate** data. In many cases, ordered pairs generated in a scientific experiment tend to cluster around a straight line when plotted. In these situations, the main question is usually to determine how closely the two quantities are related to each other. The summary statistic most often used to measure the closeness of the association between two variables is the **correlation coefficient**, which we will study in Section 2.1. When two variables are closely related to each other, it is often of interest to try to predict the value of one of them when given the value of the other. This is often done with the equation of a line known as the **least-squares line**, which we will study in Sections 2.2 and 2.3.

2.1 The Correlation Coefficient

The article "Advances in Oxygen Equivalence Equations for Predicting the Properties of Titanium Welds" (D. Harwig, W. Ittiwattana, and H. Castner, *The Welding Journal*, 2001:126s–136s) presents data concerning the chemical composition and strength characteristics of a number of titanium welds. One of the goals of the research reported in this article was to discover factors that could be used to predict the strength of welds. Figure 2.1a (page 38) is a plot of the yield strength (in thousands of pounds per square

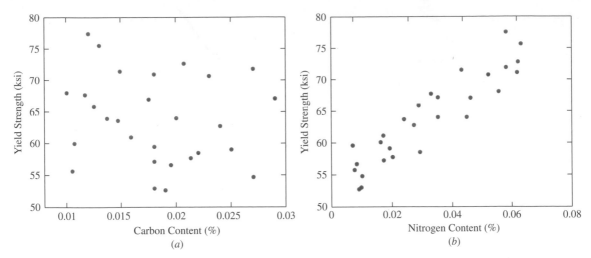

FIGURE 2.1 *(a)* A scatterplot showing that there is not much of a relationship between carbon content and yield strength for a certain group of welds. *(b)* A scatterplot showing that for these same welds, higher nitrogen content is associated with higher yield strength.

inch [ksi]) versus carbon content (in %) for some of these welds. Figure 2.1b is a plot of the yield strength (in ksi) versus nitrogen content (in %) for the same welds. Figures 2.1a and 2.1b are examples of **scatterplots**. When data consist of ordered pairs, a scatterplot is constructed simply by plotting each point on a two-dimensional coordinate system.

The scatterplot of yield strength versus nitrogen content (Figure 2.1b) shows some clear structure—the points seem to be following a line from lower left to upper right. In this way, the scatterplot illustrates a relationship between nitrogen content and yield strength: Welds with higher nitrogen content tend to have higher yield strength. This scatterplot might lead investigators to try to predict strength from nitrogen content. In contrast, there does not seem to be much structure to the scatterplot of yield strength versus carbon content (Figure 2.1a), and thus there is no evidence of a relationship between these two quantities. This scatterplot would discourage investigators from trying to predict strength from carbon content.

Looking again at Figure 2.1b, it is apparent that the points tend to follow a line with positive slope. Figure 2.2 presents the scatterplot with a line drawn through it. It is clear that the points tend to cluster around this line. This line is a special line known as the **least-squares line**. It is the line that fits the data best, in a sense to be described later. We will learn how to compute the least-squares line in Section 2.2.

In general, the degree to which the points in a scatterplot tend to cluster around a line reflects the strength of the linear relationship between *x* and *y*. We would like a way to measure the strength of this linear relationship. The visual impression of a scatterplot can be misleading in this regard, because changing the scale of the axes can make the clustering appear tighter or looser. For this reason, we define the **correlation coefficient**, which is a numerical measure of the strength of the linear relationship between two

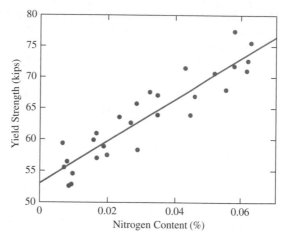

FIGURE 2.2 Yield strength versus nitrogen content for a sample of welds. The least-squares line is superimposed.

variables. The correlation coefficient is usually denoted by the letter r. There are several equivalent formulas for r. They are all a bit complicated, and it is not immediately obvious how they work. We will present the formulas, then show how they work.

Let $(x_1, y_1), \ldots, (x_n, y_n)$ represent n points on a scatterplot. To compute the correlation, first compute the deviations of the xs and ys, that is, $x_i - \bar{x}$ and $y_i - \bar{y}$, for each x_i and y_i. The correlation coefficient is given by

$$r = \frac{\sum_{i=1}^{n}(x_i - \bar{x})(y_i - \bar{y})}{\sqrt{\sum_{i=1}^{n}(x_i - \bar{x})^2}\sqrt{\sum_{i=1}^{n}(y_i - \bar{y})^2}} \tag{2.1}$$

By performing some algebra on the numerator and denominator of Equation (2.1), we obtain an equivalent formula that is often easiest to use when computing by hand:

$$r = \frac{\sum_{i=1}^{n} x_i y_i - n\bar{x}\bar{y}}{\sqrt{\sum_{i=1}^{n} x_i^2 - n\bar{x}^2}\ \sqrt{\sum_{i=1}^{n} y_i^2 - n\bar{y}^2}} \tag{2.2}$$

It is a mathematical fact that the correlation coefficient is always between -1 and 1. Positive values of the correlation coefficient indicate that the least-squares line has a positive slope, which means that greater values of one variable are associated with greater values of the other. Negative values of the correlation coefficient indicate that the least-squares line has a negative slope, which means that greater values of one variable are associated with lesser values of the other. Values of the correlation coefficient close to 1 or to -1 indicate a strong linear relationship; values close to 0 indicate a weak linear relationship. The correlation coefficient is equal to 1 (or to -1) only when the points in the scatterplot lie exactly on a straight line of positive (or negative) slope—in other words, when there is a perfect linear relationship. As a technical note, if the points lie exactly on a horizontal or a vertical line, the correlation coefficient is undefined, because the denominator in the formula is equal to zero. Finally, a bit of terminology: Whenever $r \neq 0$, x and y are said to be **correlated**. If $r = 0$, x and y are said to be **uncorrelated**.

The correlation between nitrogen content and yield strength in Figure 2.1b is $r = 0.92$. The correlation between carbon content and yield strength in Figure 2.1a is $r = -0.16$. We can interpret these results by saying that nitrogen content and yield strength have a strong positive linear relationship, while carbon content and yield strength have a weak negative linear relationship.

How the Correlation Coefficient Works

How does the formula (Equation 2.1) for the correlation coefficient r measure the strength of the linear association between two variables? First, note that the denominator of the expression in Equation (2.1) is always positive, so the sign of the correlation coefficient is the same as the sign of the numerator, which is $\sum_{i=1}^{n}(x_i - \overline{x})(y_i - \overline{y})$, the sum of the products of the deviations. Figure 2.3 now illustrates how the correlation coefficient works. In this scatterplot, the origin is placed at the point of averages $(\overline{x}, \overline{y})$. Therefore, in the first quadrant, the deviations $(x_i - \overline{x})$ and $(y_i - \overline{y})$ are both positive, so their product is positive as well. Thus, each point in the first quadrant contributes a positive amount to the sum in Equation (2.1). In the second quadrant, the deviations for the x-coordinates of the points are negative, while the deviations for the y-coordinates are positive. Therefore, the products of the deviations are negative, so each point in the second quadrant contributes a negative amount to the sum in Equation (2.1). Similarly, points in the third quadrant contribute positive amounts, and points in the fourth quadrant contribute negative amounts. Clearly, in Figure 2.3 there are more points in the first and third quadrants than in the second and fourth, so the correlation will be positive. If the plot had a negative slope, there would be more points in the second and fourth quadrants, and the correlation coefficient would be negative.

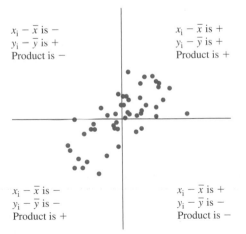

FIGURE 2.3 How the correlation coefficient works.

The Correlation Coefficient Is Unitless

Examination of Equation (2.1) shows that the numerator and the denominator have the same units—both are units of x multiplied by units of y. For this reason, the correlation coefficient is unitless. This fact is important to the usefulness of r, because it allows us to compare the strengths of the linear relationships between two plots even when the units of the quantities on one plot differ from the units of the quantities on the other.

Another important property of the correlation coefficient is that it is unaffected by changes in scale. This means that if each x_i or each y_i is multiplied by a positive constant, the value of the correlation coefficient will be unchanged. As a result, the value of the correlation coefficient does not depend on the units of x or of y. In addition, the value of the correlation coefficient is unchanged if the same constant value is added to each x_i, or to each y_i, or to both. The reason is that the correlation coefficient depends only on the deviations $x_i - \bar{x}$ and $y_i - \bar{y}$. Finally, it can be seen from Equation (2.1) that interchanging the values of x and y does not change the value of the correlation coefficient.

Summary

The correlation coefficient remains unchanged under each of the following operations:

- Multiplying each value of a variable by a positive constant.
- Adding a constant to each value of a variable.
- Interchanging the values of x and y.

Figure 2.4 (page 42) presents plots of mean temperatures for the months of April and October for several U.S. cities. Whether the temperatures are measured in °C or °F, the correlation is the same. The reason is that converting from °C to °F involves multiplying by 1.8 and adding 32.

The Correlation Coefficient Measures Only *Linear* Association

An object is fired upward from the ground with an initial velocity of 64 ft/s. At each of several times $x_1, \ldots, x_n$, the heights $y_1, \ldots, y_n$ of the object above the surface of the earth are measured. In the absence of friction, and assuming that there is no measurement error, the scatterplot of the points $(x_1, y_1), \ldots, (x_n, y_n)$ will look like Figure 2.5 (page 42). The plot shows a strong relationship between x and y; in fact, the value of y is determined by x through the equation $y = 64x - 16x^2$. Yet the correlation between x and y is equal to 0. Is something wrong? No. The value of 0 for the correlation indicates that there is no *linear* relationship between x and y, which is true. The relationship is purely *quadratic*. The lesson of this example is that the correlation coefficient should not be used when the points on a scatterplot tend to follow a curve. The results can then be misleading.

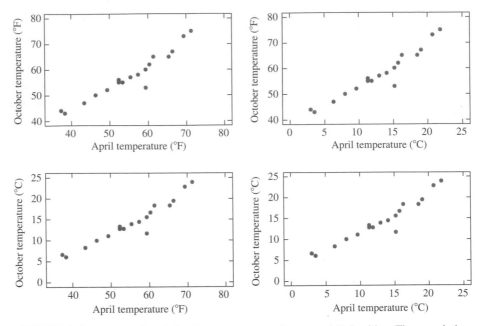

FIGURE 2.4 Mean April and October temperatures for several U.S. cities. The correlation coefficient is 0.96 for each plot; the choice of units does not matter.

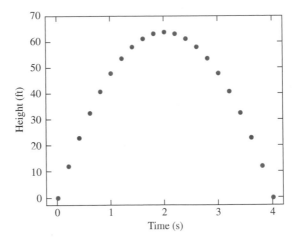

FIGURE 2.5 The relationship between the height of a free falling object with a positive initial velocity and the time in free fall is quadratic. The correlation is equal to 0.

Outliers

In Figure 2.6, the point $(0, 3)$ is an outlier, a point that is detached from the main body of the data. The correlation for this scatterplot is $r = 0.26$, which indicates a weak linear relationship. Yet 10 of the 11 points have a perfect linear relationship. Outliers

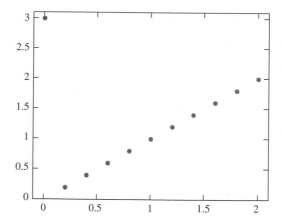

FIGURE 2.6 The correlation is 0.26. Because of the outlier, the correlation coefficient is misleading.

can greatly distort the correlation coefficient, especially in small data sets, and present a serious problem for data analysts. Some outliers are caused by data-recording errors or by failure to follow experimental protocol. It is appropriate to correct these outliers when possible or to delete them if they cannot be corrected. It is tempting to delete outliers from a plot without cause, simply to make the plot easier to interpret. This is not appropriate, because it results in an underestimation of the variability of the process that generated the data. Interpreting data that contain outliers can be difficult, because there are few easy rules to follow.

Correlation Is Not Causation

For children, vocabulary size is strongly correlated with shoe size. However, learning new words does not cause feet to grow, nor do growing feet cause one's vocabulary to increase. There is a third factor—age—that is correlated with both shoe size and vocabulary. Older children tend to have both larger shoe sizes and larger vocabularies, and this results in a positive correlation between vocabulary and shoe size. This phenomenon is known as **confounding**. Confounding occurs when there is a third variable that is correlated with both of the variables of interest, resulting in a correlation between them.

To restate this example in more detail: Individuals with larger ages tend to have larger shoe sizes. Individuals with larger ages also tend to have larger vocabularies. It follows that individuals with larger shoe sizes will tend to have larger vocabularies. In other words, because both shoe size and vocabulary are positively correlated with age, they are positively correlated with each other.

In this example, the confounding was easy to spot. In many cases, it is not so easy. The example shows that simply because two variables are correlated with each other, we cannot assume that a change in one will tend to cause a change in the other. Before we can conclude that two variables have a causal relationship, we must rule out the possibility of confounding.

When planning a scientific experiment, it is important to design the experiment so as to reduce the possibility of confounding by as much as is practical. The topic of **experimental design** (see Chapter 9) is largely concerned with this topic. Here is a simple example.

Example
2.1

An environmental scientist is studying the rate of absorption of a certain chemical into skin. She places differing volumes of the chemical on different pieces of skin and allows the skin to remain in contact with the chemical for varying lengths of time. She then measures the volume of chemical absorbed into each piece of skin. She obtains the results shown in the following table.

Volume (mL)	Time (h)	Percent Absorbed
0.05	2	48.3
0.05	2	51.0
0.05	2	54.7
2.00	10	63.2
2.00	10	67.8
2.00	10	66.2
5.00	24	83.6
5.00	24	85.1
5.00	24	87.8

The scientist plots the percent absorbed against both volume and time, as shown in the following figure. She calculates the correlation between volume and absorption and obtains $r = 0.988$. She concludes that increasing the volume of the chemical causes the percentage absorbed to increase. She then calculates the correlation between time and absorption, obtaining $r = 0.987$. She concludes that increasing the time that the skin is in contact with the chemical causes the percentage absorbed to increase as well. Are these conclusions justified?

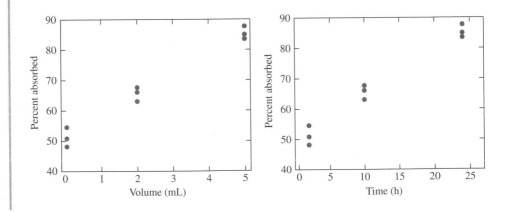

Solution

No. The scientist should look at the plot of time versus volume, presented in the following figure. The correlation between time and volume is $r = 0.999$, so these two variables are almost completely confounded. If *either* time or volume affects the percentage absorbed, *both* will appear to do so, because they are highly correlated with each other. For this reason, it is impossible to determine whether it is the time or the volume that is having an effect. This relationship between time and volume resulted from the design of the experiment and should have been avoided.

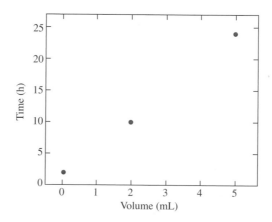

Example 2.2

The scientist in Example 2.1 has repeated the experiment, this time with a new design. The results are presented in the following table.

Volume (mL)	Time (h)	Percent Absorbed
0.05	2	49.2
0.05	10	51.0
0.05	24	84.3
2.00	2	54.1
2.00	10	68.7
2.00	24	87.2
5.00	2	47.7
5.00	10	65.1
5.00	24	88.4

The scientist plots the percent absorbed against both volume and time, as shown in the following figure.

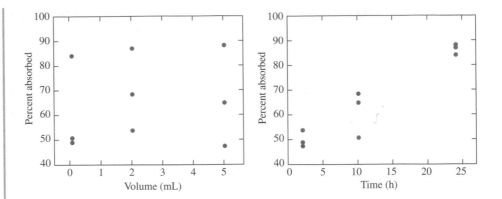

She then calculates the correlation between volume and absorption and obtains $r = 0.121$. She concludes that increasing the volume of the chemical has little or no effect on the percentage absorbed. She then calculates the correlation between time and absorption and obtains $r = 0.952$. She concludes that increasing the time that the skin is in contact with the chemical will cause the percentage absorbed to increase. Are these conclusions justified?

Solution

These conclusions are much better justified than the ones in Example 2.1. To see why, look at the plot of time versus volume in the following figure. This experiment has been designed so that time and volume are uncorrelated. It now appears that the time, but not the volume, has an effect on the percentage absorbed. Before making a final conclusion that increasing the time actually causes the percentage absorbed to increase, the scientist must make sure that no other potential confounders are around. For example, if the ambient temperature varied with each replication of the experiment and was highly correlated with time, then it might be the case that the temperature, rather than the time, was causing the percentage absorbed to vary.

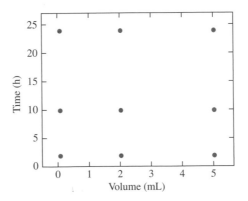

Controlled Experiments Reduce the Risk of Confounding

In Examples 2.1 and 2.2, the experimenter was able to reduce confounding by choosing values for volume and time so that these two variables were uncorrelated. This is a controlled experiment, because the experimenter could choose the values for these factors. In controlled experiments, confounding can often be avoided by choosing values for factors in a way so that the factors are uncorrelated.

Observational studies are studies in which the values of factors cannot be chosen by the experimenter. Studies involving public health issues, such as the effect of environmental pollutants on human health, are usually observational, because experimenters cannot deliberately expose people to high levels of pollution. In these studies, confounding is often difficult to avoid. For example, people who live in areas with higher levels of pollution may tend to have lower socioeconomic status, which may affect their health. Because confounding is difficult to avoid, observational studies must generally be repeated a number of times, under a variety of conditions, before reliable conclusions can be drawn.

Exercises for Section 2.1

1. Compute the correlation coefficient for the following data set.

x	1	2	3	4	5
y	2	1	4	3	7

2. For each of the following data sets, explain why the correlation coefficient is the same as for the data set in Exercise 1.

   a.

x	1	2	3	4	5
y	4	2	8	6	14

   b.

x	13	23	33	43	53
y	4	2	8	6	14

   c.

x	3	2	5	4	8
y	2	4	6	8	10

3. For each of the following scatterplots, state whether the correlation coefficient is an appropriate summary, and explain briefly.

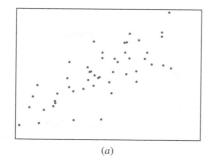

(a)

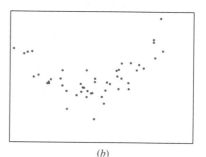

(b)

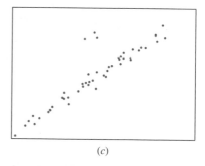

(c)

4. True or false, and explain briefly:

   a. If the correlation coefficient is positive, then above-average values of one variable are associated with above-average values of the other.

   b. If the correlation coefficient is negative, then below-average values of one variable are associated with below-average values of the other.

c. If the correlation between x and y is positive, then x is usually greater than y.

5. An investigator collected data on heights and weights of college students. The correlation between height and weight for men was about 0.6, and for women it was about the same. If men and women are taken together, will the correlation between height and weight be more than 0.6, less than 0.6, or about equal to 0.6? It might be helpful to make a rough scatterplot.

6. In a laboratory test of a new engine design, the emissions rate (in mg/s of oxides of nitrogen, NO_x) was measured as a function of engine speed (in rpm). The results are presented in the following table.

Speed	1671.8	1532.8	1595.6	1487.5	1432.3	1622.2	1577.7
Emissions	323.3	94.6	149.6	60.5	26.3	158.7	106.0

a. Compute the correlation coefficient between emissions and speed.
b. Construct a scatterplot for these data.
c. Is the correlation coefficient an appropriate summary for these data? Explain why or why not.

7. A chemical engineer is studying the effect of temperature and stirring rate on the yield of a certain product. The process is run 16 times, at the settings indicated in the following table. The units for yield are percent of a theoretical maximum.

Temperature (°C)	Stirring Rate (rpm)	Yield (%)
110	30	70.27
110	32	72.29
111	34	72.57
111	36	74.69
112	38	76.09
112	40	73.14
114	42	75.61
114	44	69.56
117	46	74.41
117	48	73.49
122	50	79.18
122	52	75.44
130	54	81.71
130	56	83.03
143	58	76.98
143	60	80.99

a. Compute the correlation between temperature and yield, between stirring rate and yield, and between temperature and stirring rate.
b. Do these data provide good evidence that increasing the temperature causes the yield to increase, within the range of the data? Or might the result be due to confounding? Explain.
c. Do these data provide good evidence that increasing the stirring rate causes the yield to increase, within the range of the data? Or might the result be due to confounding? Explain.

8. Another chemical engineer is studying the same process as in Exercise 7 and uses the following experimental matrix.

Temperature (°C)	Stirring Rate (rpm)	Yield (%)
110	30	70.27
110	40	74.95
110	50	77.91
110	60	82.69
121	30	73.43
121	40	73.14
121	50	78.27
121	60	74.89
132	30	69.07
132	40	70.83
132	50	79.18
132	60	78.10
143	30	73.71
143	40	77.70
143	50	74.31
143	60	80.99

a. Compute the correlation between temperature and yield, between stirring rate and yield, and between temperature and stirring rate.
b. Do these data provide good evidence that the yield is unaffected by temperature, within the range of the data? Or might the result be due to confounding? Explain.
c. Do these data provide good evidence that increasing the stirring rate causes the yield to increase, within the range of the data? Or might the result be due to confounding? Explain.
d. Which experimental design is better, this one or the one in Exercise 7? Explain.

2.2 The Least-Squares Line

When two variables have a linear relationship, the scatterplot tends to be clustered around a line known as the least-squares line (see Figure 2.2 in Section 2.1). In this section, we will learn how to compute the least-squares line and how it can be used to describe bivariate data.

Table 2.1 presents the nitrogen content (in percent) and yield strength (in ksi) for 28 welds. These data were first presented in Figure 2.1b in Section 2.1.

TABLE 2.1 Nitrogen Content and Yield Strength

Nitrogen Content x	Yield Strength (ksi) y	Nitrogen Content x	Yield Strength (ksi) y
0.0088	52.67	0.0169	61.05
0.0200	57.67	0.0286	65.85
0.0350	64.00	0.0430	71.45
0.0460	67.00	0.0580	77.45
0.0617	71.00	0.0630	75.55
0.0620	72.67	0.0099	54.70
0.0074	55.67	0.0190	59.05
0.0160	60.00	0.0270	62.75
0.0237	63.67	0.0350	67.10
0.0327	67.67	0.0520	70.65
0.0555	68.00	0.0580	71.80
0.0447	64.00	0.0290	58.50
0.0068	59.50	0.0170	57.15
0.0081	56.60	0.0096	52.95

Figure 2.7 (page 50) presents the scatterplot of y versus x with the least-squares line superimposed. We write the equation of the line as

$$y = \hat{\beta}_0 + \hat{\beta}_1 x \tag{2.3}$$

The quantities $\hat{\beta}_0$ and $\hat{\beta}_1$ are called the **least-squares coefficients**. The coefficient $\hat{\beta}_1$ is the slope of the least-squares line, and the coefficient $\hat{\beta}_0$ is the y-intercept. The variable represented by x, in this case nitrogen content, is called the **independent variable**. The variable represented by y, in this case yield strength, is called the **dependent variable**.

The least-squares line is the line that fits the data "best." We now define what we mean by "best." For each data point (x_i, y_i), the vertical distance to the point $(x_i, \hat{y}_i)$ on the least-squares line is $e_i = y_i - \hat{y}_i$ (see Figure 2.7). The quantity $\hat{y}_i = \hat{\beta}_0 + \hat{\beta}_1 x_i$ is called the **fitted value**, and the quantity e_i is called the **residual** associated with the point (x_i, y_i). The residual e_i is the difference between the value y_i observed in the data and the fitted value $\hat{y}_i$ predicted by the least-squares line. This is the vertical distance from the point to the line. Points above the least-squares line have positive residuals, and points below the least-squares line have negative residuals. The closer the residuals are to 0, the closer the fitted values are to the observations and the better the line fits the data. We define the least-squares line to be the line for which the sum of the squared residuals

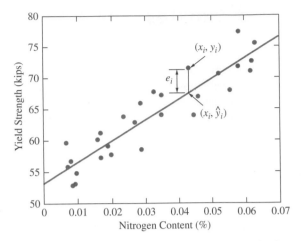

FIGURE 2.7 Plot of yield strengths versus nitrogen content. The least-squares line $y = \hat{\beta}_0 + \hat{\beta}_1 x$ is superimposed. The vertical distance from a data point (x_i, y_i) to the point $(x_i, \hat{y}_i)$ on the line is the ith residual e_i. The least-squares line is the line that minimizes the sum of the squared residuals.

$\sum_{i=1}^{n} e_i^2$ is minimized. In this sense, the least-squares line fits the data better than any other line.

In the weld example, there is only one independent variable (nitrogen content). In other cases, we may use several independent variables. For example, to predict the yield of a certain crop, we might need to know the amount of fertilizer used, the amount of water applied, and various measurements of chemical properties of the soil. Linear models with only one independent variable are called **simple linear regression** models. Linear models with more than one independent variable are called **multiple regression** models. This chapter covers simple linear regression. Multiple regression is covered in Chapter 8.

Computing the Equation of the Least-Squares Line

To compute the equation of the least-squares line, we must determine the values for the slope $\hat{\beta}_1$ and the intercept $\hat{\beta}_0$ that minimize the sum of the squared residuals $\sum_{i=1}^{n} e_i^2$. To do this, we first express e_i in terms of $\hat{\beta}_0$ and $\hat{\beta}_1$:

$$e_i = y_i - \hat{y}_i = y_i - \hat{\beta}_0 - \hat{\beta}_1 x_i \tag{2.4}$$

Therefore, $\hat{\beta}_0$ and $\hat{\beta}_1$ are the quantities that minimize the sum

$$S = \sum_{i=1}^{n} e_i^2 = \sum_{i=1}^{n} (y_i - \hat{\beta}_0 - \hat{\beta}_1 x_i)^2 \tag{2.5}$$

It can be shown that these quantities are

$$\hat{\beta}_1 = \frac{\sum_{i=1}^{n}(x_i - \bar{x})(y_i - \bar{y})}{\sum_{i=1}^{n}(x_i - \bar{x})^2} \tag{2.6}$$

$$\hat{\beta}_0 = \bar{y} - \hat{\beta}_1\bar{x} \tag{2.7}$$

Computing Formulas

The quantities $\sum_{i=1}^{n}(x_i - \bar{x})^2$ and $\sum_{i=1}^{n}(x_i - \bar{x})(y_i - \bar{y})$ need to be computed in order to determine the equation of the least-squares line, and, as we will soon see, the quantity $\sum_{i=1}^{n}(y_i - \bar{y})^2$ needs to be computed in order to determine how well the line fits the data. When computing these quantities by hand, there are alternate formulas that are often easier to use. They are given in the following box.

Computing Formulas

The expressions on the right are equivalent to those on the left and are often easier to compute:

$$\sum_{i=1}^{n}(x_i - \bar{x})^2 = \sum_{i=1}^{n}x_i^2 - n\bar{x}^2 \tag{2.8}$$

$$\sum_{i=1}^{n}(y_i - \bar{y})^2 = \sum_{i=1}^{n}y_i^2 - n\bar{y}^2 \tag{2.9}$$

$$\sum_{i=1}^{n}(x_i - \bar{x})(y_i - \bar{y}) = \sum_{i=1}^{n}x_i y_i - n\bar{x}\,\bar{y} \tag{2.10}$$

Example

2.3

Using the weld data in Table 2.1, compute the least-squares estimates of the slope and intercept of the least-squares line.

Solution
The slope is $\hat{\beta}_1$, and the the intercept is $\hat{\beta}_0$. From Table 2.1 we compute

$$\bar{x} = 0.031943 \qquad \bar{y} = 63.7900$$

$$\sum_{i=1}^{n}(x_i - \bar{x})^2 = \sum_{i=1}^{n}x_i^2 - n\bar{x}^2 = 0.100203$$

$$\sum_{i=1}^{n}(x_i - \bar{x})(y_i - \bar{y}) = \sum_{i=1}^{n}x_i y_i - n\bar{x}\,\bar{y} = 3.322921$$

Using Equations (2.6) and (2.7), we compute

$$\hat{\beta}_1 = \frac{3.322921}{0.100203} = 331.62$$

$$\hat{\beta}_0 = 63.7900 - (331.62)(0.031943) = 53.197$$

The equation of the least-squares line is $y = \hat{\beta}_0 + \hat{\beta}_1 x$. Substituting the computed values for $\hat{\beta}_0$ and $\hat{\beta}_1$, we obtain

$$y = 53.197 + 331.62x$$

Using the equation of the least-squares line, we can compute the fitted values $\hat{y}_i = \hat{\beta}_0 + \hat{\beta}_1 x_i$ and the residuals $e_i = y_i - \hat{y}_i$ for each point (x_i, y_i) in the data set. The results are presented in Table 2.2. The point whose residual is shown in Figure 2.7 is the one where $x = 0.0430$.

TABLE 2.2 Yield strengths of welds with various nitrogen contents, with fitted values and residuals

Nitrogen x	Strength y	Fitted Value $\hat{y}$	Residual e	Nitrogen x	Strength y	Fitted Value $\hat{y}$	Residual e
0.0088	52.67	56.12	−3.45	0.0169	61.05	58.80	2.25
0.0200	57.67	59.83	−2.16	0.0286	65.85	62.68	3.17
0.0350	64.00	64.80	−0.80	0.0430	71.45	67.46	3.99
0.0460	67.00	68.45	−1.45	0.0580	77.45	72.43	5.02
0.0617	71.00	73.66	−2.66	0.0630	75.55	74.09	1.46
0.0620	72.67	73.76	−1.09	0.0099	54.70	56.48	−1.78
0.0074	55.67	55.65	0.02	0.0190	59.05	59.50	−0.45
0.0160	60.00	58.50	1.50	0.0270	62.75	62.15	0.60
0.0237	63.67	61.06	2.61	0.0350	67.10	64.80	2.30
0.0327	67.67	64.04	3.63	0.0520	70.65	70.44	0.21
0.0555	68.00	71.60	−3.60	0.0580	71.80	72.43	−0.63
0.0447	64.00	68.02	−4.02	0.0290	58.50	62.81	−4.31
0.0068	59.50	55.45	4.05	0.0170	57.15	58.83	−1.68
0.0081	56.60	55.88	0.72	0.0096	52.95	56.38	−3.43

Often we know the value of the independent variable and want to predict the value of the dependent variable. When the data follow a linear pattern, the quantity $\hat{y} = \hat{\beta}_0 + \hat{\beta}_1 x$ is a prediction of the value of the dependent variable. Examples 2.4 and 2.5 illustrate this.

Example 2.4

Using the weld data, predict the yield strength for a weld whose nitrogen content is 0.05%.

Solution

In Example 2.3, the equation of the least-squares line was computed to be $y = 53.197 + 331.62x$. Using the value $x = 0.05$, we estimate the yield strength of a weld whose nitrogen content is 0.05% to be

$$\hat{y} = 53.197 + (331.62)(0.05) = 69.78 \text{ ksi}$$

Example

2.5

Using the weld data, predict the yield strength for a weld whose nitrogen content is 0.02%.

Solution
The estimate is $\hat{y} = 53.197 + (331.62)(0.02) = 59.83$ ksi.

In Example 2.5, note that the data set contains a weld whose nitrogen content was 0.02%. The yield strength of that weld was 57.67 ksi (see Table 2.2). It might seem reasonable to predict that another weld with a nitrogen content of 0.02% would have the same strength as that of the weld already observed. But the least-squares estimate of 59.83 ksi is a better prediction, because it is based on all the data.

Summary

Given points $(x_1, y_1), \ldots, (x_n, y_n)$:

■ The least-squares line is $\hat{y} = \hat{\beta}_0 + \hat{\beta}_1 x$.

■ $\hat{\beta}_1 = \dfrac{\sum_{i=1}^{n}(x_i - \overline{x})(y_i - \overline{y})}{\sum_{i=1}^{n}(x_i - \overline{x})^2}$

■ $\hat{\beta}_0 = \overline{y} - \hat{\beta}_1 \overline{x}$

■ For any x, $\hat{y} = \hat{\beta}_0 + \hat{\beta}_1 x$ is a prediction of the value of the dependent variable for an item whose independent variable value is x.

Exercises for Section 2.2

1. Each month for several months, the average temperature in °C (x) and the amount of steam in kg (y) consumed by a certain chemical plant were measured. The least-squares line computed from the resulting data is $y = 111.74 + 0.51x$.

a. Predict the amount of steam consumed in a month where the average temperature is 65°C.

b. If two months differ in their average temperatures by 5°C, by how much do you predict the amount of steam consumed to differ?

2. In a study of the relationship between the Brinell hardness (x) and tensile strength in ksi (y) of specimens of cold drawn copper, the least-squares line was $y = -196.32 + 2.42x$.

a. Predict the tensile strength of a specimen whose Brinell hardness is 102.7.

b. If two specimens differ in their Brinell hardness by 3, by how much do you predict their tensile strengths to differ?

3. One of the earliest uses of the correlation coefficient was by Sir Francis Galton, who computed the correlation between height and forearm length for a sample of adult men. Assume that the least-squares line for predicting forearm length (y) from height (x) is $y = -0.2967 + 0.2738x$. Both forearm length and height are measured in inches in this equation.

a. Predict the forearm length of a man whose height is 70 in.

b. How tall must a man be so that we would predict his forearm length to be 19 in.?

c. All the men in a certain group have heights greater than the height computed in part (b). Can you conclude that all their forearms will be at least 19 in. long? Explain.

4. In a study relating the degree of warping, in mm, of a copper plate (y) to temperature in °C (x), the following summary statistics were calculated: $n = 40$,

$\sum_{i=1}^{n}(x_i - \bar{x})^2 = 98,775, \sum_{i=1}^{n}(y_i - \bar{y})^2 = 19.10,$
$\bar{x} = 26.36, \bar{y} = 0.5188, \sum_{i=1}^{n}(x_i - \bar{x})(y_i - \bar{y}) = 826.94.$

a. Compute the correlation r between the degree of warping and the temperature.

b. Compute the least-squares line for predicting warping from temperature.

c. Predict the warping at a temperature of 40°C.

d. At what temperature will we predict the warping to be 0.5 mm?

e. Assume it is important that the warping not exceed 0.5 mm. An engineer suggests that if the temperature is kept below the level computed in part (d), we can be sure that the warping will not exceed 0.5 mm. Is this a correct conclusion? Explain.

5. Inertial weight (in tons) and fuel economy (in mi/gal) were measured for a sample of seven diesel trucks. The results are presented in the following table. (From "In-Use Emissions from Heavy-Duty Diesel Vehicles," J. Yanowitz, Ph.D. thesis, Colorado School of Mines, 2001.)

Weight	Mileage
8.00	7.69
24.50	4.97
27.00	4.56
14.50	6.49
28.50	4.34
12.75	6.24
21.25	4.45

a. Construct a scatterplot of mileage (y) versus weight (x).

b. Compute the least-squares line for predicting mileage from weight.

c. If two trucks differ in weight by 5 tons, by how much would you predict their mileages to differ?

d. Predict the mileage for trucks with a weight of 15 tons.

e. What are the units of the estimated slope $\hat{\beta}_1$?

f. What are the units of the estimated intercept $\hat{\beta}_0$?

6. The processing of raw coal involves "washing," in which coal ash (nonorganic, incombustible material) is removed. The article "Quantifying Sampling Pre-cision for Coal Ash Using Gy's Discrete Model of the Fundamental Error" (*Journal of Coal Quality*, 1989:33–39) provides data relating the percentage of ash to the density of a coal particle. The average percentage ash for five densities of coal particles was measured. The data are presented in the following table:

Density (g/cm³)	Percent ash
1.25	1.93
1.325	4.63
1.375	8.95
1.45	15.05
1.55	23.31

a. Construct a scatterplot of percent ash (y) versus density (x).

b. Compute the least-squares line for predicting percent ash from density.

c. If two coal particles differed in density by 0.1 g/cm³, by how much would you predict their percent ash to differ?

d. Predict the percent ash for particles with density 1.40 g/cm³.

e. Compute the fitted values.

f. Compute the residuals. Which point has the residual with the largest magnitude?

g. Compute the correlation between density and percent ash.

7. In tests designed to measure the effect of a certain additive on the drying time of paint, the following data were obtained.

Concentration of Additive (%)	Drying Time (h)
4.0	8.7
4.2	8.8
4.4	8.3
4.6	8.7
4.8	8.1
5.0	8.0
5.2	8.1
5.4	7.7
5.6	7.5
5.8	7.2

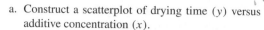

a. Construct a scatterplot of drying time (y) versus additive concentration (x).

b. Compute the least-squares line for predicting drying time from additive concentration.

c. Compute the fitted value and the residual for each point.

d. If the concentration of the additive is increased by 0.1%, by how much would you predict the drying time to increase or decrease?

e. Predict the drying time for a concentration of 4.4%.

f. For what concentration would you predict a drying time of 8.2 hours?

8. The article "Polyhedral Distortions in Tourmaline" (A. Ertl, J. Hughes, et al., *Canadian Mineralogist*, 2002:153–162) presents a model for calculating bond-length distortion in vanadium-bearing tourmaline. To check the accuracy of the model, several calculated values (x) were compared with directly observed values (y). The results (read from a graph) are presented in the following table.

Observed Value	Calculated Value	Observed Value	Calculated Value
0.33	0.36	0.74	0.78
0.36	0.36	0.79	0.86
0.54	0.58	0.97	0.97
0.56	0.64	1.03	1.11
0.66	0.64	1.10	1.06
0.66	0.67	1.13	1.08
0.74	0.58	1.14	1.17

a. Compute the least-squares line $y = \hat{\beta}_0 + \hat{\beta}_1 x$.

b. Predict the value y when the calculated value is $x = 1.0$.

c. For which values of x will the predicted value y be less than the calculated value x?

9. Measurements of the absorbance of a solution are made at various concentrations. The results are presented in the following table.

Concentration (mol/cm³)	1.00	1.20	1.50	1.70	2.00
Absorbance (L/cm³)	0.99	1.13	1.52	1.73	1.96

a. Let $A = \hat{\beta}_0 + \hat{\beta}_1 C$ be the equation of the least-squares line for predicting absorbance (A) from concentration (C). Compute the values of $\hat{\beta}_0$ and $\hat{\beta}_1$.

b. Predict the absorbance for a concentration $C = 1.3$ mol/cm³.

10. For a sample of 12 trees, the volume of lumber (in m³) and the diameter (in cm) at a fixed height above ground level was measured. The results were as follows.

Diameter	Volume	Diameter	Volume
35.1	0.81	33.8	0.80
48.4	1.39	45.3	1.69
47.9	1.31	25.2	0.30
35.3	0.67	28.5	0.19
47.3	1.46	30.1	0.63
26.4	0.47	30.0	0.64

a. Construct a scatterplot of volume (y) versus diameter (x).

b. Compute the least-squares line for predicting volume from diameter.

c. Compute the fitted value and the residual for each point.

d. If two trees differ in diameter by 8 cm, by how much would you predict their volumes to differ?

e. Predict the volume for a tree whose diameter is 44 cm.

f. For what diameter would you predict a volume of 1 m³?

11. A sample of 10 households was monitored for one year. The household income (in $1000s) and the amount of energy consumed (in 10^{10} joules) were determined. The results follow.

Income	Energy	Income	Energy
31	16.0	96	98.3
40	40.2	70	93.8
28	29.8	100	77.1
48	45.6	145	114.8
195	184.6	78	67.0

a. Construct a scatterplot of energy consumption (y) versus income (x).

b. Compute the least-squares line for predicting energy consumption from income.

c. If two families differ in income by $12,000, by how much would you predict their energy consumptions to differ?

d. Predict the energy consumption for a family whose income is $50,000.

e. For what income would you predict an energy consumption of 100?

12. The following table presents measurements of hardness (in durometers) for tire treads containing various percentages of reclaimed rubber.

a. Construct a scatterplot of hardness (y) versus percent reclaimed rubber (x).

b. Compute the least-squares line for predicting hardness from percent reclaimed rubber.

c. If two tires differ in the amount of reclaimed rubber by 15%, by how much would you predict their hardnesses to differ?

d. The table shows that a tire that had 30% reclaimed rubber had a hardness of 67.3. If another tire was manufactured with 30% reclaimed rubber, what would you predict its hardness to be?

e. For what percent reclaimed rubber would you predict a hardness of 65?

Percent	Hardness	Percent	Hardness	Percent	Hardness
0	61.6	22	62.3	42	67.9
2	61.5	24	67.2	44	71.9
4	60.9	26	65.5	46	70.3
6	61.4	28	67.1	48	68.9
8	63.2	30	67.3	50	70.9
10	61.7	32	67.4	52	69.8
12	64.5	34	66.4	54	69.9
14	64.6	36	66.9	56	71.8
16	63.0	38	68.1	58	73.9
18	64.2	40	67.3	60	73.2
20	63.3				

2.3 Features and Limitations of the Least-Squares Line

Don't Extrapolate Outside the Range of the Data

The nitrogen contents of the welds in the data set presented in Table 2.1 in Section 2.2 range from 0.0068% to 0.0630%. Should we use the least-squares line to estimate the yield strength of a weld whose nitrogen content is 0.100%, which is outside the range of the data? The answer is no. We can compute the least-squares estimate, which is $53.197 + 331.62(0.100) = 86.359$ ksi. However, because the value of the independent variable is outside the range of the data, this estimate is unreliable.

Although the data follow a linear pattern for nitrogen contents within the range 0.0068% to 0.0630%, this does not guarantee that a weld with a nitrogen content outside this range would follow the same pattern. For many variables, linear relationships hold within a certain range, but not outside it. If we extrapolate a least-squares line outside the range of the data, therefore, there is no guarantee that it will properly describe the relationship. If we want to predict the yield strength of a weld with a nitrogen content of 0.100%, we must include welds with nitrogen content 0.100% or more in the data set.

Summary

Do not extrapolate a fitted line (such as the least-squares line) outside the range of the data. The linear relationship may not hold there.

Don't Use the Least-Squares Line When the Data Aren't Linear

In Section 2.1, we learned that the correlation coefficient should be used only when the relationship between x and y is linear. The same holds true for the least-squares line. When the scatterplot follows a curved pattern, it does not make sense to summarize it with a straight line. To illustrate this, Figure 2.8 presents a plot of the relationship between the height y of an object released from a height of 256 ft and the time x since its release. The relationship between x and y is nonlinear. The least-squares line does not fit the data well.

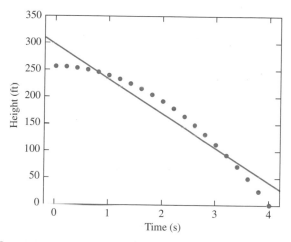

FIGURE 2.8 The relationship between the height of a free-falling object and the time in free fall is not linear. The least-squares line does not fit the data well and should not be used to predict the height of the object at a given time.

Outliers and Influential Points

Outliers are points that are detached from the bulk of the data. Scatterplots should always be examined for outliers. The first thing to do with an outlier is to try to determine why it is different from the rest of the points. Sometimes outliers are caused by data-recording errors or equipment malfunction. In these cases, the outliers may be deleted from the data set. But many times the cause for an outlier cannot be determined with certainty. Deleting the outlier is then unwise, because it results in underestimating the variability of the process that generated the data.

When there is no justification for deleting outliers, one approach is first to fit the line to the whole data set, and then to remove each outlier in turn, fitting the line to the data set with the one outlier deleted. If none of the outliers upon removal make a noticeable difference to the least-squares line or to the estimated standard deviations of the slope and intercept, then use the fit with the outliers included. If one or more of the outliers does make a difference when removed, then the range of values for the least-squares coefficients should be reported.

An outlier that makes a considerable difference to the least-squares line when removed is called an **influential point**. Figure 2.9 presents an example of an influential point, along with an outlier that is not influential. In general, outliers with unusual x values are more likely to be influential than those with unusual y values, but every outlier should be checked. Many software packages identify potentially influential points. Further information on treatment of outliers and influential points can be found in Draper and Smith (1998); Belsey, Kuh, and Welsch (2004); and Cook and Weisberg (1994).

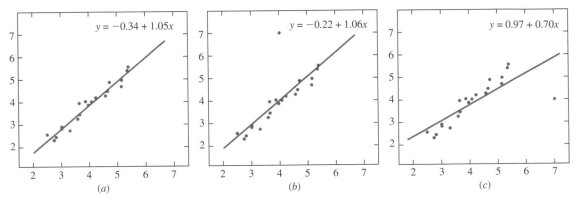

FIGURE 2.9 *(a)* Scatterplot with no outliers. *(b)* An outlier is added to the plot. There is little change in the least-squares line, so this point is not influential. *(c)* An outlier is added to the plot. There is a considerable change in the least-squares line, so this point is influential.

Finally, we remark that some authors restrict the definition of outliers to points that have unusually large residuals. Under this definition, a point that is far from the bulk of the data, yet near the least-squares line, is not an outlier.

Another Look at the Least-Squares Line

The expression (2.6) for $\hat{\beta}_1$ can be rewritten in a way that provides a useful interpretation. Starting with the definition of the correlation coefficient r (Equation 2.1 in Section 2.1) and multiplying both sides by $\sqrt{\sum_{i=1}^{n}(y_i - \overline{y})^2}/\sqrt{\sum_{i=1}^{n}(x_i - \overline{x})^2} = s_y/s_x$ yields the result

$$\hat{\beta}_1 = r\frac{s_y}{s_x} \tag{2.11}$$

Equation (2.11) allows us to interpret the slope of the least-squares line in terms of the correlation coefficient. The units of $\hat{\beta}_1$, the slope of the least-squares line, must be units of y/x. The correlation coefficient r is a unitless number that measures the strength of the linear relationship between x and y. Equation (2.11) shows that the slope $\hat{\beta}_1$ is proportional to the correlation coefficient, where the constant of proportionality is the quantity s_y/s_x that adjusts for the units in which x and y are measured.

Using Equation (2.11), we can write the equation of the least-squares line in a useful form: Substituting $\overline{y} - \hat{\beta}_1\overline{x}$ for $\hat{\beta}_0$ in the equation for the least-squares line $\hat{y} = \hat{\beta}_0 + \hat{\beta}_1 x$ and rearranging terms yields

$$\hat{y} - \overline{y} = \hat{\beta}_1(x - \overline{x}) \tag{2.12}$$

Combining Equations (2.11) and (2.12) yields

$$\hat{y} - \overline{y} = r\frac{s_y}{s_x}(x - \overline{x}) \tag{2.13}$$

Thus the least-squares line is the line that passes through the center of mass of the scatterplot $(\overline{x}, \overline{y})$, with slope $\hat{\beta}_1 = r(s_y/s_x)$.

Measuring Goodness-of-Fit

A goodness-of-fit statistic is a quantity that measures how well a model explains a given set of data. A linear model fits well if there is a strong linear relationship between x and y. We mentioned in Section 2.1 that the correlation coefficient r measures the strength of the linear relationship between x and y. Therefore r is a goodness-of-fit statistic for the linear model. We will now describe how r measures the goodness-of-fit. Figure 2.10 presents the weld data. The points on the scatterplot are (x_i, y_i) where x_i is the nitrogen content of the ith weld and y_i is the yield strength. There are two lines superimposed on the scatterplot. One is the least-squares line, and the other is the horizontal line $y = \overline{y}$. Now imagine that we must predict the yield strength of one of the welds. If we have no knowledge of the nitrogen content, we must predict the yield strength to be the mean, $\overline{y}$. Our prediction error is $y_i - \overline{y}$. If we predict the strength of each weld this way, the sum of squared prediction errors will be $\sum_{i=1}^{n}(y_i - \overline{y})^2$. If, on the other hand, we know

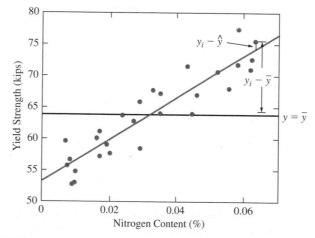

FIGURE 2.10 Nitrogen content and yield strengths of welds. The least-squares line and the horizontal line $y = \overline{y}$ are superimposed.

the nitrogen content of each weld before predicting the yield strength, we can use the least-squares line, and we will predict the ith yield strength to be $\hat{y}_i$. The prediction error will be the residual $y_i - \hat{y}_i$, and the sum of squared prediction errors is $\sum_{i=1}^{n}(y_i - \hat{y}_i)^2$. The strength of the linear relationship can be measured by computing the reduction in sum of squared prediction errors obtained by using $\hat{y}_i$ rather than $\bar{y}$. This is the difference $\sum_{i=1}^{n}(y_i - \bar{y})^2 - \sum_{i=1}^{n}(y_i - \hat{y}_i)^2$. The bigger this difference is, the more tightly clustered the points are around the least-squares line, and the stronger the linear relationship is between x and y. Thus, $\sum_{i=1}^{n}(y_i - \bar{y})^2 - \sum_{i=1}^{n}(y_i - \hat{y}_i)^2$ is a goodness-of-fit statistic.

There is a problem with using $\sum_{i=1}^{n}(y_i - \bar{y})^2 - \sum_{i=1}^{n}(y_i - \hat{y}_i)^2$ as a goodness-of-fit statistic, however. This quantity has units—namely, the squared units of y. We could not use this statistic to compare the goodness-of-fit of two models fit to different data sets, since the units would be different. We need to use a goodness-of-fit statistic that is unitless, so that we can measure goodness-of-fit on an absolute scale.

This is where the correlation coefficient r comes in. It can be shown that

$$r^2 = \frac{\sum_{i=1}^{n}(y_i - \bar{y})^2 - \sum_{i=1}^{n}(y_i - \hat{y}_i)^2}{\sum_{i=1}^{n}(y_i - \bar{y})^2} \tag{2.14}$$

The quantity r^2, the square of the correlation coefficient, is called the **coefficient of determination**. It is the reduction in the sum of the squared prediction errors obtained by using $\hat{y}_i$ rather than $\bar{y}$, expressed as a fraction of the sum of squared prediction errors $\sum_{i=1}^{n}(y_i - \bar{y})^2$ obtained by using $\bar{y}$. This interpretation of r^2 is important to know. In Chapter 8, we will see how it can be generalized to provide a measure of the goodness-of-fit of linear relationships involving several variables.

For a visual interpretation of r^2, look at Figure 2.10. For each point (x_i, y_i) on the scatterplot, the quantity $y_i - \bar{y}$ is the vertical distance from the point to the horizontal line $y = \bar{y}$, and the quantity $y_i - \hat{y}_i$ is the vertical distance from the point to the least-squares line. Thus, the quantity $\sum_{i=1}^{n}(y_i - \bar{y})^2$ measures the overall spread of the points around the line $y = \bar{y}$, and the quantity $\sum_{i=1}^{n}(y_i - \hat{y}_i)^2$ measures the overall spread of the points around the least-squares line. The quantity $\sum_{i=1}^{n}(y_i - \bar{y})^2 - \sum_{i=1}^{n}(y_i - \hat{y}_i)^2$ therefore measures the reduction in the spread of the points obtained by using the least-squares line rather than $y = \bar{y}$. The coefficient of determination r^2 expresses this reduction as a proportion of the spread around $y = \bar{y}$.

The sums of squares appearing in this discussion are used so often that statisticians have given them names. They call $\sum_{i=1}^{n}(y_i - \hat{y}_i)^2$ the **error sum of squares** and $\sum_{i=1}^{n}(y_i - \bar{y})^2$ the **total sum of squares**. Their difference $\sum_{i=1}^{n}(y_i - \bar{y})^2 - \sum_{i=1}^{n}(y_i - \hat{y}_i)^2$ is called the **regression sum of squares**. Clearly, the following relationship holds:

$$\text{Total sum of squares} = \text{Regression sum of squares} + \text{Error sum of squares}$$

Using the preceding terminology, we can write Equation (2.14) as

$$r^2 = \frac{\text{Regression sum of squares}}{\text{Total sum of squares}}$$

Since the total sum of squares is just the sample variance of the y_i without dividing by $n - 1$, statisticians (and others) often refer to r^2 as the **proportion of the variance in y explained by regression**.

Interpreting Computer Output

Nowadays, calculations involving least-squares lines are usually done on a computer. Many software packages are available that will perform these calculations. The following output (from MINITAB), which presents the results of fitting a least-squares line to the weld data, is fairly typical.

```
Regression Analysis: Length versus Weight

The regression equation is
Strength = 53.197 + 331.62 Nitrogen

Predictor          Coef      SE Coef           T           P
Constant        53.19715      1.02044      52.131       0.000
Nitrogen        331.6186     27.4872       12.064       0.000

S = 2.75151        R-Sq = 84.8%         R-Sq(adj) = 84.3%
```

In the output, the equation of the least-squares line appears at the top, labeled as the "regression equation." The least-squares coefficients appear in the column labeled "Coef." The quantity labeled "R-sq" is the goodness-of-fit statistic r^2. The remaining values in the output are used for statistical inference. We will discuss them in Section 8.1.

Exercises for Section 2.3

1. Ultimate strengths (in MPa) were measured at several temperatures (in °C) for three types of plastic. The results follow.

Type 1		Type 2		Type 3	
Temperature	Strength	Temperature	Strength	Temperature	Strength
10	3	10	5	10	8
12	11	12	7	11	11
14	11	14	7	12	11
16	16	16	9	16	15
18	19	18	11	19	16
20	21	20	13	22	21
22	25	22	17	23	24
24	25	24	18	25	26
26	29	26	22	26	30
28	26	28	18	29	27
30	31	30	25	31	32
32	33	32	28	31	32
34	32	34	26	33	32
36	39	36	37	35	38
38	38	38	36	35	38
40	47	40	52	60	41

a. Construct a scatterplot for each of the three data sets.

b. For each data set, say whether the least-squares line is appropriate for summarizing the data, and explain why.

2. Following are measurements of shear strain (in %) and shear strain (in MPa) for a sample of copper bars.

Stress	Strain	Stress	Strain
8.2	77.6	13.4	80.9
9.6	78.8	13.5	81.3
11.0	78.9	14.7	82.2
12.0	80.2	14.9	82.7

a. Compute the least-squares line using strain as the dependent variable (y) and stress as the independent variable (x).

b. Compute the least-squares line using stress as the dependent variable (y) and strain as the independent variable (x). Rewrite the equation of this line to express strain in terms of stress.

c. Are the lines in parts (a) and (b) the same?

d. Compute the correlation coefficient r. Express the ratio of the slopes of the two lines in terms of r.

3. In an experiment to determine corrosion rates, steel plates were placed in a solution of 10% hydrochloric acid (HCl) for various periods of time (in hours), and the weight loss (in mg) was measured. Following are the results.

Time	2	4	6	8	10	12
Weight loss	0.7	6.9	7.2	11.9	16.2	22.5

a. Compute the least-squares line for predicting the weight loss (y) at a given time (x).

b. Should the line be used to predict the weight loss after 1 hour? If so, predict the weight loss. If not, explain why not.

c. Should the line be used to predict the weight loss after 5 hours? If so, predict the weight loss. If not, explain why not.

d. Should the line be used to predict the weight loss after 20 hours? If so, predict the weight loss. If not, explain why not.

4. For a certain scatterplot, the value of the correlation coefficient is $r = 0.5$. Which of the following statements is correct?

i. The total sum of squares is half as much as the regression sum of squares.

ii. The regression sum of squares is half as much as the total sum of squares.

iii. The total sum of squares is one-fourth as much as the regression sum of squares.

iv. The regression sum of squares is one-fourth as much as the total sum of squares.

5. A least-squares line is fit to a set of points. If the total sum of squares is $\sum(y_i - \overline{y})^2 = 9615$, and the error sum of squares is $\sum(y_i - \hat{y}_i)^2 = 1450$, compute the coefficient of determination r^2.

6. A least-squares line is fit to a set of points. If the total sum of squares is $\sum(y_i - \overline{y})^2 = 181.2$, and the error sum of squares is $\sum(y_i - \hat{y}_i)^2 = 33.9$, compute the coefficient of determination r^2.

7. Moisture content in percent by volume (x) and conductivity in mS/m (y) were measured for 50 soil specimens. The means and standard deviations were $\overline{x} = 8.1, s_x = 1.2, \overline{y} = 30.4, s_y = 1.9$. The correlation between conductivity and moisture was computed to be $r = 0.85$. Find the equation of the least-squares line for predicting soil conductivity from moisture content.

8. Curing times in days (x) and compressive strengths in MPa (y) were recorded for several concrete specimens. The means and standard deviations of the x and y values were $\overline{x} = 5, s_x = 2, \overline{y} = 1350, s_y = 100$. The correlation between curing time and compressive strength was computed to be $r = 0.7$. Find the equation of the least-squares line to predict compressive strength from curing time.

9. Varying amounts of pectin were added to canned jellies, to study the relationship between pectin concentration in percent (x) and a firmness index (y). The means and standard deviations of the x and y values were $\overline{x} = 3, s_x = 0.5, \overline{y} = 50, s_y = 10$. The correlation between curing time and firmness was computed to be $r = 0.5$. Find the equation of the least-squares line to predict firmness from pectin concentration.

10. An engineer wants to predict the value for y when $x = 4.5$, using the following data set.

x	y	$z = \ln y$	x	y	$z = \ln y$
1	0.2	−1.61	6	2.3	0.83
2	0.3	−1.20	7	2.9	1.06
3	0.5	−0.69	8	4.5	1.50
4	0.5	−0.69	9	8.7	2.16
5	1.3	0.26	10	12.0	2.48

a. Construct a scatterplot of the points (x, y).

b. Should the least-squares line be used to predict the value of y when $x = 4.5$? If so, compute the least-squares line and the predicted value. If not, explain why not.

c. Construct a scatterplot of the points (x, z), where $z = \ln y$.

d. Use the least-squares line to predict the value of z when $x = 4.5$. Is this an appropriate method of prediction? Explain why or why not.

e. Let $\hat{z}$ denote the predicted value of z computed in part (d). Let $\hat{y} = e^{\hat{z}}$. Explain why $\hat{y}$ is a reasonable predictor of the value of y when $x = 4.5$.

Supplementary Exercises for Chapter 2

1. A simple random sample of 100 men aged 25–34 averaged 70 inches in height and had a standard deviation of 3 inches. Their incomes averaged $47,500 and had a standard deviation of $18,300. Fill in the blank: From the least-squares line, we would predict that the income of a man 70 inches tall would be _____

   i. less than $47,500.

  ii. greater than $47,500.

 iii. equal to $47,500.

 iv. we cannot tell unless we know the correlation.

2. In an accelerated life test, units are operated under extreme conditions until failure. In one such test, 12 motors were operated under high temperature conditions. The ambient temperatures (in °C) and lifetimes (in hours) are presented in the following table:

Temperature	Lifetime
40	851
45	635
50	764
55	708
60	469
65	661
70	586
75	371
80	337
85	245
90	129
95	158

a. Construct a scatterplot of lifetime (y) versus temperature (x). Verify that a linear model is appropriate.

b. Compute the least-squares line for predicting lifetime from temperature.

c. Compute the fitted value and the residual for each point.

d. If the temperature is increased by 5°C, by how much would you predict the lifetime to increase or decrease?

e. Predict the lifetime for a temperature of 73°C.

f. Should the least-squares line be used to predict the lifetime for a temperature of 120°C? If so, predict the lifetime. If not, explain why not.

g. For what temperature would you predict a lifetime of 500 hours?

3. In a sample of adults, would the correlation between age and the year graduated from high school be closest to −1, −0.5, 0, 0.5, or 1? Explain.

4. A mixture of sucrose and water was heated on a hot plate, and the temperature (in °C) was recorded each minute for 20 minutes by three thermocouples. The results are shown in the following table.

Time	T_1	T_2	T_3
0	20	18	21
1	18	22	11
2	29	22	26
3	32	25	35
4	37	37	33
5	36	46	35
6	46	45	44
7	46	44	43
8	56	54	63
9	58	64	68
10	64	69	62
11	72	65	65
12	79	80	80
13	84	74	75
14	82	87	78
15	87	93	88
16	98	90	91
17	103	100	103
18	101	98	109
19	103	103	107
20	102	103	104

a. Compute the least-squares line for estimating the temperature as a function of time, using T_1 as the value for temperature.

b. Compute the least-squares line for estimating the temperature as a function of time, using T_2 as the value for temperature.

c. Compute the least-squares line for estimating the temperature as a function of time, using T_3 as the value for temperature.

d. It is desired to compute a single line to estimate temperature as a function of time. One person suggests averaging the three slope estimates to obtain a single slope estimate, and averaging the three intercept estimates to obtain a single intercept estimate. Find the equation of the line that results from this method.

e. Someone else suggests averaging the three temperature measurements at each time to obtain $\overline{T} = (T_1 + T_2 + T_3)/3$. Compute the least-squares line using $\overline{T}$ as the value for temperature.

f. Are the results of parts (d) and (e) different?

5. A scatterplot contains three points: $(-1,-1)$, $(0,0)$, $(1,1)$. A fourth point, $(2, y)$, is to be added to the plot. Let r represent the correlation between x and y.

a. Find the value of y so that $r = 1$.

b. Find the value of y so that $r = 0$.

c. Explain why there is no value of y for which $r = -1$.

6. The article "Experimental Measurement of Radiative Heat Transfer in Gas–Solid Suspension Flow System" (G. Han, K. Tuzla, and J. Chen, *AIChe Journal*, 2002:1910–1916) discusses the calibration of a radiometer. Several measurements were made on the electromotive force readings of the radiometer (in volts) and the radiation flux (in kilowatts per square meter). The results (read from a graph) are presented in the following table.

Heat flux (y)	15	31	51	55	67	89
Signal output (x)	1.08	2.42	4.17	4.46	5.17	6.92

a. Compute the least-squares line for predicting heat flux from the signal output.

b. If the radiometer reads 3.00 V, predict the heat flux.

c. If the radiometer reads 8.00 V, should the heat flux be predicted? If so, predict it. If not, explain why.

7. The article "Estimating Population Abundance in Plant Species With Dormant Life-Stages: Fire and the Endangered Plant *Grevillea caleye* R. Br." (T. Auld and J. Scott, *Ecological Management and Restoration*, 2004:125–129) presents estimates of population sizes of a certain rare shrub in areas burnt by fire. The following table presents population counts and areas (in m^2) for several patches containing the plant.

Area	Population	Area	Population
3739	3015	2521	707
5277	1847	213	113
400	17	11958	1392
345	142	1200	157
392	40	12000	711
7000	2878	10880	74
2259	223	841	1720
81	15	1500	300
33	18	228	31
1254	229	228	17
1320	351	10	4
1000	92		

a. Compute the least-square line for predicting population (y) from area (x).

b. Use the least-squares line computed in part (a) to predict the population in an area of 5000 m^2.

c. Compute the least-square line for predicting ln y from ln x.

d. Use the least-squares line computed in part (c) to predict the population in an area of 5000 m^2.

e. Construct a scatterplot of population (y) versus area (x).

f. Construct a scatterplot of ln y versus ln x.

g. Which of the two scatterplots appears to be more appropriately summarized by the least-squares line?

h. Which prediction is more reliable, the one in part (b) or the one in part (d)? Explain.

8. In a test of military ordnance, a large number of bombs were dropped on a target from various heights. The initial velocity of the bombs in the direction of the ground was 0. Let y be the height in meters from which a bomb is dropped, let x be the time in seconds for the bomb to strike the ground, let $u = x^2$, and let $v = \sqrt{y}$. The relationship between x and y is given by $y = 4.9x^2$. For each of the following pairs of variables, state whether the correlation coefficient is an appropriate summary.

a. x and y

b. u and y

c. x and v

d. u and v

e. ln x and ln y

3

Probability

Introduction

The development of the theory of probability was financed by seventeenth-century gamblers, who hired some of the leading mathematicians of the day to calculate the correct odds for certain games of chance. Later, people realized that scientific processes involve chance as well, and since then the methods of probability have been used to study the physical world.

Probability is now an extensive branch of mathematics. Many books are devoted to the subject, and many researchers have dedicated their professional careers to its further development. In this chapter we present an introduction to the ideas of probability that are most important to the study of statistics.

3.1 Basic Ideas

To make a systematic study of probability, we need some terminology. An **experiment** is a process that results in an outcome that cannot be predicted in advance with certainty. Tossing a coin, rolling a die, measuring the diameter of a bolt, weighing the contents of a box of cereal, and measuring the breaking strength of a length of fishing line are all examples of experiments. To discuss an experiment in probabilistic terms, we must specify its possible outcomes:

Definition

The set of all possible outcomes of an experiment is called the **sample space** for the experiment.

For tossing a coin, we can use the set {Heads, Tails} as the sample space. For rolling a six-sided die, we can use the set {1, 2, 3, 4, 5, 6}. These sample spaces are finite. Some experiments have sample spaces with an infinite number of outcomes. For example, imagine that a punch with a diameter of 10 mm punches holes in sheet metal.

Because of variations in the angle of the punch and slight movements in the sheet metal, the diameters of the holes vary between 10.0 and 10.2 mm. For the experiment of punching a hole, then, a reasonable sample space is the interval $(10.0, 10.2)$, or in set notation, $\{x \mid 10.0 < x < 10.2\}$. This set obviously contains an infinite number of outcomes.

When discussing experiments, we are often interested in a particular subset of outcomes. For example, we might be interested in the probability that a die comes up an even number. The sample space for the experiment is $\{1, 2, 3, 4, 5, 6\}$, and coming up even corresponds to the subset $\{2, 4, 6\}$. In the hole punch example, we might be interested in the probability that a hole has a diameter less than 10.1 mm. This corresponds to the subset $\{x \mid 10.0 < x < 10.1\}$. There is a special name for a subset of a sample space:

Definition

A subset of a sample space is called an **event**.

Note that for any sample space, the empty set $\emptyset$ is an event, as is the entire sample space. A given event is said to have occurred if the outcome of the experiment is one of the outcomes in the event. For example, if a die comes up 2, the events $\{2, 4, 6\}$ and $\{1, 2, 3\}$ have both occurred, along with every other event that contains the outcome "2."

Example
3.1

An engineer has a box containing four bolts and another box containing four nuts. The diameters of the bolts are 4, 6, 8, and 10 mm, and the diameters of the nuts are 6, 10, 12, and 14 mm. One bolt and one nut are chosen. Let A be the event that the bolt diameter is less than 8, let B be the event that the nut diameter is greater than 10, and let C be the event that the bolt and the nut have the same diameter. Find a sample space for this experiment, and specify the subsets corresponding to the events A, B, and C.

Solution

A good sample space for this experiment is the set of ordered pairs in which the first component is the diameter of the bolt and the second component is the diameter of the nut. We will denote this sample space by $\mathcal{S}$.

$$\mathcal{S} = \{(4, 6), (4, 10), (4, 12), (4, 14), (6, 6), (6, 10), (6, 12), (6, 14),$$
$$(8, 6), (8, 10), (8, 12), (8, 14), (10, 6), (10, 10), (10, 12), (10, 14)\}$$

The events A, B, and C are given by

$A = \{(4, 6), (4, 10), (4, 12), (4, 14), (6, 6), (6, 10), (6, 12), (6, 14)\}$
$B = \{(4, 12), (4, 14), (6, 12), (6, 14), (8, 12), (8, 14), (10, 12), (10, 14)\}$
$C = \{(6, 6), (10, 10)\}$

Combining Events

We often construct events by combining simpler events. Because events are subsets of sample spaces, it is traditional to use the notation of sets to describe events constructed in this way. We review the necessary notation here.

- The **union** of two events A and B, denoted $A \cup B$, is the set of outcomes that belong to A, to B, or to both. In words, $A \cup B$ means "A or B." Thus the event $A \cup B$ occurs whenever either A or B (or both) occurs.

- The **intersection** of two events A and B, denoted $A \cap B$, is the set of outcomes that belong both to A and to B. In words, $A \cap B$ means "A and B." Thus the event $A \cap B$ occurs whenever both A and B occur.

- The **complement** of an event A, denoted A^c, is the set of outcomes that do not belong to A. In words, A^c means "not A." Thus the event A^c occurs whenever A does *not* occur.

Mutually Exclusive Events

There are some events that can never occur together. For example, it is impossible that a coin can come up both heads and tails, and it is impossible that a steel pin can be both too long and too short. Events like this are said to be **mutually exclusive**.

Definition

- The events A and B are said to be **mutually exclusive** if they have no outcomes in common.
- More generally, a collection of events $A_1, A_2, \ldots, A_n$ is said to be mutually exclusive if no two of them have any outcomes in common.

Refer to Example 3.1. If the experiment is performed, is it possible for events A and B both to occur? How about B and C? A and C? Which pair of events is mutually exclusive?

Solution
If the outcome is (4, 12), (4, 14), (6, 12), or (6, 14), then events A and B both occur. If the outcome is (6, 6), then both A and C occur. It is impossible for B and C both to occur, because these events are mutually exclusive, having no outcomes in common.

Probabilities

Each event in a sample space has a **probability** of occurring. Intuitively, the probability is a quantitative measure of how likely the event is to occur. Formally speaking, there are several interpretations of probability; the one we shall adopt is that the probability of an event is the proportion of times the event would occur in the long run, if the experiment were to be repeated over and over again.

We often use the letter P to stand for probability. Thus when tossing a coin, the notation "$P(\text{heads}) = 1/2$" means that the probability that the coin lands heads is equal to $1/2$.

Summary

Given any experiment and any event A:

- The expression $P(A)$ denotes the probability that the event A occurs.
- $P(A)$ is the proportion of times that event A would occur in the long run, if the experiment were to be repeated over and over again.

In many situations, the only way to estimate the probability of an event is to repeat the experiment many times and determine the proportion of times that the event occurs. For example, if it is desired to estimate the probability that a printed circuit board manufactured by a certain process is defective, it is usually necessary to produce a number of boards and test them to determine the proportion that are defective. In some cases, probabilities can be determined through knowledge of the physical nature of an experiment. For example, if it is known that the shape of a die is nearly a perfect cube and that its mass is distributed nearly uniformly, it may be assumed that each of the six faces is equally likely to land upward when the die is rolled.

Once the probabilities of some events have been found through scientific knowledge or experience, the probabilities of other events can be computed mathematically. For example, if it has been estimated through experimentation that the probability that a printed circuit board is defective is 0.10, an estimate of the probability that a board is not defective can be calculated to be 0.90. As another example, assume that steel pins manufactured by a certain process can fail to meet a length specification either by being too short or too long. By measuring a large number of pins, it is estimated that the probability that a pin is too short is 0.02, and the probability that a pin is too long is 0.03. It can then be estimated that the probability that a pin fails to meet the specification is 0.05.

In practice, scientists and engineers estimate the probabilities of some events on the basis of scientific understanding and experience, and then use mathematical rules to compute estimates of the probabilities of other events. In the rest of this section, we will explain some of these rules and show how to use them.

Axioms of Probability

The subject of probability is based on three commonsense rules, known as axioms. They are:

The Axioms of Probability

1. Let $\mathcal{S}$ be a sample space. Then $P(\mathcal{S}) = 1$.
2. For any event A, $0 \leq P(A) \leq 1$.
3. If A and B are mutually exclusive events, then $P(A \cup B) = P(A) + P(B)$. More generally, if $A_1, A_2, \ldots$ are mutually exclusive events, then
$$P(A_1 \cup A_2 \cup \cdots) = P(A_1) + P(A_2) + \cdots.$$

With a little thought, it is easy to see that the three axioms do indeed agree with common sense. The first axiom says that the outcome of an experiment is always in the sample space. This is obvious, because by definition the sample space contains all the possible outcomes of the experiment. The second axiom says that the long-run frequency of any event is always between 0 and 100%. For an example illustrating the third axiom, we previously discussed a process that manufactures steel pins, in which the probability that a pin is too short is 0.02 and the probability that a pin is too long is 0.03. The third axiom says that the probability that the pin is either too short or too long is $0.02 + 0.03 = 0.05$.

We now present two simple rules that are helpful in computing probabilities. These rules are intuitively obvious, and it can be shown that they follow logically from the axioms.

For any event A,

$$P(A^c) = 1 - P(A) \tag{3.1}$$

Let $\emptyset$ denote the empty set. Then

$$P(\emptyset) = 0 \tag{3.2}$$

Equation (3.1) says that the probability that an event does not occur is equal to 1 minus the probability that it does occur. For example, if there is a 40% chance of rain, there is a 60% chance that it does not rain. Equation (3.2) says that it is impossible for an experiment to have no outcome.

Example 3.3

A target on a test firing range consists of a bull's-eye with two concentric rings around it. A projectile is fired at the target. The probability that it hits the bull's-eye is 0.10, the probability that it hits the inner ring is 0.25, and the probability that it hits the outer ring is 0.45. What is the probability that the projectile hits the target? What is the probability that it misses the target?

Solution

Hitting the bull's-eye, hitting the inner ring, and hitting the outer ring are mutually exclusive events, since it is impossible for more than one of these events to occur. Therefore, using Axiom 3,

$$P(\text{hits target}) = P(\text{bull's-eye}) + P(\text{inner ring}) + P(\text{outer ring})$$
$$= 0.10 + 0.25 + 0.45$$
$$= 0.80$$

We can now compute the probability that the projectile misses the target by using Equation (3.1):

$$P(\text{misses target}) = 1 - P(\text{hits target})$$
$$= 1 - 0.80$$
$$= 0.20$$

Sample Spaces with Equally Likely Outcomes

For some experiments, a sample space can be constructed in which all the outcomes are equally likely. A simple example is the roll of a fair die, in which the sample space is $\{1, 2, 3, 4, 5, 6\}$ and each of these outcomes has probability 1/6. If a sample space contains N equally likely outcomes, the probability of each outcome is $1/N$. This is so because the probability of the whole sample space must be 1, and this probability is equally divided among the N outcomes. If A is an event that contains k outcomes, then $P(A)$ can be found by summing the probabilities of the k outcomes, so $P(A) = k/N$.

If $\mathcal{S}$ is a sample space containing N equally likely outcomes, and if A is an event containing k outcomes, then

$$P(A) = \frac{k}{N} \qquad (3.3)$$

Example 3.4

An extrusion die is used to produce aluminum rods. Specifications are given for the length and the diameter of the rods. For each rod, the length is classified as too short, too long, or OK, and the diameter is classified as too thin, too thick, or OK. In a population of 1000 rods, the number of rods in each class is as follows:

	Diameter		
Length	Too Thin	OK	Too Thick
Too Short	10	3	5
OK	38	900	4
Too Long	2	25	13

A rod is sampled at random from this population. What is the probability that it is too short?

Solution

We can think of each of the 1000 rods as an outcome in a sample space. Each of the 1000 outcomes is equally likely. We'll solve the problem by counting the number of outcomes that correspond to the event. The number of rods that are too short is $10 + 3 + 5 = 18$. Since the total number of rods is 1000,

$$P(\text{too short}) = \frac{18}{1000}$$

The Addition Rule

If A and B are mutually exclusive events, then $P(A \cup B) = P(A) + P(B)$. This rule can be generalized to cover the case where A and B are not mutually exclusive. Example 3.5 illustrates the reasoning.

Example
3.5

Refer to Example 3.4. If a rod is sampled at random, what is the probability that it is either too short or too thick?

Solution

First we'll solve this problem by counting the number of outcomes that correspond to the event. In the following table the numbers of rods that are too thick are circled, and the numbers of rods that are too short have rectangles around them. Note that 5 of the rods are both too short and too thick.

		Diameter		
		Too Thin	OK	Too Thick
	Too Short	$\boxed{10}$	$\boxed{3}$	$\boxed{⑤}$
Length	OK	38	900	④
	Too Long	2	25	⑬

Of the 1000 outcomes, the number that are either too short or too thick is $10 + 3 + 5 + 4 + 13 = 35$. Therefore

$$P(\text{too short or too thick}) = \frac{35}{1000}$$

Now we will solve the problem in a way that leads to a more general method. In the sample space, there are $10 + 3 + 5 = 18$ rods that are too short and $5 + 4 + 13 = 22$ rods that are too thick. But if we try to find the number of rods that are either too short or too thick by adding $18 + 22$, we get too large a number (40 instead of 35). The reason is that there are five rods that are both too short and too thick, and these are counted twice. We can still solve the problem by adding 18 and 22, but we must then subtract 5 to correct for the double counting.

We restate this reasoning, using probabilities:

$$P(\text{too short}) = \frac{18}{1000}, \quad P(\text{too thick}) = \frac{22}{1000}, \quad P(\text{too short and too thick}) = \frac{5}{1000}$$

$$P(\text{too short or too thick}) = P(\text{too short}) + P(\text{too thick}) - P(\text{too short and too thick})$$

$$= \frac{18}{1000} + \frac{22}{1000} - \frac{5}{1000}$$

$$= \frac{35}{1000}$$

The method of Example 3.5 holds for any two events in any sample space. In general, to find the probability that either of two events occurs, add the probabilities of the events and then subtract the probability that they both occur.

Summary

Let A and B be any events. Then

$$P(A \cup B) = P(A) + P(B) - P(A \cap B) \qquad (3.4)$$

Exercises for Section 3.1

1. The probability that a bearing fails during the first month of use is 0.12. What is the probability that it does not fail during the first month?

2. An die (six faces) has the number 1 painted on three of its faces, the number 2 painted on two of its faces, and the number 3 painted on one face. Assume that each face is equally likely to come up.

   a. Find a sample space for this experiment.

   b. Find $P(\text{odd number})$.

   c. If the die were loaded so that the face with the 3 on it were twice as likely to come up as each of the other five faces, would this change the sample space? Explain.

   d. If the die were loaded so that the face with the 3 on it were twice as likely to come up as each of the other five faces, would this change the value of $P(\text{odd number})$? Explain.

3. Silicon wafers are used in the manufacture of integrated circuits. Of the wafers manufactured by a certain process, 10% have resistances below specification, and 5% have resistances above specification.

   a. What is the probability that the resistance of a randomly chosen wafer does not meet the specification?

   b. If a randomly chosen wafer has a resistance that does not meet the specification, what is the probability that it is too low?

4. A system contains two components, A and B. The system will function so long as either A or B functions. The probability that A functions is 0.95, the probability that B functions is 0.90, and the probability that both function is 0.88. What is the probability that the system functions?

5. A system contains two components, A and B. The system will function only if both components function. The probability that A functions is 0.98, the probability that B functions is 0.95, and the probability that either A or B functions is 0.99. What is the probability that the system functions?

6. Human blood may contain either or both of two antigens, A and B. Blood that contains only the A antigen is called type A, blood that contains only the B antigen is called type B, blood that contains both antigens is called type AB, and blood that contains neither antigen is called type O. At a certain blood bank, 35% of the blood donors have type A blood, 10% have type B, and 5% have type AB.

   a. What is the probability that a randomly chosen blood donor is type O?

   b. A recipient with type A blood may safely receive blood from a donor whose blood does not contain the B antigen. What is the probability that a randomly chosen blood donor may donate to a recipient with type A blood?

7. Sixty percent of large purchases made at a certain computer retailer are personal computers, 30% are laptop computers, and 10% are peripheral devices such as printers. As part of an audit, one purchase record is sampled at random.

   a. What is the probability that it is a personal computer?

   b. What is the probability that it is either a personal computer or a laptop computer?

8. A flywheel is attached to a crankshaft by 12 bolts, numbered 1 through 12. Each bolt is checked to determine whether it is torqued correctly. Let A be the event that

all the bolts are torqued correctly, let B be the event that the #3 bolt is not torqued correctly, let C be the event that exactly one bolt is not torqued correctly, and let D be the event that bolts #5 and #8 are torqued correctly.

State whether each of the following pairs of events is mutually exclusive.

a. A and B c. C and D

b. B and D d. B and C

3.2 Conditional Probability and Independence

A sample space contains all the possible outcomes of an experiment. Sometimes we obtain some additional information about an experiment that tells us that the outcome comes from a certain part of the sample space. In this case, the probability of an event is based on the outcomes in that part of the sample space. A probability that is based on a part of a sample space is called a **conditional probability**. We explore this idea through some examples.

In Example 3.4 (in Section 3.1) we discussed a population of 1000 aluminum rods. For each rod, the length is classified as too short, too long, or OK, and the diameter is classified as too thin, too thick, or OK. These 1000 rods form a sample space in which each rod is equally likely to be sampled. The number of rods in each category is presented in Table 3.1. Of the 1000 rods, 928 meet the diameter specification. Therefore, if a rod is sampled, $P(\text{diameter OK}) = 928/1000 = 0.928$. This probability is called the **unconditional probability**, since it is based on the entire sample space. Now assume that a rod is sampled and found to meet the length specification. What is the probability that the rod also meets the diameter specification? The key to computing this probability is to realize that knowledge that the length meets the specification reduces the sample space from which the rod is drawn. Table 3.2 presents this idea. Once we know that the length specification is met, we know that the rod will be one of the 942 rods in the sample space presented in Table 3.2.

TABLE 3.1 Sample space containing 1000 aluminum rods

	Diameter		
Length	**Too Thin**	**OK**	**Too Thick**
Too Short	10	3	5
OK	38	900	4
Too Long	2	25	13

TABLE 3.2 Reduced sample space containing 942 aluminum rods that meet the length specification

	Diameter		
Length	**Too Thin**	**OK**	**Too Thick**
Too Short	—	—	—
OK	38	900	4
Too Long	—	—	—

Of the 942 rods in this sample space, 900 of them meet the diameter specification. Therefore, if we know that the rod meets the length specification, the probability that the rod meets the diameter specification is 900/942. We say that the **conditional probability** that the rod meets the diameter specification **given** that it meets the length specification is equal to 900/942, and we write P(diameter OK | length OK) $= 900/942 = 0.955$. Note that the conditional probability P(diameter OK | length OK) differs from the unconditional probability P(diameter OK), which was computed from the full sample space (Table 3.1) to be 0.928.

Example 3.6

Compute the conditional probability P(diameter OK | length too long). Is this the same as the unconditional probability P(diameter OK)?

Solution

The conditional probability P(diameter OK | length too long) is computed under the assumption that the rod is too long. This reduces the sample space to the 40 items indicated in boldface in the following table.

	Diameter		
Length	Too Thin	OK	Too Thick
Too Short	10	3	5
OK	38	900	4
Too Long	2	25	13

Of the 40 outcomes, 25 meet the diameter specification. Therefore

$$P(\text{diameter OK} \mid \text{length too long}) = \frac{25}{40} = 0.625$$

The unconditional probability P(diameter OK) is computed on the basis of all 1000 outcomes in the sample space and is equal to $928/1000 = 0.928$. In this case, the conditional probability differs from the unconditional probability.

Let's look at the solution to Example 3.6 more closely. We found that

$$P(\text{diameter OK} \mid \text{length too long}) = \frac{25}{40}$$

In the answer 25/40, the denominator, 40, represents the number of outcomes that satisfy the condition that the rod is too long, while the numerator, 25, represents the number of outcomes that satisfy both the condition that the rod is too long and that its diameter is OK. If we divide both the numerator and denominator of this answer by the number of outcomes in the full sample space, which is 1000, we obtain

$$P(\text{diameter OK} \mid \text{length too long}) = \frac{25/1000}{40/1000}$$

Now 40/1000 represents the *probability* of satisfying the condition that the rod is too long. That is,

$$P(\text{length too long}) = \frac{40}{1000}$$

The quantity 25/1000 represents the *probability* of satisfying both the condition that the rod is too long and that the diameter is OK. That is,

$$P(\text{diameter OK and length too long}) = \frac{25}{1000}$$

We can now express the conditional probability as

$$P(\text{diameter OK} \mid \text{length too long}) = \frac{P(\text{diameter OK and length too long})}{P(\text{length too long})}$$

This reasoning can be extended to construct a definition of conditional probability that holds for any sample space:

Definition

Let A and B be events with $P(B) \neq 0$. The conditional probability of A given B is

$$P(A|B) = \frac{P(A \cap B)}{P(B)} \tag{3.5}$$

Figure 3.1 presents Venn diagrams to illustrate the idea of conditional probability.

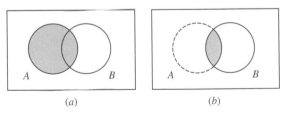

(a) (b)

FIGURE 3.1 *(a)* The diagram represents the unconditional probability $P(A)$. $P(A)$ is illustrated by considering the event A in proportion to the entire sample space, which is represented by the rectangle. *(b)* The diagram represents the conditional probability $P(A|B)$. Since the event B is known to occur, the event B now becomes the sample space. For the event A to occur, the outcome must be in the intersection $A \cap B$. The conditional probability $P(A|B)$ is therefore illustrated by considering the intersection $A \cap B$ in proportion to the entire event B.

Independent Events

Sometimes the knowledge that one event has occurred does not change the probability that another event occurs. In this case the conditional and unconditional probabilities are the same, and the events are said to be **independent**. We present an example.

If an aluminum rod is sampled from the sample space presented in Table 3.1, find $P(\text{too long})$ and $P(\text{too long} \mid \text{too thin})$. Are these probabilities different?

Solution

$$P(\text{too long}) = \frac{40}{1000} = 0.040$$

$$P(\text{too long} \mid \text{too thin}) = \frac{P(\text{too long and too thin})}{P(\text{too thin})}$$

$$= \frac{2/1000}{50/1000}$$

$$= 0.040$$

The conditional probability and the unconditional probability are the same. The information that the rod is too thin does not change the probability that the rod is too long.

Example 3.7 shows that knowledge that an event occurs sometimes does not change the probability that another event occurs. In these cases, the two events are said to be **independent**. The event that a rod is too long and the event that a rod is too thin are independent. We now give a more precise definition of the term, both in words and in symbols.

Definition

Two events A and B are **independent** if the probability of each event remains the same whether or not the other occurs.

In symbols: If $P(A) \neq 0$ and $P(B) \neq 0$, then A and B are independent if

$$P(B|A) = P(B) \quad \text{or, equivalently,} \quad P(A|B) = P(A) \qquad (3.6)$$

If either $P(A) = 0$ or $P(B) = 0$, then A and B are independent.

If A and B are independent, then the following pairs of events are also independent: A and B^c, A^c and B, and A^c and B^c. The proof of this fact is left as an exercise.

The concept of independence can be extended to more than two events:

Definition

Events $A_1, A_2, \ldots, A_n$ are independent if the probability of each remains the same no matter which of the others occur.

In symbols: Events $A_1, A_2, \ldots, A_n$ are independent if for each A_i, and each collection $A_{j1}, \ldots, A_{jm}$ of events with $P(A_{j1} \cap \cdots \cap A_{jm}) \neq 0$,

$$P(A_i | A_{j1} \cap \cdots \cap A_{jm}) = P(A_i) \tag{3.7}$$

The Multiplication Rule

Sometimes we know $P(A|B)$ and we wish to find $P(A \cap B)$. We can obtain a result that is useful for this purpose by multiplying both sides of Equation (3.5) by $P(B)$. This leads to the multiplication rule.

If A and B are two events with $P(B) \neq 0$, then

$$P(A \cap B) = P(B)P(A|B) \tag{3.8}$$

If A and B are two events with $P(A) \neq 0$, then

$$P(A \cap B) = P(A)P(B|A) \tag{3.9}$$

If $P(A) \neq 0$ and $P(B) \neq 0$, then Equations (3.8) and (3.9) both hold.

When two events are independent, then $P(A|B) = P(A)$ and $P(B|A) = P(B)$, so the multiplication rule simplifies:

If A and B are independent events, then

$$P(A \cap B) = P(A)P(B) \tag{3.10}$$

This result can be extended to any number of events. If $A_1, A_2, \ldots, A_n$ are independent events, then for each collection $A_{j1}, \ldots, A_{jm}$ of events

$$P(A_{j1} \cap A_{j2} \cap \cdots \cap A_{jm}) = P(A_{j1})P(A_{j2}) \cdots P(A_{jm}) \tag{3.11}$$

In particular,

$$P(A_1 \cap A_2 \cap \cdots \cap A_n) = P(A_1)P(A_2) \cdots P(A_n) \tag{3.12}$$

Example
3.8

It is known that 5% of the cars and 10% of the light trucks produced by a certain manufacturer require warranty service. If someone purchases both a car and a light truck from this manufacturer, then, assuming the vehicles function independently, what is the probability that both will require warranty service?

Solution
We need to compute P(car requires service and truck requires service). Since the vehicles function independently, we may use Equation (3.10):

$$P(\text{car requires service and truck requires service})$$
$$= P(\text{car requires service})\, P(\text{truck requires service})$$
$$= (0.05)(0.10)$$
$$= 0.005$$

Equations (3.10) and (3.11) tell us how to compute probabilities when we know that events are independent, but they are usually not much help when it comes to deciding whether two events really *are* independent. In most cases, the best way to determine whether events are independent is through an understanding of the process that produces the events. Here are a few illustrations:

- A die is rolled twice. It is reasonable to believe that the outcome of the second roll is not affected by the outcome of the first roll. Therefore, knowing the outcome of the first roll does not help to predict the outcome of the second roll. The two rolls are independent.

- A certain chemical reaction is run twice, using different equipment each time. It is reasonable to believe that the yield of one reaction will not affect the yield of the other. In this case the yields are independent.

- A chemical reaction is run twice in succession, using the same equipment. In this case, it might not be wise to assume that the yields are independent. For example, a low yield on the first run might indicate that there is more residue than usual left behind. This might tend to make the yield on the next run higher. Thus knowing the yield on the first run could help to predict the yield on the second run.

- The items in a simple random sample may be treated as independent, unless the population is finite and the sample comprises more than about 5% of the population (see the discussion of independence in Section 1.1).

Application to Reliability Analysis

Reliability analysis is the branch of engineering concerned with estimating the failure rates of systems. While some problems in reliability analysis require advanced mathematical methods, there are many problems that can be solved with the methods we have learned so far. We begin with an example illustrating the computation of the reliability of a system consisting of two components connected *in series*.

xample
3.9

A system contains two components, A and B, connected in series as shown in the following diagram.

The system will function only if both components function. The probability that A functions is given by $P(A) = 0.98$, and the probability that B functions is given by $P(B) = 0.95$. Assume that A and B function independently. Find the probability that the system functions.

Solution
Since the system will function only if both components function, it follows that

$$P(\text{system functions}) = P(A \cap B)$$
$$= P(A)P(B) \text{ by the assumption of independence}$$
$$= (0.98)(0.95)$$
$$= 0.931$$

Example 3.10 illustrates the computation of the reliability of a system consisting of two components connected *in parallel*.

*xample*
3.10

A system contains two components, C and D, connected in parallel as shown in the following diagram.

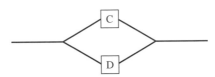

The system will function if either C or D functions. The probability that C functions is 0.90, and the probability that D functions is 0.85. Assume C and D function independently. Find the probability that the system functions.

Solution
Since the system will function so long as either of the two components functions, it follows that

$$P(\text{system functions}) = P(C \cup D)$$
$$= P(C) + P(D) - P(C \cap D)$$
$$= P(C) + P(D) - P(C)P(D)$$
$$\qquad \text{by the assumption of independence}$$
$$= 0.90 + 0.85 - (0.90)(0.85)$$
$$= 0.985$$

The reliability of more complex systems can often be determined by decomposing the system into a series of subsystems, each of which contains components connected either in series or in parallel. Example 3.11 illustrates the method.

Example

3.11

The thesis "Dynamic, Single-Stage, Multiperiod, Capacitated Production Sequencing Problem with Multiple Parallel Resources" (D. Ott, M.S. thesis, Colorado School of Mines, 1998) describes a production method used in the manufacture of aluminum cans. The following schematic diagram, slightly simplified, depicts the process.

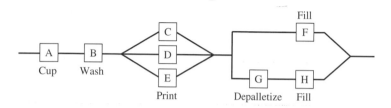

The initial input into the process consists of coiled aluminum sheets, approximately 0.25 mm thick. In a process known as "cupping," these sheets are uncoiled and shaped into can bodies, which are cylinders that are closed on the bottom and open on top. These can bodies are then washed and sent to the printer, which prints the label on the can. In practice there are several printers on a line; the diagram presents a line with three printers. The printer deposits the cans onto pallets, which are wooden structures that hold 7140 cans each. The cans next go to be filled. Some fill lines can accept cans directly from the pallets, but others can accept them only from cell bins, which are large containers holding approximately 100,000 cans each. To use these fill lines, the cans must be transported from the pallets to cell bins, in a process called depalletizing. In practice, there are several fill lines; the diagram presents a case where there are two fill lines, one of which will accept cans from the pallets, and the other of which will not. In the filling process the cans are filled, and the can top is seamed on. The cans are then packaged and shipped to distributors.

It is desired to estimate the probability that the process will function for one day without failing. Assume that the cupping process has probability 0.995 of functioning successfully for one day. Since this component is denoted by "A" in the diagram, we will express this probability as $P(A) = 0.995$. Assume that the other process components have the following probabilities of functioning successfully during a one-day period: $P(B) = 0.99$, $P(C) = P(D) = P(E) = 0.95$, $P(F) = 0.90$, $P(G) = 0.90$, $P(H) = 0.98$. Assume the components function independently. Find the probability that the process functions successfully for one day.

Solution

We can solve this problem by noting that the entire process can be broken down into subsystems, each of which consists of simple series or parallel component systems. Specifically, subsystem 1 consists of the cupping and washing components, which are connected in series. Subsystem 2 consists of the printers, which are connected in parallel.

Subsystem 3 consists of the fill lines, which are connected in parallel, with one of the two lines consisting of two components connected in series.

We compute the probabilities of successful functioning for each subsystem, denoting the probabilities p_1, p_2, and p_3.

$$\begin{aligned} P(\text{subsystem 1 functions}) = p_1 &= P(A \cap B) \\ &= P(A)P(B) \\ &= (0.995)(0.990) \\ &= 0.985050 \end{aligned}$$

$$\begin{aligned} P(\text{subsystem 2 functions}) = p_2 &= 1 - P(\text{subsystem 2 fails}) \\ &= 1 - P(C^c \cap D^c \cap E^c) \\ &= 1 - P(C^c)P(D^c)P(E^c) \\ &= 1 - (0.05)^3 \\ &= 0.999875 \end{aligned}$$

Subsystem 3 functions if F functions, or if both G and H function. Therefore,

$$\begin{aligned} P(\text{subsystem 3 functions}) = p_3 &= P(F \cup (G \cap H)) \\ &= P(F) + P(G \cap H) - P(F \cap G \cap H) \\ &= P(F) + P(G)P(H) - P(F)P(G)P(H) \\ &= (0.90) + (0.90)(0.98) - (0.90)(0.90)(0.98) \\ &= 0.988200 \end{aligned}$$

The entire process consists of the three subsystems connected in series. Therefore, for the process to function, all three subsystems must function. We conclude that

$$\begin{aligned} P(\text{system functions}) &= P(\text{systems 1, 2, and 3 all function}) \\ &= p_1 p_2 p_3 \\ &= (0.985050)(0.999875)(0.988200) \\ &= 0.973 \end{aligned}$$

We remark that the assumption that the components function independently is crucial in the solutions of Examples 3.9, 3.10, and 3.11. When this assumption is not met, it can be very difficult to make accurate reliability estimates. If the assumption of independence is used without justification, reliability estimates may be misleading.

Exercises for Section 3.2

1. Suppose that start-up companies in the area of biotechnology have probability 0.2 of becoming profitable and that those in the area of information technology have probability 0.15 of becoming profitable. A venture capitalist invests in one firm of each type. Assume the companies function independently.

   a. What is the probability that both companies become profitable?

   b. What is the probability that neither company become profitable?

   c. What is the probability that at least one of the two companies become profitable?

2. A drag racer has two parachutes, a main and a backup, that are designed to bring the vehicle to a stop after the end of a run. Suppose that the main chute deploys with probability 0.99 and that if the main fails to deploy, the backup deploys with probability 0.98.

   a. What is the probability that one of the two parachutes deploys?

   b. What is the probability that the backup parachute deploys?

3. A population of 600 semiconductor wafers contains wafers from three lots. The wafers are categorized by lot and by whether they conform to a thickness specification. The following table presents the number of wafers in each category. A wafer is chosen at random from the population.

Lot	Conforming	Nonconforming
A	88	12
B	165	35
C	260	40

   a. If the wafer is from Lot A, what is the probability that it is conforming?

   b. If the wafer is conforming, what is the probability that it is from Lot A?

   c. If the wafer is conforming, what is the probability that it is not from Lot C?

   d. If the wafer is not from Lot C, what is the probability that it is conforming?

4. Refer to Exercise 3. Let E_1 be the event that the wafer comes from Lot A, and let E_2 be the event that the wafer is conforming. Are E_1 and E_2 independent? Explain.

5. A geneticist is studying two genes. Each gene can be either dominant or recessive. A sample of 100 individuals is categorized as follows.

	Gene 2	
Gene 1	Dominant	Recessive
Dominant	56	24
Recessive	14	6

   a. What is the probability that a randomly sampled individual, Gene 1 is dominant?

   b. What is the probability that a randomly sampled individual, Gene 2 is dominant?

   c. Given that Gene 1 is dominant, what is the probability that Gene 2 is dominant?

   d. These genes are said to be in linkage equilibrium if the event that Gene 1 is dominant is independent of the event that Gene 2 is dominant. Are these genes in linkage equilibrium?

6. A system consists of four components connected as shown in the following diagram:

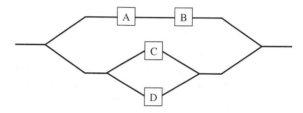

   Assume A, B, C, and D function independently. If the probabilities that A, B, C, and D fail are 0.10, 0.05, 0.10, and 0.20, respectively, what is the probability that the system functions?

7. A system consists of four components connected as shown in the following diagram:

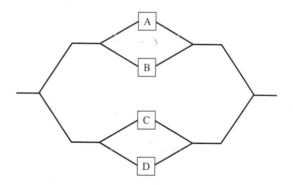

   Assume A, B, C, and D function independently. If the probabilities that A, B, C, and D fail are 0.1, 0.2, 0.05, and 0.3, respectively, what is the probability that the system functions?

8. A system contains two components, A and B, connected in series, as shown in the diagram.

   Assume A and B function independently. For the system to function, both components must function.

   a. If the probability that A fails is 0.05, and the probability that B fails is 0.03, find the probability that the system functions.

   b. If both A and B have probability p of failing, what must the value of p be so that the probability that the system functions is 0.90?

   c. If three components are connected in series, and each has probability p of failing, what must the value of p be so that the probability that the system functions is 0.90?

9. A system contains two components, C and D, connected in parallel as shown in the diagram.

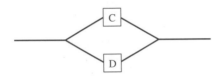

   Assume C and D function independently. For the system to function, either C or D must function.

   a. If the probability that C fails is 0.08 and the probability that D fails is 0.12, find the probability that the system functions.

   b. If both C and D have probability p of failing, what must the value of p be so that the probability that the system functions is 0.99?

   c. If three components are connected in parallel, function independently, and each has probability p of failing, what must the value of p be so that the probability that the system functions is 0.99?

   d. If components function independently, and each component has probability 0.5 of failing, what is the minimum number of components that must be connected in parallel so that the probability that the system functions is at least 0.99?

10. If A and B are independent events, prove that the following pairs of events are independent: A^c and B, A and B^c, and A^c and B^c.

11. An electronic message consists of a string of bits (0s and 1s). The message must pass through two relays before being received. At each relay the probability is 0.1 that the bit will be reversed before being relayed (i.e., a 1 will be changed to a 0, or a 0 to a 1). Find the probability that the value of a bit received at its final destination is the same as the value of the bit that was sent.

3.3   Random Variables

In many situations, it is desirable to assign a numerical value to each outcome of an experiment. Such an assignment is called a **random variable**. To make the idea clear, we present an example. Suppose that three machines are available to cut concrete blocks. Each machine needs to be brought down from time to time for maintenance. At any given moment, the probability that a machine is down is 0.10, and the probability that it is functioning is 0.90. Assume the machines function independently.

There are eight possible outcomes regarding which of the machines are up and which are down. Their probabilities are easy to compute. For example, the probability that Machine 1 is up, Machine 2 is down, and Machine 3 is up is $(0.9)(0.1)(0.9) = 0.081$.

The sample space that lists these outcomes, along with their probabilities, is presented in the following table.

	Machine			
1	**2**	**3**	**Probability**
Up	Up	Up	$(0.9)(0.9)(0.9) = 0.729$
Up	Up	Down	$(0.9)(0.9)(0.1) = 0.081$
Up	Down	Up	$(0.9)(0.1)(0.9) = 0.081$
Up	Down	Down	$(0.9)(0.1)(0.1) = 0.009$
Down	Up	Up	$(0.1)(0.9)(0.9) = 0.081$
Down	Up	Down	$(0.1)(0.9)(0.1) = 0.009$
Down	Down	Up	$(0.1)(0.1)(0.9) = 0.009$
Down	Down	Down	$(0.1)(0.1)(0.1) = 0.001$

Now let us suppose that what is important is the number of machines that are up, not specifically which machines are up. We will therefore assign to each outcome the quantity representing the number of machines that are up. This assignment, represented by the letter X, is presented in the following table.

	Machine				
1	**2**	**3**	X	**Probability**
Up	Up	Up	3	$(0.9)(0.9)(0.9) = 0.729$
Up	Up	Down	2	$(0.9)(0.9)(0.1) = 0.081$
Up	Down	Up	2	$(0.9)(0.1)(0.9) = 0.081$
Up	Down	Down	1	$(0.9)(0.1)(0.1) = 0.009$
Down	Up	Up	2	$(0.1)(0.9)(0.9) = 0.081$
Down	Up	Down	1	$(0.1)(0.9)(0.1) = 0.009$
Down	Down	Up	1	$(0.1)(0.1)(0.9) = 0.009$
Down	Down	Down	0	$(0.1)(0.1)(0.1) = 0.001$

The function X, which assigns a numerical value to each outcome in the sample space, is a random variable.

A **random variable** assigns a numerical value to each outcome in a sample space.

It is customary to denote random variables with uppercase letters. The letters X, Y, and Z are most often used.

We can compute probabilities for random variables in an obvious way. In the example just presented, the event $X = 2$ corresponds to the event {(Up, Up, Down), (Up, Down, Up), (Down, Up, Up)} of the sample space. Therefore $P(X = 2) = P(\{(\text{Up}, \text{Up}, \text{Down}), (\text{Up}, \text{Down}, \text{Up}), (\text{Down}, \text{Up}, \text{Up})\}) = 0.081 + 0.081 + 0.081 = 0.243$.

Example

3.12

List the possible values of the random variable X, and find the probability of each of them.

Solution

The possible values are 0, 1, 2, and 3. To find the probability of one of these values, we add the probabilities of the outcomes in the sample space that correspond to the value. The results are given in the following table.

x	$P(X = x)$
0	0.001
1	0.027
2	0.243
3	0.729

The table of probabilities in Example 3.12 contains all the information needed to compute any probability regarding the random variable X. Note that the outcomes of the sample space are not presented in the table. When the probabilities pertaining to a random variable are known, we usually do not think about the sample space; we just focus on the probabilities.

There are two important types of random variables, **discrete** and **continuous**. A discrete random variable is one whose possible values form a discrete set. This means that the values can be ordered and there are gaps between adjacent values. The random variable X described in Example 3.12 is discrete. In contrast, the possible values of a continuous random variable always contain an interval—that is, all the points between some two numbers. We will provide precise definitions of these types of random variables later in this section.

We present some more examples of random variables.

Example

3.13

For a certain type of printed-circuit board, 50% contain no defects, 25% contain one defect, 12% contain two defects, 8% contain three defects, and the remaining 5% contain four defects. Let Y represent the number of defects in a randomly chosen board. What are the possible values for Y? Is Y discrete or continuous? Find $P(Y = y)$ for each possible value y.

Solution

The possible values for Y are the integers 0, 1, 2, 3, and 4. The random variable Y is discrete, because it takes on only integer values. Fifty percent of the outcomes in the sample space are assigned the value 0. Therefore $P(Y = 0) = 0.5$. Similarly $P(Y = 1) = 0.25$, $P(Y = 2) = 0.12$, $P(Y = 3) = 0.08$, and $P(Y = 4) = 0.05$.

Example 3.14

A certain type of magnetic disk must function in an environment where it is exposed to corrosive gases. It is known that 10% of all such disks have lifetimes less than or equal to 100 hours, 50% have lifetimes greater than 100 hours but less than or equal to 500 hours, and 40% have lifetimes greater than 500 hours. Let Z represent the number of hours in the lifetime of a randomly chosen disk. Is Z continuous or discrete? Find $P(Z \leq 500)$. Can we compute all the probabilities for Z? Explain.

Solution

The lifetime of a component is not limited to a list of discretely spaced values; Z is continuous. Of all the components, 60% have lifetimes less than or equal to 500 hours. Therefore $P(Z \leq 500) = 0.60$. We do not have enough information to compute all the probabilities for Z. We can compute some of them, for example, $P(Z \leq 100) = 0.10$, $P(100 < Z \leq 500) = 0.50$, and $P(Z > 500) = 0.40$. But we do not know, for example, the proportion of components that have lifetimes between 100 and 200 hours, or between 200 and 300 hours, so we cannot find the probability that the random variable Z falls into either of these intervals. To compute all the probabilities for Z, we would need to be able to compute the probability for every possible interval—for example, $P(200 < Z \leq 300)$, $P(200 < Z \leq 201)$, $P(200 < Z \leq 200.1)$, and so on. We will see how this can be done later in this section, when we discuss continuous random variables.

Random Variables and Populations

It is often useful to think of a value of a random variable as having been sampled from a population. For example, consider the random variable Y described in Example 3.13. Observing a value for this random variable is like sampling a value from a population consisting of the integers 0, 1, 2, 3 in the following proportions: 0s, 50%; 1s, 25%; 2s, 12%; 3s, 8%; and 4s, 5%. For a continuous random variable, it is appropriate to imagine an infinite population containing all the possible values of the random variable. For example, for the random variable Z in Example 3.14 we would imagine a population containing all the positive numbers, with 10% of the population values less than or equal to 100, 50% greater than 100 but less than or equal to 500, and 40% greater than 500. The proportion of population values in any interval would be equal to the probability that the variable Z is in that interval.

Methods for working with random variables differ somewhat between discrete and continuous random variables. We begin with the discrete case.

Discrete Random Variables

We begin by reviewing the definition of a discrete random variable.

Definition

A random variable is **discrete** if its possible values form a discrete set. This means that if the possible values are arranged in order, there is a gap between each value and the next one. The set of possible values may be infinite; for example, the set of all integers and the set of all positive integers are both discrete sets.

It is common for the possible values of a discrete random variable to be a set of integers. For any discrete random variable, if we specify the list of its possible values along with the probability that the random variable takes on each of these values, then we have completely described the population from which the random variable is sampled. We illustrate with an example.

The number of flaws in a 1-inch length of copper wire manufactured by a certain process varies from wire to wire. Overall, 48% of the wires produced have no flaws, 39% have one flaw, 12% have two flaws, and 1% have three flaws. Let X be the number of flaws in a randomly selected piece of wire. Then

$$P(X = 0) = 0.48 \quad P(X = 1) = 0.39 \quad P(X = 2) = 0.12 \quad P(X = 3) = 0.01$$

The possible values 0, 1, 2, 3, along with the probabilities for each, provide a complete description of the population from which X is drawn. This description has a name—the **probability mass function**.

Definition

The **probability mass function** of a discrete random variable X is the function $p(x) = P(X = x)$. The probability mass function is sometimes called the **probability distribution**.

Thus for the random variable X representing the number of flaws in a length of wire, $p(0) = 0.48$, $p(1) = 0.39$, $p(2) = 0.12$, $p(3) = 0.01$, and $p(x) = 0$ for any value of x other than 0, 1, 2, or 3. Note that if the values of the probability mass function are added over all the possible values of X, the sum is equal to 1. This is true for any probability mass function. The reason is that summing the values of a probability mass function over all the possible values of the corresponding random variable produces the probability that the random variable is equal to one of its possible values, and this probability is always equal to 1.

The probability mass function can be represented by a graph in which a vertical line is drawn at each of the possible values of the random variable. The heights of the lines are equal to the probabilities of the corresponding values. The physical interpretation of this graph is that each line represents a mass equal to its height. Figure 3.2 presents a graph of the probability mass function of the random variable X.

The Cumulative Distribution Function of a Discrete Random Variable

The probability mass function specifies the probability that a random variable is equal to a given value. A function called the **cumulative distribution function** specifies the probability that a random variable is less than or equal to a given value. The cumulative distribution function of the random variable X is the function $F(x) = P(X \leq x)$.

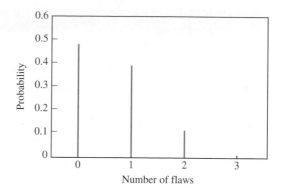

FIGURE 3.2 Probability mass function of X, the number of flaws in a randomly chosen piece of wire.

Example

3.15

Let $F(x)$ denote the cumulative distribution function of the random variable X that represents the number of flaws in a randomly chosen wire. Find $F(2)$. Find $F(1.5)$.

Solution

Since $F(2) = P(X \leq 2)$, we need to find $P(X \leq 2)$. We do this by summing the probabilities for the values of X that are less than or equal to 2, namely, 0, 1, and 2. Thus

$$F(2) = P(X \leq 2)$$
$$= P(X = 0) + P(X = 1) + P(X = 2)$$
$$= 0.48 + 0.39 + 0.12$$
$$= 0.99$$

Now $F(1.5) = P(X \leq 1.5)$. Therefore, to compute $F(1.5)$ we must sum the probabilities for the values of X that are less than or equal to 1.5, which are 0 and 1. Thus

$$F(1.5) = P(X \leq 1.5)$$
$$= P(X = 0) + P(X = 1)$$
$$= 0.48 + 0.39$$
$$= 0.87$$

In general, for any discrete random variable X, the cumulative distribution function $F(x)$ can be computed by summing the probabilities of all the possible values of X that are less than or equal to x. Note that $F(x)$ is defined for any number x, not just for the possible values of X.

Summary

Let X be a discrete random variable. Then

- The probability mass function of X is the function $p(x) = P(X = x)$.
- The cumulative distribution function of X is the function
  $F(x) = P(X \le x)$.
- $F(x) = \sum_{t \le x} p(t) = \sum_{t \le x} P(X = t)$.

- $\sum_x p(x) = \sum_x P(X = x) = 1$, where the sum is over all the possible
  values of X.

xample

3.16

Plot the cumulative distribution function $F(x)$ of the random variable X that represents the number of flaws in a randomly chosen wire.

Solution

First we compute $F(x)$ for each of the possible values of X, which are 0, 1, 2, and 3.

$$F(0) = P(X \le 0) = 0.48$$
$$F(1) = P(X \le 1) = 0.48 + 0.39 = 0.87$$
$$F(2) = P(X \le 2) = 0.48 + 0.39 + 0.12 = 0.99$$
$$F(3) = P(X \le 3) = 0.48 + 0.39 + 0.12 + 0.01 = 1$$

For any value x, we compute $F(x)$ by summing the probabilities of all the possible values of X that are less than or equal to x. For example, if $1 \le x < 2$, the possible values of X that are less than or equal to x are 0 and 1, so $F(x) = P(X = 0) + P(X = 1) = F(1) = 0.87$. Therefore

$$F(x) = \begin{cases} 0 & x < 0 \\ 0.48 & 0 \le x < 1 \\ 0.87 & 1 \le x < 2 \\ 0.99 & 2 \le x < 3 \\ 1 & x \ge 3 \end{cases}$$

A plot of $F(x)$ is presented in the following figure.

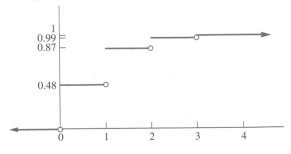

For a discrete random variable, the graph of $F(x)$ consists of a series of horizontal lines (called "steps") with jumps at each of the possible values of X. Note that the size of the jump at any point x is equal to the value of the probability mass function $p(x) = P(X = x)$.

Mean and Variance for Discrete Random Variables

The **population mean** of a discrete random variable can be thought of as the mean of a hypothetical sample that follows the probability distribution perfectly. To make this idea concrete, assume that the number of flaws in a wire, X, has the probability mass function given previously, with $P(X = 0) = 0.48$, $P(X = 1) = 0.39$, $P(X = 2) = 0.12$, and $P(X = 3) = 0.01$. Now imagine that we have a sample of 100 wires and that the sample follows this distribution perfectly, so that exactly 48 of the wires have 0 flaws, 39 have 1 flaw, 12 have 2 flaws, and 1 has 3 flaws. The sample mean is the total number of flaws divided by 100:

$$\text{Mean} = \frac{0(48) + 1(39) + 2(12) + 3(1)}{100} = 0.66$$

This can be rewritten as

$$\text{Mean} = 0(0.48) + 1(0.39) + 2(0.12) + 3(0.01) = 0.66$$

This shows that the mean of a perfect sample can be obtained by multiplying each possible value of X by its probability, and summing the products. This is the definition of the population mean of a discrete random variable. The population mean of a random variable X may also be called the **expectation**, or **expected value**, of X, and can be denoted by μ_X, by $E(X)$, or simply by μ. Sometimes we will drop the word "population," and simply refer to the population mean as the mean.

Definition

Let X be a discrete random variable with probability mass function $p(x) = P(X = x)$.

The mean of X is given by

$$\mu_X = \sum_x x P(X = x) \tag{3.13}$$

where the sum is over all possible values of X.

The mean of X is sometimes called the expectation, or expected value, of X and may also be denoted by $E(X)$ or by μ.

Example

3.17

A certain industrial process is brought down for recalibration whenever the quality of the items produced falls below specifications. Let X represent the number of times the process is recalibrated during a week, and assume that X has the following probability mass function.

x	0	1	2	3	4
$p(x)$	0.35	0.25	0.20	0.15	0.05

Find the mean of X.

Solution

Using Equation (3.13), we compute

$$\mu_X = 0(0.35) + 1(0.25) + 2(0.20) + 3(0.15) + 4(0.05) = 1.30$$

The population mean has an important physical interpretation. It is the horizontal component of the center of mass of the probability mass function, that is, it is the point on the horizontal axis at which the graph of the probability mass would balance if supported there. Figure 3.3 illustrates this property for the probability mass function described in Example 3.17, where the population mean is $\mu = 1.30$.

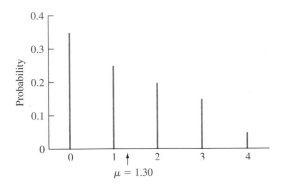

FIGURE 3.3 The graph of a probability mass function will balance if supported at the population mean.

We will describe the **population variance** of a discrete random variable by making an analogy with the sample variance. The sample variance was discussed in Section 1.2. Recall that for a sample $X_1, \ldots, X_n$, the sample variance is given by $\sum(X_i - \overline{X})^2/(n-1)$. The sample variance is thus essentially the average of the squared differences between the sample points and the sample mean (except that we divide by $n - 1$ instead of n).

By analogy, the population variance of a discrete random variable X is a weighted average of the squared differences $(x - \mu_X)^2$ where x ranges through all the possible values of the random variable X. This weighted average is computed by multiplying each squared difference $(x - \mu_X)^2$ by the probability $P(X = x)$ and summing the results.

The population variance of a random variable X can be denoted by σ_X^2, by $V(X)$, or simply by σ^2. The population variance is given by the formula

$$\sigma_X^2 = \sum_x (x - \mu_X)^2 P(X = x)$$

By performing some algebra, an alternate formula can be obtained.

$$\sigma_X^2 = \sum_x x^2 P(X = x) - \mu_X^2$$

We also define the **population standard deviation** to be the square root of the population variance. We denote the population standard deviation of a random variable X by σ_X or simply by σ. As with the mean, we will sometimes drop the word "population," and simply refer to the population variance and population standard deviation as the variance and standard deviation, respectively.

Summary

Let X be a discrete random variable with probability mass function $p(x) = P(X = x)$. Then

■ The variance of X is given by

$$\sigma_X^2 = \sum_x (x - \mu_X)^2 P(X = x) \qquad (3.14)$$

■ An alternate formula for the variance is given by

$$\sigma_X^2 = \sum_x x^2 P(X = x) - \mu_X^2 \qquad (3.15)$$

■ The variance of X may also be denoted by $V(X)$ or by σ^2.
■ The standard deviation is the square root of the variance: $\sigma_X = \sqrt{\sigma_X^2}$.

Example **3.18**

Find the variance and standard deviation for the random variable X described in Example 3.17, representing the number of times a process is recalibrated.

Solution
In Example 3.17 we computed the mean of X to be $\mu_X = 1.30$. We compute the variance by using Equation (3.14):

$$\sigma_X^2 = (0 - 1.30)^2 P(X = 0) + (1 - 1.30)^2 P(X = 1) + (2 - 1.30)^2 P(X = 2)$$
$$+ (3 - 1.30)^2 P(X = 3) + (4 - 1.30)^2 P(X = 4)$$
$$= (1.69)(0.35) + (0.09)(0.25) + (0.49)(0.20) + (2.89)(0.15) + (7.29)(0.05)$$
$$= 1.51$$

The standard deviation is $\sigma_X = \sqrt{1.51} = 1.23$.

Example 3.19

Use the alternate formula, Equation (3.15), to compute the variance of X, the number of times a process is recalibrated.

Solution

In Example 3.17 the mean was computed to be $\mu_X = 1.30$. The variance is therefore

$$
\begin{aligned}
\sigma_X^2 &= 0^2 P(X=0) + 1^2 P(X=1) + 2^2 P(X=2) + 3^2 P(X=3) \\
&\quad + 4^2 P(X=4) - (1.30)^2 \\
&= (0)(0.35) + (1)(0.25) + (4)(0.20) + (9)(0.15) + (16)(0.05) - (1.30)^2 \\
&= 1.51
\end{aligned}
$$

The Probability Histogram

When the possible values of a discrete random variable are evenly spaced, the probability mass function can be represented by a histogram, with rectangles centered at the possible values of the random variable. The area of a rectangle centered at a value x is equal to $P(X = x)$. Such a histogram is called a **probability histogram**, because the areas represent probabilities. Figure 3.2 presented the graph of the probability mass function of a random variable X representing the number of flaws in a wire. Figure 3.4 presents a probability histogram for this random variable.

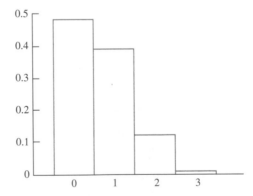

FIGURE 3.4 Probability histogram for X, the number of flaws in a randomly chosen piece of wire. Compare with Figure 3.2.

The probability that the value of a random variable falls into a given interval is given by an area under the probability histogram. Example 3.20 illustrates the idea.

Example 3.20

Use the probability histogram in Figure 3.4 to find the probability that a randomly chosen wire has more than one flaw. Indicate this probability as an area under the probability histogram.

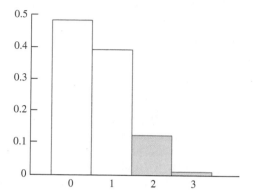

FIGURE 3.5 Probability histogram for X, the number of flaws in a randomly chosen piece of wire. The area corresponding to values of X greater than 1 is shaded. This area is equal to $P(X > 1)$.

Solution

We wish to find $P(X > 1)$. Since no wire has more than three flaws, the proportion of wires that have more than one flaw can be found by adding the proportion that have two flaws to the proportion that have three flaws. In symbols, $P(X > 1) = P(X = 2) + P(X = 3)$. The probability mass function specifies that $P(X = 2) = 0.12$ and $P(X = 3) = 0.01$. Therefore $P(X > 1) = 0.12 + 0.01 = 0.13$.

This probability is given by the area under the probability histogram corresponding to those rectangles centered at values greater than 1 (see Figure 3.5). There are two such rectangles; their areas are $P(X = 2) = 0.12$ and $P(X = 3) = 0.01$. This is another way to show that $P(X > 1) = 0.12 + 0.01 = 0.13$.

In Chapter 4, we will see that probabilities for discrete random variables can sometimes be approximated by computing the area under a curve. Representing the discrete probabilities with a probability histogram will make it easier to understand how this is done.

Continuous Random Variables

Figure 1.8 (in Section 1.3) presents a histogram for the emissions, in grams of particulates per gallon of fuel consumed, of a sample of 62 vehicles. Note that emissions is a continuous variable, because its possible values are not restricted to some discretely spaced set. The class intervals are chosen so that each interval contains a reasonably large number of vehicles. If the sample were larger, we could make the intervals narrower. In particular, if we had information on the entire population, containing millions of vehicles, we could make the intervals extremely narrow. The histogram would then look quite smooth and could be approximated with a curve, which might look like Figure 3.6 (page 96).

If a vehicle were chosen at random from this population to have its emissions measured, the emissions level X would be a random variable. The probability that X falls between any two values a and b is equal to the area under the histogram between a and b. Because the histogram in this case is represented by a curve, the probability would be found by computing an integral.

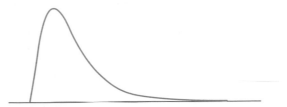

FIGURE 3.6 The histogram for a large continuous population could be drawn with extremely narrow rectangles and might look like this curve.

The random variable X described here is an example of a **continuous random variable**. A continuous random variable is defined to be a random variable whose probabilities are represented by areas under a curve. This curve is called the **probability density function**. Because the probability density function is a curve, the computations of probabilities involve integrals, rather than the sums that are used in the discrete case.

Definition

A random variable is **continuous** if its probabilities are given by areas under a curve. The curve is called a **probability density function** for the random variable.

The probability density function is sometimes called the **probability distribution**.

Computing Probabilities with the Probability Density Function

Let X be a continuous random variable. Let the function $f(x)$ be the probability density function of X. Let a and b be any two numbers, with $a < b$.

The proportion of the population whose values of X lie between a and b is given by $\int_a^b f(x)\,dx$, the area under the probability density function between a and b. This is the probability that the random variable X takes on a value between a and b. Note that the area under the curve does not depend on whether the endpoints a and b are included in the interval. Therefore, probabilities involving X do not depend on whether endpoints are included.

Summary

Let X be a continuous random variable with probability density function $f(x)$. Let a and b be any two numbers, with $a < b$. Then

$$P(a \leq X \leq b) = P(a \leq X < b) = P(a < X \leq b) = P(a < X < b) = \int_a^b f(x)\,dx$$

In addition,

$$P(X \leq b) = P(X < b) = \int_{-\infty}^b f(x)\,dx \qquad (3.16)$$

$$P(X \geq a) = P(X > a) = \int_a^\infty f(x)\,dx \qquad (3.17)$$

If $f(x)$ is the probability density function of a random variable X, then the area under the entire curve from $-\infty$ to ∞ is the probability that the value of X is between $-\infty$ and ∞. This probability must be equal to 1, because the value of X is always between $-\infty$ and ∞. Therefore the area under the entire curve $f(x)$ is equal to 1.

Summary

Let X be a continuous random variable with probability density function $f(x)$. Then

$$\int_{-\infty}^{\infty} f(x)\,dx = 1$$

Example
3.21

A hole is drilled in a sheet-metal component, and then a shaft is inserted through the hole. The shaft clearance is equal to the difference between the radius of the hole and the radius of the shaft. Let the random variable X denote the clearance, in millimeters. The probability density function of X is

$$f(x) = \begin{cases} 1.25(1 - x^4) & 0 < x < 1 \\ 0 & \text{otherwise} \end{cases}$$

Components with clearances larger than 0.8 mm must be scrapped. What proportion of components are scrapped?

Solution

Figure 3.7 presents the probability density function of X. Note that the density $f(x)$ is 0 for $x \leq 0$ and for $x \geq 1$. This indicates that the clearances are always between 0 and 1 mm. The proportion of components that must be scrapped is $P(X > 0.8)$, which is equal to the area under the probability density function to the right of 0.8.

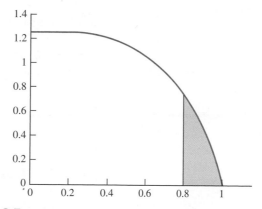

FIGURE 3.7 Graph of the probability density function of X, the clearance of a shaft. The area shaded is equal to $P(X > 0.8)$.

This area is given by

$$P(X > 0.8) = \int_{0.8}^{\infty} f(x)\, dx$$

$$= \int_{0.8}^{1} 1.25(1 - x^4)\, dx$$

$$= 1.25 \left(x - \frac{x^5}{5} \right) \Big|_{0.8}^{1}$$

$$= 0.0819$$

The Cumulative Distribution Function of a Continuous Random Variable

The cumulative distribution function of a continuous random variable X is $F(x) = P(X \le x)$, just as it is for a discrete random variable. For a discrete random variable, $F(x)$ can be found by summing values of the probability mass function. For a continuous random variable, the value of $F(x)$ is obtained by integrating the probability density function. Since $F(x) = P(X \le x)$, it follows from Equation (3.16) that $F(x) = \int_{-\infty}^{x} f(t)\, dt$, where $f(t)$ is the probability density function.

Definition

Let X be a continuous random variable with probability density function $f(x)$. The cumulative distribution function of X is the function

$$F(x) = P(X \le x) = \int_{-\infty}^{x} f(t)\, dt \qquad (3.18)$$

Example 3.22

Refer to Example 3.21. Find the cumulative distribution function $F(x)$ and plot it.

Solution

The probability density function of X is given by $f(t) = 0$ if $t \le 0$, $f(t) = 1.25(1 - t^4)$ if $0 < t < 1$, and $f(t) = 0$ if $t \ge 1$. The cumulative distribution function is given by $F(x) = \int_{-\infty}^{x} f(t)\, dt$. Since $f(t)$ is defined separately on three different intervals, the computation of the cumulative distribution function involves three separate cases.

If $x \le 0$:

$$F(x) = \int_{-\infty}^{x} f(t)\, dt$$

$$= \int_{-\infty}^{x} 0\, dt$$

$$= 0$$

If $0 < x < 1$:

$$F(x) = \int_{-\infty}^{x} f(t)\, dt$$

$$= \int_{-\infty}^{0} f(t)\, dt + \int_{0}^{x} f(t)\, dt$$

$$= \int_{-\infty}^{0} 0\, dt + \int_{0}^{x} 1.25(1 - t^4)\, dt$$

$$= 0 + 1.25 \left(t - \frac{t^5}{5} \right) \Bigg|_{0}^{x}$$

$$= 1.25 \left(x - \frac{x^5}{5} \right)$$

If $x > 1$:

$$F(x) = \int_{-\infty}^{x} f(t)\, dt$$

$$= \int_{-\infty}^{0} f(t)\, dt + \int_{0}^{1} f(t)\, dt + \int_{1}^{x} f(t)\, dt$$

$$= \int_{-\infty}^{0} 0\, dt + \int_{0}^{1} 1.25(1 - t^4)\, dt + \int_{1}^{x} 0\, dt$$

$$= 0 + 1.25 \left(t - \frac{t^5}{5} \right) \Bigg|_{0}^{1} + 0$$

$$= 0 + 1 + 0$$

$$= 1$$

Therefore

$$F(x) = \begin{cases} 0 & x \le 0 \\ 1.25 \left(x - \dfrac{x^5}{5} \right) & 0 < x < 1 \\ 1 & x \ge 1 \end{cases}$$

A plot of $F(x)$ is presented here.

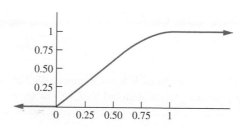

Note that the cumulative distribution function $F(x)$ in Example 3.22 is a continuous function—there are no jumps in its graph. This is characteristic of continuous random variables. The cumulative distribution function of a continuous random variable will always be continuous, while the cumulative distribution function of a discrete random variable will never be continuous.

Example 3.23

Refer to Example 3.21. Use the cumulative distribution function to find the probability that the shaft clearance is less than 0.5 mm.

Solution

Let X denote the shaft clearance. We need to find $P(X \leq 0.5)$. This is equivalent to finding $F(0.5)$, where $F(x)$ is the cumulative distribution function. Using the results of Example 3.22, $F(0.5) = 1.25(0.5 - 0.5^5/5) = 0.617$.

Mean and Variance for Continuous Random Variables

The population mean and variance of a continuous random variable are defined in the same way as those of a discrete random variable, except that the probability density function is used instead of the probability mass function. Specifically, if X is a continuous random variable, its population mean is defined to be the center of mass of its probability density function, and its population variance is the moment of inertia around a vertical axis through the population mean. The formulas are analogous to Equations (3.13) through (3.15), with the sums replaced by integrals.

As was the case with discrete random variables, we will sometimes drop the word "population" and refer to the population mean, population variance, and population standard deviation more simply as the mean, variance, and standard deviation, respectively.

Definition

Let X be a continuous random variable with probability density function $f(x)$. Then the mean of X is given by

$$\mu_X = \int_{-\infty}^{\infty} x f(x)\, dx \tag{3.19}$$

The mean of X is sometimes called the expectation, or expected value, of X and may also be denoted by $E(X)$ or by μ.

Definition

Let X be a continuous random variable with probability density function $f(x)$. Then

- The variance of X is given by

$$\sigma_X^2 = \int_{-\infty}^{\infty} (x - \mu_X)^2 f(x)\, dx \qquad (3.20)$$

- An alternate formula for the variance is given by

$$\sigma_X^2 = \int_{-\infty}^{\infty} x^2 f(x)\, dx - \mu_X^2 \qquad (3.21)$$

- The variance of X may also be denoted by $V(X)$ or by σ^2.
- The standard deviation is the square root of the variance: $\sigma_X = \sqrt{\sigma_X^2}$.

Example **3.24**

Refer to Example 3.21. Find the mean clearance and the variance of the clearance.

Solution

Using Equation (3.19), we can find the mean clearance:

$$\mu_X = \int_{-\infty}^{\infty} x f(x)\, dx$$

$$= \int_0^1 x[1.25(1 - x^4)]\, dx$$

$$= 1.25 \left(\frac{x^2}{2} - \frac{x^6}{6} \right) \Big|_0^1$$

$$= 0.4167$$

Having computed $\mu_X = 0.4167$, we can now compute σ_X^2. It is easiest to use the alternate formula, Equation (3.21):

$$\sigma_X^2 = \int_{-\infty}^{\infty} x^2 f(x)\, dx - \mu_X^2$$

$$= \int_0^1 x^2[1.25(1 - x^4)]\, dx - (0.4167)^2$$

$$= 1.25 \left(\frac{x^3}{3} - \frac{x^7}{7} \right) \Big|_0^1 - (0.4167)^2$$

$$= 0.0645$$

Exercises for Section 3.3

1. Determine whether each of the following random variables is discrete or continuous.

   a. The number of heads in 100 tosses of a coin.

   b. The length of a rod randomly chosen from a day's production.

   c. The final exam score of a randomly chosen student from last semester's engineering statistics class.

   d. The age of a randomly chosen Colorado School of Mines student.

   e. The age that a randomly chosen Colorado School of Mines student will be on his or her next birthday.

2. Computer chips often contain surface imperfections. For a certain type of computer chip, the probability mass function of the number of defects X is presented in the following table.

x	0	1	2	3	4
$p(x)$	0.4	0.3	0.15	0.10	0.05

   a. Find $P(X \leq 2)$.

   b. Find $P(X > 1)$.

   c. Find μ_X.

   d. Find σ_X^2.

3. A chemical supply company ships a certain solvent in 10-gallon drums. Let X represent the number of drums ordered by a randomly chosen customer. Assume X has the following probability mass function:

x	1	2	3	4	5
$p(x)$	0.4	0.2	0.2	0.1	0.1

   a. Find the mean number of drums ordered.

   b. Find the variance of the number of drums ordered.

   c. Find the standard deviation of the number of drums ordered.

   d. Let Y be the number of gallons ordered. Find the probability mass function of Y.

   e. Find the mean number of gallons ordered.

   f. Find the variance of the number of gallons ordered.

   g. Find the standard deviation of the number of gallons ordered.

4. The element titanium has five stable occurring isotopes, differing from each other in the number of neutrons an atom contains. If X is the number of neutrons in a randomly chosen titanium atom, the probability mass function of X is given as follows:

x	24	25	26	27	28
$p(x)$	0.0825	0.0744	0.7372	0.0541	0.0518

   a. Find μ_X.

   b. Find σ_X.

5. A certain type of component is packaged in lots of four. Let X represent the number of properly functioning components in a randomly chosen lot. Assume that the probability that exactly x components function is proportional to x; in other words, assume that the probability mass function of X is given by

   $$p(x) = \begin{cases} cx & x = 1, 2, 3, \text{ or } 4 \\ 0 & \text{otherwise} \end{cases}$$

   where c is a constant.

   a. Find the value of the constant c so that $p(x)$ is a probability mass function.

   b. Find $P(X = 2)$.

   c. Find the mean number of properly functioning components.

   d. Find the variance of the number of properly functioning components.

   e. Find the standard deviation of the number of properly functioning components.

6. After manufacture, computer disks are tested for errors. Let X be the number of errors detected on a randomly chosen disk. The following table presents values of the cumulative distribution function $F(x)$ of X.

x	$F(x)$
0	0.41
1	0.72
2	0.83
3	0.95
4	1.00

   a. What is the probability that two or fewer errors are detected?

   b. What is the probability that more than three errors are detected?

c. What is the probability that exactly one error is detected?

d. What is the probability that no errors are detected?

e. What is the most probable number of errors to be detected?

7. Resistors labeled $100\,\Omega$ have true resistances that are between $80\,\Omega$ and $120\,\Omega$. Let X be the mass of a randomly chosen resistor. The probability density function of X is given by

$$f(x) = \begin{cases} \dfrac{x - 80}{800} & 80 < x < 120 \\ 0 & \text{otherwise} \end{cases}$$

a. What proportion of resistors have resistances less than $90\,\Omega$?

b. Find the mean resistance.

c. Find the standard deviation of the resistances.

d. Find the cumulative distribution function of the resistances.

8. Elongation (in %) of steel plates treated with aluminum are random with probability density function

$$f(x) = \begin{cases} \dfrac{x}{250} & 20 < x < 30 \\ 0 & \text{otherwise} \end{cases}$$

a. What proportion of steel plates have elongations greater than 25%?

b. Find the mean elongation.

c. Find the variance of the elongations.

d. Find the standard deviation of the elongations.

e. Find the cumulative distribution function of the elongations.

f. A particular plate elongates 28%. What proportion of plates elongate more than this?

9. The lifetime of a transistor in a certain application has a lifetime that is random with probability density function

$$f(t) = \begin{cases} 0.1e^{-0.1t} & t > 0 \\ 0 & t \leq 0 \end{cases}$$

a. Find the mean lifetime.

b. Find the standard deviation of the lifetimes.

c. Find the cumulative distribution function of the lifetime.

d. Find the probability that the lifetime will be less than 12 months.

10. The diameter of a rivet (in mm) is a random variable with probability density function

$$f(x) = \begin{cases} 6(x - 12)(13 - x) & 12 < x \leq 13 \\ 0 & \text{otherwise} \end{cases}$$

a. What is the probability that the diameter is less than 12.5 mm?

b. Find the mean diameter.

c. Find the standard deviation of the diameters.

d. Find the cumulative distribution function of the diameters.

e. The specification for the diameter is 12.3 to 12.7 mm. What is the probability that the specification is met?

11. The concentration of a reactant is a random variable with probability density function

$$f(x) = \begin{cases} 1.2(x + x^2) & 0 < x < 1 \\ 0 & \text{otherwise} \end{cases}$$

a. What is the probability that the concentration is greater than 0.5?

b. Find the mean concentration.

c. Find the probability that the concentration is within ± 0.1 of the mean.

d. Find the standard deviation σ of the concentrations.

e. Find the probability that the concentration is within $\pm 2\sigma$ of the mean.

f. Find the cumulative distribution function of the concentration.

12. The error in the length of a part (absolute value of the difference between the actual length and the target length), in mm, is a random variable with probability density function

$$f(x) = \begin{cases} \dfrac{e^{-x}}{1 - e^{-1}} & 0 < x < 1 \\ 0 & \text{otherwise} \end{cases}$$

a. What is the probability that the error is less than 0.2 mm?

b. Find the mean error.

c. Find the variance of the error.

d. Find the cumulative distribution function of the error.

e. The specification for the error is 0 to 0.3 mm. What is the probability that the specification is met?

13. The thickness of a washer (in mm) is a random variable with probability density function

$$f(x) = \begin{cases} (3/52)x(6-x) & 2 < x < 4 \\ 0 & \text{otherwise} \end{cases}$$

a. What is the probability that the thickness is less than 2.5 m?

b. What is the probability that the thickness is between 2.5 and 3.5 m?

c. Find the mean thickness.

d. Find the standard deviation σ of the thicknesses.

e. Find the probability that the thickness is within $\pm\sigma$ of the mean.

f. Find the cumulative distribution function of the thickness.

14. Particles are a major component of air pollution in many areas. It is of interest to study the sizes of contaminating particles. Let X represent the diameter, in micrometers, of a randomly chosen particle. Assume that in a certain area, the probability density function of X is inversely proportional to the volume of the particle; that is, assume that

$$f(x) = \begin{cases} \dfrac{c}{x^3} & x \geq 1 \\ 0 & x < 1 \end{cases}$$

where c is a constant.

a. Find the value of c so that $f(x)$ is a probability density function.

b. Find the mean particle diameter.

c. Find the cumulative distribution function of the particle diameter.

d. The term PM_{10} refers to particles 10 μm or less in diameter. What proportion of the contaminating particles are PM_{10}?

e. The term $PM_{2.5}$ refers to particles 2.5 μm or less in diameter. What proportion of the contaminating particles are $PM_{2.5}$?

f. What proportion of the PM_{10} particles are $PM_{2.5}$?

15. The level of impurity (in %) in the product of a certain chemical process is a random variable with probability density function

$$f(x) = \begin{cases} (3/64)x^2(4-x) & 0 < x < 4 \\ 0 & \text{otherwise} \end{cases}$$

a. What is the probability that the impurity level is greater than 3%?

b. What is the probability that the impurity level is between 2% and 3%?

c. Find the mean impurity level.

d. Find the variance of the impurity levels.

e. Find the cumulative distribution function of the impurity level.

16. The reading given by a thermometer calibrated in ice water (actual temperature 0°C) is a random variable with probability density function

$$f(x) = \begin{cases} k(1-x^2) & -1 < x < 1 \\ 0 & \text{otherwise} \end{cases}$$

where k is a constant.

a. Find the value of k.

b. What is the probability that the thermometer reads above 0°C?

c. What is the probability that the reading is within 0.25°C of the actual temperature?

d. What is the mean reading?

e. What is the standard deviation of the readings?

3.4 Functions of Random Variables

In practice we often construct new random variables by performing arithmetic operations on other random variables. For example, we might add a constant to a random variable, multiply a random variable by a constant, or add two or more random variables together. In this section, we describe how to compute means and variances of random variables constructed in these ways, and we present some practical examples.

Adding a Constant

When a constant is added to a random variable, the mean is increased by the value of the constant, but the variance and standard deviation are unchanged. For example, assume that steel rods produced by a certain machine have a mean length of 5.0 in. and a variance of $\sigma^2 = 0.003$ in². Each rod is attached to a base that is exactly 1.0 in. long. The mean length of the assembly will be $5.0 + 1.0 = 6.0$ in. Since each length is increased by the same amount, the spread in the lengths does not change, so the variance remains the same. To put this in statistical terms, let X be the length of a randomly chosen rod, and let $Y = X + 1$ be the length of the assembly. Then $\mu_Y = \mu_{X+1} = \mu_X + 1$, and $\sigma_Y^2 = \sigma_{X+1}^2 = \sigma_X^2$. In general, when a constant is added to a random variable, the mean is shifted by that constant, and the variance is unchanged.

Summary

If X is a random variable and b is a constant, then

$$\mu_{X+b} = \mu_X + b \tag{3.22}$$

$$\sigma_{X+b}^2 = \sigma_X^2 \tag{3.23}$$

Multiplying by a Constant

Often we need to multiply a random variable by a constant. This might be done, for example, to convert to a more convenient set of units. We continue the example of steel rod production to show how multiplication by a constant affects the mean, variance, and standard deviation of a random variable.

If we measure the lengths of the rods described earlier in centimeters rather than inches, the mean length will be $(2.54 \text{ cm/in.})(5.0 \text{ in.}) = 12.7$ cm. In statistical terms, let the random variable X be the length in inches of a randomly chosen rod, and let $Y = 2.54X$ be the length in centimeters. Then $\mu_Y = 2.54\mu_X$. In general, when a random variable is multiplied by a constant, its mean is multiplied by the same constant.

Summary

If X is a random variable and a is a constant, then

$$\mu_{aX} = a\mu_X \tag{3.24}$$

When the length X of a rod is measured in inches, the variance σ_X^2 must have units of in². If $Y = 2.54X$ is the length in centimeters, then σ_Y^2 must have units of cm². Therefore we obtain σ_Y^2 by multiplying σ_X^2 by 2.54^2, which is the conversion factor from in² to cm². In general, when a random variable is multiplied by a constant, its variance is multiplied by the *square* of the constant.

Summary

If X is a random variable and a is a constant, then

$$\sigma_{aX}^2 = a^2\sigma_X^2 \qquad (3.25)$$

$$\sigma_{aX} = |a|\sigma_X \qquad (3.26)$$

If a random variable is multiplied by a constant and then added to another constant, the effect on the mean and variance can be determined by combining Equations (3.22) and (3.24) and Equations (3.23) and (3.25). The results are presented in the following summary.

Summary

If X is a random variable, and a and b are constants, then

$$\mu_{aX+b} = a\mu_X + b \qquad (3.27)$$

$$\sigma_{aX+b}^2 = a^2\sigma_X^2 \qquad (3.28)$$

$$\sigma_{aX+b} = |a|\sigma_X \qquad (3.29)$$

Note that Equations (3.27) through (3.29) are analogous to results for the sample mean and standard deviation presented in Section 1.2.

Example 3.25

The molarity of a solute in solution is defined to be the number of moles of solute per liter of solution (1 mole $= 6.02 \times 10^{23}$ molecules). If the molarity of a stock solution of concentrated sulfuric acid (H_2SO_4) is X, and if one part of the solution is mixed with N parts water, the molarity Y of the dilute solution is given by $Y = X/(N + 1)$. Assume that the stock solution is manufactured by a process that produces a molarity with mean 18 and standard deviation 0.1. If 100 mL of stock solution is added to 300 mL of water, find the mean and standard deviation of the molarity of the dilute solution.

Solution
The molarity of the dilute solution is $Y = 0.25X$. The mean and standard deviation of X are $\mu_X = 18$ and $\sigma_X = 0.1$, respectively. Therefore

$$
\begin{aligned}
\mu_Y &= \mu_{0.25X} \\
&= 0.25\mu_X \qquad \text{(using Equation 3.24)} \\
&= 0.25(18.0) \\
&= 4.5
\end{aligned}
$$

Also,

$$\sigma_Y = \sigma_{0.25X}$$
$$= 0.25\sigma_X \qquad \text{(using Equation 3.26)}$$
$$= 0.25(0.1)$$
$$= 0.025$$

Means of Linear Combinations of Random Variables

Consider the case of adding two random variables. For example, assume that two machines fabricate a certain metal part. The mean daily production of machine A is 100 parts, and the mean daily production of machine B is 150 parts. Clearly the mean daily production from the two machines together is 250 parts. Putting this in mathematical notation, let X be the number of parts produced on a given day by machine A, and let Y be the number of parts produced on the same day by machine B. The total number of parts is $X + Y$, and we have that $\mu_{X+Y} = \mu_X + \mu_Y$.

This idea extends to any number of random variables.

If $X_1, X_2, \ldots, X_n$ are random variables, then the mean of the sum $X_1 + X_2 + \cdots + X_n$ is given by

$$\mu_{X_1+X_2+\cdots+X_n} = \mu_{X_1} + \mu_{X_2} + \cdots + \mu_{X_n} \qquad (3.30)$$

The sum $X_1 + X_2 + \cdots + X_n$ is a special case of a **linear combination**:

If $X_1, \ldots, X_n$ are random variables and $c_1, \ldots, c_n$ are constants, then the random variable

$$c_1 X_1 + \cdots + c_n X_n$$

is called a **linear combination** of $X_1, \ldots, X_n$.

To find the mean of a linear combination of random variables, we can combine Equations (3.24) and (3.30):

If X and Y are random variables, and a and b are constants, then

$$\mu_{aX+bY} = \mu_{aX} + \mu_{bY} = a\mu_X + b\mu_Y \qquad (3.31)$$

More generally, if $X_1, X_2, \ldots, X_n$ are random variables and $c_1, c_2, \ldots, c_n$ are constants, then the mean of the linear combination $c_1 X_1 + c_2 X_2 + \cdots + c_n X_n$ is given by

$$\mu_{c_1 X_1 + c_2 X_2 + \cdots + c_n X_n} = c_1 \mu_{X_1} + c_2 \mu_{X_2} + \cdots + c_n \mu_{X_n} \qquad (3.32)$$

Independent Random Variables

The notion of independence for random variables is very much like the notion of independence for events. Two random variables are independent if knowledge concerning one of them does not affect the probabilities of the other. When two events are independent, the probability that both occur is found by multiplying the probabilities for each event (see Equations 3.10 and 3.11 in Section 3.2). There are analogous formulas for independent random variables. The notation for these formulas is as follows. Let X be a random variable, and let S be a set of numbers. The notation "$X \in S$" means that the value of the random variable X is in the set S.

Definition

If X and Y are **independent** random variables, and S and T are sets of numbers, then

$$P(X \in S \text{ and } Y \in T) = P(X \in S)P(Y \in T) \tag{3.33}$$

More generally, if $X_1, \ldots, X_n$ are independent random variables, and $S_1, \ldots, S_n$ are sets, then

$$P(X_1 \in S_1 \text{ and } X_2 \in S_2 \text{ and } \cdots \text{ and } X_n \in S_n) =$$
$$P(X_1 \in S_1)P(X_2 \in S_2) \cdots P(X_n \in S_n) \tag{3.34}$$

Example 3.26

Cylindrical cans have specifications regarding their height and radius. Let H be the length and R be the radius, in millimeters, of a randomly sampled can. The probability mass function of H is given by $P(H = 119) = 0.2$, $P(H = 120) = 0.7$, and $P(H = 121) = 0.1$. The probability mass function of R is given by $P(R = 30) = 0.6$ and $P(R = 31) = 0.4$. The volume of a can is given by $V = \pi R^2 H$. Assume R and H are independent. Find the probability that the area is 108,000 mm^3.

Solution
The volume will be equal to 108,000 if $H = 120$ and $R = 30$. Therefore

$$
\begin{aligned}
P(V = 108{,}000) &= P(H = 120 \text{ and } R = 30) \\
&= P(H = 120)P(R = 30) \qquad \text{since } H \text{ and } R \text{ are independent} \\
&= (0.7)(0.6) \\
&= 0.42
\end{aligned}
$$

Variances of Linear Combinations of Independent Random Variables

We have seen that the mean of a sum of random variables is always equal to the sum of the means (Equation 3.30). In general, the formula for the variance of a sum of random variables is a little more complicated than this. But when random variables are *independent*, the result is simple: the variance of the sum is the sum of the variances.

If $X_1, X_2, \ldots, X_n$ are *independent* random variables, then the variance of the sum $X_1 + X_2 + \cdots + X_n$ is given by

$$\sigma^2_{X_1+X_2+\cdots+X_n} = \sigma^2_{X_1} + \sigma^2_{X_2} + \cdots + \sigma^2_{X_n} \qquad (3.35)$$

To find the variance of a linear combination of random variables, we can combine Equations (3.35) and (3.25):

If $X_1, X_2, \ldots, X_n$ are *independent* random variables and $c_1, c_2, \ldots, c_n$ are constants, then the variance of the linear combination $c_1 X_1 + c_2 X_2 + \cdots + c_n X_n$ is given by

$$\sigma^2_{c_1 X_1+c_2 X_2+\cdots+c_n X_n} = c_1^2\sigma^2_{X_1} + c_2^2\sigma^2_{X_2} + \cdots + c_n^2\sigma^2_{X_n} \qquad (3.36)$$

Two frequently encountered linear combinations are the sum and the difference of two random variables. Interestingly enough, when the random variables are independent, the variance of the sum is the same as the variance of the difference.

If X and Y are *independent* random variables with variances σ^2_X and σ^2_Y, then the variance of the sum $X + Y$ is

$$\sigma^2_{X+Y} = \sigma^2_X + \sigma^2_Y \qquad (3.37)$$

The variance of the difference $X - Y$ is

$$\sigma^2_{X-Y} = \sigma^2_X + \sigma^2_Y \qquad (3.38)$$

The fact that the variance of the difference is the *sum* of the variances may seem counterintuitive. However, it follows from Equation (3.36) by setting $c_1 = 1$ and $c_2 = -1$.

Example 3.27

An object with initial temperature T_0 is placed in an environment with ambient temperature T_a. According to Newton's law of cooling, the temperature T of the object is given by $T = cT_0 + (1-c)T_a$, where c is a constant that depends on the physical properties of the object and on the elapsed time. Assume that T_0 has mean 25°C and standard deviation 2°C, and T_a has mean 5°C and standard deviation 1°C. Find the mean of T at the time when $c = 0.25$. Assuming T_0 and T_a to be independent, find the standard deviation of T at that time.

Solution
Using Equation (3.32), we find that the temperature T has mean

$$\mu_T = \mu_{0.25T_0+0.75T_a}$$
$$= 0.25\mu_{T_0} + 0.75\mu_{T_a}$$
$$= 6.25 + 3.75$$
$$= 10$$

Since X_1 and X_2 are independent, we can use Equation (3.36) to find the standard deviation σ_C:

$$\sigma_C = \sqrt{\sigma_{0.25T_0 + 0.75T_a}^2}$$

$$= \sqrt{(0.25)^2 \sigma_{T_0}^2 + (0.75)^2 \sigma_{T_a}^2}$$

$$= \sqrt{0.0625(2)^2 + 0.5625(1)^2}$$

$$= 0.9014$$

Independence and Simple Random Samples

When a simple random sample of numerical values is drawn from a population, each item in the sample can be thought of as a random variable. The items in a simple random sample may be treated as independent, except when the sample is a large proportion (more than 5%) of a finite population (see the discussion of independence in Section 1.1). From here on, unless explicitly stated to the contrary, we will assume this exception has not occurred, so that the values in a simple random sample may be treated as independent random variables.

Summary

If $X_1, X_2, \ldots, X_n$ is a simple random sample, then $X_1, X_2, \ldots, X_n$ may be treated as independent random variables, all with the same distribution.

When $X_1, \ldots, X_n$ are independent random variables, all with the same distribution, it is sometimes said that $X_1, \ldots, X_n$ are **independent and identically distributed (i.i.d.)**.

The Mean and Variance of a Sample Mean

The most frequently encountered linear combination is the sample mean. Specifically, if $X_1, \ldots, X_n$ is a simple random sample from a population with mean μ and variance σ^2, then the sample mean $\overline{X}$ is the linear combination

$$\overline{X} = \frac{1}{n}X_1 + \cdots + \frac{1}{n}X_n$$

From this fact we can compute the mean and variance of $\overline{X}$.

$$\mu_{\overline{X}} = \mu_{\frac{1}{n}X_1 + \cdots + \frac{1}{n}X_n}$$

$$= \frac{1}{n}\mu_{X_1} + \cdots + \frac{1}{n}\mu_{X_n} \qquad \text{(using Equation 3.32)}$$

$$= \frac{1}{n}\mu + \cdots + \frac{1}{n}\mu$$

$$= (n)\left(\frac{1}{n}\right)\mu$$

$$= \mu$$

As discussed previously, the items in a simple random sample may be treated as independent random variables. Therefore

$$\sigma_{\overline{X}}^2 = \sigma_{\frac{1}{n}X_1 + \cdots + \frac{1}{n}X_n}^2$$

$$= \frac{1}{n^2}\sigma_{X_1}^2 + \cdots + \frac{1}{n^2}\sigma_{X_n}^2 \qquad \text{(using Equation 3.36)}$$

$$= \frac{1}{n^2}\sigma^2 + \cdots + \frac{1}{n^2}\sigma^2$$

$$= (n)\left(\frac{1}{n^2}\right)\sigma^2$$

$$= \frac{\sigma^2}{n}$$

Summary

If $X_1, \ldots, X_n$ is a simple random sample from a population with mean μ and variance σ^2, then the sample mean $\overline{X}$ is a random variable with

$$\mu_{\overline{X}} = \mu \qquad (3.39)$$

$$\sigma_{\overline{X}}^2 = \frac{\sigma^2}{n} \qquad (3.40)$$

The standard deviation of $\overline{X}$ is

$$\sigma_{\overline{X}} = \frac{\sigma}{\sqrt{n}} \qquad (3.41)$$

Example
3.28

The lifetime of a light bulb in a certain application has mean 700 hours and standard deviation 20 hours. The light bulbs are packaged 12 to a box. Assuming that the light bulbs in a box are a simple random sample of light bulbs, find the mean and standard deviation of the average lifetime of the light bulbs in a box.

Solution
Let $T_1, \ldots, T_{12}$ represent the lifetimes of the 12 light bulbs in a box. This is a simple random sample from a population with mean $\mu = 700$ and standard deviation $\sigma = 20$. The average lifetime is $\overline{T} = (T_1 + \cdots + T_{12})/12$. Using Equation (3.39),

$$\mu_{\overline{T}} = \mu = 700$$

Using Equation (3.41),

$$\sigma_{\overline{T}} = \frac{\sigma}{\sqrt{12}} = 5.77$$

Standard Deviations of Nonlinear Functions of Random Variables

The examples we have seen so far involve estimating standard deviations of linear functions of random variables. In many cases we wish to estimate the standard deviation of a nonlinear function of a random variable. For example, suppose the radius R of a circle is measured to be 5.43 cm, and the area is therefore calculated to be $A = \pi 5.43^2 = 92.6\, \text{cm}^2$. Furthermore, suppose that due to measurement error, it is appropriate to model the measured value as a random variable with standard deviation 0.01 cm. What is the standard deviation of the calculated area A? In statistical terms, we know that the standard deviation σ_R is 0.01 cm, and we must calculate the standard deviation of A, where A is the function of R given by $A = \pi R^2$.

In general, the type of problem we wish to solve is this: Given a random variable X, with known standard deviation σ_X, and given a function $U = U(X)$, how do we compute the standard deviation σ_U? If U is a linear function, we can use Equation (3.29). If U is not linear, we can still approximate σ_U, by multiplying σ_X by the absolute value of the derivative dU/dX. The approximation will be good so long as σ_X is small.

If X is a random variable whose standard deviation σ_X is small, and if U is a function of X, then

$$\sigma_U \approx \left| \frac{dU}{dX} \right| \sigma_X \qquad (3.42)$$

In practice, we evaluate the derivative dU/dX at the observed value of X.

Equation (3.42) is known as the **propagation of error** formula.

Propagation of Error Uncertainties Are Only Approximate

Standard deviations computed by using Equation (3.42) are often only rough approximations. For this reason, these standard deviations should be expressed with no more than two significant digits. Indeed, some authors suggest using only one significant digit.

Example 3.29

The radius R of a circle is measured to be 5.43 cm, with a standard deviation of 0.01 cm. Estimate the area of the circle, and find the standard deviation of this estimate.

Solution
The area A is given by $A = \pi R^2$. The estimate of A is $\pi(5.43\ \text{cm})^2 = 92.6\ \text{cm}^2$. Now $\sigma_R = 0.01$ cm, and $dA/dR = 2\pi R = 10.86\pi$ cm. We can now find the standard deviation of A:

$$\sigma_A = \left| \frac{dA}{dR} \right| \sigma_R$$
$$= (10.86\pi\ \text{cm})(0.01\ \text{cm})$$
$$= 0.34\ \text{cm}^2$$

We estimate the area of the circle to be $92.6\ \text{cm}^2$, with a standard deviation of $0.3\ \text{cm}^2$.

Often we need to estimate a quantity as a function of several random variables. For example, we might measure the mass m and volume V of a rock and compute the density as $D = m/V$. In practice, we might need to estimate the standard deviation σ_D in terms of σ_m and σ_V.

The general problem is to estimate the standard deviation of a quantity that is a function of several *independent* random variables. The basic formula is given here.

If $X_1, X_2, \ldots, X_n$ are *independent* random variables whose standard deviations $\sigma_{X_1}, \sigma_{X_2}, \ldots, \sigma_{X_n}$ are small, and if $U = U(X_1, X_2, \ldots, X_n)$ is a function of $X_1, X_2, \ldots, X_n$, then

$$\sigma_U \approx \sqrt{\left(\frac{\partial U}{\partial X_1}\right)^2 \sigma_{X_1}^2 + \left(\frac{\partial U}{\partial X_2}\right)^2 \sigma_{X_2}^2 + \cdots + \left(\frac{\partial U}{\partial X_n}\right)^2 \sigma_{X_n}^2} \qquad (3.43)$$

In practice, we evaluate the partial derivatives at the point $(X_1, X_2, \ldots, X_n)$.

Equation (3.43) is the **multivariate propagation of error formula**. It is important to note that it is valid only when the random variables $X_1, X_2, \ldots, X_n$ are independent. As in the case of one random variable, the standard deviations computed by the propagation of error formula are often only rough approximations.

Example 3.30

Assume the mass of a rock is measured to be $m = 674.0 \pm 1.0\,\text{g}$, and the volume of the rock is measured to be $V = 261.0 \pm 0.1\,\text{mL}$. The density D of the rock is given by $D = m/V$. Estimate the density and find the standard deviation of the estimate.

Solution

Substituting $m = 674.0\,\text{g}$ and $V = 261.0\,\text{mL}$, the estimate of the density D is $674.0/261.0 = 2.582\,\text{g/mL}$. Since $D = m/V$, the partial derivatives of D are

$$\frac{\partial D}{\partial m} = \frac{1}{V} = 0.003831\,\text{mL}^{-1}$$

$$\frac{\partial D}{\partial V} = \frac{-m}{V^2} = -0.009894\,\text{g/mL}^2$$

The standard deviation of D is therefore

$$\sigma_D = \sqrt{\left(\frac{\partial D}{\partial m}\right)^2 \sigma_m^2 + \left(\frac{\partial D}{\partial V}\right)^2 \sigma_V^2}$$

$$= \sqrt{(0.003831)^2(1.0)^2 + (-0.009894)^2(0.1)^2}$$

$$= 0.0040\,\text{g/mL}$$

We estimate the density of the rock to be $2.582\,\text{g/mL}$, with a standard deviation of $0.004\,\text{g/mL}$.

One of the great benefits of the multivariate propagation of error formula is that it enables one to determine which measurement errors are most responsible for the random variation in the final result. Example 3.31 illustrates this.

Example
3.31

The density of the rock in Example 3.30 is to be estimated again with different equipment, in order to improve the precision. Which would improve the precision of the density estimate more: decreasing the standard deviation of the mass estimate to 0.5 g, or decreasing the standard deviation of the volume estimate to 0.05 mL?

Solution

From Example 3.30, $\sigma_D = \sqrt{(0.003831)^2\sigma_m^2 + (-0.009894)^2\sigma_V^2}$. We have a choice between having $\sigma_m = 0.5$ and $\sigma_V = 0.1$, or having $\sigma_m = 1.0$ and $\sigma_V = 0.05$. The first choice results in $\sigma_D = 0.002$ g/mL, while the second choice results in $\sigma_D = 0.004$ g/mL. It is better to reduce σ_m to 0.5 g.

Exercises for Section 3.4

1. If X and Y are independent random variables with means $\mu_X = 9.5$ and $\mu_Y = 6.8$, and standard deviations $\sigma_X = 0.4$ and $\sigma_Y = 0.1$, find the means and standard deviations of the following:

   a. $3X$

   b. $Y - X$

   c. $X + 4Y$

2. The bottom of a cylindrical container has an area of 10 cm^2. The container is filled to a height whose mean is 5 cm and whose standard deviation is 0.1 cm. Let V denote the volume of fluid in the container.

   a. Find μ_V.

   b. Find σ_V.

3. A process that fills plastic bottles with a beverage has a mean fill volume of 2.013 L and a standard deviation of 0.005 L. A case contains 24 bottles. Assuming that the bottles in a case are a simple random sample of bottles filled by this method, find the mean and standard deviation of the average volume per bottle in a case.

4. The force, in N, exerted by gravity on a mass of m kg is given by $F = 9.8m$. Objects of a certain type have mass whose mean is 2.3 kg with a standard deviation of 0.2 kg. Find the mean and standard deviation of F.

5. A piece of plywood is composed of five layers. The layers are a simple random sample from a popula-

   tion whose thickness has mean 3.50 mm and standard deviation 0.10 mm.

   a. Find the mean thickness of a piece of plywood.

   b. Find the standard deviation of the thickness of a piece of plywood.

6. The period of a pendulum is estimated by measuring the starting and stopping times and taking their difference. If the starting and stopping times are measured independently, each with standard deviation 0.2 s, what is the standard deviation of the estimated period?

7. The molarity of a solute in solution is defined to be the number of moles of solute per liter of solution (1 mole $= 6.02 \times 10^{23}$ molecules). If X is the molarity of a solution of magnesium chloride ($MgCl_2$), and Y is the molarity of a solution of ferric chloride ($FeCl_3$), the molarity of chloride ion (Cl^-) in a solution made of equal parts of the solutions of $MgCl_2$ and $FeCl_3$ is given by $M = X + 1.5Y$. Assume that X has mean 0.125 and standard deviation 0.05, and that Y has mean 0.350 and standard deviation 0.10.

   a. Find μ_M.

   b. Assuming X and Y to be independent, find σ_M.

8. Two independent measurements are made of the lifetime of a charmed strange meson. Each measurement has a standard deviation of 7×10^{-15} seconds. The

lifetime of the meson is estimated by averaging the two measurements. What is the standard deviation of this estimate?

9. The four sides of a picture frame consist of two pieces selected from a population whose mean length is 30 cm with standard deviation 0.1 cm, and two pieces selected from a population whose mean length is 45 cm with standard deviation 0.3 cm.

   a. Find the mean perimeter.

   b. Assuming the four pieces are chosen independently, find the standard deviation of the perimeter.

10. A gas station earns $2.60 in revenue for each gallon of regular gas it sells, $2.75 for each gallon of midgrade gas, and $2.90 for each gallon of premium gas. Let X_1, X_2, and X_3 denote the numbers of gallons of regular, midgrade, and premium gasoline sold in a day. Assume that X_1, X_2, and X_3 have means $\mu_1 = 1500$, $\mu_2 = 500$, and $\mu_3 = 300$, and standard deviations $\sigma_1 = 180$, $\sigma_2 = 90$, and $\sigma_3 = 40$, respectively.

   a. Find the mean daily revenue.

   b. Assuming X_1, X_2, and X_3 to be independent, find the standard deviation of the daily revenue.

11. The number of miles traveled per gallon of gasoline for a certain car has a mean of 25 and a standard deviation of 2. The tank holds 20 gallons.

   a. Find the mean number of miles traveled per tank.

   b. Assume the distances traveled are independent for each gallon of gas. Find the standard deviation of the number of miles traveled per tank.

   c. The car owner travels X miles on 20 gallons of gas and estimates her gas mileage as $X/20$. Find the mean of the estimated gas mileage.

d. Assuming the distances traveled are independent for each gallon of gas, find the standard deviation of the estimated gas mileage.

12. The Needleman-Wunsch method for aligning DNA sequences assigns 1 point whenever a mismatch occurs, and 3 points whenever a gap (insertion or deletion) appears in a sequence. Assume that under certain conditions, the number of mismatches has mean 5 and standard deviation 2, and the number of gaps has mean 2 and standard deviation 1.

   a. Find the mean of the Needleman-Wunsch score.

   b. Assume the number of gaps is independent of the number of mismatches. Find the variance of the Needleman-Wunsch score.

13. The acceleration g due to gravity is estimated by dropping an object and measuring the time it takes to travel a certain distance. Assume the distance s is known to be exactly 5 m, and the time is measured to be $t = 1.01 \pm 0.02$ s. Estimate g, and find the standard deviation of the estimate. (Note that $g = 2s/t^2$.)

14. One way to measure the water content of a soil is to weigh the soil both before and after drying it in an oven. The water content is $W = (M_1 - M_2)/M_1$, where M_1 is the mass before drying and M_2 is the mass after drying. Assume that $M_1 = 1.32 \pm 0.01$ kg and $M_2 = 1.04 \pm 0.01$ kg.

   a. Estimate W, and find the standard deviation of the estimate.

   b. Which would provide a greater reduction in the standard deviation of W: reducing the standard deviation of M_1 to 0.005 kg or reducing the standard deviation of M_2 to 0.005 kg?

Supplementary Exercises for Chapter 3

1. Among the cast aluminum parts manufactured on a certain day, 80% were flawless, 15% had only minor flaws, and 5% had major flaws. Find the probability that a randomly chosen part

   a. has a flaw (major or minor).

   b. has no major flaw.

2. In a certain type of automobile engine, the cylinder head is fastened to the block by 10 bolts, each of which

should be torqued to 60 N·m. Assume that the torques of the bolts are independent. If each bolt is torqued correctly with probability 0.99, what is the probability that all the bolts on a cylinder head are torqued correctly?

3. Nuclear power plants have redundant components in important systems to reduce the chance of catastrophic failure. Assume that a plant has two gauges to measure

the level of coolant in the reactor core and that each gauge has probability 0.01 of failing. Assume that one potential cause of gauge failure is that the electric cables leading from the core to the control room where the gauges are located may burn up in a fire. Someone wishes to estimate the probability that both gauges fail and makes the following calculation:

P(both gauges fail)

$\quad = P$(first gauge fails) P(second gauge fails)

$\quad = (0.01)(0.01)$

$\quad = 0.0001$

a. What assumption is being made in this calculation?

b. Explain why this assumption is probably not justified in the present case.

c. Is the probability of 0.0001 likely to be too high or too low? Explain.

4. The lifetime, in years, of a certain type of fuel cell is a random variable with probability density function

$$f(x) = \begin{cases} \dfrac{81}{(x+3)^4} & x > 0 \\ 0 & x \le 0 \end{cases}$$

a. What is the probability that a fuel cell lasts more than 3 years?

b. What is the probability that a fuel cell lasts between 1 and 3 years?

c. Find the mean lifetime.

d. Find the variance of the lifetimes.

e. Find the cumulative distribution function of the lifetime.

5. A system consists of four components, connected as shown in the diagram. Suppose that the components function independently, and that the probabilities of failure are 0.05 for A, 0.03 for B, 0.07 for C, and 0.14 for D. Find the probability that the system functions.

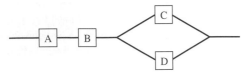

6. Let A and B be events with $P(A) = 0.4$ and $P(A \cap B) = 0.2$. For what value of $P(B)$ will A and B be independent?

7. Let X and Y be independent random variables with $\mu_X = 2$, $\sigma_X = 1$, $\mu_Y = 2$, and $\sigma_Y = 3$. Find the means and variances of the following quantities.

a. $3X$

b. $X + Y$

c. $X - Y$

d. $2X + 6Y$

8. The article "Traps in Mineral Valuations—Proceed With Care" (W. Lonegan, *Journal of the Australasian Institute of Mining and Metallurgy*, 2001:18–22) models the value (in millions of dollars) of a mineral deposit yet to be mined as a random variable X with probability mass function $p(x)$ given by $p(10) = 0.40$, $p(60) = 0.50$, $p(80) = 0.10$, and $p(x) = 0$ for values of x other than 10, 60, or 80.

a. Is this article treating the value of a mineral deposit as a discrete or a continuous random variable?

b. Compute μ_X.

c. Compute σ_X.

d. The project will be profitable if the value is more than $50 million. What is the probability that the project is profitable?

9. The period T of a simple pendulum is given by $T = 2\pi\sqrt{L/g}$, where L is the length of the pendulum and g is the acceleration due to gravity. Assume $g = 9.80 \text{ m/s}^2$ exactly, and L is measured to be 0.742 m with a standard deviation of 0.005 m. Estimate T, and find the standard deviation of the estimate.

10. The pressure P, temperature T, and volume V of one mole of an ideal gas are related by the equation $T = 0.1203PV$, when P is measured in kilopascals, T is measured in kelvins, and V is measured in liters.

a. Assume that P is measured to be 242.52 kPa with a standard deviation of 0.03 kPa and V is measured to be 10.103 L with a standard deviation of 0.002 L. Estimate T, and find the standard deviation of the estimate.

b. Which would provide a greater reduction in the standard deviation of T: reducing the standard deviation of P to 0.01 kPa or reducing the standard deviation of V to 0.001 L?

11. The repair time (in hours) for a certain machine is a random variable with probability density function

$$f(x) = \begin{cases} xe^{-x} & x > 0 \\ 0 & x \le 0 \end{cases}$$

a. What is the probability that the repair time is less than 2 hours?

b. What is the probability that the repair time is between 1.5 and 3 hours?

c. Find the mean repair time.

d. Find the cumulative distribution function of the repair times.

12. A process that manufactures piston rings produces rings whose diameters (in centimeters) vary according to the probability density function

$$f(x) = \begin{cases} 3[1 - 16(x - 10)^2] & 9.75 < x < 10.25 \\ 0 & \text{otherwise} \end{cases}$$

a. Find the mean diameter of rings manufactured by this process.

b. Find the standard deviation of the diameters of rings manufactured by this process. (*Hint*: Equation 3.20 may be easier to use than Equation 3.21.)

c. Find the cumulative distribution function of piston ring diameters.

d. What proportion of piston rings have diameters less than 9.75 cm?

e. What proportion of piston rings have diameters between 9.75 and 10.25 cm?

13. Refer to Exercise 12. A competing process produces rings whose diameters (in centimeters) vary according to the probability density function

$$f(x) = \begin{cases} 15[1 - 25(x - 10.05)^2]/4 & 9.85 < x < 10.25 \\ 0 & \text{otherwise} \end{cases}$$

Specifications call for the diameter to be 10.0 ± 0.1 cm. Which process is better, this one or the one in Exercise 12? Explain.

14. An environmental scientist is concerned with the rate at which a certain toxic solution is absorbed into the skin. Let X be the volume in microliters of solution absorbed by 1 in^2 of skin in 1 min. Assume that the probability density function of X is well approximated

by the function $f(x) = (2\sqrt{2\pi})^{-1}e^{-(x-10)^2/8}$, defined for $-\infty < x < \infty$.

a. Find the mean volume absorbed in 1 min.

b. (Hard) Find the standard deviation of the volume absorbed in 1 min.

15. In the article "An Investigation of the Ca–CO$_3$–CaF$_2$–K$_2$SiO$_3$–SiO$_2$–Fe Flux System Using the Submerged Arc Welding Process on HSLA-100 and AISI-1018 Steels" (G. Fredrickson, M.S. thesis, Colorado School of Mines, 1992), the carbon equivalent P of a weld metal is defined to be a linear combination of the weight percentages of carbon (C), manganese (Mn), copper (Cu), chromium (Cr), silicon (Si), nickel (Ni), molybdenum (Mo), vanadium (V), and boron (B). The carbon equivalent is given by

$$P = C + \frac{Mn + Cu + Cr}{20} + \frac{Si}{30} + \frac{Ni}{60} + \frac{Mo}{15} + \frac{V}{10} + 5B$$

Means and standard deviations of the weight percents of these chemicals were estimated from measurements on 45 weld metals produced on HSLA-100 steel base metal. Assume the means and standard deviations (SD) are as given in the following table.

	Mean	SD
C	0.0695	0.0018
Mn	1.0477	0.0269
Cu	0.8649	0.0225
Cr	0.7356	0.0113
Si	0.2171	0.0185
Ni	2.8146	0.0284
Mo	0.5913	0.0031
V	0.0079	0.0006
B	0.0006	0.0002

a. Find the mean carbon equivalent of weld metals produced from HSLA-1000 steel base metal.

b. Assuming the weight percents to be independent, find the standard deviation of the carbon equivalent of weld metals produced from HSLA-1000 steel base metal.

16. The oxygen equivalence number of a weld is a number that can be used to predict properties such as hardness, strength, and ductility. The article "Advances in

Oxygen Equivalence Equations for Predicting the Properties of Titanium Welds" (D. Harwig, W. Ittiwattana, and H. Castner, *The Welding Journal*, 2001:126s–136s) presents several equations for computing the oxygen equivalence number of a weld. One equation, designed to predict the hardness of a weld, is $X = O + 2N + (2/3)C$, where X is the oxygen equivalence, and O, N, and C are the amounts of oxygen, nitrogen, and carbon, respectively, in weight percent, in the weld. Suppose that for welds of a certain type, $\mu_O = 0.1668$, $\mu_N = 0.0255$, $\mu_C = 0.0247$, $\sigma_O = 0.0340$, $\sigma_N = 0.0194$, and $\sigma_C = 0.0131$.

a. Find μ_X.

b. Suppose the weight percents of O, N, and C are independent. Find σ_X.

Chapter 4

Commonly Used Distributions

Introduction

Statistical inference involves drawing a sample from a population and analyzing the sample data to learn about the population. In many situations, one has some knowledge about the probability mass function or probability density function of the population. In particular, the probability mass or density function can often be well approximated by one of several standard families of curves, or functions. In this chapter, we describe some of these standard families, and for each one we describe some conditions under which it is an appropriate model.

4.1 The Binomial Distribution

Imagine an experiment that can result in one of two outcomes. One outcome is labeled "success" and the other outcome is labeled "failure." The probability of success is denoted by p. The probability of failure is therefore $1 - p$. Such a trial is called a **Bernoulli trial** with success probability p. The simplest Bernoulli trial is the toss of a coin. The two outcomes are heads and tails. If we define heads to be the success outcome, then p is the probability that the coin comes up heads. For a fair coin, $p = 1/2$. Another example of a Bernoulli trial is the selection of a component from a population of components, some of which are defective. If we define "success" as a defective component, then p is the proportion of defective components in the population.

Many experiments consist of performing a sequence of Bernoulli trials. For example, we might sample several components from a very large lot and count the number of defectives among them. This amounts to conducting several independent Bernoulli trials and counting the number of successes. The number of successes is then a random variable, which is said to have a **binomial distribution**.

We now present a formal description of the binomial distribution. Assume that a series of n Bernoulli trials is conducted, each with the same success probability p. Assume further that the trials are *independent*—that is, that the outcome of one trial does not influence the outcomes of any of the other trials. Let the random variable X equal the number of successes in these n trials. Then X is said to have the **binomial distribution** with parameters n and p. The notation is $X \sim \text{Bin}(n, p)$. X is a discrete random variable, and its possible values are $0, 1, \ldots, n$.

Summary

If a total of n Bernoulli trials are conducted, and

- The trials are independent
- Each trial has the same success probability p
- X is the number of successes in the n trials

then X has the binomial distribution with parameters n and p, denoted $X \sim \text{Bin}(n, p)$.

Example 4.1

A fair coin is tossed 10 times. Let X be the number of heads that appear. What is the distribution of X?

Solution
There are 10 independent Bernoulli trials, each with success probability $p = 0.5$. The random variable X is equal to the number of successes in the 10 trials. Therefore $X \sim \text{Bin}(10, 0.5)$.

Probability Mass Function of a Binomial Random Variable

We now derive the probability mass function of a binomial random variable by considering an example. A biased coin has probability 0.6 of coming up heads. The coin is tossed three times. Let X be the number of heads. Then $X \sim \text{Bin}(3, 0.6)$. We will compute $P(X = 2)$.

There are three arrangements of two heads in three tosses of a coin, HHT, HTH, and THH. We first compute the probability of HHT. The event HHT is a sequence of independent events: H on the first toss, H on the second toss, T on the third toss. We know the probabilities of each of these events separately:

$P(\text{H on the first toss}) = 0.6$, $P(\text{H on the second toss}) = 0.6$, $P(\text{T on the third toss}) = 0.4$

Since the events are independent, the probability that they all occur is equal to the product of their probabilities (Equation 3.11 in Section 3.2). Thus

$$P(\text{HHT}) = (0.6)(0.6)(0.4) = (0.6)^2(0.4)^1$$

Similarly, $P(\text{HTH}) = (0.6)(0.4)(0.6) = (0.6)^2(0.4)^1$, and $P(\text{THH}) = (0.4)(0.6)(0.6) = (0.6)^2(0.4)^1$. It is easy to see that all the different arrangements of two heads and one tail have the same probability. Now

$$P(X = 2) = P(\text{HHT or HTH or THH})$$

$$= P(\text{HHT}) + P(\text{HTH}) + P(\text{THH})$$

$$= (0.6)^2(0.4)^1 + (0.6)^2(0.4)^1 + (0.6)^2(0.4)^1$$

$$= 3(0.6)^2(0.4)^1$$

Examining this result, we see that the number 3 represents the number of arrangements of two successes (heads) and one failure (tails), 0.6 is the success probability p, the exponent 2 is the number of successes, 0.4 is the failure probability $1 - p$, and the exponent 1 is the number of failures.

We can now generalize this result to produce a formula for the probability of x successes in n independent Bernoulli trials with success probability p, in terms of x, n, and p. In other words, we can compute $P(X = x)$ where $X \sim \text{Bin}(n, p)$. We can see that

$$P(X = x) = (\text{number of arrangements of } x \text{ successes in } n \text{ trials}) \cdot p^x(1 - p)^{n-x}$$

$$(4.1)$$

All we need to do now is to provide an expression for the number of arrangements of x successes in n trials. To describe this number, we need factorial notation. For any positive integer n, the quantity $n!$ (read "n factorial") is the number

$$n! = (n)(n - 1)(n - 2) \cdots (3)(2)(1)$$

We also define $0! = 1$. The number of arrangements of x successes in n trials is $n!/[x!(n - x)!]$. We can now define the probability mass function for a binomial random variable.

If $X \sim \text{Bin}(n, p)$, the probability mass function of X is

$$p(x) = P(X = x) = \begin{cases} \dfrac{n!}{x!(n-x)!} p^x(1 - p)^{n-x} & x = 0, 1, \ldots, n \\ 0 & \text{otherwise} \end{cases}$$

$$(4.2)$$

Figure 4.1 (page 122) presents probability histograms for the $\text{Bin}(10, 0.4)$ and $\text{Bin}(20, 0.1)$ probability mass functions.

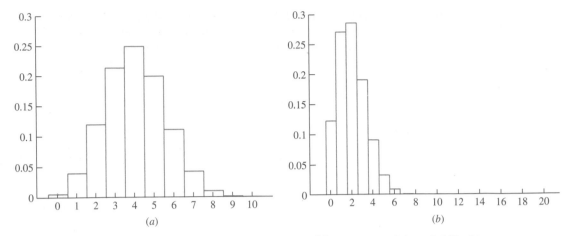

FIGURE 4.1 *(a)* The Bin(10, 0.4) probability histogram. *(b)* The Bin(20, 0.1) probability histogram.

xample 4.2

The probability that a newborn baby is a girl is approximately 0.49. Find the probability that of the next five single births in a certain hospital, no more than two are girls.

Solution

Each birth is a Bernoulli trial with success probability 0.49. Let X denote the number of girls in 5 births. Then $X \sim \text{Bin}(5, 0.49)$. We need to find $P(X \leq 2)$. Using the probability mass function,

$$
\begin{aligned}
P(X \leq 2) &= P(X = 0 \text{ or } X = 1 \text{ or } X = 2) \\
&= P(X = 0) + P(X = 1) + P(X = 2) \\
&= \frac{5!}{0!(5-0)!}(0.49)^0(0.51)^{5-0} + \frac{5!}{1!(5-1)!}(0.49)^1(0.51)^{5-1} \\
&\quad + \frac{5!}{2!(5-2)!}(0.49)^2(0.51)^{5-2} \\
&= 0.0345 + 0.1657 + 0.3185 \\
&= 0.5187
\end{aligned}
$$

Sampling from a Finite Population

A manufacturing process produces 10,000 stainless steel rivets during the course of a day. It is known that 8% of the rivets manufactured by this process fail to meet a strength specification. A simple random sample of 20 rivets will be packaged for shipment to a customer. A quality engineer wants to know the probability that the shipment will contain exactly three defective rivets.

To compute this probability, we can use the binomial distribution. Each of the 20 rivets in the shipment is a trial. A "success" occurs if the rivet is defective. It is clear that each trial has a success probability of 0.08. We want to know the probability that we observe three successes in 20 trials. In order to use the binomial distribution, we must be sure that the trials are independent. To see that this is so, recall from the discussion of independence in Section 1.1 that when drawing a sample from a finite population, the sample items may be treated as independent if the sample comprises less than 5% of the population. Therefore, if X represents the number of defective rivets in the sample of 20,

$$P(X = 3) = \frac{20!}{3!17!}(0.08)^3(0.92)^{17} = 0.1414$$

Summary

Assume that a finite population contains items of two types, successes and failures, and that a simple random sample is drawn from the population. Then if the sample size is no more than 5% of the population, the binomial distribution may be used to model the number of successes.

Example 4.3

A lot contains several thousand components, 10% of which are defective. Nine components are sampled from the lot. Let X represent the number of defective components in the sample. Find the probability that exactly two are defective.

Solution
Since the sample size is small compared to the population (i.e., less than 5%), the number of successes in the sample approximately follows a binomial distribution. Therefore we model X with the Bin(9, 0.1) distribution. The probability that exactly two are defective is

$$P(X = 2) = \frac{9!}{2!(9 - 2)!}(0.1)^2(0.9)^7 = 0.1722$$

Using the Binomial Table

Table A.1 (in Appendix A) presents binomial probabilities of the form $P(X \leq x)$ for $n \leq 20$ and selected values of p. Examples 4.4 and 4.5 illustrate the use of this table.

Example 4.4

Of all the new vehicles of a certain model that are sold, 20% require repairs to be done under warranty during the first year of service. A particular dealership sells 14 such vehicles. What is the probability that fewer than five of them require warranty repairs?

Solution
Let X represent the number of vehicles that require warranty repairs. Then $X \sim \text{Bin}(14, 0.2)$. The probability that fewer than five vehicles require warranty repairs is $P(X \leq 4)$. We consult Table A.1 with $n = 14$, $p = 0.2$, and $x = 4$. We find that $P(X \leq 4) = 0.870$.

Sometimes the best way to compute the probability of an event is to compute the probability that the event does not occur, and then subtract from 1. Example 4.5 provides an illustration.

Example
4.5

Refer to Example 4.4. What is the probability that more than 2 of the 14 vehicles require warranty repairs?

Solution
Let X represent the number of invoices in the sample that receive discounts. We wish to compute the probability $P(X > 2)$. Table A.1 presents probabilities of the form $P(X \leq x)$. We note that $P(X > 2) = 1 - P(X \leq 2)$. Consulting the table with $n = 14$, $p = 0.2$, $x = 2$, we find that $P(X \leq 2) = 0.448$. Therefore $P(X > 2) = 1 - 0.448 = 0.552$.

The Mean and Variance of a Binomial Random Variable

With a little thought, it is easy to see how to compute the mean of a binomial random variable. For example, if a fair coin is tossed 10 times, we expect on the average to see five heads. The number 5 comes from multiplying the success probability (0.5) by the number of trials (10). This method works in general. If we perform n Bernoulli trials, each with success probability p, the mean number of successes is np. Therefore, if $X \sim \text{Bin}(n, p)$, then $\mu_X = np$. The variance of a binomial random variable is not so intuitive; it turns out to be $np(1 - p)$.

Summary

If $X \sim \text{Bin}(n, p)$, then the mean and variance of X are given by

$$\mu_X = np \tag{4.3}$$

$$\sigma_X^2 = np(1 - p) \tag{4.4}$$

Exercises for Section 4.1

1. Let $X \sim \text{Bin}(10, 0.6)$. Find

a. $P(X = 3)$

b. $P(X = 6)$

c. $P(X \leq 4)$

d. $P(X > 8)$

e. μ_X

f. σ_X^2

2. Let $X \sim \text{Bin}(5, 0.3)$. Find

a. $P(X < 3)$

b. $P(X \geq 1)$

c. $P(1 \leq X \leq 3)$

d. $P(2 < X < 5)$

e. $P(X = 0)$

f. $P(X = 3)$

g. μ_X

h. σ_X^2

3. Find the following probabilities:

    a. $P(X = 7)$ when $X \sim \text{Bin}(13, 0.4)$

    b. $P(X \geq 2)$ when $X \sim \text{Bin}(8, 0.4)$

    c. $P(X < 5)$ when $X \sim \text{Bin}(6, 0.7)$

    d. $P(2 \leq X \leq 4)$ when $X \sim \text{Bin}(7, 0.1)$

4. Ten percent of the items in a large lot are defective. A sample of six items is drawn from this lot.

    a. Find the probability that none of the sampled items are defective.

    b. Find the probability that one or more of the sampled items is defective.

    c. Find the probability that exactly one of the sampled items is defective.

    d. Find the probability that fewer than two of the sampled items are defective.

5. Of all the weld failures in a certain assembly, 85% of them occur in the weld metal itself, and the remaining 15% occur in the base metal. A sample of 20 weld failures is examined.

    a. What is the probability that exactly five of them are base metal failures?

    b. What is the probability that fewer than four of them are base metal failures?

    c. What is the probability that none of them are base metal failures?

    d. Find the mean number of base metal failures.

    e. Find the standard deviation of the number of base metal failures.

6. A large industrial firm allows a discount on any invoice that is paid within 30 days. Of all invoices, 10% receive the discount. In a company audit, 12 invoices are sampled at random.

    a. What is the probability that exactly four of them receive the discount?

    b. What is the probability that fewer than three of them receive the discount?

    c. What is the probability that none of them receive the discount?

    d. Find the mean number that receive the discount.

    e. Find the standard deviation of the number that receive the discount.

7. A fair coin is tossed eight times.

    a. What is the probability of obtaining exactly five heads?

    b. Find the mean number of heads obtained.

    c. Find the variance of the number of heads obtained.

    d. Find the standard deviation of the number of heads obtained.

8. In a large shipment of automobile tires, 10% have a flaw. Four tires are chosen at random to be installed on a car.

    a. What is the probability that none of the tires have a flaw?

    b. What is the probability that exactly one of the tires has a flaw?

    c. What is the probability that one or more of the tires has a flaw?

9. Of the bolts manufactured for a certain application, 85% meet the length specification and can be used immediately, 10% are too long and can be used after being cut, and 5% are too short and must be scrapped.

    a. Find the probability that a randomly selected bolt can be used (either immediately or after being cut).

    b. Find the probability that fewer than 9 out of a sample of 10 bolts can be used (either immediately or after being cut).

10. A distributor receives a large shipment of components. The distributor would like to accept the shipment if 10% or fewer of the components are defective and to return it if more than 10% of the components are defective. She decides to sample 10 components and to return the shipment if more than 1 of the 10 is defective. If the proportion of defectives in the batch is in fact 10%, what is the probability that she will return the shipment?

11. A k out of n system is one in which there is a group of n components, and the system will function if at least k of the components function. Assume the components function independently of one another.

    a. In a 3 out of 5 system, each component has probability 0.9 of functioning. What is the probability that the system will function?

    b. In a 3 out of n system, in which each component has probability 0.9 of functioning, what is the smallest value of n needed so that the probability that the system functions is at least 0.90?

12. Refer to Exercise 11 for the definition of a k out of n system. For a certain 4 out of 6 system, assume that on a rainy day each component has probability 0.7 of functioning and that on a nonrainy day each component has probability 0.9 of functioning.

    a. What is the probability that the system functions on a rainy day?

    b. What is the probability that the system functions on a nonrainy day?

    c. Assume that the probability of rain tomorrow is 0.20. What is the probability that the system will function tomorrow?

13. A certain large shipment comes with a guarantee that it contains no more than 15% defective items. If the proportion of defective items in the shipment is greater than 15%, the shipment may be returned. You draw a random sample of 10 items. Let X be the number of defective items in the sample.

    a. If in fact 15% of the items in the shipment are defective (so that the shipment is good, but just barely), what is $P(X \geq 7)$?

    b. Based on the answer to part (a), if 15% of the items in the shipment are defective, would 7 defectives in a sample of size 10 be an unusually large number?

    c. If you found that 7 of the 10 sample items were defective, would this be convincing evidence that the shipment should be returned? Explain.

    d. If in fact 15% of the items in the shipment are defective, what is $P(X \geq 2)$?

    e. Based on the answer to part (d), if 15% of the items in the shipment are defective, would 2 defectives in a sample of size 10 be an unusually large number?

    f. If you found that 2 of the 10 sample items were defective, would this be convincing evidence that the shipment should be returned? Explain.

14. An insurance company offers a discount to homeowners who install smoke detectors in their homes. A company representative claims that 80% or more of policyholders have smoke detectors. You draw a random sample of eight policyholders. Let X be the number of policyholders in the sample who have smoke detectors.

    a. If exactly 80% of the policyholders have smoke detectors (so the representative's claim is true, but just barely), what is $P(X \leq 1)$?

    b. Based on the answer to part (a), if 80% of the policyholders have smoke detectors, would one policyholder with a smoke detector in a sample of size 8 be an unusually small number?

    c. If you found that exactly one of the eight sample policyholders had a smoke detector, would this be convincing evidence that the claim is false? Explain.

    d. If exactly 80% of the policyholders have smoke detectors, what is $P(X \leq 6)$?

    e. Based on the answer to part (d), if 80% of the policyholders have smoke detectors, would six policyholders with smoke detectors in a sample of size 8 be an unusually small number?

    f. If you found that exactly six of the eight sample policyholders had smoke detectors, would this be convincing evidence that the claim is false? Explain.

15. A message consists of a string of bits (0s and 1s). Due to noise in the communications channel, each bit has probability 0.3 of being reversed (i.e., a 1 will be changed to a 0 or a 0 to a 1). To improve the accuracy of the communication, each bit is sent five times, so, for example, 0 is sent as 00000. The receiver assigns the value 0 if three or more of the bits are decoded as 0, and 1 if three or more of the bits are decoded as 1. Assume that errors occur independently.

    a. A 0 is sent (as 00000). What is the probability that the receiver assigns the correct value of 0?

    b. Assume that each bit is sent n times, where n is an odd number, and that the receiver assigns the value decoded in the majority of the bits. What is the minimum value of n necessary so that the probability that the correct value is assigned is at least 0.90?

16. One design for a system requires the installation of two identical components. The system will work if at least one of the components works. An alternative design requires four of these components, and the system will work if at least two of the four components work. If the probability that a component works is 0.9, and if the components function independently, which design has the greater probability of functioning?

4.2 The Poisson Distribution

The Poisson distribution arises frequently in scientific work. We will introduce this distribution by describing a classic application in physics: the number of particles emitted by a radioactive mass.

A mass contains 10,000 atoms of a radioactive substance. The probability that a given atom will decay in a one-minute time period is 0.0002. Let X represent the number of atoms that decay in one minute. Now each atom can be thought of as a Bernoulli trial, where success occurs if the atom decays. Thus X is the number of successes in 10,000 independent Bernoulli trials, each with success probability 0.0002, so the distribution of X is Bin(10,000, 0.0002).

Now assume that we want to compute the probability that exactly three atoms from this mass decay in one minute. Using the binomial probability mass function, we would compute as follows:

$$P(X = 3) = \frac{10,000!}{3! \, 9997!} (0.0002)^3 (0.9998)^{9997} = 0.18047$$

Computing the binomial probability involves some large factorials and large exponents. There is a simpler calculation that yields almost exactly the same result. Specifically, if n is large and p is small, and we let $\lambda = np$, it can be shown by advanced methods that for any non-negative integer x,

$$\frac{n!}{x!(n-x)!} p^x (1-p)^{n-x} \approx e^{-\lambda} \frac{\lambda^x}{x!} \tag{4.5}$$

We are led to define a new probability mass function, called the Poisson probability mass function. The Poisson probability mass function is defined by

$$p(x) = P(X = x) = \begin{cases} e^{-\lambda} \dfrac{\lambda^x}{x!} & \text{if } x \text{ is a non-negative integer} \\ 0 & \text{otherwise} \end{cases} \tag{4.6}$$

If X is a random variable whose probability mass function is given by Equation (4.6), then X is said to have the **Poisson distribution** with parameter λ. The notation is $X \sim \text{Poisson}(\lambda)$.

For the radioactive mass just described, we would use the Poisson mass function to approximate $P(X = x)$ by substituting $\lambda = (10,000)(0.0002) = 2$ into Equation (4.6). The result is

$$P(X = 3) = e^{-2} \frac{2^3}{3!} = 0.18045$$

The Poisson probability agrees with the binomial probability to four decimal places.

4.6

If $X \sim$ Poisson(5), compute $P(X = 3)$, $P(X = 8)$, $P(X = 0)$, $P(X = -1)$, and $P(X = 0.5)$.

Solution

Using the probability mass function (4.6), with $\lambda = 5$, we obtain

$$P(X = 3) = e^{-5}\frac{5^3}{3!} = 0.1404$$

$$P(X = 8) = e^{-5}\frac{5^8}{8!} = 0.0653$$

$$P(X = 0) = e^{-5}\frac{5^0}{0!} = 0.0067$$

$$P(X = -1) = 0 \qquad\qquad \text{because } -1 \text{ is not a non-negative integer}$$

$$P(X = 0.5) = 0 \qquad\qquad \text{because } 0.5 \text{ is not a non-negative integer}$$

Example
4.7

If $X \sim$ Poisson(4), compute $P(X \le 2)$ and $P(X > 1)$.

Solution

$$P(X \le 2) = P(X = 0) + P(X = 1) + P(X = 2)$$

$$= e^{-4}\frac{4^0}{0!} + e^{-4}\frac{4^1}{1!} + e^{-4}\frac{4^2}{2!}$$

$$= 0.0183 + 0.0733 + 0.1465$$

$$= 0.2381$$

To find $P(X > 1)$, we might try to start by writing

$$P(X > 1) = P(X = 2) + P(X = 3) + \cdots$$

This leads to an infinite sum that is difficult to compute. Instead, we write

$$P(X > 1) = 1 - P(X \le 1)$$

$$= 1 - [P(X = 0) + P(X = 1)]$$

$$= 1 - \left(e^{-4}\frac{4^0}{0!} + e^{-4}\frac{4^1}{1!}\right)$$

$$= 1 - (0.0183 + 0.0733)$$

$$= 0.9084$$

Summary

If $X \sim \text{Poisson}(\lambda)$, then

- X is a discrete random variable whose possible values are the non-negative integers.
- The parameter λ is a positive constant.
- The probability mass function of X is

$$p(x) = P(X = x) = \begin{cases} e^{-\lambda}\dfrac{\lambda^x}{x!} & \text{if } x \text{ is a non-negative integer} \\ 0 & \text{otherwise} \end{cases}$$

- The Poisson probability mass function is very close to the binomial probability mass function when n is large, p is small, and $\lambda = np$.

The Mean and Variance of a Poisson Random Variable

The mean and variance of a Poisson random variable can be computed by using the probability mass function along with the definitions given by Equations (3.13) and (3.14) (in Section 3.3). It turns out that the mean and variance are both equal to λ.

Mean and Variance of a Poisson Random Variable

If $X \sim \text{Poisson}(\lambda)$, then the mean and variance of X are given by

$$\mu_X = \lambda \tag{4.7}$$

$$\sigma_X^2 = \lambda \tag{4.8}$$

Figure 4.2 presents probability histograms for the Poisson(1) and Poisson(10) probability mass functions.

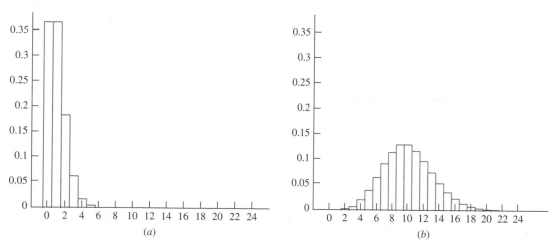

FIGURE 4.2 *(a)* The Poisson(1) probability histogram. *(b)* The Poisson(10) probability histogram.

One of the earliest industrial uses of the Poisson distribution involved an application to the brewing of beer. A crucial step in the brewing process is the addition of yeast culture to prepare mash for fermentation. The living yeast cells are kept suspended in a liquid medium. Because the cells are alive, their concentration in the medium changes over time. Therefore, just before the yeast is added, it is necessary to estimate the concentration of yeast cells per unit volume of suspension, so as to be sure to add the right amount.

Up until the early part of the twentieth century, this posed a problem for brewers. They estimated the concentration in the obvious way, by withdrawing a small volume of the suspension and counting the yeast cells in it under a microscope. Of course, the estimates determined this way were subject to random variation, but no one knew how to measure the likely size of this variation. Therefore no one knew by how much the concentration in the sample was likely to differ from the actual concentration.

William Sealy Gosset, a young man in his mid-twenties who was employed by the Guinness Brewing Company of Dublin, Ireland, discovered in 1904 that the number of yeast cells in a sampled volume of suspension follows a Poisson distribution. He was then able to develop methods to measure the concentration of yeast cells with greater precision. Gosset's discovery not only enabled Guinness to produce a more consistent product, it showed that the Poisson distribution could have important applications in many situations. Gosset wanted to publish his result, but his managers at Guinness considered his discovery to be proprietary information. They required him to use a pseudonym, so that competing breweries would not realize how useful the results could be. Gosset chose the pseudonym "Student."

In Example 4.8, we will follow a train of thought that leads to Student's result. Before we get to it though, we will mention that shortly after publishing this result, Student made another discovery that solved one of the most important outstanding problems in statistics, which has profoundly influenced work in virtually all fields of science ever since. We will discuss this result in Section 5.4.

Example 4.8

Particles (e.g., yeast cells) are suspended in a liquid medium at a concentration of 10 particles per mL. A large volume of the suspension is thoroughly agitated, and then 1 mL is withdrawn. What is the probability that exactly eight particles are withdrawn?

Solution
So long as the volume withdrawn is a small fraction of the total, the solution to this problem does not depend on the total volume of the suspension but only on the concentration of particles in it. Let V be the total volume of the suspension, in mL. Then the total number of particles in the suspension is $10V$. Think of each of the $10V$ particles as a Bernoulli trial. A particle "succeeds" if it is withdrawn. Now 1 mL out of the total of V mL is to be withdrawn. Therefore, the amount to be withdrawn is $1/V$ of the total, so it follows that each particle has probability $1/V$ of being withdrawn. Let X denote the number of particles withdrawn. Then X represents the number of successes in $10V$ Bernoulli trials, each with probability $1/V$ of success. Therefore $X \sim \text{Bin}(10V, \ 1/V)$. Since V is large, $10V$ is large and $1/V$ is small. Thus, to a very close approximation, $X \sim \text{Poisson}(10)$. We compute $P(X = 8)$ with the Poisson probability mass function:
$P(X = 8) = e^{-10}10^8/8! = 0.1126$.

In Example 4.8, λ had the value 10 because the mean number of particles in 1 mL of suspension (the volume withdrawn) was 10.

Example 4.9

Particles are suspended in a liquid medium at a concentration of 6 particles per mL. A large volume of the suspension is thoroughly agitated, and then 3 mL are withdrawn. What is the probability that exactly 15 particles are withdrawn?

Solution

Let X represent the number of particles withdrawn. The mean number of particles in a 3 mL volume is 18. Therefore $X \sim \text{Poisson}(18)$. The probability that exactly 15 particles are withdrawn is

$$P(X = 15) = e^{-18}\frac{18^{15}}{15!}$$
$$= 0.0786$$

Note that for the solutions to Examples 4.8 and 4.9 to be correct, it is important that the amount of suspension withdrawn not be too large a fraction of the total. For example, if the total volume in Example 4.9 was 3 mL, so that the entire amount was withdrawn, it would be certain that all 18 particles would be withdrawn, so the probability of withdrawing 15 particles would be zero.

Example 4.10

Grandma bakes chocolate chip cookies in batches of 100. She puts 300 chips into the dough. When the cookies are done, she gives you one. What is the probability that your cookie contains no chocolate chips?

Solution

This is another instance of particles in a suspension. Let X represent the number of chips in your cookie. The mean number of chips is 3 per cookie, so $X \sim \text{Poisson}(3)$. It follows that $P(X = 0) = e^{-3}3^{0}/0! = 0.0498$.

Examples 4.8 and 4.9 show that for particles distributed uniformly at random throughout a medium, the number of particles that happen to fall in a small portion of the medium follows a Poisson distribution. In these examples, the particles were actual particles, and the medium was spatial in nature. In many cases, however, the "particles" represent events, and the medium is time. We saw such an example previously, where the number of radioactive decay events in a fixed time interval turned out to follow a Poisson distribution. Example 4.11 presents another.

Example 4.11

The number of email messages received by a computer server follows a Poisson distribution with a mean of 6 per minute. Find the probability that exactly 20 messages will be received in the next 3 minutes.

Solution

Let X be the number of messages received in 3 minutes. The mean number of messages received in 3 minutes is $(6)(3) = 18$, so $X \sim$ Poisson(18). Using the Poisson(18) probability mass function, we find that

$$P(X = 20) = e^{-18} \frac{18^{20}}{20!}$$

$$= 0.0798$$

Exercises for Section 4.2

1. Let $X \sim$ Poisson(3). Find

    a. $P(X = 2)$
    b. $P(X = 0)$
    c. $P(X < 3)$
    d. $P(X > 2)$
    e. μ_X
    f. σ_X

    a. $P(X = 5)$
    b. $P(X = 0)$
    c. $P(X < 2)$
    d. $P(X > 1)$
    e. μ_X
    f. σ_X

2. The concentration of particles in a suspension is 4 per mL. The suspension is thoroughly agitated, and then 2 mL is withdrawn. Let X represent the number of particles that are withdrawn. Find

    a. $P(X = 6)$
    b. $P(X \leq 3)$
    c. $P(X > 2)$
    d. μ_X
    e. σ_X

3. Suppose that 0.2% of diodes in a certain application fail within the first month of use. Let X represent the number of diodes in a random sample of 1000 that fail within the first month. Find

    a. $P(X = 4)$
    b. $P(X \leq 1)$
    c. $P(1 \leq X < 4)$
    d. μ_X
    e. σ_X

4. The number of flaws in a given area of aluminum foil follows a Poisson distribution with a mean of 3 per m². Let X represent the number of flaws in a 1 m² sample of foil.

5. The number of hits on a certain website follows a Poisson distribution with a mean rate of 4 per minute.

    a. What is the probability that five messages are received in a given minute?
    b. What is the probability that 9 messages are received in 1.5 minutes?
    c. What is the probability that fewer than three messages are received in a period of 30 seconds?

6. One out of every 5000 individuals in a population carries a certain defective gene. A random sample of 1000 individuals is studied.

    a. What is the probability that exactly one of the sample individuals carries the gene?
    b. What is the probability that none of the sample individuals carries the gene?
    c. What is the probability that more than two of the sample individuals carry the gene?
    d. What is the mean of the number of sample individuals that carry the gene?
    e. What is the standard deviation of the number of sample individuals that carry the gene?

7. A sensor network consists of a large number of microprocessors spread out over an area, in

communication with each other and with a base station. In a certain network, the probability that a message will fail to reach the base station is 0.005. Assume that during a particular day, 1000 messages are sent.

a. What is the probability that exactly 3 of the messages fail to reach the base station?

b. What is the probability that fewer than 994 of the messages reach the base station?

c. What is the mean of the number of messages that fail to reach the base station?

d. What is the standard deviation of the number of messages that fail to reach the base station?

8. Geologists estimate the time since the most recent cooling of a mineral by counting the number of uranium fission tracks on the surface of the mineral. A certain mineral specimen is of such an age that there should be an average of 6 tracks per cm^2 of surface area. Assume the number of tracks in an area follows a Poisson distribution. Let X represent the number of tracks counted in 1 cm^2 of surface area. Find

a. $P(X = 7)$

b. $P(X \geq 3)$

c. $P(2 < X < 7)$

d. μ_X

e. σ_X

9. A random variable X has a binomial distribution, and a random variable Y has a Poisson distribution. Both X and Y have means equal to 3. Is it possible to determine which random variable has the larger variance? Choose one of the following answers:

i. Yes, X has the larger variance.

ii. Yes, Y has the larger variance.

iii. No, we need to know the number of trials, n, for X.

iv. No, we need to know the success probability, p, for X.

v. No, we need to know the value of λ for Y.

10. You have received a radioactive mass that is claimed to have a mean decay rate of at least 1 particle per

second. If the mean decay rate is less than 1 per second, you may return the product for a refund. Let X be the number of decay events counted in 10 seconds.

a. If the mean decay rate is exactly 1 per second (so that the claim is true, but just barely), what is $P(X \leq 1)$?

b. Based on the answer to part (a), if the mean decay rate is 1 particle per second, would one event in 10 seconds be an unusually small number?

c. If you counted one decay event in 10 seconds, would this be convincing evidence that the product should be returned? Explain.

d. If the mean decay rate is exactly 1 per second, what is $P(X \leq 8)$?

e. Based on the answer to part (d), if the mean decay rate is 1 particle per second, would eight events in 10 seconds be an unusually small number?

f. If you counted eight decay events in 10 seconds, would this be convincing evidence that the product should be returned? Explain.

11. Someone claims that a certain suspension contains at least seven particles per mL. You sample 1 mL of solution. Let X be the number of particles in the sample.

a. If the mean number of particles is exactly seven per mL (so that the claim is true, but just barely), what is $P(X \leq 1)$?

b. Based on the answer to part (a), if the suspension contains seven particles per mL, would one particle in a 1 mL sample be an unusually small number?

c. If you counted one particle in the sample, would this be convincing evidence that the claim is false? Explain.

d. If the mean number of particles is exactly 7 per mL, what is $P(X \leq 6)$?

e. Based on the answer to part (d), if the suspension contains seven particles per mL, would six particles in a 1 mL sample be an unusually small number?

f. If you counted six particles in the sample, would this be convincing evidence that the claim is false? Explain.

4.3 The Normal Distribution

The **normal distribution** (also called the **Gaussian distribution**) is by far the most commonly used distribution in statistics. This distribution provides a good model for many, although not all, continuous populations. Part of the reason for this is the Central Limit Theorem, which we shall discuss in Section 4.8.

The normal distribution is continuous rather than discrete. The mean of a normal random variable may have any value, and the variance may have any positive value. The probability density function of a normal random variable with mean μ and variance σ^2 is given by

$$f(x) = \frac{1}{\sigma\sqrt{2\pi}} e^{-(x-\mu)^2/(2\sigma^2)} \tag{4.9}$$

If X is a random variable whose probability density function is normal with mean μ and variance σ^2, we write $X \sim N(\mu, \sigma^2)$.

Summary

If $X \sim N(\mu, \sigma^2)$, then the mean and variance of X are given by

$$\mu_X = \mu$$
$$\sigma_X^2 = \sigma^2$$

Figure 4.3 presents a plot of the normal probability density function with mean μ and standard deviation σ. The normal probability density function is sometimes called the **normal curve**. It is also the case that for any normal population

■ About 68% of the population is in the interval $\mu \pm \sigma$.
■ About 95% of the population is in the interval $\mu \pm 2\sigma$.
■ About 99.7% of the population is in the interval $\mu \pm 3\sigma$.

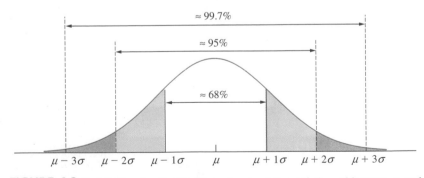

FIGURE 4.3 Probability density function of a normal population with mean μ and variance σ^2.

The proportion of a normal population that is within a given number of standard deviations of the mean is the same for any normal population. For this reason, when dealing with normal populations, we often convert from the units in which the population items were originally measured to **standard units**. Standard units tell how many standard deviations an observation is from the population mean.

Example
4.12

Resistances in a population of wires are normally distributed with mean $20\,\mathrm{m\Omega}$ and standard deviation $3\,\mathrm{m\Omega}$. The resistance of two randomly chosen wires are $23\,\mathrm{m\Omega}$ and $16\,\mathrm{m\Omega}$. Convert these amounts to standard units.

Solution

A resistance of $23\,\mathrm{m\Omega}$ is $3\,\mathrm{m\Omega}$ more than the mean of 20, and $3\,\mathrm{m\Omega}$ is equal to one standard deviation. So $23\,\mathrm{m\Omega}$ is one standard deviation above the mean and is thus equivalent to one standard unit. A resistance of $16\,\mathrm{m\Omega}$ is 1.33 standard deviations below the mean, so $16\,\mathrm{m\Omega}$ is equivalent to -1.33 standard units.

In general, we convert to standard units by subtracting the mean and dividing by the standard deviation. Thus, if x is an item sampled from a normal population with mean μ and variance σ^2, the standard unit equivalent of x is the number z, where

$$z = \frac{x - \mu}{\sigma} \tag{4.10}$$

The number z is sometimes called the "z-score" of x. The z-score is an item sampled from a normal population with mean 0 and standard deviation 1. This normal population is called the **standard normal population**.

Example
4.13

The yield, in grams, of a dye manufacturing process is normally distributed with mean 1500 and standard deviation 50. The yield of a particular run is 1568 grams. Find the z-score.

Solution

The quantity 1568 is an observation from a normal population with mean $\mu = 1500$ and standard deviation $\sigma = 50$. Therefore

$$z = \frac{1568 - 1500}{50}$$
$$= 1.36$$

Example
4.14

Refer to Example 4.13. The yield of a certain run has a z-score of -1.4. Find the yield in the original units of grams.

Solution

We use Equation (4.10), substituting -1.4 for z and solving for x. We obtain

$$-1.4 = \frac{x - 1500}{50}$$

Solving for x yields $x = 1430$. The yield is 1430 grams.

The proportion of a normal population that lies within a given interval is equal to the area under the normal probability density above that interval. This would suggest that we compute these proportions by integrating the normal probability density given in Equation (4.9). Interestingly enough, areas under this curve cannot be found by the method, taught in elementary calculus, of computing the antiderivative of the function and plugging in the limits of integration. This is because the antiderivative of this function is an infinite series and cannot be written down exactly. Instead, areas under this curve must be approximated numerically.

Areas under the standard normal curve (mean 0, variance 1) have been extensively tabulated. A typical such table, called a **standard normal table**, or **z table**, is given as Table A.2 (in Appendix A). To find areas under a normal curve with a different mean and variance, we convert to standard units and use the z table. Table A.2 provides areas in the left-hand tail of the curve for values of z. Other areas can be calculated by subtraction or by using the fact that the total area under the curve is equal to 1. We now present several examples to illustrate the use of the z table.

Example 4.15

Find the area under the normal curve to the left of $z = 0.47$.

Solution
From the z table, the area is 0.6808. See Figure 4.4.

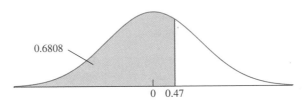

FIGURE 4.4 Solution to Example 4.15.

Example 4.16

Find the area under the normal curve to the right of $z = 1.38$.

Solution
From the z table, the area to the *left* of $z = 1.38$ is 0.9162. Therefore the area to the right is $1 - 0.9162 = 0.0838$. See Figure 4.5.

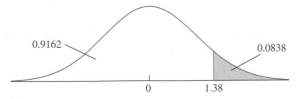

FIGURE 4.5 Solution to Example 4.16.

Example
4.17

Find the area under the normal curve between $z = 0.71$ and $z = 1.28$.

Solution
From the z table, the area to the left of $z = 1.28$ is 0.8997. The area to the left of $z = 0.71$ is 0.7611. The area between $z = 0.71$ and $z = 1.28$ is therefore $0.8997 - 0.7611 = 0.1386$. See Figure 4.6.

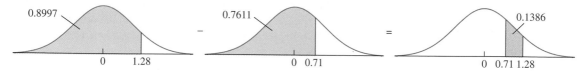

FIGURE 4.6 Solution to Example 4.17.

Recall that the pth percentile of a sample is the value that divides the sample so that p percent of the sample values are less than the pth percentile (see Section 1.2). This idea extends to populations as well. The pth percentile of a population is the value that divides the sample so that p percent of the population values are less than the pth percentile. For a normal population, therefore, the pth percentile is the value for which p percent of the area under the normal curve is to the left. The median is the 50th percentile. For a normal population, the median is equal to the mean.

Example
4.18

What z-score corresponds to the 75th percentile of a normal distribution? The 25th percentile? The median?

Solution
To answer this question, we use the z table in reverse. We need to find the z-score for which 75% of the area of the curve is to the left. From the body of the table, the closest area to 75% is 0.7486, corresponding to a z-score of 0.67. Therefore the 75th percentile is approximately 0.67. By the symmetry of the curve, the 25th percentile is $z = -0.67$ (this can also be looked up in the table directly). See Figure 4.7. The median is $z = 0$.

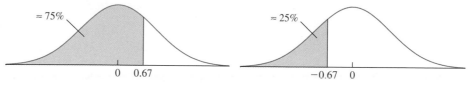

FIGURE 4.7 Solution to Example 4.18.

Example
4.19

Lifetimes of batteries in a certain application are normally distributed with mean 50 hours and standard deviation 5 hours. Find the probability that a randomly chosen battery lasts between 42 and 52 hours.

Solution

Let X represent the lifetime of a randomly chosen battery. Then $X \sim N(50, 5^2)$. Figure 4.8 presents the probability density function of the $N(50, 5^2)$ population. The shaded area represents $P(42 < X < 52)$, the probability that a randomly chosen battery has a lifetime between 42 and 52 hours. To compute the area, we will use the z table. First we need to convert the quantities 42 and 52 to standard units. We have

$$z = \frac{42 - 50}{5} = -1.60 \qquad z = \frac{52 - 50}{5} = 0.40$$

From the z table, the area to the left of $z = -1.60$ is 0.0548, and the area to the left of $z = 0.40$ is 0.6554. The probability that a battery has a lifetime between 42 and 52 hours is $0.6554 - 0.0548 = 0.6006$.

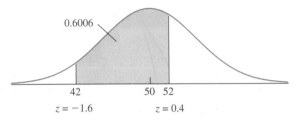

FIGURE 4.8 Solution to Example 4.19.

Refer to Example 4.19. Find the 40th percentile of battery lifetimes.

Solution

From the z table, the closest area to 0.4000 is 0.4013, corresponding to a z-score of -0.25. The population of lifetimes has mean 50 and standard deviation 5. The 40th percentile is the point 0.25 standard deviations below the mean. We find this value by converting the z-score to a raw score, using Equation (4.10):

$$-0.25 = \frac{x - 50}{5}$$

Solving for x yields $x = 48.75$. The 40th percentile of battery lifetimes is 48.75 hours. See Figure 4.9.

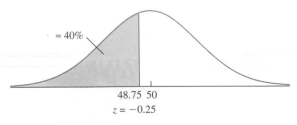

FIGURE 4.9 Solution to Example 4.20.

Example
4.21

A process manufactures ball bearings whose diameters are normally distributed with mean 2.505 cm and standard deviation 0.008 cm. Specifications call for the diameter to be in the interval 2.5 ± 0.01 cm. What proportion of the ball bearings will meet the specification?

Solution

Let X represent the diameter of a randomly chosen ball bearing. Then $X \sim N(2.505, 0.008^2)$. Figure 4.10 presents the probability density function of the $N(2.505, 0.008^2)$ population. The shaded area represents $P(2.49 < X < 2.51)$, which is the proportion of ball bearings that meet the specification.

We compute the z-scores of 2.49 and 2.51:

$$z = \frac{2.49 - 2.505}{0.008} = -1.88 \qquad z = \frac{2.51 - 2.505}{0.008} = 0.63$$

The area to the left of $z = -1.88$ is 0.0301. The area to the left of $z = 0.63$ is 0.7357. The area between $z = 0.63$ and $z = -1.88$ is $0.7357 - 0.0301 = 0.7056$. Approximately 70.56% of the diameters will meet the specification.

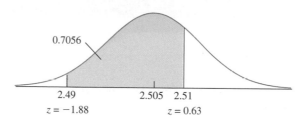

FIGURE 4.10 Solution to Example 4.21.

Linear Functions of Normal Random Variables

If a normal random variable is multiplied by a nonzero constant or has a constant added to it, the resulting random variable is also normal, with a mean and variance that are determined by the original mean and variance and the constants. Specifically,

---

Summary

Let $X \sim N(\mu, \sigma^2)$, and let $a \neq 0$ and b be constants. Then

$$aX + b \sim N(a\mu + b, a^2\sigma^2). \tag{4.11}$$

Example
4.22

A chemist measures the temperature of a solution in °C. The measurement is denoted C and is normally distributed with mean 40°C and standard deviation 1°C. The measurement is converted to °F by the equation $F = 1.8C + 32$. What is the distribution of F?

Solution

Since C is normally distributed, so is F. Now $\mu_C = 40$, so $\mu_F = 1.8(40) + 32 = 104$, and $\sigma_C^2 = 1$, so $\sigma_F^2 = 1.8^2(1) = 3.24$. Therefore $F \sim N(104, 3.24)$.

Linear Combinations of Independent Normal Random Variables

One of the remarkable features of the normal distribution is that linear combinations of independent normal random variables are themselves normal random variables. To be specific, suppose that $X_1 \sim N(\mu_1, \sigma_1^2)$, $X_2 \sim N(\mu_2, \sigma_2^2)$, ..., $X_n \sim N(\mu_n, \sigma_n^2)$ are independent normal random variables. Note that the means and variances of these random variables can differ from one another. Let $c_1, c_2, \ldots, c_n$ be constants. Then the linear combination $c_1 X_1 + c_2 X_2 + \cdots + c_n X_n$ is a normally distributed random variable. The mean and variance of the linear combination are $c_1\mu_1 + c_2\mu_2 + \cdots + c_n\mu_n$ and $c_1^2\sigma_1^2 + c_2^2\sigma_2^2 + \cdots + c_n^2\sigma_n^2$, respectively (see Equations 3.32 and 3.36 in Section 3.4).

Summary

Let $X_1, X_2, \ldots, X_n$ be independent and normally distributed with means $\mu_1, \mu_2, \ldots, \mu_n$ and variances $\sigma_1^2, \sigma_2^2, \ldots, \sigma_n^2$. Let $c_1, c_2, \ldots, c_n$ be constants, and $c_1 X_1 + c_2 X_2 + \cdots + c_n X_n$ be a linear combination. Then

$$c_1 X_1 + c_2 X_2 + \cdots + c_n X_n \sim N(c_1\mu_1 + c_2\mu_2 + \cdots + c_n\mu_n, c_1^2\sigma_1^2 + c_2^2\sigma_2^2 + \cdots + c_n^2\sigma_n^2)$$

(4.12)

In the article "Advances in Oxygen Equivalent Equations for Predicting the Properties of Titanium Welds" (D. Harwig, W. Ittiwattana, and H. Castner, *The Welding Journal*, 2001:126s–136s), the authors propose an oxygen equivalence equation to predict the strength, ductility, and hardness of welds made from nearly pure titanium. The equation is $E = 2C + 3.5N + O$, where E is the oxygen equivalence, and C, N, and O are the proportions by weight, in parts per million, of carbon, nitrogen, and oxygen, respectively (a constant term involving iron content has been omitted). Assume that for a particular grade of commercially pure titanium, the quantities C, N, and O are approximately independent and normally distributed with means $\mu_C = 150$, $\mu_N = 200$, $\mu_O = 1500$, and standard deviations $\sigma_C = 30$, $\sigma_N = 60$, $\sigma_O = 100$. Find the distribution of E. Find $P(E > 3000)$.

Solution

Since E is a linear combination of independent normal random variables, its distribution is normal. We must now find the mean and variance of E. Using Equation (4.12), we compute

$$\mu_E = 2\mu_C + 3.5\mu_N + 1\mu_O$$
$$= 2(150) + 3.5(200) + 1(1500)$$
$$= 2500$$
$$\sigma_E^2 = 2^2\sigma_C^2 + 3.5^2\sigma_N^2 + 1^2\sigma_O^2$$
$$= 2^2(30^2) + 3.5^2(60^2) + 1^2(100^2)$$
$$= 57{,}700$$

We conclude that $E \sim N(2500, 57{,}700)$.

To find $P(E > 3000)$, we compute the z-score: $z = (3000 - 2500)/\sqrt{57{,}700} = 2.08$. The area to the right of $z = 2.08$ under the standard normal curve is 0.0188. So $P(E > 3000) = 0.0188$.

If $X_1, \ldots, X_n$ is a random sample from any population with mean μ and variance σ^2, then the sample mean $\overline{X}$ has mean $\mu_{\overline{X}} = \mu$ and variance $\sigma_{\overline{X}}^2 = \sigma^2/n$. If the population is normal, then $\overline{X}$ is normal as well, because it is a linear combination of $X_1, \ldots, X_n$ with coefficients $c_1 = \cdots = c_n = 1/n$.

Summary

Let $X_1, \ldots, X_n$ be independent and normally distributed with mean μ and variance σ^2. Then

$$\overline{X} \sim N\left(\mu, \frac{\sigma^2}{n}\right) \tag{4.13}$$

Other important linear combinations are the sum and difference of two random variables. If X and Y are independent normal random variables, the sum $X + Y$ and the difference $X - Y$ are linear combinations. The distributions of $X + Y$ and $X - Y$ can be determined by using Equation (4.12) with $c_1 = 1$, $c_2 = 1$ for $X + Y$ and $c_1 = 1$, $c_2 = -1$ for $X - Y$.

Summary

Let X and Y be independent, with $X \sim N(\mu_X, \sigma_X^2)$ and $Y \sim N(\mu_Y, \sigma_Y^2)$. Then

$$X + Y \sim N(\mu_X + \mu_Y, \sigma_X^2 + \sigma_Y^2) \tag{4.14}$$
$$X - Y \sim N(\mu_X - \mu_Y, \sigma_X^2 + \sigma_Y^2) \tag{4.15}$$

How Can I Tell Whether My Data Come from a Normal Population?

In practice, we often have a sample from some population, and we must use the sample to decide whether the population distribution is approximately normal. If the sample is reasonably large, the sample histogram may give a good indication. Large samples from normal populations have histograms that look something like the normal density function—peaked in the center, and decreasing more or less symmetrically on either side. Probability plots, which will be discussed in Section 4.7, provide another good way of determining whether a reasonably large sample comes from a population that is approximately normal. For small samples, it can be difficult to tell whether the normal distribution is appropriate. One important fact is this: *Samples from normal populations rarely contain outliers.* Therefore the normal distribution should generally not be used for data sets that contain outliers. This is especially true when the sample size is small. Unfortunately, for small data sets that do not contain outliers, it is difficult to determine whether the population is approximately normal. In general, some knowledge of the process that generated the data is needed.

Exercises for Section 4.3

1. Find the area under the normal curve

   a. To the right of $z = -0.75$.

   b. Between $z = 0.40$ and $z = 1.15$.

   c. Between $z = -0.60$ and $z = 0.70$.

   d. Outside $z = 0.75$ to $z = 1.30$.

2. Find the area under the normal curve

   a. To the right of $z = 0.73$.

   b. Between $z = 2.12$ and $z = 2.57$.

   c. Outside $z = 0.96$ to $z = 1.62$.

   d. Between $z = -0.13$ and $z = 0.70$.

3. Let $Z \sim N(0, 1)$. Find a constant c for which

   a. $P(Z \leq c) = 0.8413$

   b. $P(0 \leq Z \leq c) = 0.3051$

   c. $P(-c \leq Z \leq c) = 0.8664$

   d. $P(c \leq Z \leq 0) = 0.4554$

   e. $P(|Z| \geq c) = 0.1470$

4. If $X \sim N(3, 4)$, compute

   a. $P(X \geq 3)$

   b. $P(1 \leq X < 8)$

   c. $P(-1.5 \leq X < 1)$

   d. $P(-2 \leq X - 3 < 4)$

5. Scores on a standardized test are approximately normally distributed with a mean of 460 and a standard deviation of 80.

   a. What proportion of the scores are above 550?

   b. What is the 35th percentile of the scores?

   c. If someone's score is 600, what percentile is she on?

   d. What proportion of the scores are between 420 and 520?

6. Weights of female cats of a certain breed are normally distributed with mean 4.1 kg and standard deviation 0.6 kg.

   a. What proportion of female cats have weights between 3.7 and 4.4 kg?

   b. A certain female cat has a weight that is 0.5 standard deviations above the mean. What proportion of female cats are heavier than this one?

   c. How heavy is a female cat whose weight is on the 80th percentile?

   d. A female cat is chosen at random. What is the probability that she weighs more than 4.5 kg?

   e. Six female cats are chosen at random. What is the probability that exactly one of them weighs more than 4.5 kg?

7. The lifetime of a light bulb in a certain application is normally distributed with mean $\mu = 1400$ hours and standard deviation $\sigma = 200$ hours.

   a. What is the probability that a light bulb will last more than 1800 hours?

   b. Find the 10th percentile of the lifetimes.

   c. A particular battery lasts 1645 hours. What percentile is its lifetime on?

   d. What is the probability that the lifetime of a battery is between 1350 and 1550 hours?

8. At a certain university, math SAT scores for the entering freshman class averaged 650 and had a standard deviation of 100. The maximum possible score is 800. Is it possible that the scores of these freshmen are normally distributed? Explain.

9. The strength of an aluminum alloy is normally distributed with mean 10 gigapascals (GPa) and standard deviation 1.4 GPa.

   a. What is the probability that a specimen of this alloy will have a strength greater than 12 GPa?

   b. Find the first quartile of the strengths of this alloy.

   c. Find the 95th percentile of the strengths of this alloy.

10. The temperature recorded by a certain thermometer when placed in boiling water (true temperature 100°C) is normally distributed with mean $\mu = 99.8$°C and standard deviation 0.1°C.

    a. What is the probability that the thermometer reading is greater than 100°C?

    b. What is the probability that the thermometer reading is within ±0.05°C of the true temperature?

11. Penicillin is produced by the *Penicillium* fungus, which is grown in a broth whose sugar content must be carefully controlled. The optimum sugar concentration is 4.9 mg/mL. If the concentration exceeds 6.0 mg/mL, the fungus dies and the process must be shut down for the day.

    a. If sugar concentration in batches of broth is normally distributed with mean 4.9 mg/mL and standard deviation 0.6 mg/mL, on what proportion of days will the process shut down?

    b. The supplier offers to sell broth with a sugar content that is normally distributed with mean 5.2 mg/mL and standard deviation 0.4 mg/mL. Will this broth result in fewer days of production lost? Explain.

12. The quality assurance program for a certain adhesive formulation process involves measuring how well the adhesive sticks a piece of plastic to a glass surface. When the process is functioning correctly, the adhesive strength X is normally distributed with a mean of 200 N and a standard deviation of 10 N. Each hour, you make one measurement of the adhesive strength. You are supposed to inform your supervisor if your measurement indicates that the process has strayed from its target distribution.

    a. Find $P(X \leq 160)$, under the assumption that the process is functioning correctly.

    b. Based on your answer to part (a), if the process is functioning correctly, would a strength of 160 N be unusually small? Explain.

    c. If you observed an adhesive strength of 160 N, would this be convincing evidence that the process was no longer functioning correctly? Explain.

    d. Find $P(X \geq 203)$, under the assumption that the process is functioning correctly.

    e. Based on your answer to part (d), if the process is functioning correctly, would a strength of 203 N be unusually large? Explain.

    f. If you observed an adhesive strength of 203 N, would this be convincing evidence that the process was no longer functioning correctly? Explain.

    g. Find $P(X \leq 195)$, under the assumption that the process is functioning correctly.

    h. Based on your answer to part (g), if the process is functioning correctly, would a strength of 195 N be unusually small? Explain.

    i. If you observed an adhesive strength of 195 N, would this be convincing evidence that the process was no longer functioning correctly? Explain.

4.4 The Lognormal Distribution

For data that are highly skewed or that contain outliers, the normal distribution is generally not appropriate. The **lognormal** distribution, which is related to the normal distribution, is often a good choice for these data sets. The lognormal distribution is derived from the normal distribution as follows: If X is a normal random variable with mean μ and variance σ^2, then the random variable $Y = e^X$ is said to have the **lognormal distribution** with parameters μ and σ^2. Note that if Y has the lognormal distribution with parameters μ and σ^2, then $X = \ln Y$ has the normal distribution with mean μ and variance σ^2.

Summary

- If $X \sim N(\mu, \ \sigma^2)$, then the random variable $Y = e^X$ has the lognormal distribution with parameters μ and σ^2.
- If Y has the lognormal distribution with parameters μ and σ^2, then the random variable $X = \ln Y$ has the $N(\mu, \ \sigma^2)$ distribution.

The probability density function of a lognormal random variable with parameters μ and σ is

$$f(x) = \begin{cases} \dfrac{1}{\sigma x \sqrt{2\pi}} \ \exp\left[-\dfrac{1}{2\sigma^2}(\ln x - \mu)^2\right] & \text{if } x > 0 \\ 0 & \text{if } x \leq 0 \end{cases} \qquad (4.16)$$

Figure 4.11 presents a graph of the lognormal density function with parameters $\mu = 0$ and $\sigma = 1$.

Note that the density function is highly skewed. This is the reason that the lognormal distribution is often used to model processes that tend to produce occasional large values, or outliers.

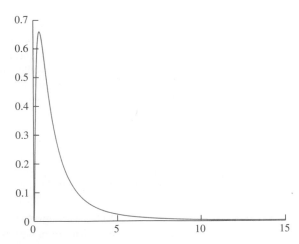

FIGURE 4.11 The probability density function of the lognormal distribution with parameters $\mu = 0$ and $\sigma = 1$.

It can be shown by advanced methods that if Y is a lognormal random variable with parameters μ and σ^2, then the mean $E(Y)$ and variance $V(Y)$ are given by

$$E(Y) = e^{\mu + \sigma^2/2} \qquad V(Y) = e^{2\mu + 2\sigma^2} - e^{2\mu + \sigma^2} \qquad (4.17)$$

Note that if Y has the lognormal distribution, the parameters μ and σ^2 *do not* refer to the mean and variance of Y. They refer instead to the mean and variance of the normal random variable $\ln Y$. In Equation (4.17) we used the notation $E(Y)$ instead of μ_Y and $V(Y)$ instead of σ_Y^2, in order to avoid confusion with μ and σ.

To compute probabilities involving lognormal random variables, take logs and use the z table (Table A.2). Example 4.24 illustrates the method.

Example 4.24

When a pesticide comes into contact with skin, a certain percentage of it is absorbed. The percentage that is absorbed during a given time period is often modeled with a lognormal distribution. Assume that for a given pesticide, the amount that is absorbed (in percent) within two hours of application is lognormally distributed with $\mu = 1.5$ and $\sigma = 0.5$. Find the probability that more than 5% of the pesticide is absorbed within two hours.

Solution

Let Y represent the percent of pesticide that is absorbed. We need to find $P(Y > 5)$. We cannot use the z table for Y, because Y is not normally distributed. However, $\ln Y$ is normally distributed; specifically, $\ln Y \sim N(1.5, 0.5^2)$. We express $P(Y > 5)$ as a probability involving $\ln Y$:

$$P(Y > 5) = P(\ln Y > \ln 5) = P(\ln Y > 1.609)$$

The z-score of 1.609 is

$$z = \frac{1.609 - 1.500}{0.5}$$

$$= 0.22$$

From the z table, we find that $P(\ln Y > 1.609) = 0.4129$. We conclude that the probability that more than 5% of the pesticide is absorbed is approximately 0.41.

How Can I Tell Whether My Data Come from a Lognormal Population?

As stated previously, samples from normal populations rarely contain outliers. In contrast, samples from lognormal populations often contain outliers in the right-hand tail. That is, the samples often contain a few values that are much larger than the rest of the data. This of course reflects the long right-hand tail in the lognormal density function (Figure 4.11). For samples with outliers on the right, we *transform* the data, by taking the natural logarithm (or any logarithm) of each value. We then try to determine whether these logs come from a normal population, by plotting them on a histogram or on a probability plot. Probability plots will be discussed in Section 4.7.

Exercises for Section 4.4

1. The lifetime (in days) of a certain electronic component that operates in a high-temperature environment is lognormally distributed with $\mu = 1.2$ and $\sigma = 0.4$.

   a. Find the mean lifetime.

   b. Find the probability that a component lasts between three and six days.

   c. Find the median lifetime.

   d. Find the 90th percentile of the lifetimes.

2. The article "Assessment of Dermopharmacokinetic Approach in the Bioequivalence Determination of Topical Tretinoin Gel Products" (L. Pershing, J. Nelson, et al., *Journal of The American Academy of Dermatology*, 2003:740–751) reports that the amount of a certain antifungal ointment that is absorbed into the skin can be modeled with a lognormal distribution. Assume that the amount (in ng/cm^2) of active ingredient in the skin two hours after application is lognormally distributed with $\mu = 2.2$ and $\sigma = 2.1$.

   a. Find the mean amount absorbed.

   b. Find the median amount absorbed.

   c. Find the probability that the amount absorbed is more than 100 ng/cm^2.

   d. Find the probability that the amount absorbed is less than 50 ng/cm^2.

   e. Find the 80th percentile of the amount absorbed.

   f. Find the standard deviation of the amount absorbed.

3. The body mass index (BMI) of a person is defined to be the person's body mass divided by the square of the person's height. The article "Influences of Parameter Uncertainties within the ICRP 66 Respiratory Tract Model: Particle Deposition" (W. Bolch, E. Farfan, et al., *Health Physics*, 2001:378–394) states that body mass index (in kg/m^2) in men aged 25–34 is lognormally distributed with parameters $\mu = 3.215$ and $\sigma = 0.157$.

   a. Find the mean BMI for men aged 25–34.

   b. Find the standard deviation of BMI for men aged 25–34.

   c. Find the median BMI for men aged 25–34.

   d. What proportion of men aged 25–34 have a BMI less than 22?

   e. Find the 75th percentile of BMI for men aged 25–34.

4. The article "Stochastic Estimates of Exposure and Cancer Risk from Carbon Tetrachloride Released to the Air from the Rocky Flats Plant" (A. Rood, P. McGavran, et al., *Risk Analysis*, 2001:675–695) models the increase in the risk of cancer due to exposure to carbon tetrachloride as lognormal with $\mu = -15.65$ and $\sigma = 0.79$.

   a. Find the mean risk.

   b. Find the median risk.

   c. Find the standard deviation of the risk.

   d. Find the 5th percentile.

   e. Find the 95th percentile.

5. The prices of stocks or other financial instruments are often modeled with a lognormal distribution. An investor is considering purchasing stock in one of two companies, A or B. The price of a share of stock today is $1 for both companies. For company A, the value of the stock one year from now is modeled as lognormal with parameters $\mu = 0.05$ and $\sigma = 0.1$. For company B, the value of the stock one year from now is modeled as lognormal with parameters $\mu = 0.02$ and $\sigma = 0.2$.

   a. Find the mean of the price of one share of company A one year from now.

   b. Find the probability that the price of one share of company A one year from now will be greater than $1.20.

   c. Find the mean of the price of one share of company B one year from now.

   d. Find the probability that the price of one share of company B one year from now will be greater than $1.20.

6. A manufacturer claims that the tensile strength of a certain composite (in MPa) has the lognormal distribution with $\mu = 5$ and $\sigma = 0.5$. Let X be the strength of a randomly sampled specimen of this composite.

   a. If the claim is true, what is $P(X < 20)$?

   b. Based on the answer to part (a), if the claim is true, would a strength of 20 MPa be unusually small?

   c. If you observed a tensile strength of 20 MPa, would this be convincing evidence that the claim is false? Explain.

d. If the claim is true, what is $P(X < 130)$?

e. Based on the answer to part (d), if the claim is true, would a strength of 130 MPa be unusually small?

f. If you observed a tensile strength of 130 MPa, would this be convincing evidence that the claim is false? Explain.

4.5 The Exponential Distribution

The **exponential distribution** is a continuous distribution that is sometimes used to model the time that elapses before an event occurs. Such a time is often called a **waiting time**. A common example is the lifetime of a component, which is the time that elapses before the component fails. In addition, there is a close connection between the exponential distribution and the Poisson distribution.

The probability density function of the exponential distribution involves a parameter, which is a positive constant λ whose value determines the density function's location and shape.

Definition

The probability density function of the exponential distribution with parameter $\lambda > 0$ is

$$f(x) = \begin{cases} \lambda e^{-\lambda x} & x > 0 \\ 0 & x \le 0 \end{cases} \qquad (4.18)$$

Figure 4.12 presents the probability density function of the exponential distribution for various values of λ. If X is a random variable whose distribution is exponential with parameter λ, we write $X \sim \text{Exp}(\lambda)$.

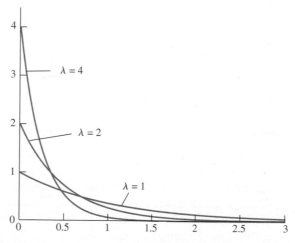

FIGURE 4.12 Plots of the exponential probability density function for various values of λ.

The cumulative distribution function of the exponential distribution is easy to compute. For $x \leq 0$, $F(x) = P(X \leq x) = 0$. For $x > 0$, the cumulative distribution function is

$$F(x) = P(X \leq x) = \int_0^x \lambda e^{-\lambda t} \, dt = 1 - e^{-\lambda x}$$

Summary

If $X \sim \text{Exp}(\lambda)$, the cumulative distribution function of X is

$$F(x) = P(X \leq x) = \begin{cases} 1 - e^{-\lambda x} & x > 0 \\ 0 & x \leq 0 \end{cases} \qquad (4.19)$$

The mean and variance of an exponential random variable can be computed by using integration by parts. The results follow.

If $X \sim \text{Exp}(\lambda)$, then

$$\mu_X = \frac{1}{\lambda} \qquad (4.20)$$

$$\sigma_X^2 = \frac{1}{\lambda^2} \qquad (4.21)$$

4.25

If $X \sim \text{Exp}(4)$, find μ_X, σ_X^2, and $P(X \leq 0.2)$.

Solution

We compute μ_X and σ_X^2 from Equations (4.20) and (4.21), substituting $\lambda = 2$. We obtain $\mu_X = 0.25$, $\sigma_X^2 = 0.0625$. Using Equation (4.19), we find that

$$P(X \leq 0.2) = 1 - e^{-4(0.2)} = 0.5507$$

The Poisson Process

Recall from Section 4.2 that the Poisson distribution is often used to model the number of events that occur in a given region of time or space. We say that events follow a **Poisson process** with rate parameter λ when the numbers of events in disjoint intervals of time or space are independent, and the number X of events that occur in any interval of length t has a Poisson distribution with mean λt—in other words, when $X \sim \text{Poisson}(\lambda t)$.

We mentioned that the exponential distribution is sometimes used to model the waiting time to an event. It turns out that the exponential distribution is the correct model for waiting times whenever the events follow a Poisson process. The connection between the exponential distribution and the Poisson process is as follows:

If events follow a Poisson process with rate parameter λ, and if T represents the waiting time from any starting point until the next event, then $T \sim \text{Exp}(\lambda)$.

Example 4.26

A radioactive mass emits particles according to a Poisson process at a mean rate of 15 particles per minute. At some point, a clock is started. What is the probability that more than 5 seconds will elapse before the next emission? What is the mean waiting time until the next particle is emitted?

Solution

We will measure time in seconds. Let T denote the time in seconds that elapses before the next particle is emitted. The mean rate of emissions is 0.25 per second, so the rate parameter is $\lambda = 0.25$, and $T \sim \text{Exp}(0.25)$. The probability that more than 5 seconds will elapse before the next emission is equal to

$$
\begin{aligned}
P(T > 5) &= 1 - P(T \le 5) \\
&= 1 - (1 - e^{-0.25(5)}) \\
&= e^{-1.25} \\
&= 0.2865
\end{aligned}
$$

The mean waiting time is $\mu_T = \dfrac{1}{0.25} = 4$ seconds.

Lack of Memory Property

The exponential distribution has a property known as the lack of memory property, which we illustrate with Examples 4.27 and 4.28.

Example 4.27

The lifetime of a transistor in a particular circuit has an exponential distribution with mean 1.25 years. Find the probability that the circuit lasts longer than two years.

Solution

Let T represent the lifetime of the circuit. Since $\mu_T = 1.25$, $\lambda = 1/1.25 = 0.8$. We need to find $P(T > 2)$.

$$
\begin{aligned}
P(T > 2) &= 1 - P(T \le 2) \\
&= 1 - (1 - e^{-0.8(2)}) \\
&= e^{-1.6} \\
&= 0.202
\end{aligned}
$$

Example 4.28

Refer to Example 4.27. Assume the transistor is now three years old and is still functioning. Find the probability that it functions for more than two additional years. Compare this probability with the probability that a new transistor functions for more than two years, which was calculated in Example 4.27.

Solution

We are given that the lifetime of the transistor will be more than three years, and we must compute the probability that the lifetime will be more than $3 + 2 = 5$ years. The probability is given by

$$P(T > 5|T > 3) = \frac{P(T > 5 \text{ and } T > 3)}{P(T > 3)}$$

If $T > 5$, then $T > 3$ as well. Therefore $P(T > 5 \text{ and } T > 3) = P(T > 5)$. It follows that

$$P(T > 5|T > 3) = \frac{P(T > 5)}{P(T > 3)}$$

$$= \frac{e^{-0.8(5)}}{e^{-0.8(3)}}$$

$$= e^{-0.8(2)}$$

$$= e^{-1.6}$$

$$= 0.202$$

The probability that a 3-year-old transistor lasts 2 additional years is exactly the same as the probability that a new transistor lasts 2 years.

Examples 4.27 and 4.28 illustrate the lack of memory property. The probability that we must wait an additional t units, given that we have already waited s units, is the same as the probability that we must wait t units from the start. The exponential distribution does not "remember" how long we have been waiting. In particular, if the lifetime of a component follows the exponential distribution, then the probability that a component that is s time units old will last an additional t time units is the same as the probability that a new component will last t time units. In other words, a component whose lifetime follows an exponential distribution does not show any effects of age or wear.

The calculations in Examples 4.27 and 4.28 could be repeated for any values s and t in place of 3 and 2, and for any value of λ in place of 0.8. We now state the lack of memory property in its general form:

Lack of Memory Property

If $T \sim \text{Exp}(\lambda)$, and t and s are positive numbers, then

$$P(T > t + s \mid T > s) = P(T > t)$$

Exercises for Section 4.5

1. Let $T \sim \text{Exp}(0.5)$. Find

  a. μ_T

  b. σ_T^2

  c. $P(T > 5)$

  d. The median of T

2. The time between requests to a web server is exponentially distributed with mean 0.5 seconds.

  a. What is the value of the parameter λ?

  b. What is the median time between requests?

  c. What is the standard deviation?

d. What is the 80th percentile?

e. Find the probability that more than one second elapses between requests.

f. If there have been no requests for the past two seconds, what is the probability that more than one additional second will elapse before the next request?

3. A catalyst researcher states that the diameters, in microns, of the pores in a new product she has made follow the exponential distribution with parameter $\lambda = 0.25$.

a. What is the mean pore diameter?

b. What is the standard deviation of the pore diameters?

c. What proportion of the pores are less than 3 microns in diameter?

d. What proportion of the pores are greater than 11 microns in diameter?

e. What is the median pore diameter?

f. What is the third quartile of the pore diameters?

g. What is the 99th percentile of the pore diameters?

4. The distance between flaws on a long cable is exponentially distributed with mean 12 m.

a. Find the probability that the distance between two flaws is greater than 15 m.

b. Find the probability that the distance between two flaws is between 8 and 20 m.

c. Find the median distance.

d. Find the standard deviation of the distances.

e. Find the 65th percentile of the distances.

5. A certain type of component can be purchased new or used. Fifty percent of all new components last more than five years, but only 30% of used components last more than five years. Is it possible that the lifetimes of new components are exponentially distributed? Explain.

6. A radioactive mass emits particles according to a Poisson process at a mean rate of 2 per second. Let T be the waiting time, in seconds, between emissions.

a. What is the mean waiting time?

b. What is the median waiting time?

c. Find $P(T > 2)$.

d. Find $P(T < 0.1)$.

e. Find $P(0.3 < T < 1.5)$.

f. If 3 seconds have elapsed with no emission, what is the probability that there will be an emission within the next second?

4.6 Some Other Continuous Distributions

The Uniform Distribution

The continuous uniform distribution, which we will sometimes refer to just as the uniform distribution, is the simplest of the continuous distributions. It often plays an important role in computer simulation studies. The uniform distribution has two parameters, a and b, with $a < b$.

Definition

The probability density function of the continuous uniform distribution with parameters a and b is

$$f(x) = \begin{cases} \dfrac{1}{b - a} & a < x < b \\ 0 & \text{otherwise} \end{cases} \tag{4.22}$$

If X is a random variable with probability density function $f(x)$, we say that X is uniformly distributed on the interval (a, b).

Since the probability density function is constant on the interval (a, b), we can think of the probability as being distributed "uniformly" on the interval. If X is a random variable whose distribution is uniform on the interval (a, b), we write $X \sim U(a, b)$.

The mean and variance of a uniform random variable can easily be computed from the definitions (Equations 3.19, 3.20, and 3.21 in Section 3.3).

Let $X \sim U(a, b)$. Then

$$\mu_X = \frac{a + b}{2} \tag{4.23}$$

$$\sigma_X^2 = \frac{(b - a)^2}{12} \tag{4.24}$$

The Gamma Distribution

The **gamma distribution** is a continuous distribution, one of whose purposes is to extend the usefulness of the exponential distribution in modeling waiting times. It involves a certain integral known as the **gamma function**. We define the gamma function and state some of its properties.

Definition

For $r > 0$, the gamma function is defined by

$$\Gamma(r) = \int_0^\infty t^{r-1} e^{-t} \, dt \tag{4.25}$$

The gamma function has the following properties:

1. If r is an integer, then $\Gamma(r) = (r - 1)!$.
2. For any r, $\Gamma(r + 1) = r\Gamma(r)$.
3. $\Gamma(1/2) = \sqrt{\pi}$.

The gamma function is used to define the probability density function of the gamma distribution. The gamma probability density function has two parameters, r and λ, both of which are positive constants.

Definition

The probability density function of the gamma distribution with parameters $r > 0$ and $\lambda > 0$ is

$$f(x) = \begin{cases} \dfrac{\lambda^r x^{r-1} e^{-\lambda x}}{\Gamma(r)} & x > 0 \\ 0 & x \le 0 \end{cases} \tag{4.26}$$

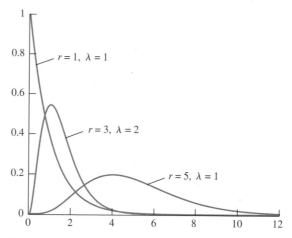

FIGURE 4.13 The gamma probability density function for various values of r and λ.

If X is a random variable whose probability density function is gamma with parameters r and λ, we write $X \sim \Gamma(r, \lambda)$. Note that when $r = 1$, the gamma distribution is the same as the exponential. In symbols, $\Gamma(1, \lambda) = \text{Exp}(\lambda)$. Figure 4.13 presents plots of the gamma probability density function for various values of r and λ.

The mean and variance of a gamma random variable can be computed from the probability density function. The results are as follows.

If $X \sim \Gamma(r, \lambda)$, then

$$\mu_X = \frac{r}{\lambda} \tag{4.27}$$

$$\sigma_X^2 = \frac{r}{\lambda^2} \tag{4.28}$$

A gamma distribution for which the parameter r is a positive integer is sometimes called an **Erlang distribution**. If $r = k/2$ where k is a positive integer, the $\Gamma(r, 1/2)$ distribution is called a **chi-square distribution with k degrees of freedom**. The chi-square distribution is widely used in statistical inference. We will discuss some of its uses in Section 6.5.

The Weibull Distribution

The Weibull distribution is a continuous distribution that is used in a variety of situations. A common application of the Weibull distribution is to model the lifetimes of components such as bearings, ceramics, capacitors, and dielectrics. The Weibull probability density function has two parameters, both positive constants, that determine its location and

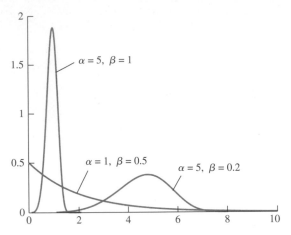

FIGURE 4.14 The Weibull probability density function for various choices of α and β.

shape. We denote these parameters α and β. The probability density function of the Weibull distribution is

$$f(x) = \begin{cases} \alpha\beta^{\alpha}x^{\alpha-1}e^{-(\beta x)^{\alpha}} & x > 0 \\ 0 & x \le 0 \end{cases} \qquad (4.29)$$

If X is a random variable whose probability density function is Weibull with parameters α and β, we write $X \sim$ Weibull(α, β). Note that when $\alpha = 1$, the Weibull distribution is the same as the exponential distribution with parameter $\lambda = \beta$. In symbols, Weibull$(1, \beta) = $ Exp(β).

Figure 4.14 presents plots of the Weibull(α, β) probability density function for several choices of the parameters α and β. By varying the values of α and β, a wide variety of curves can be generated. Because of this, the Weibull distribution can be made to fit a wide variety of data sets. This is the main reason for the usefulness of the Weibull distribution.

The Weibull cumulative distribution function can be computed by integrating the probability density function:

$$F(x) = P(X \le x) = \begin{cases} \displaystyle\int_{0}^{x} \alpha\beta^{\alpha}t^{\alpha-1}e^{-(\beta t)^{\alpha}}\,dt = 1 - e^{-(\beta x)^{\alpha}} & x > 0 \\ 0 & x \le 0 \end{cases} \qquad (4.30)$$

This integral is not as hard as it looks. Just substitute $u = (\beta t)^{\alpha}$, and $du = \alpha\beta^{\alpha}t^{\alpha-1}\,dt$.

The mean and variance of the Weibull distribution are expressed in terms of the gamma function.

If $X \sim$ Weibull(α, β), then

$$\mu_X = \frac{1}{\beta} \Gamma \left(1 + \frac{1}{\alpha} \right) \tag{4.31}$$

$$\sigma_X^2 = \frac{1}{\beta^2} \left\{ \Gamma \left(1 + \frac{2}{\alpha} \right) - \left[\Gamma \left(1 + \frac{1}{\alpha} \right) \right]^2 \right\} \tag{4.32}$$

In the special case that $1/\alpha$ is an integer, then

$$\mu_X = \frac{1}{\beta} \left[\left(\frac{1}{\alpha} \right)! \right] \qquad \sigma_X^2 = \frac{1}{\beta^2} \left\{ \left(\frac{2}{\alpha} \right)! - \left[\left(\frac{1}{\alpha} \right)! \right]^2 \right\}$$

If the quantity $1/\alpha$ is an integer, then $1 + 1/\alpha$ and $1 + 2/\alpha$ are integers, so Property 1 of the gamma function can be used to compute μ_X and σ_X^2 exactly. If the quantity $1/\alpha$ is of the form $n/2$, where n is an integer, then in principle μ_X and σ_X^2 can be computed exactly through repeated applications of Properties 2 and 3 of the gamma function. For other values of α, μ_X and σ_X^2 must be approximated. Many computer packages can do this.

Example 4.29

In the article "Snapshot: A Plot Showing Program through a Device Development Laboratory" (D. Lambert, J. Landwehr, and M. Shyu, *Statistical Case Studies for Industrial Process Improvement*, ASA-SIAM 1997), the authors suggest using a Weibull distribution to model the duration of a bake step in the manufacture of a semiconductor. Let T represent the duration in hours of the bake step for a randomly chosen lot. If $T \sim$ Weibull(0.3, 0.1), what is the probability that the bake step takes longer than four hours? What is the probability that it takes between two and seven hours?

Solution

We use the cumulative distribution function, Equation (4.30). Substituting 0.3 for α and 0.1 for β, we have

$$P(T \leq t) = 1 - e^{-(0.1t)^{0.3}}$$

Therefore

$$\begin{aligned}
P(T > 4) &= 1 - P(T \leq 4) \\
&= 1 - \left(1 - e^{-[(0.1)(4)]^{0.3}} \right) \\
&= e^{-(0.4)^{0.3}} \\
&= e^{-0.7597} \\
&= 0.468
\end{aligned}$$

The probability that the step takes between two and seven hours is

$$
\begin{aligned}
P(2 < T < 7) &= P(T \le 7) - P(T \le 2) \\
&= \left(1 - e^{-[(0.1)(7)]^{0.3}}\right) - \left(1 - e^{-[(0.1)(2)]^{0.3}}\right) \\
&= e^{-[(0.1)(2)]^{0.3}} - e^{-[(0.1)(7)]^{0.3}} \\
&= e^{-(0.2)^{0.3}} - e^{-(0.7)^{0.3}} \\
&= e^{-0.6170} - e^{-0.8985} \\
&= 0.132
\end{aligned}
$$

Exercises for Section 4.6

1. A person arrives at a certain bus stop each morning. The waiting time, in minutes, for a bus to arrive is uniformly distributed on the interval $(0, 10)$.

   a. Find the mean waiting time.

   b. Find the standard deviation of the waiting times.

2. Resistors are labeled 100 Ω. In fact, the actual resistances are uniformly distributed on the interval $(95, 103)$.

   a. Find the mean resistance.

   b. Find the standard deviation of the resistances.

3. Let $T \sim \Gamma(6, 2)$.

   a. Find μ_T.

   b. Find σ_T.

4. Let $T \sim \Gamma(r, \lambda)$. If $\mu_T = 8$ and $\sigma_T^2 = 16$, find r and λ.

5. Let $T \sim$ Weibull$(0.5, 2)$.

   a. Find μ_T.

   b. Find σ_T^2.

   c. Find $P(T \le 2)$.

   d. Find $P(T > 3)$.

   e. Find $P(1 < T \le 2)$.

6. If T is a continuous random variable that is always positive (such as a waiting time), with probability density function $f(t)$ and cumulative distribution function $F(t)$, then the **hazard function** is defined to be the function

   $$
   h(t) = \frac{f(t)}{1 - F(t)}
   $$

   The hazard function is the rate of failure per unit time, expressed as a proportion of the items that have not failed.

   a. If $T \sim$ Weibull(α, β), find $h(t)$.

   b. For what values of α is the hazard rate increasing with time? For what values of α is it decreasing?

7. In the article "Parameter Estimation with Only One Complete Failure Observation" (W. Pang, P. Leung, et al., *International Journal of Reliability, Quality, and Safety Engineering*, 2001:109–122), the lifetime, in hours, of a certain type of bearing is modeled with the Weibull distribution with parameters $\alpha = 2.25$ and $\beta = 4.474 \times 10^{-4}$.

   a. Find the probability that a bearing lasts more than 1000 hours.

   b. Find the probability that a bearing lasts less than 2000 hours.

   c. Find the median lifetime of a bearing.

   d. The hazard function is defined in Exercise 6. What is the hazard at $t = 2000$ hours?

8. The lifetime of a certain battery is modeled with the Weibull distribution with $\alpha = 2$ and $\beta = 0.1$.

   a. What proportion of batteries will last longer than 10 hours?

   b. What proportion of batteries will last less than 5 hours?

   c. What proportion of batteries will last longer than 20 hours?

   d. The hazard function is defined in Exercise 6. What is the hazard at $t = 10$ hours?

9. The lifetime of a cooling fan, in hours, that is used in a computer system has the Weibull distribution with $\alpha = 1.5$ and $\beta = 0.0001$.

a. What is the probability that a fan lasts more than 10,000 hours?

b. What is the probability that a fan lasts less than 5000 hours?

c. What is the probability that a fan lasts between 3000 and 9000 hours?

10. Someone suggests that the lifetime T (in days) of a certain component can be modeled with the Weibull distribution with parameters $\alpha = 3$ and $\beta = 0.01$.

   a. If this model is correct, what is $P(T \leq 1)$?

b. Based on the answer to part (a), if the model is correct, would one day be an unusually short lifetime? Explain.

c. If you observed a component that lasted one day, would you find this model to be plausible? Explain.

d. If this model is correct, what is $P(T \leq 90)$?

e. Based on the answer to part (d), if the model is correct, would 90 days be an unusually short lifetime? An unusually long lifetime? Explain.

f. If you observed a component that lasted 90 days, would you find this model to be plausible? Explain.

4.7 Probability Plots

Scientists and engineers frequently work with data that can be thought of as a random sample from some population. In many such cases, it is important to determine a probability distribution that approximately describes that population. In some cases, knowledge of the process that generated the data can guide the decision. More often, though, the only way to determine an appropriate distribution is to examine the sample to find a probability distribution that fits.

Probability plots provide a good way to do this. Given a random sample $X_1, \ldots, X_n$, a probability plot can determine whether the sample might plausibly have come from some specified population. We will present the idea behind probability plots with a simple example. A random sample of size 5 is drawn, and we want to determine whether the population from which it came might have been normal. The sample, arranged in increasing order, is

$$3.01, \ 3.35, \ 4.79, \ 5.96, \ 7.89$$

Denote the values, in increasing order, by $X_1, \ldots, X_n$ ($n = 5$ in this case). The first thing to do is to assign increasing, evenly spaced values between 0 and 1 to the X_i. There are several acceptable ways to do this; perhaps the simplest is to assign the value $(i - 0.5)/n$ to X_i. The following table shows the assignment for the given sample.

i	X_i	$(i - 0.5)/5$
1	3.01	0.1
2	3.35	0.3
3	4.79	0.5
4	5.96	0.7
5	7.89	0.9

The value $(i - 0.5)/n$ is chosen to reflect the position of X_i in the ordered sample. There are $i - 1$ values less than X_i, and i values less than or equal to X_i. The quantity $(i - 0.5)/n$ is a compromise between the proportions $(i - 1)/n$ and i/n.

The goal is to determine whether the sample might have come from a normal population. The most plausible normal distribution is the one whose mean and standard deviation are the same as the sample mean and standard deviation. The sample mean is $\overline{X} = 5.00$, and the sample standard deviation is $s = 2.00$. We will therefore determine whether this sample might have come from a $N(5, 2^2)$ distribution. The first step is to compute the $100(i - 0.5)/5$ percentiles (i.e., the 10th, 30th, 50th, 70th, and 90th percentiles) of the $N(5, 2^2)$ distribution. We could approximate these values by looking up the z-scores corresponding to these percentiles and then converting to raw scores. In practice, the Q_i are invariably calculated by a computer software package. The following table presents the X_i and the Q_i for this example.

i	X_i	Q_i
1	3.01	2.44
2	3.35	3.95
3	4.79	5.00
4	5.96	6.05
5	7.89	7.56

The **probability plot** consists of the points (X_i, Q_i). Since the distribution that generated the Q_i was a normal distribution, this is called a **normal probability plot**. If $X_1, \ldots, X_n$ do in fact come from the distribution that generated the Q_i, the points should lie close to a straight line.

Figure 4.15 presents a normal probability plot for the sample $X_1, \ldots, X_5$. A straight line is superimposed on the plot, to make it easier to judge whether or not the points lie

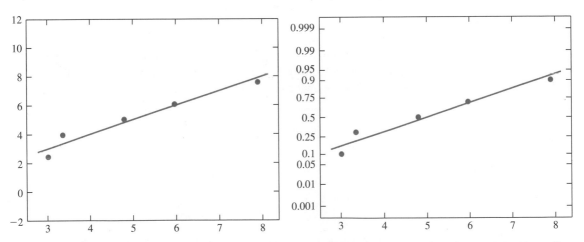

FIGURE 4.15 Normal probability plots for the sample $X_1, \ldots, X_5$. The plots are identical, except for the scaling on the vertical axis. The sample points lie approximately on a straight line, so it is plausible that they came from a normal population.

close to a straight line. Two versions of the plot are presented; they are identical except for the scaling on the vertical axis. In the plot on the left, the values on the vertical axis represent the Q_i. In the plot on the right, the values on the vertical axis represent the percentiles (as decimals, so 0.1 is the 10th percentile) of the Q_i. For example, the 10th percentile of $N(5, 2^2)$ is 2.44, so the value 0.1 on the right-hand plot corresponds to the value 2.44 on the left-hand plot. The 50th percentile, or median, is 5, so the value 0.5 on the right-hand plot corresponds to the value 5 on the left-hand plot. Computer packages often scale the vertical axis like the plot on the right. In Figure 4.15, the sample points are close to the line, so it is quite plausible that the sample came from a normal distribution.

We remark that the points $Q_1, \ldots, Q_n$ are called **quantiles** of the distribution from which they are generated. Sometimes the sample points $X_1, \ldots, X_n$ are called **empirical quantiles**. For this reason the probability plot is sometimes called a quantile–quantile plot, or QQ plot.

In this example, we used a sample of only five points to make the calculations clear. In practice, probability plots work better with larger samples. A good rule of thumb is to require at least 30 points before relying on a probability plot. Probability plots can still be used for smaller samples, but they will detect only fairly large departures from normality.

Figure 4.16 shows two normal probability plots. The plot in Figure 4.16a is of the monthly productions of 255 gas wells. These data do not lie close to a straight line and thus do not come from a population that is close to normal. The plot in Figure 4.16b is of the natural logs of the monthly productions. These data lie much closer to a straight line, although some departure from normality can be detected.

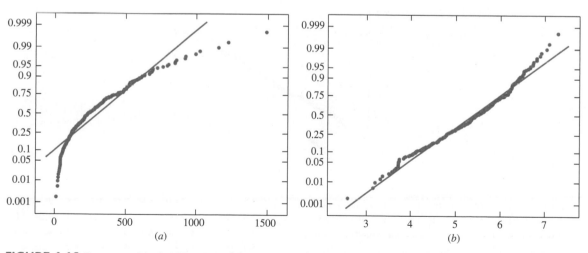

FIGURE 4.16 Two normal probability plots. *(a)* A plot of the monthly productions of 255 gas wells. These data do not lie close to a straight line, and thus do not come from a population that is close to normal. *(b)* A plot of the natural logs of the monthly productions. These data lie much closer to a straight line, although some departure from normality can be detected.

Interpreting Probability Plots

It's best not to use hard-and-fast rules when interpreting a probability plot. Judge the straightness of the plot by eye. When deciding whether the points on a probability plot lie close to a straight line, do not pay too much attention to the points at the very ends (high or low) of the sample, unless they are quite far from the line. It is common for a few points at either end to stray from the line somewhat. However, a point that is very far from the line when most other points are close is an outlier, and deserves attention.

Exercises for Section 4.7

1. Each of three samples has been plotted on a normal probability plot. For each, say whether the sample appears to have come from an approximately normal population.

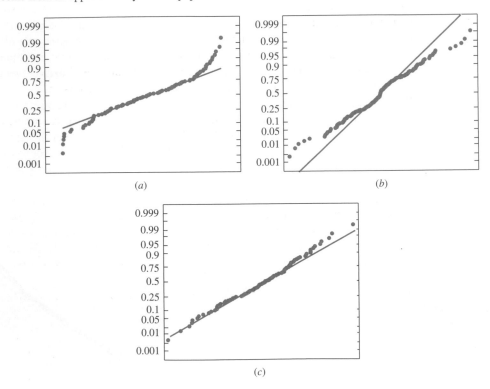

2. Construct a normal probability plot for the soapbars data in Exercise 1 in Section 1.3 (page 30). Do these data appear to come from an approximately normal distribution?

3. Below are the durations (in minutes) of 40 eruptions of the geyser Old Faithful in Yellowstone National Park.

4.1	1.8	3.2	1.9	4.6	2.0	4.5	3.9	4.3	2.3
3.8	1.9	4.6	1.8	4.7	1.8	4.6	1.9	3.5	4.0
3.7	3.7	4.3	3.6	3.8	3.8	3.8	2.5	4.5	4.1
3.7	3.8	3.4	4.0	2.3	4.4	4.1	4.3	3.3	2.0

Construct a normal probability plot for these data. Do the data appear to come from an approximately normal distribution?

4. Below are the durations (in minutes) of 40 time intervals between eruptions of the geyser Old Faithful in Yellowstone National Park.

```
91  51  79  53  82  51  76  82  84  53
86  51  85  45  88  51  80  49  82  75
73  67  68  86  72  75  75  66  84  70
79  60  86  71  67  81  76  83  76  55
```

Construct a normal probability plot for these data. Do they appear to come from an approximately normal distribution?

5. Construct a normal probability plot for the PM data in Table 1.2 (page 17). Do the PM data appear to come from a normal population?

6. Construct a normal probability plot for the logs of the PM data in Table 1.2. Do the logs of the PM data appear to come from a normal population?

7. Can the plot in Exercise 6 be used to determine whether the PM data appear to come from a lognormal population? Explain.

4.8 The Central Limit Theorem

The **Central Limit Theorem** is by far the most important result in statistics. Many commonly used statistical methods rely on this theorem for their validity. The Central Limit Theorem says that if we draw a large enough sample from a population, then the distribution of the sample mean is approximately normal, no matter what population the sample was drawn from. This allows us to compute probabilities for sample means using the z table, even though the population from which the sample was drawn is not normal. We now explain this more fully.

Let $X_1, \ldots, X_n$ be a simple random sample from a population with mean μ and variance σ^2. Let $\overline{X} = (X_1 + \cdots + X_n)/n$ be the sample mean. Now imagine drawing many such samples and computing their sample means. If one could draw every possible sample of size n from the original population and compute the sample mean for each one, the resulting collection would be the population of sample means. One could construct the probability density function of this population. One might think that the shape of this probability density function would depend on the shape of the population from which the sample was drawn. The surprising thing is that if the sample size is sufficiently large, this is not so. If the sample size is large enough, the distribution of the sample mean is approximately normal, no matter what the distribution of the population from which the sample was drawn.

The Central Limit Theorem

Let $X_1, \ldots, X_n$ be a simple random sample from a population with mean μ and variance σ^2.

Let $\overline{X} = \dfrac{X_1 + \cdots + X_n}{n}$ be the sample mean.

Let $S_n = X_1 + \cdots + X_n$ be the sum of the sample observations.

Then if n is sufficiently large,

$$\overline{X} \sim N\left(\mu, \frac{\sigma^2}{n}\right) \qquad \text{approximately} \qquad (4.33)$$

and

$$S_n \sim N(n\mu, n\sigma^2) \qquad \text{approximately} \qquad (4.34)$$

The Central Limit Theorem says that $\overline{X}$ and S_n are approximately normally distributed, if the sample size n is large enough. The natural question to ask is: How large is large enough? The answer depends on the shape of the underlying population. If the sample is drawn from a nearly symmetric distribution, the normal approximation can be good even for a fairly small value of n. However, if the population is heavily skewed, a fairly large n may be necessary. Empirical evidence suggests that for most populations, a sample size of 30 or more is large enough for the normal approximation to be adequate.

> For most populations, if the sample size is greater than 30, the Central Limit Theorem approximation is good.

*E*xample
4.30

Let X denote the number of flaws in a 1 in. length of copper wire. The probability mass function of X is presented in the following table.

x	P(X = x)
0	0.48
1	0.39
2	0.12
3	0.01

One hundred wires are sampled from this population. What is the probability that the average number of flaws per wire in this sample is less than 0.5?

Solution
The population mean number of flaws is $\mu = 0.66$, and the population variance is $\sigma^2 = 0.5244$. (These quantities can be calculated using Equations 3.13 and 3.14 in Section 3.3.) Let $X_1, \ldots, X_{100}$ denote the number of flaws in the 100 wires sampled from this population. We need to find $P(\overline{X} < 0.5)$. Now the sample size is $n = 100$, which is a large sample. It follows from the Central Limit Theorem (expression 4.33) that $\overline{X} \sim N(0.66, 0.005244)$. The z-score of 0.5 is therefore

$$z = \frac{0.5 - 0.66}{\sqrt{0.005244}} = -2.21$$

From the z table, the area to the left of -2.21 is 0.0136. Therefore $P(\overline{X} < 0.5) = 0.0136$, so only 1.36% of samples of size 100 will have fewer than 0.5 flaws per wire. See Figure 4.17.

Note that in Example 4.30 we needed to know only the mean and variance of the population, not the probability mass function.

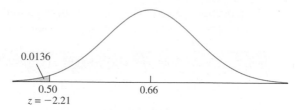

FIGURE 4.17 Solution to Example 4.30

Example

4.31

At a large university, the mean age of the students is 22.3 years, and the standard deviation is 4 years. A random sample of 64 students is drawn. What is the probability that the average age of these students is greater than 23 years?

Solution

Let $X_1, \ldots, X_{64}$ be the ages of the 64 students in the sample. We wish to find $P(\overline{X} > 23)$. Now the population from which the sample was drawn has mean $\mu = 22.3$ and variance $\sigma^2 = 16$. The sample size is $n = 64$. It follows from the Central Limit Theorem (expression 4.33) that $\overline{X} \sim N(22.3, 0.25)$. The z-score for 23 is

$$z = \frac{23 - 22.3}{\sqrt{0.25}} = 1.40$$

From the z table, the area to the right of 1.40 is 0.0808. Therefore $P(\overline{X} > 23) = 0.0808$. See Figure 4.18.

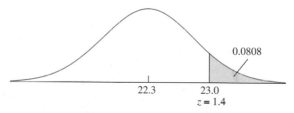

FIGURE 4.18 Solution to Example 4.31

Normal Approximation to the Binomial

Recall from Section 4.1 that if $X \sim \text{Bin}(n, p)$ then X represents the number of successes in n independent trials, each of which has success probability p. Furthermore, X has mean np and variance $np(1 - p)$. If the number of trials is large enough, the Central Limit Theorem says that $X \sim N(np, np(1 - p))$, approximately. Thus we can use the normal distribution to compute probabilities concerning X.

Again the question arises: How large a sample is large enough? In the binomial case, the accuracy of the normal approximation depends on the mean number of successes np and on the mean number of failures $n(1 - p)$. The larger the values of np and $n(1 - p)$, the better the approximation. A common rule of thumb is to use the normal approximation whenever $np > 5$ and $n(1 - p) > 5$. A better and more conservative rule is to use the normal approximation whenever $np > 10$ and $n(1 - p) > 10$.

Summary

If $X \sim \text{Bin}(n, p)$, and if $np > 10$ and $n(1 - p) > 10$, then

$$X \sim N(np, np(1 - p)) \qquad \text{approximately} \qquad (4.35)$$

To illustrate the accuracy of the normal approximation to the binomial, Figure 4.19 presents the $\text{Bin}(100, 0.2)$ probability histogram with the $N(20, 16)$ probability density function superimposed. While a slight degree of skewness can be detected in the binomial distribution, the normal approximation is quite close.

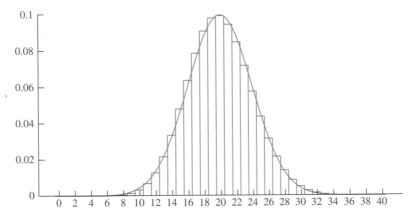

FIGURE 4.19 The $\text{Bin}(100, 0.2)$ probability histogram, with the $N(20, 16)$ probability density function superimposed.

The Continuity Correction

The binomial distribution is discrete, while the normal distribution is continuous. The **continuity correction** is an adjustment, made when approximating a discrete distribution with a continuous one, that can improve the accuracy of the approximation. To see how it works, imagine that a fair coin is tossed 100 times. Let X represent the number of heads. Then $X \sim \text{Bin}(100, 0.5)$. Imagine that we wish to compute the probability that X is between 45 and 55. This probability will differ depending on whether the endpoints, 45 and 55, are included or excluded. Figure 4.20 illustrates the case where the endpoints are included—that is, where we wish to compute $P(45 \leq X \leq 55)$. The exact probability is given by the total area of the rectangles of the binomial probability histogram corresponding to the integers 45 to 55 inclusive. The approximating normal curve is superimposed. To get the best approximation, we should compute the area under the normal curve between 44.5 and 55.5. In contrast, Figure 4.21 illustrates the case where we wish to compute $P(45 < X < 55)$. Here the endpoints are excluded. The

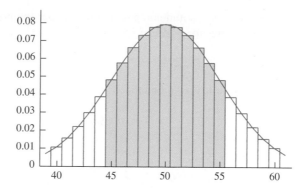

FIGURE 4.20 To compute $P(45 \leq X \leq 55)$, the areas of the rectangles corresponding to 45 and to 55 should be included. To approximate this probability with the normal curve, compute the area under the curve between 44.5 and 55.5.

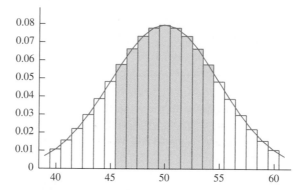

FIGURE 4.21 To compute $P(45 < X < 55)$, the areas of the rectangles corresponding to 45 and to 55 should be excluded. To approximate this probability with the normal curve, compute the area under the curve between 45.5 and 54.5.

exact probability is given by the total area of the rectangles of the binomial probability histogram corresponding to the integers 46 to 54. The best normal approximation is found by computing the area under the normal curve between 45.5 and 54.5.

In summary, to apply the continuity correction, determine precisely which rectangles of the discrete probability histogram you wish to include, and then compute the area under the normal curve corresponding to those rectangles.

Example

4.32

If a fair coin is tossed 100 times, use the normal curve to approximate the probability that the number of heads is between 45 and 55 *inclusive*.

Solution
This situation is illustrated in Figure 4.20. Let X be the number of heads obtained. Then $X \sim \text{Bin}(100, 0.5)$. Substituting $n = 100$ and $p = 0.5$ into Equation (4.35), we obtain the normal approximation $X \sim N(50, 25)$. Since the endpoints 45 and 55 are to be

included, we should compute the area under the normal curve between 44.5 and 55.5. The z-scores for 44.5 and 55.5 are

$$z = \frac{44.5 - 50}{5} = -1.1, \qquad z = \frac{55.5 - 50}{5} = 1.1$$

From the z table we find that the probability is 0.7286. See Figure 4.22.

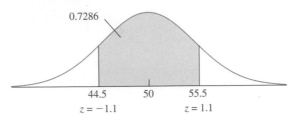

FIGURE 4.22 Solution to Example 4.32.

If a fair coin is tossed 100 times, use the normal curve to approximate the probability that the number of heads is between 45 and 55 *exclusive*.

Solution

This situation is illustrated in Figure 4.21. Let X be the number of heads obtained. As in Example 4.32, $X \sim \text{Bin}(100, 0.5)$, and the normal approximation is $X \sim N(50, 25)$. Since the endpoints 45 and 55 are to be excluded, we should compute the area under the normal curve between 45.5 and 54.5. The z-scores for 45.5 and 54.5 are

$$z = \frac{45.5 - 50}{5} = -0.9, \qquad z = \frac{54.5 - 50}{5} = 0.9$$

From the z table we find that the probability is 0.6318. See Figure 4.23.

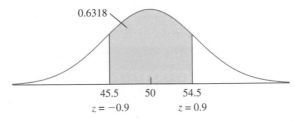

FIGURE 4.23 Solution to Example 4.33.

Accuracy of the Continuity Correction

The continuity correction improves the accuracy of the normal approximation to the binomial distribution in most cases. For binomial distributions with large n and small p, however, when computing a probability that corresponds to an area in the tail of the distribution, use of the continuity correction can in some cases reduce the accuracy of the normal approximation somewhat. This results from the fact that the normal approximation is not perfect; it fails to account for a small degree of skewness in these distributions. In summary, use of the continuity correction makes the normal approximation to the binomial distribution better in most cases, but not all.

Normal Approximation to the Poisson

Recall that if $X \sim \text{Poisson}(\lambda)$, then X is approximately binomial with n large and $np = \lambda$. Recall also that $\mu_X = \lambda$ and $\sigma_X^2 = \lambda$. It follows that if λ is sufficiently large, i.e., $\lambda > 10$, then X is approximately binomial, with $np > 10$. It follows from the Central Limit Theorem that X is also approximately normal, with mean and variance both equal to λ. Thus we can use the normal distribution to approximate the Poisson.

Summary

If $X \sim \text{Poisson}(\lambda)$, where $\lambda > 10$, then

$$X \sim N(\lambda, \lambda) \qquad \text{approximately} \qquad (4.36)$$

Continuity Correction for the Poisson Distribution

Since the Poisson distribution is discrete, the continuity correction can in principle be applied when using the normal approximation. For areas that include the central part of the curve, the continuity correction generally improves the normal approximation, but for areas in the tails the continuity correction sometimes makes the approximation worse. We will not use the continuity correction for the Poisson distribution.

Example
4.34

The number of hits on a website follows a Poisson distribution, with a mean of 27 hits per hour. Find the probability that there will be 90 or more hits in three hours.

Solution

Let X denote the number of hits on the website in three hours. The mean number of hits in three hours is 81, so $X \sim \text{Poisson}(81)$. Using the normal approximation, we find that $X \sim N(81, 81)$. We wish to find $P(X \geq 90)$. We compute the z-score of 90, which is

$$z = \frac{90 - 81}{\sqrt{81}} = 1.00$$

Using the z table, we find that $P(X \geq 90) = 0.1587$. See Figure 4.24.

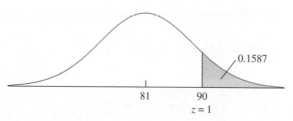

FIGURE 4.24 Solution to Example 4.34.

Exercises for Section 4.8

1. Bottles filled by a certain machine are supposed to contain 12 oz of liquid. In fact, the fill volume is random, with mean 12.01 oz and standard deviation 0.2 oz.

   a. What is the probability that the mean volume of a random sample of 144 bottles is less than 12 oz?

   b. If the population mean fill volume is increased to 12.03 oz what is the probability that the mean volume of a sample of size 144 will be less than 12 oz?

2. Among all the income tax forms filed in a certain year, the mean tax paid was $2000, and the standard deviation was $500. In addition, for 10% of the forms, the tax paid was greater than $3000. A random sample of 625 tax forms is drawn.

   a. What is the probability that the average tax paid on the sample forms is greater than $1980?

   b. What is the probability that more than 60 of the sampled forms have a tax of greater than $3000?

3. In a galvanized coating process for pipes, the mean coating weight is 125 lb, with a standard deviation of 10 lb. A random sample of 50 pipes is drawn.

   a. What is the probability that the sample mean coating weight is greater than 128 lb?

   b. Find the 90th percentile of the sample mean coating weight.

   c. How large a sample size is needed so that the probability is 0.02 that the sample mean is less than 123?

4. A 500-page book contains 250 sheets of paper. The thickness of the paper used to manufacture the book has mean 0.08 mm and standard deviation 0.01 mm.

   a. What is the probability that a randomly chosen book is more than 20.2 mm thick (not including the covers)?

   b. What is the 10th percentile of book thicknesses?

   c. Someone wants to know the probability that a randomly chosen page is more than 0.1 mm thick. Is enough information given to compute this probability? If so, compute the probability. If not, explain why not.

5. In a process that manufactures bearings, 90% of the bearings meet a thickness specification. In a sample of 500 bearings, what is the probability that more than 440 meet the specification?

6. Among the adults in a large city, 30% have a college degree. A simple random sample of 100 adults is chosen. What is the probability that more than 35 of them have a college degree?

7. The breaking strength (in kg/mm) for a certain type of fabric has mean 1.86 and standard deviation 0.27. A random sample of 80 pieces of fabric is drawn.

   a. What is the probability that the sample mean breaking strength is less than 1.8 kg/mm?

   b. Find the 80th percentile of the sample mean breaking strength.

   c. How large a sample size is needed so that the probability is 0.01 that the sample mean is less than 1.8?

8. The amount of warpage in a type of wafer used in the manufacture of integrated circuits has mean 1.3 mm and standard deviation 0.1 mm. A random sample of 200 wafers is drawn.

   a. What is the probability that the sample mean warpage exceeds 1.305 mm?

   b. Find the 25th percentile of the sample mean.

   c. How many wafers must be sampled so that the probability is 0.05 that the sample mean exceeds 1.305?

9. A sample of 225 wires is drawn from the population of wires described in Example 4.30. Find the probability that fewer than 110 of these wires have no flaws.

10. A battery manufacturer claims that the lifetime of a certain type of battery has a population mean of 40 hours and a standard deviation of 5 hours. Let $\overline{X}$ represent the mean lifetime of the batteries in a simple random sample of size 100.

   a. If the claim is true, what is $P(\overline{X} \leq 36.7)$?

   b. Based on the answer to part (a), if the claim is true, is a sample mean lifetime of 36.7 hours unusually short?

   c. If the sample mean lifetime of the 100 batteries were 36.7 hours, would you find the manufacturer's claim to be plausible? Explain.

   d. If the claim is true, what is $P(\overline{X} \leq 39.8)$?

e. Based on the answer to part (d), if the claim is true, is a sample mean lifetime of 39.8 hours unusually short?

f. If the sample mean lifetime of the 100 batteries were 39.8 hours, would you find the manufacturer's claim to be plausible? Explain.

11. A new process has been designed to make ceramic tiles. The goal is to have no more than 5% of the tiles be nonconforming due to surface defects. A random sample of 1000 tiles is inspected. Let X be the number of nonconforming tiles in the sample.

a. If 5% of the tiles produced are nonconforming, what is $P(X \geq 75)$?

b. Based on the answer to part (a), if 5% of the tiles are nonconforming, is 75 nonconforming tiles out of 1000 an unusually large number?

c. If 75 of the sample tiles were nonconforming, would it be plausible that the goal had been reached? Explain.

d. If 5% of the tiles produced are nonconforming, what is $P(X \geq 53)$?

e. Based on the answer to part (d), if 5% of the tiles are nonconforming, is 53 nonconforming tiles out of 1000 an unusually large number?

f. If 53 of the sample tiles were nonconforming, would it be plausible that the goal had been reached? Explain.

Supplementary Exercises for Chapter 4

1. An airplane has 100 seats for passengers. Assume that the probability that a person holding a ticket appears for the flight is 0.90. If the airline sells 105 tickets, what is the probability that everyone who appears for the flight will get a seat?

2. The number of large cracks in a length of pavement along a certain street has a Poisson distribution with a mean of 1 crack per 100 m.

a. What is the probability that there will be exactly 8 cracks in a 500 m length of pavement?

b. What is the probability that there will be no cracks in a 100 m length of pavement?

c. Let T be the distance in meters between two successive cracks. What is the probability density function of T?

d. What is the probability that the distance between two successive cracks will be more than 50 m?

3. Pea plants contain two genes for seed color, each of which may be Y (for yellow seeds) or G (for green seeds). Plants that contain one of each type of gene are called heterozygous. According to the Mendelian theory of genetics, if two heterozygous plants are crossed, each of their offspring will have probability 0.75 of having yellow seeds and probability 0.25 of having green seeds.

a. Out of 10 offspring of heterozygous plants, what is the probability that exactly 3 have green seeds?

b. Out of 10 offspring of heterozygous plants, what is the probability that more than 2 have green seeds?

c. Out of 100 offspring of heterozygous plants, what is the probability that more than 30 have green seeds?

d. Out of 100 offspring of heterozygous plants, what is the probability that between 30 and 35 inclusive have green seeds?

e. Out of 100 offspring of heterozygous plants, what is the probability that fewer than 80 have yellow seeds?

4. A simple random sample $X_1, \ldots, X_n$ is drawn from a population, and the quantities $\ln X_1, \ldots, \ln X_n$ are plotted on a normal probability plot. The points approximately follow a straight line. True or false:

a. $X_1, \ldots, X_n$ come from a population that is approximately lognormal.

b. $X_1, \ldots, X_n$ come from a population that is approximately normal.

c. $\ln X_1, \ldots, \ln X_n$ come from a population that is approximately lognormal.

d. $\ln X_1, \ldots, \ln X_n$ come from a population that is approximately normal.

5. The Environmental Protection Agency (EPA) has contracted with your company for equipment to monitor water quality for several lakes in your water district. A total of 10 devices will be used. Assume that each

device has a probability of 0.01 of failure during the course of the monitoring period.

a. What is the probability that none of the devices fail?

b. What is the probability that two or more devices fail?

c. If the EPA requires the probability that none of the devices fail to be at least 0.95, what is the largest individual failure probability allowable?

6. In the article "Occurrence and Distribution of Ammonium in Iowa Groundwater" (K. Schilling, *Water Environment Research*, 2002:177–186), ammonium concentrations (in mg/L) were measured at a large number of wells in Iowa. The mean concentration was 0.71, the median was 0.22, and the standard deviation was 1.09. Is it possible to determine whether these concentrations are approximately normally distributed? If so, say whether they are normally distributed, and explain how you know. If not, describe the additional information you would need to determine whether they are normally distributed.

7. Medication used to treat a certain condition is administered by syringe. The target dose in a particular application is μ. Because of the variations in the syringe, in reading the scale, and in mixing the fluid suspension, the actual dose administered is normally distributed with mean μ and variance σ^2.

a. What is the probability that the dose administered differs from the mean μ by less than σ?

b. If X represents the dose administered, find the value of z so that $P(X < \mu + z\sigma) = 0.90$.

c. If the mean dose is 10 mg, the variance is 2.6 mg^2, and a clinical overdose is defined as a dose larger than 15 mg, what is the probability that a patient will receive an overdose?

8. You have a large box of resistors whose resistances are normally distributed with mean 10 Ω and standard deviation 1 Ω.

a. What proportion of the resistors have resistances between 9.3 and 10.7 Ω?

b. If you sample 100 resistors, what is the probability that 50 or more of them will have resistances between 9.3 and 10.7 Ω?

c. How many resistors must you sample so that the probability is 0.99 that 50 or more of the sam-

pled resistors will have resistances between 9.3 and 10.7 Ω?

9. The intake valve clearances on new engines of a certain type are normally distributed with mean 200 μm and standard deviation 10 μm.

a. What is the probability that the clearance is greater than 215 μm?

b. What is the probability that the clearance is between 180 and 205 μm?

c. An engine has six intake valves. What is the probability that exactly two of them have clearances greater than 215 μm?

10. The stiffness of a certain type of steel beam used in building construction has mean 30 kN/mm and standard deviation 2 kN/mm.

a. Is it possible to compute the probability that the stiffness of a randomly chosen beam is greater than 32 kN/mm? If so, compute the probability. If not, explain why not.

b. In a sample of 100 beams, is it possible to compute the probability that the sample mean stiffness of the beams is greater than 30.2 kN/mm? If so, compute the probability. If not, explain why not.

11. In a certain process, the probability of producing an oversize assembly is 0.05.

a. In a sample of 300 randomly chosen assemblies, what is the probability that fewer than 20 of them are oversize?

b. In a sample of 10 randomly chosen assemblies, what is the probability that one or more of them is oversize?

c. To what value must the probability of an oversize assembly be reduced so that only 1% of lots of 300 assemblies contain 20 or more that are oversize?

12. A process for manufacturing plate glass leaves an average of three small bubbles per 10 m^2 of glass. The number of bubbles on a piece of plate glass follows a Poisson distribution.

a. What is the probability that a piece of glass 3 × 5 m will contain more than two bubbles?

b. What is the probability that a piece of glass 4 × 6 m will contain no bubbles?

c. What is the probability that 50 pieces of glass, each 3 × 6 m, will contain more than 300 bubbles in total?

13. The lifetime of a bearing (in years) follows the Weibull distribution with parameters $\alpha = 1.5$ and $\beta = 0.8$.

    a. What is the probability that a bearing lasts more than 1 year?

    b. What is the probability that a bearing lasts less than 2 years?

14. The length of time to perform an oil change at a certain shop is normally distributed with mean 29.5 minutes and standard deviation 3 minutes. What is the probability that a mechanic can complete 16 oil changes in an eight-hour day?

15. A cereal manufacturer claims that the gross weight (including packaging) of a box of cereal labeled as weighing 12 oz has a mean of 12.2 oz and a standard deviation of 0.1 oz. You gather 75 boxes and weigh them all together. Let S denote the total weight of the 75 boxes of cereal.

    a. If the claim is true, what is $P(S \le 914.8)$?

    b. Based on the answer to part (a), if the claim is true, is 914.8 oz an unusually small total weight for a sample of 75 boxes?

    c. If the total weight of the boxes were 914.8 oz, would you be convinced that the claim was false? Explain.

    d. If the claim is true, what is $P(S \le 910.3)$?

    e. Based on the answer to part (d), if the claim is true, is 910.3 oz an unusually small total weight for a sample of 75 boxes?

    f. If the total weight of the boxes were 910.3 oz, would you be convinced that the claim was false? Explain.

16. Someone claims that the number of hits on his website has a Poisson distribution with mean 20 per hour. Let X be the number of hits in five hours.

    a. If the claim is true, what is $P(X \le 95)$?

    b. Based on the answer to part (a), if the claim is true, is 95 hits in a five-hour time period an unusually small number?

    c. If you observed 95 hits in a five-hour time period, would this be convincing evidence that the claim is false? Explain.

    d. If the claim is true, what is $P(X \le 65)$?

    e. Based on the answer to part (d), if the claim is true, is 65 hits in a five-hour time period an unusually small number?

    f. If you observed 65 hits in a five-hour time period, would this be convincing evidence that the claim is false? Explain.

Chapter 5

Point and Interval Estimation for a Single Sample

Introduction

When data are collected, it is often with the purpose of estimating some numerical characteristic of the population from which they came. For example, we might measure the diameters of a sample of bolts from a large population and compute the sample mean in order to estimate the population mean diameter. We might also compute the proportion of the sample bolts that meet a strength specification in order to estimate the proportion of bolts in the population that meet the specification.

The sample mean and sample proportion are examples of **point estimates**, because they are single numbers, or points. More useful are **interval estimates**, also called **confidence intervals**. The purpose of a confidence interval is to provide a margin of error for the point estimate, to indicate how far off the true value it is likely to be.

For example, suppose that the diameters of a sample of 100 bolts are measured, and the sample mean is 5.0 mm with a sample standard deviation of 0.2 mm. The sample mean 5.0 mm is a point estimate for the population mean diameter μ. The population mean is likely to be close to 5.0 but is not likely to be exactly equal to 5.0. A confidence interval gives an idea of how close the population mean is likely to be to the sample mean. An example of a confidence interval is 5.0 ± 0.04, or $(4.96, 5.04)$. In fact, the results of Section 5.2 will show that we can be approximately 95% confident that the true mean μ is contained in the interval $(4.96, 5.04)$.

5.1 Point Estimation

When we gather data, we often do so in order to estimate some numerical characteristic of the population from which the data came. For example, if $X_1, \ldots, X_n$ is a random sample from a population, the sample mean $\overline{X}$ is often used to estimate the population mean μ, and the sample variance s^2 is often used to estimate the population variance σ^2. As another example, if $X \sim \text{Bin}(n, p)$, the sample proportion $\hat{p} = X/n$ is often used to estimate the unknown population proportion p.

In general, a quantity calculated from data is called a statistic, and a statistic that is used to estimate an unknown constant, or parameter, is called a **point estimator** or **point estimate**. The term *point estimate* is used when a particular numerical value is specified for the data. For example, if $X \sim \text{Bin}(10, p)$, and we observe $X = 3$, then the number $\hat{p} = 3/10$ is a point estimate of the unknown parameter p. On the other hand, if no particular value is specified for X, the random quantity $\hat{p} = X/10$ is often called a point estimator of p. Often, point estimators and point estimates are simply called **estimators** and **estimates**, respectively.

Summary

- A numerical summary of a sample is called a **statistic**. The sample mean $\overline{X}$ and the sample variance s^2 are examples of statistics.
- A numerical summary of a population is called a **parameter**. The population mean μ and the population variance σ^2 are examples of parameters.
- Statistics are often used to estimate parameters. A statistic that is used to estimate a parameter is called a point estimate.

Point estimates are almost never exactly equal to the true value they are estimating. The closeness of an estimate the true value will vary from sample to sample. For example, for some samples, the sample mean $\overline{X}$ will be very close to the population mean μ, while for other samples it will not be so close. Because the closeness of an estimator varies from sample to sample, we determine how good an estimator is by measuring its closeness to the true value on the average. Specifically, there are two important measures, the **bias** and the variance. We discuss the bias first.

The bias of an estimator is the difference between the mean of the estimator and the true value being estimated. A small bias is better than a large one. A small bias indicates that on the average, the value of the estimator is close to the true value. Many good estimators are **unbiased**, which means that their bias is 0 or, in other words, that the mean of the estimator is equal to the true value.

It is important to note that the bias only describes the closeness of the mean of the estimator. Even when the bias is small, any single value of the estimate may be far from the true value being estimated.

Summary

Let θ be a parameter, and $\hat{\theta}$ an estimator of θ. The bias of $\hat{\theta}$ is the difference between the mean of $\hat{\theta}$ and the true value θ. In symbols,

$$\text{Bias} = \mu_{\hat{\theta}} - \theta \tag{5.1}$$

An estimator with a bias of 0 is said to be unbiased.

5.1

Let $X \sim \text{Bin}(n, p)$ where p is unknown. Find the bias of $\hat{p} = X/n$.

Solution

The bias is $\mu_{\hat{p}} - p$. We must therefore compute $\mu_{\hat{p}}$. From Equation 4.3 in Section 4.1, we know that $\mu_X = np$. Thus

$$\mu_{\hat{p}} = \mu_{X/n} = \mu_X/n = np/n = p$$

The bias is $\mu_{\hat{p}} - p = p - p = 0$. Therefore $\hat{p}$ is unbiased.

The second important measure of the performance of an estimator is its variance. Variance, of course, measures spread. A small variance is better than a large variance. When the variance of an estimator is small, it means that repeated values of the estimator will be close to one another. Of course, this does not mean that they will necessarily be close to the true value being estimated.

Figure 5.1 illustrates the performance of some hypothetical estimators under differing conditions regarding bias and variance. The sets of measurements in Figure 5.1(a) and (b) are fairly close together, indicating that the variance is small. The sets of measurements in Figure 5.1(a) and (c) are centered near the true value, indicating that the bias is small.

The most commonly used measure of the performance of an estimator combines both the bias and the variance to produce a single number. This measure is called the **mean squared error** (abbreviated MSE), and is equal to the variance plus the square of the bias.

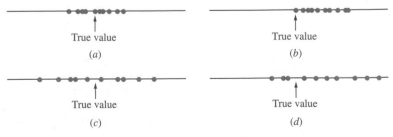

FIGURE 5.1 *(a)* Both bias and variance are small. *(b)* Bias is large; variance is small. *(c)* Bias is small; variance is large. *(d)* Both bias and variance are large.

> ### Definition
>
> Let θ be a parameter, and $\hat{\theta}$ an estimator of θ. The mean squared error (MSE) of $\hat{\theta}$ is
>
> $$\text{MSE}_{\hat{\theta}} = (\mu_{\hat{\theta}} - \theta)^2 + \sigma_{\hat{\theta}}^2 \qquad (5.2)$$
>
> An equivalent expression for the MSE is
>
> $$\text{MSE}_{\hat{\theta}} = \mu_{(\hat{\theta}-\theta)^2} \qquad (5.3)$$

Equation (5.2) says that the MSE is equal to the square of the bias, plus the variance. To interpret Equation (5.3), note that the quantity $\hat{\theta} - \theta$ is the difference between the estimated value and the true value, and it is called the error. So Equation (5.3) says that the MSE is the mean of the squared error, and indeed it is this property that gives the MSE its name.

Equations (5.2) and (5.3) yield identical results, so either may be used in any situation to compute the MSE. In practice, Equation (5.2) is often somewhat easier to use.

Let $X \sim \text{Bin}(n, p)$ where p is unknown. Find the MSE of $\hat{p} = X/n$.

Solution

We compute the bias and variance of $\hat{p}$ and use Equation (5.2). As shown in Example 5.1, the bias of $\hat{p}$ is 0. We must now compute the variance of $\hat{p}$. From Equation (4.3) in Section 4.1, we know that $\mu_X = np$. Therefore the variance is

$$\sigma_{\hat{p}}^2 = \sigma_{X/n}^2 = \sigma_X^2/n^2 = np(1 - p)/n^2 = p(1 - p)/n$$

The MSE of $\hat{p}$ is therefore $0 + p(1 - p)/n$, or $p(1 - p)/n$.

In Example 5.2 the estimator was unbiased, so the MSE was equal to the variance of the estimator.

Limitations of Point Estimates

Point estimates by themselves are of limited use. The reason for this is that they are almost never exactly equal to the true values they are estimating. They are almost always off—sometimes by a little, sometimes by a lot. For a point estimate to be useful, it is necessary to describe just how far off the true value it is likely to be. One way to do this is to report the value of the MSE along with the estimate. In practice, though, people usually use a different type of estimate altogether. These are called interval estimates or, more commonly, confidence intervals. A confidence interval consists of two numbers: a lower limit and an upper limit, constructed in such a way that it is likely that the true value is between the two limits. Along with the confidence interval comes a level of confidence, which is a number that tells us just how likely it is that the true value is contained within the interval. Most of the rest of this chapter will be devoted to the construction of confidence intervals.

Exercises for Section 5.1

1. Choose the best answer to fill in the blank. If an estimator is unbiased, then _____
   i. the estimator is equal to the true value.
   ii. the estimator is usually close to the true value.
   iii. the mean of the estimator is equal to the true value.
   iv. the mean of the estimator is usually close to the true value.

2. Choose the best answer to fill in the blank. The variance of an estimator measures _____
   i. how close the estimator is to the true value.
   ii. how close repeated values of the estimator are to each other.
   iii. how close the mean of the estimator is to the true value.
   iv. how close repeated values of the mean of the estimator are to each other.

3. Let X_1 and X_2 be independent, each with unknown mean μ and known variance $\sigma^2 = 1$.
   a. Let $\hat{\mu}_1 = \dfrac{X_1 + X_2}{2}$. Find the bias, variance, and mean squared error of $\hat{\mu}_1$.
   b. Let $\hat{\mu}_2 = \dfrac{X_1 + 2X_2}{3}$. Find the bias, variance, and mean squared error of $\hat{\mu}_2$.
   c. Let $\hat{\mu}_3 = \dfrac{X_1 + X_2}{4}$. Find the bias, variance, and mean squared error of $\hat{\mu}_3$.

4. Refer to Exercise 3. For what values of μ does $\hat{\mu}_3$ have smaller mean squared error than $\hat{\mu}_1$?

5. Refer to Exercise 3. For what values of μ does $\hat{\mu}_3$ have smaller mean squared error than $\hat{\mu}_2$?

5.2   Large-Sample Confidence Intervals for a Population Mean

We begin with an example. An important measure of the performance of an automotive battery is its cold cranking amperage, which is the current, in amperes, that the battery can provide for 30 seconds at 0°F while maintaining a specified voltage. An engineer wants to estimate the mean cold cranking amperage for batteries of a certain design. He draws a simple random sample of 100 batteries, and finds that the sample mean amperage is $\overline{X} = 185.5$ A, and the sample standard deviation is $s = 5.0$ A.

The sample mean $\overline{X} = 185.5$ is a point estimate of the population mean, which will not be exactly equal to the population mean. The population mean will actually be somewhat larger or smaller than 185.5. To get an idea how much larger or smaller it is likely to be, we construct a **confidence interval** around 185.5 that is likely to cover the population mean. We can then quantify our level of confidence that the population mean is actually covered by the interval. To see how to construct a confidence interval in this example, let μ represent the unknown population mean and let σ^2 represent the unknown population variance. Let $X_1, \ldots, X_{100}$ be the 100 amperages of the sample batteries. The observed value of the sample mean is $\overline{X} = 185.5$. Since $\overline{X}$ is the mean of a large sample, the Central Limit Theorem specifies that it comes from a normal distribution whose mean is μ and whose standard deviation is $\sigma_{\overline{X}} = \sigma/\sqrt{100}$.

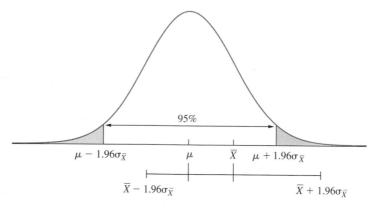

FIGURE 5.2 The sample mean $\overline{X}$ is drawn from a normal distribution with mean μ and standard deviation $\sigma_{\overline{X}} = \sigma/\sqrt{n}$. For this particular sample, $\overline{X}$ comes from the middle 95% of the distribution, so the 95% confidence interval $\overline{X} \pm 1.96\sigma_{\overline{X}}$ succeeds in covering the population mean μ.

Figure 5.2 presents a normal curve, which represents the distribution of $\overline{X}$. The middle 95% of the curve, extending a distance $1.96\sigma_{\overline{X}}$ on either side of population mean μ, is indicated. The observed value $\overline{X} = 185.5$ is a single draw from this distribution. We have no way to know from what part of the curve this particular value of $\overline{X}$ was drawn. Figure 5.2 presents one possibility, which is that the sample mean $\overline{X}$ lies within the middle 95% of the distribution. Ninety-five percent of all the samples that could have been drawn fall into this category. The horizontal line below the curve in Figure 5.2 is an interval around $\overline{X}$ that is exactly the same length as the middle 95% of the distribution, namely, the interval $\overline{X} \pm 1.96\sigma_{\overline{X}}$. This interval is a **95% confidence interval** for the population mean μ. It is clear that this interval covers the population mean μ.

In contrast, Figure 5.3 (page 178) represents a sample whose mean $\overline{X}$ lies outside the middle 95% of the curve. Only 5% of all the samples that could have been drawn fall into this category. For these more unusual samples, the 95% confidence interval $\overline{X} \pm 1.96\sigma_{\overline{X}}$ fails to cover the population mean μ.

We will now compute a 95% confidence interval $\overline{X} \pm 1.96\sigma_{\overline{X}}$ for the mean amperage. The value of $\overline{X}$ is 185.5. The population standard deviation σ and thus $\sigma_{\overline{X}} = \sigma/\sqrt{100}$ are unknown. However, in this example, since the sample size is large, we may approximate σ with the sample standard deviation $s = 5.0$. We therefore compute a 95% confidence interval for the population mean amperage μ to be $185.5 \pm (1.96)(5.0/\sqrt{100})$, or 185.5 ± 0.98, or $(184.52, 186.48)$. We can say that we are **95% confident**, or **confident at the 95% level**, that the population mean amperage lies between 184.52 and 186.48.

Does this 95% confidence interval actually cover the population mean μ? It depends on whether this particular sample happened to be one whose mean came from the middle 95% of the distribution, or whether it was a sample whose mean was unusually large or

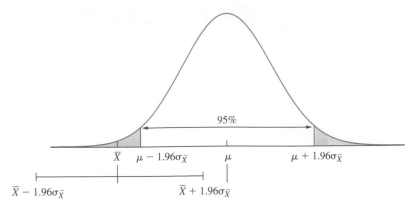

FIGURE 5.3 The sample mean $\overline{X}$ is drawn from a normal distribution with mean μ and standard deviation $\sigma_{\overline{X}} = \sigma/\sqrt{n}$. For this particular sample, $\overline{X}$ comes from the outer 5% of the distribution, so the 95% confidence interval $\overline{X} \pm 1.96\sigma_{\overline{X}}$ fails to cover the population mean μ.

small, in the outer 5% of the distribution. There is no way to know for sure into which category this particular sample falls. But imagine that the engineer were to repeat this procedure every day, drawing a large sample and computing the 95% confidence interval $\overline{X} \pm 1.96\sigma_{\overline{X}}$. In the long run, 95% of the samples he draws will have means in the middle 95% of the distribution, so 95% of the confidence intervals he computes will cover the population mean. The number 95% is called the **confidence level** of the confidence interval. The confidence level is the success rate of the procedure used to compute the confidence interval. Thus a 95% confidence interval is computed by a procedure that will succeed in covering the population mean for 95% of all the samples that might possibly be drawn.

> **Confidence level of a confidence interval**
> The confidence level is the proportion of all possible samples for which the confidence interval will cover the true value.

Let's look again at the 95% confidence interval 185.5 ± 0.98 to review how these numbers were computed. The value 185.5 is the sample mean $\overline{X}$, which is a point estimate for the population mean μ. We call the plus-or-minus number 0.98 the **margin of error**. The margin of error, 0.98, is the product of 1.96 and $\sigma_{\overline{X}}$, where $\sigma_{\overline{X}} = 0.5$. We refer to $\sigma_{\overline{X}}$, which is the standard deviation of $\overline{X}$, as the **standard error**. In general, the standard error is the standard deviation of the point estimator or an approximation of it. The number 1.96, which multiplies the standard error, is called the **critical value** for the confidence interval. The reason that 1.96 is the critical value for a 95% confidence interval is that 95% of the area under the normal curve is within -1.96 and 1.96 standard errors of the population mean μ.

> ### Summary
>
> Many confidence intervals follow the pattern just described. They have the form
>
> $$\text{point estimate} \pm \text{margin of error}$$
>
> where
>
> $$\text{margin of error} = (\text{critical value})(\text{standard error})$$

Once we have a point estimate and a standard error, we can compute an interval with any confidence level we like simply by choosing the appropriate critical value. For example, we can construct a 68% confidence interval for μ as follows. We know that the middle 68% of the normal curve corresponds to the interval extending a distance $1.0\sigma_{\overline{X}}$ on either side of the population mean μ. It follows that a critical value of 1.0 will produce a confidence interval that will cover the population mean for 68% of the samples that could possibly be drawn. Therefore a 68% confidence interval for the mean amperage of the batteries is $185.5 \pm (1.0)(0.5)$, or $(185.0, 186.0)$.

We now illustrate how to find a confidence interval with any desired level of confidence. Specifically, let α be a number between 0 and 1, and let $100(1-\alpha)\%$ denote the required confidence level. Figure 5.4 presents a normal curve representing the distribution of $\overline{X}$. Define $z_{\alpha/2}$ to be the z-score that cuts off an area of $\alpha/2$ in the right-hand tail. For example, the z table (Table A.2) indicates that $z_{.025} = 1.96$, since 2.5% of the area under the standard normal curve is to the right of 1.96. Similarly, the quantity $-z_{\alpha/2}$ cuts off an area of $\alpha/2$ in the left-hand tail. The middle $1-\alpha$ of the area under the curve corresponds to the interval $\mu \pm z_{\alpha/2}\sigma_{\overline{X}}$. By the reasoning shown in Figures 5.2 and 5.3, it follows that the interval $\overline{X} \pm z_{\alpha/2}\sigma_{\overline{X}}$ will cover the population mean μ for a proportion $1-\alpha$ of all the samples that could possibly be drawn. Therefore a **level $100(1-\alpha)\%$ confidence interval** for μ is obtained by choosing the critical value to be $z_{\alpha/2}$. The confidence interval is $\overline{X} \pm z_{\alpha/2}\sigma_{\overline{X}}$, or $\overline{X} \pm z_{\alpha/2}\sigma/\sqrt{n}$.

We note that even for large samples, the distribution of $\overline{X}$ is only *approximately* normal, rather than exactly normal. Therefore the levels stated for confidence intervals are approximate. When the sample size is large enough for the Central Limit Theorem

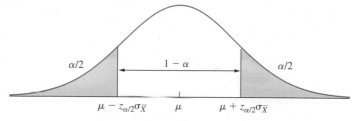

FIGURE 5.4 The sample mean $\overline{X}$ is drawn from a normal distribution with mean μ and standard deviation $\sigma_{\overline{X}} = \sigma/\sqrt{n}$. The quantity $z_{\alpha/2}$ is the z-score that cuts off an area of $\alpha/2$ in the right-hand tail. The quantity $-z_{\alpha/2}$ is the z-score that cuts off an area of $\alpha/2$ in the left-hand tail. The interval $\overline{X} \pm z_{\alpha/2}\sigma_{\overline{X}}$ will cover the population mean μ for a proportion $1-\alpha$ of all samples that could possibly be drawn. Therefore $\overline{X} \pm z_{\alpha/2}\sigma_{\overline{X}}$ is a level $100(1-\alpha)\%$ confidence interval for μ.

to be used, the distinction between approximate and exact levels is generally ignored in practice.

Summary

Let $X_1, \ldots, X_n$ be a *large* ($n > 30$) random sample from a population with mean μ and standard deviation σ, so that $\overline{X}$ is approximately normal. Then a level $100(1 - \alpha)\%$ confidence interval for μ is

$$\overline{X} \pm z_{\alpha/2}\sigma_{\overline{X}} \tag{5.4}$$

where $\sigma_{\overline{X}} = \sigma/\sqrt{n}$. When the value of σ is unknown, it can be replaced with the sample standard deviation s.

In particular,

- $\overline{X} \pm \dfrac{s}{\sqrt{n}}$ is a 68% confidence interval for μ.

- $\overline{X} \pm 1.645\dfrac{s}{\sqrt{n}}$ is a 90% confidence interval for μ.

- $\overline{X} \pm 1.96\dfrac{s}{\sqrt{n}}$ is a 95% confidence interval for μ.

- $\overline{X} \pm 2.58\dfrac{s}{\sqrt{n}}$ is a 99% confidence interval for μ.

- $\overline{X} \pm 3\dfrac{s}{\sqrt{n}}$ is a 99.7% confidence interval for μ.

Example 5.3

The sample mean and standard deviation for the cold cranking amperages of 100 batteries are $\overline{X} = 185.5$ and $s = 5.0$. Find an 85% confidence interval for the mean amperage of the batteries.

Solution

We must find the point estimate, the standard error, and the critical value. The point estimate is $\overline{X} = 185.5$. The standard error is $\sigma_{\overline{X}}$, which we approximate with $\sigma_{\overline{X}} \approx s/\sqrt{n} = 0.5$. To find the critical value for an 85% confidence interval, set $1 - \alpha = 0.85$ to obtain $\alpha = 0.15$ and $\alpha/2 = 0.075$. The critical value is $z_{.075}$, the z-score that cuts off 7.5% of the area in the right-hand tail. From the z table, we find $z_{.075} = 1.44$. The margin of error is therefore $(1.44)(0.5) = 0.72$. So the 85% confidence interval is 185.5 ± 0.72 or, equivalently, $(184.78, 186.22)$.

Example 5.4

The article "Study on the Life Distribution of Microdrills" (Z. Yang, Y. Chen, and Y. Yang, *Journal of Engineering Manufacture*, 2002:301–305) reports that in a sample of 50 microdrills drilling a low-carbon alloy steel, the average lifetime (expressed as the number of holes drilled before failure) was 12.68 with a standard deviation of 6.83. Find a 95% confidence interval for the mean lifetime of microdrills under these conditions.

Solution

First let's translate the problem into statistical language. We have a simple random sample $X_1, \ldots, X_{50}$ of lifetimes. The sample mean and standard deviation are $\overline{X} = 12.68$ and $s = 6.83$. The population mean is unknown, and denoted by μ. The confidence interval has the form $\overline{X} \pm z_{\alpha/2}\sigma_{\overline{X}}$, as specified in expression (5.4). Since we want a 95% confidence interval, the confidence level $1 - \alpha$ is equal to 0.95. Thus $\alpha = 0.05$, and the critical value is $z_{\alpha/2} = z_{.025} = 1.96$. We approximate σ with $s = 6.83$, and obtain the standard error $\sigma_{\overline{X}} \approx 6.83/\sqrt{50} = 0.9659$. Therefore the 95% confidence interval is $12.68 \pm (1.96)(0.9659)$. This can be written as 12.68 ± 1.89, or as $(10.79, \ 14.57)$.

The following computer output (from MINITAB) presents the 95% confidence interval calculated in Example 5.4.

```
One-Sample Z

The assumed standard deviation = 6.830000

 N      Mean    SE Mean           95% CI
50   12.680000  0.965908   (10.786821, 14.573179)
```

Most of the output is self-explanatory. The quantity labeled "SE Mean" is the standard error $\sigma_{\overline{X}}$, approximated by $s/\sqrt{n}$. ("SE Mean" stands for standard error of the mean.)

Example 5.5

Use the data in Example 5.4 to find an 80% confidence interval.

Solution

To find an 80% confidence interval, set $1-\alpha = 0.80$ to obtain $\alpha = 0.20$. The critical value is $z_{.10}$, the z-score that cuts off 10% of the area in the right-hand tail. From the z table, we find that this value is $z_{.10} = 1.28$. In Example 5.4, we found the point estimate to be $\overline{X} = 12.68$ and the standard error to be $\sigma_{\overline{X}} = 0.9659$. The 80% confidence interval is therefore $12.68 \pm (1.28)(0.9659)$. This can be written as 12.68 ± 1.24, or as $(11.44, 13.92)$.

We have seen how to compute a confidence interval with a given confidence level. It is also possible to compute the level of a given confidence interval. The next example illustrates the method.

Example 5.6

Based on the microdrill lifetime data presented in Example 5.4, an engineer reported a confidence interval of $(11.09, 14.27)$ but neglected to specify the level. What is the level of this confidence interval?

Solution

The confidence interval has the form $\overline{X} \pm z_{\alpha/2}s/\sqrt{n}$. We will solve for $z_{\alpha/2}$, then consult the z table to determine the value of α. Now $\overline{X} = 12.68$, $s = 6.83$, and $n = 50$. The upper confidence limit of 14.27 therefore satisfies the equation $14.27 = 12.68 + z_{\alpha/2}(6.83/\sqrt{50})$. It follows that $z_{\alpha/2} = 1.646$. From the z table, we determine that $\alpha/2$, the area to the right of 1.646, is approximately 0.05. The level is $100(1 - \alpha)\%$, or 90%.

More about Confidence Levels

The confidence level of an interval measures the reliability of the method used to compute the interval. A level $100(1 - \alpha)\%$ confidence interval is one computed by a method that in the long run will succeed in covering the population mean a proportion $1 - \alpha$ of all the times that it is used. Deciding what level of confidence to use involves a trade-off, because intervals with greater confidence levels have greater margins of error and therefore are less precise. For example, a 68% confidence interval specifies the population mean to within $\pm 1.0\sigma_{\overline{X}}$, while a 95% confidence interval specifies it only to within $\pm 1.96\sigma_{\overline{X}}$ and therefore has only about half the precision of the 68% confidence interval. Figure 5.5 illustrates the trade-off between confidence and precision. One hundred samples were drawn from a population with mean μ. Figure 5.5b presents one hundred 95% confidence intervals, each based on one of these samples. The confidence intervals are all different, because each sample has a different mean $\overline{X}$. (They also have different values of s with which to approximate σ, but this has a much smaller effect.) About 95% of these intervals cover the population mean μ. Figure 5.5a presents 68% confidence intervals based on the same samples. These intervals are more precise (narrower), but many of them fail to cover the population mean. Figure 5.5c presents 99.7% confidence intervals. These intervals are very reliable. In the long run, only 3 in 1000 of these intervals will fail to cover the population mean. However, they are less precise (wider) and thus do not convey as much information.

The level of confidence most often used in practice is 95%. For many applications, this level provides a good compromise between reliability and precision. Confidence levels below 90% are rarely used. For some quality assurance applications, where product reliability is extremely important, intervals with very high confidence levels, such as 99.7%, are used.

Probability versus Confidence

In the amperage example discussed at the beginning of this section, a 95% confidence interval for the population mean μ was computed to be (184.52, 186.48). It is tempting to say that the probability is 95% that μ is between 184.52 and 186.48. This, however, is not correct. The term *probability* refers to random events, which can come out differently when experiments are repeated. The numbers 184.52 and 186.48 are fixed, not random. The population mean is also fixed. The mean amperage is either in the interval 184.52 to 186.48 or it is not. There is no randomness involved. Therefore we say that we have 95% *confidence* (not probability) that the population mean is in this interval.

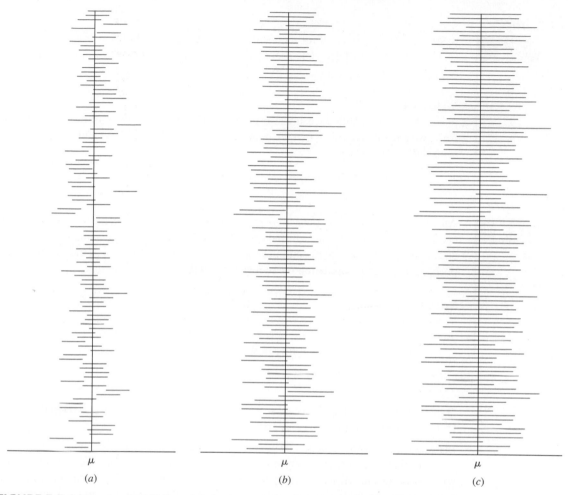

FIGURE 5.5 *(a)* One hundred 68% confidence intervals for a population mean, each computed from a different sample. Although precise, they fail to cover the population mean 32% of the time. This high failure rate makes the 68% confidence interval unacceptable for practical purposes. *(b)* One hundred 95% confidence intervals computed from these samples. This represents a good compromise between reliability and precision for many purposes. *(c)* One hundred 99.7% confidence intervals computed from these samples. These intervals fail to cover the population mean only three times in 1000. They are extremely reliable, but imprecise.

On the other hand, let's say that we are discussing a *method* used to compute a 95% confidence interval. The method will succeed in covering the population mean 95% of the time and fail the other 5% of the time. In this case, whether the population mean is covered or not is a random event, because it can vary from experiment to experiment. Therefore it *is* correct to say that a *method* for computing a 95% confidence interval has probability 95% of covering the population mean.

Example

A 90% confidence interval for the mean resistance (in Ω) of resistors is computed to be (1.43, 1.56). True or false: The probability is 90% that the mean resistance of this type of resistor is between 1.43 and 1.56.

Solution

False. A specific confidence interval is given. The mean is either in the interval or it isn't. We are 90% confident that the population mean is between 1.43 and 1.56. The term *probability* is inappropriate.

Example

5.8

An engineer plans to compute a 90% confidence interval for the mean resistance of a certain type of resistor. She will measure the resistances of a large sample of resistors, compute $\overline{X}$ and s, and then compute the interval $\overline{X} \pm 1.645s/\sqrt{n}$. True or false: The probability that the population mean resistance will be in this interval is 90%.

Solution

True. What is described here is a method for computing a confidence interval, rather than a specific numerical value. It is correct to say that a method for computing a 90% confidence interval has probability 90% of covering the population mean.

Determining the Sample Size Needed for a Confidence Interval of Specified Width

In Example 5.4, a 95% confidence interval was given by 12.68 ± 1.89, or (10.79, 14.57). This interval specifies the mean to within ± 1.89. Now assume that this interval is too wide to be useful. Assume that it is desirable to produce a 95% confidence interval that specifies the mean to within ± 0.50. To do this, the sample size must be increased. We show how to calculate the sample size needed to obtain a confidence interval of any specified width.

The width of a confidence interval is $\pm z_{\alpha/2}\sigma/\sqrt{n}$. If the desired width is denoted by $\pm w$, then $w = z_{\alpha/2}\sigma/\sqrt{n}$. Solving this equation for n yields $n = z_{\alpha/2}^2\sigma^2/w^2$. This equation may be used to find the sample size n needed to construct a level $100(1-\alpha)\%$ confidence interval of width $\pm w$.

Summary

The sample size n needed to construct a level $100(1-\alpha)\%$ confidence interval of width $\pm w$ is

$$n = \frac{z_{\alpha/2}^2\sigma^2}{w^2} \tag{5.5}$$

In the amperage example discussed earlier in this section, the sample standard deviation of amperages from 100 batteries was $s = 5.0\,\text{A}$. How many batteries must be sampled to obtain a 99% confidence interval of width $\pm 1.0\,\text{A}$?

Solution

The level is 99%, so $1 - \alpha = 0.99$. Therefore $\alpha = 0.01$ and $z_{\alpha/2} = 2.58$. The value of σ is estimated with $s = 5.0$. The necessary sample size is found by substituting these values, along with $w = 1.0$, into Equation (5.5). We obtain $n \approx 167$.

One-Sided Confidence Intervals

The confidence intervals discussed so far have been **two-sided**, in that they specify both a lower and an upper confidence bound. Occasionally we are interested only in one of these bounds. In these cases, one-sided confidence intervals are appropriate. For example, assume that a reliability engineer wants to estimate the mean crushing strength of a certain type of concrete block, to determine the sorts of applications for which it will be appropriate. The engineer will probably be interested only in a lower bound for the strength, since specifications for various applications will generally specify only a minimum strength.

Assume that a large sample has sample mean $\overline{X}$ and standard deviation $\sigma_{\overline{X}}$. Figure 5.6 shows how the idea behind the two-sided confidence interval can be adapted to produce a one-sided confidence interval for the population mean μ. The normal curve represents the distribution of $\overline{X}$. For 95% of all the samples that could be drawn, $\overline{X} < \mu + 1.645\sigma_{\overline{X}}$; therefore the interval $(\overline{X} - 1.645\sigma_{\overline{X}}, \infty)$ covers μ. This interval will fail to cover μ only if the sample mean is in the upper 5% of its distribution. The interval $(\overline{X} - 1.645\sigma_{\overline{X}}, \infty)$ is a 95% one-sided confidence interval for μ, and the quantity $\overline{X} - 1.645\sigma_{\overline{X}}$ is a 95% lower confidence bound for μ.

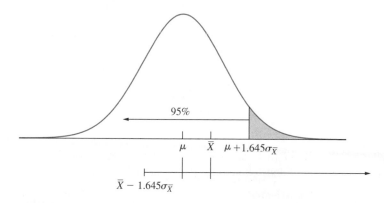

FIGURE 5.6 The sample mean $\overline{X}$ is drawn from a normal distribution with mean μ and standard deviation $\sigma_{\overline{X}} = \sigma/\sqrt{n}$. For this particular sample, $\overline{X}$ comes from the lower 95% of the distribution, so the 95% one-sided confidence interval $(\overline{X} - 1.645\sigma_{\overline{X}}, \infty)$ succeeds in covering the population mean μ.

By constructing a figure like Figure 5.6 with the lower 5% tail shaded, it can be seen that the quantity $\overline{X} + 1.645\sigma_{\overline{X}}$ is a 95% upper confidence bound for μ. We now generalize the method to produce one-sided confidence intervals of any desired level. Define z_α to be the z-score that cuts off an area α in the right-hand tail of the normal curve. For example, $z_{.05} = 1.645$. A level $100(1 - \alpha)\%$ lower confidence bound for μ is given by $\overline{X} - z_\alpha\sigma_{\overline{X}}$, and a level $1 - \alpha$ upper confidence bound for μ is given by $\overline{X} + z_\alpha\sigma_{\overline{X}}$.

Note that the point estimate $(\overline{X})$ and the standard error $(\sigma_{\overline{X}})$ are the same for both one- and two-sided confidence intervals. A one-sided interval differs from a two-sided interval in two ways. First, the critical value for a one-sided interval is z_α rather than $z_{\alpha/2}$. Second, we add the margin of error to the point estimate to obtain an upper confidence bound or subtract it to obtain a lower confidence bound, rather than both adding and subtracting as we do to obtain a two-sided confidence interval.

Summary

Let $X_1, \ldots, X_n$ be a *large* $(n > 30)$ random sample from a population with mean μ and standard deviation σ, so that $\overline{X}$ is approximately normal. Then level $100(1 - \alpha)\%$ lower confidence bound for μ is

$$\overline{X} - z_\alpha\sigma_{\overline{X}} \tag{5.6}$$

and level $100(1 - \alpha)\%$ upper confidence bound for μ is

$$\overline{X} + z_\alpha\sigma_{\overline{X}} \tag{5.7}$$

where $\sigma_{\overline{X}} = \sigma/\sqrt{n}$. When the value of σ is unknown, it can be replaced with the sample standard deviation s.

In particular,

■ $\overline{X} + 1.28\dfrac{s}{\sqrt{n}}$ is a 90% upper confidence bound for μ.

■ $\overline{X} + 1.645\dfrac{s}{\sqrt{n}}$ is a 95% upper confidence bound for μ.

■ $\overline{X} + 2.33\dfrac{s}{\sqrt{n}}$ is a 99% upper confidence bound for μ.

The corresponding lower bounds are found by replacing the "+" with "−."

Example 5.10

Refer to Example 5.4. Find both a 95% lower confidence bound and a 99% upper confidence bound for the mean lifetime of the microdrills.

Solution
The sample mean and standard deviation are $\overline{X} = 12.68$ and $s = 6.83$, respectively. The sample size is $n = 50$. We estimate $\sigma_{\overline{X}} \approx s/\sqrt{n} = 0.9659$. The 95% lower confidence bound is $\overline{X} - 1.645\sigma_{\overline{X}} = 11.09$, and the 99% upper confidence bound is $\overline{X} + 2.33\sigma_{\overline{X}} = 14.93$.

Exercises for Section 5.2

1. Find the value of $z_{\alpha/2}$ to use in expression (5.4) to construct a confidence interval with level

   a. 90%

   b. 83%

   c. 99.5%

   d. 75%

2. Find the levels of the confidence intervals that have the following values of $z_{\alpha/2}$.

   a. $z_{\alpha/2} = 1.96$

   b. $z_{\alpha/2} = 2.81$

   c. $z_{\alpha/2} = 2.17$

   d. $z_{\alpha/2} = 2.33$

3. As the confidence level goes up, the reliability goes _____, and the precision goes _____. *Options:* up, down.

4. Interpolation methods are used to estimate heights above sea level for locations where direct measurements are unavailable. In the article "Transformation of Ellipsoid Heights to Local Leveling Heights" (M. Yanalak and O. Baykal, *Journal of Surveying Engineering*, 2001:90–103), a second-order polynomial method of interpolation for estimating heights from GPS measurements is evaluated. In a sample of 74 locations, the errors made by the method averaged 3.8 cm, with a standard deviation of 4.8 cm.

   a. Find a 95% confidence interval for the mean error made by this method.

   b. Find a 98% confidence interval for the mean error made by this method.

   c. A surveyor claims that the mean error is between 3.2 and 4.4 cm. With what level of confidence can this statement be made?

   d. Approximately how many locations must be sampled so that a 95% confidence interval will specify the mean to within ± 0.7 cm?

   e. Approximately how many locations must be sampled so that a 98% confidence interval will specify the mean to within ± 0.7 cm?

5. The capacities (in ampere-hours) were measured for a sample of 120 batteries. The average was 178, and the standard deviation was 14.

   a. Find a 95% confidence interval for the mean capacity of batteries produced by this method.

   b. Find a 99% confidence interval for the mean capacity of batteries produced by this method.

   c. An engineer claims that the mean capacity is between 176 and 180 ampere-hours. With what level of confidence can this statement be made?

   d. Approximately how many batteries must be sampled so that a 95% confidence interval will specify the mean to within ± 2 ampere-hours?

   e. Approximately how many batteries must be sampled so that a 99% confidence interval will specify the mean to within ± 2 ampere-hours?

6. Resistance measurements were made on a sample of 81 wires of a certain type. The sample mean resistance was 17.3 mΩ, and the standard deviation was 1.2 mΩ.

   a. Find a 95% confidence interval for the mean resistance of this type of wire.

   b. Find a 98% confidence interval for the mean resistance of this type of wire.

   c. What is the level of the confidence interval (17.1, 17.5)?

   d. How many wires must be sampled so that a 98% confidence interval will specify the mean to within ± 0.1 mΩ?

   e. How many wires must be sampled so that a 95% confidence interval will specify the mean to within ± 0.1 mΩ?

7. In a sample of 100 boxes of a certain type, the average compressive strength was 6230 N, and the standard deviation was 221 N.

   a. Find a 95% confidence interval for the mean compressive strength of boxes of this type.

   b. Find a 99% confidence interval for the mean compressive strength of boxes of this type.

   c. An engineer claims that the mean strength is between 6205 and 6255 N. With what level of confidence can this statement be made?

   d. Approximately how many boxes must be sampled so that a 95% confidence interval will specify the mean to within ± 25 N?

e. Approximately how many boxes must be sampled so that a 99% confidence interval will specify the mean to within ±25 N?

8. Polychlorinated biphenyls (PCBs) are a group of synthetic oil-like chemicals that were at one time widely used as insulation in electrical equipment and were discharged into rivers. They were discovered to be a health hazard and were banned in the 1970s. Since then, much effort has gone into monitoring PCB concentrations in waterways. Suppose that a random sample of 63 water samples drawn from waterway has a sample mean of 1.69 ppb and a sample standard deviation of 0.25 ppb.

   a. Find a 90% confidence interval for the PCB concentration.

   b. Find a 95% confidence interval for the PCB concentration.

   c. An environmental scientist says that the PCB concentration is between 1.65 and 1.73 ppb. With what level of confidence can that statement be made?

   d. Estimate the sample size needed so that a 90% confidence interval will specify the population mean to within ±0.02 ppb.

   e. Estimate the sample size needed so that a 95% confidence interval will specify the population mean to within ±0.02 ppb.

9. In a sample of 80 ten-penny nails, the average weight was 1.56 g, and the standard deviation was 0.1 g.

   a. Find a 95% confidence interval for the mean weight of this type of nail.

   b. Find a 98% confidence interval for the mean weight of this type of nail.

   c. What is the confidence level of the interval (1.54, 1.58)?

   d. How many nails must be sampled so that a 95% confidence interval specifies the mean to within ±0.01 g?

   e. Approximately how many nails must be sampled so that a 98% confidence interval will specify the mean to within ±0.01 g?

10. One step in the manufacture of a certain metal clamp involves the drilling of four holes. In a sample of 150 clamps, the average time needed to complete this step was 72 seconds and the standard deviation was 10 seconds.

   a. Find a 95% confidence interval for the mean time needed to complete the step.

   b. Find a 99.5% confidence interval for the mean time needed to complete the step.

   c. What is the confidence level of the interval (71, 73)?

   d. How many clamps must be sampled so that a 95% confidence interval specifies the mean to within ±1.5 seconds?

   e. How many clamps must be sampled so that a 99.5% confidence interval specifies the mean to within ±1.5 seconds?

11. A supplier sells synthetic fibers to a manufacturing company. A simple random sample of 81 fibers is selected from a shipment. The average breaking strength of these is 29 lb, and the standard deviation is 9 lb.

   a. Find a 95% confidence interval for the mean breaking strength of all the fibers in the shipment.

   b. Find a 99% confidence interval for the mean breaking strength of all the fibers in the shipment.

   c. What is the confidence level of the interval (27.5, 30.5)?

   d. How many fibers must be sampled so that a 95% confidence interval specifies the mean to within ±1 lb?

   e. How many fibers must be sampled so that a 99% confidence interval specifies the mean to within ±1 lb?

12. Refer to Exercise 5.

   a. Find a 95% lower confidence bound for the mean capacity of this type of battery.

   b. An engineer claims that the mean capacity is greater than 175 hours. With what level of confidence can this statement be made?

13. Refer to Exercise 6.

   a. Find a 98% lower confidence bound for the mean resistance.

   b. An engineer says that the mean resistance is greater than 17 mΩ. With what level of confidence can this statement be made?

14. Refer to Exercise 7.

   a. Find a 95% lower confidence bound for the mean compressive strength of this type of box.

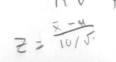

b. An engineer claims that the mean compressive strength is greater than 6220 N. With what level of confidence can this statement be made?

15. Refer to Exercise 8.

   a. Find a 99% upper confidence bound for the concentration.

   b. The claim is made that the concentration is less than 1.75 ppb. With what level of confidence can this statement be made?

16. Refer to Exercise 9.

   a. Find a 90% upper confidence bound for the mean weight.

   b. Someone says that the mean weight is less than 1.585 g. With what level of confidence can this statement be made?

17. Refer to Exercise 10.

   a. Find a 98% lower confidence bound for the mean time to complete the step.

   b. An efficiency expert says that the mean time is greater than 70 seconds. With what level of confidence can this statement be made?

18. Refer to Exercise 11.

   a. Find a 95% upper confidence bound for the mean breaking strength.

   b. The supplier claims that the mean breaking strength is greater than 28 lb. With what level of confidence can this statement be made?

5.3 Confidence Intervals for Proportions

The methods of Section 5.2, in particular expression (5.4), can be used to find a confidence interval for the mean of any population from which a large sample has been drawn. In this section, we show how to adapt these methods to construct confidence intervals for population proportions. We illustrate with an example.

In Example 5.4 (in Section 5.2), a confidence interval was constructed for the mean lifetime of a certain type of microdrill when drilling a low-carbon alloy steel. Now assume that a specification has been set that a drill should have a minimum lifetime of 10 holes drilled before failure. A sample of 144 microdrills is tested, and 120, or 83.3%, meet this specification. Let p represent the proportion of microdrills in the population that will meet the specification. We wish to find a 95% confidence interval for p. To do this we need a point estimate, a critical value, and a standard error.

There are two methods for constructing a point estimate: a traditional one that has a long history, and a more modern one that has been found to be somewhat more accurate. It's better to use the modern approach, but it's important to understand the traditional one as well, since it is still often used.

We begin with the traditional approach. To construct a point estimate for p, let X represent the number of drills in the sample that meet the specification. Then $X \sim \text{Bin}(n, p)$, where $n = 144$ is the sample size. The estimate for p is $\hat{p} = X/n$. In this example, $X = 120$, so $\hat{p} = 120/144 = 0.833$.

Since the sample size is large, it follows from the Central Limit Theorem (Equation 4.35 in Section 4.8) that

$$X \sim N\left(np,\ np(1-p)\right)$$

Since $\hat{p} = X/n$, it follows that

$$\hat{p} \sim N\left(p,\ \frac{p(1-p)}{n}\right)$$

In particular, the standard error of $\hat{p}$ is $\sigma_{\hat{p}} = \sqrt{p(1-p)/n}$. We can't use this value in the confidence interval because it contains the unknown p. The traditional approach is to replace p with $\hat{p}$, obtaining $\sqrt{\hat{p}(1-\hat{p})/n}$. Since the point estimate $\hat{p}$ is approximately normal, the critical value for a 95% confidence interval is 1.96 (obtained from the z table), so the traditional 95% confidence interval is $\hat{p} \pm 1.96\sqrt{\hat{p}(1-\hat{p})/n}$.

Recent research, involving simulation studies, has shown that this interval can be improved by modifying both n and $\hat{p}$ slightly. Specifically, one should add 4 to the number of trials and 2 to the number of successes. So in place of n we use $\tilde{n} = n + 4$, and in place of $\hat{p}$ we use $\tilde{p} = (X + 2)/\tilde{n}$. A 95% confidence interval for p is thus given by $\tilde{p} \pm 1.96\sqrt{\tilde{p}(1-\tilde{p})/\tilde{n}}$. In this example, $\tilde{n} = 148$ and $\tilde{p} = 122/148 = 0.8243$, so the 95% confidence interval is 0.8243 ± 0.0613, or $(0.763, 0.886)$.

We justified this confidence interval on the basis of the Central Limit Theorem, which requires n to be large. However, this method of computing confidence intervals is appropriate for any sample size n. When used with small samples, it may occasionally happen that the lower limit is less than 0 or that the upper limit is greater than 1. Since $0 < p < 1$, a lower limit less than 0 should be replaced with 0, and an upper limit greater than 1 should be replaced with 1.

Summary

Let X be the number of successes in n independent Bernoulli trials with success probability p, so that $X \sim \text{Bin}(n, p)$.

Define $\tilde{n} = n + 4$, and $\tilde{p} = \dfrac{X + 2}{\tilde{n}}$. Then a level $100(1 - \alpha)\%$ confidence interval for p is

$$\tilde{p} \pm z_{\alpha/2}\sqrt{\frac{\tilde{p}(1 - \tilde{p})}{\tilde{n}}} \tag{5.8}$$

If the lower limit is less than 0, replace it with 0. If the upper limit is greater than 1, replace it with 1.

The confidence interval given by expression (5.8) is sometimes called the *Agresti–Coull* interval, after Alan Agresti and Brent Coull, who developed it. For more information on this confidence interval, consult the article "Approximate Is Better Than 'Exact' for Interval Estimation of Binomial Proportions" (A. Agresti and B. Coull, *The American Statistician*, 1998:119–126).

*E*xample
5.11

Interpolation methods are used to estimate heights above sea level for locations where direct measurements are unavailable. In the article "Transformation of Ellipsoid Heights to Local Leveling Heights" (M. Yanalak and O. Baykal, *Journal of Surveying Engineering*, 2001:90–103), a weighted-average method of interpolation for estimating heights from GPS measurements is evaluated. The method made "large" errors (errors whose magnitude was above a commonly accepted threshold) at 26 of the 74 sample test locations. Find a 90% confidence interval for the proportion of locations at which this method will make large errors.

Solution
The number of successes is $X = 26$, and the number of trials is $n = 74$. We therefore compute $\tilde{n} = 74 + 4 = 78$, $\tilde{p} = (26 + 2)/78 = 0.3590$, and $\sqrt{\tilde{p}(1 - \tilde{p})/\tilde{n}} = \sqrt{(0.3590)(0.6410)/78} = 0.0543$. For a 90% confidence interval, the value of $\alpha/2$ is 0.05, so $z_{\alpha/2} = 1.645$. The 90% confidence interval is therefore $0.3590 \pm (1.645)(0.0543)$, or $(0.270, 0.448)$.

One-sided confidence intervals can be computed for proportions as well. They are analogous to the one-sided intervals for a population mean (Equations 5.6 and 5.7 in Section 5.2). The levels for one-sided confidence intervals are only roughly approximate for small samples.

Summary

Let X be the number of successes in n independent Bernoulli trials with success probability p, so that $X \sim \text{Bin}(n, p)$.

Define $\tilde{n} = n + 4$, and $\tilde{p} = \dfrac{X + 2}{\tilde{n}}$. Then a level $100(1 - \alpha)\%$ lower confidence bound for p is

$$\tilde{p} - z_{\alpha}\sqrt{\frac{\tilde{p}(1 - \tilde{p})}{\tilde{n}}} \qquad (5.9)$$

and level $100(1 - \alpha)\%$ upper confidence bound for p is

$$\tilde{p} + z_{\alpha}\sqrt{\frac{\tilde{p}(1 - \tilde{p})}{\tilde{n}}} \qquad (5.10)$$

If the lower bound is less than 0, replace it with 0. If the upper bound is greater than 1, replace it with 1.

Determining the Sample Size Needed for a Confidence Interval of Specified Width

The width of a confidence interval for a proportion is $\pm z_{\alpha/2}\sqrt{\tilde{p}(1 - \tilde{p})/\tilde{n}}$. If the desired width is denoted by $\pm w$, then $w = z_{\alpha/2}\sqrt{\tilde{p}(1 - \tilde{p})/\tilde{n}}$. Solving this equation for n yields $n = z_{\alpha/2}^2 \tilde{p}(1 - \tilde{p})/w^2$. If a preliminary value of $\tilde{p}$ is known, this equation may be used to find the approximate sample size n needed to construct a level $100(1 - \alpha)\%$ confidence interval of width $\pm w$. If no value of $\tilde{p}$ is available, we may set $\tilde{p} = 0.5$, to obtain $n = z_{\alpha/2}^2/4w^2$. The reason for setting $\tilde{p} = 0.5$ is that the quantity $\tilde{p}(1 - \tilde{p})$ is maximized when $\tilde{p} = 0.5$. Therefore the value of n obtained in this way is conservative, in that it is guaranteed to produce a width no greater than w.

Summary

The sample size n needed to construct a level $100(1 - \alpha)\%$ confidence interval of width $\pm w$ is

$$n = \frac{z_{\alpha/2}^2 \tilde{p}(1 - \tilde{p})}{w^2} \quad \text{if an estimate of } \tilde{p} \text{ is available} \tag{5.11}$$

$$n = \frac{z_{\alpha/2}^2}{4w^2} \quad \text{if no estimate of } \tilde{p} \text{ is available} \tag{5.12}$$

Example 5.12

In Example 5.11, what sample size is needed to obtain a 95% confidence interval with width ± 0.08?

Solution

Since the desired level is 95%, the value of $z_{\alpha/2}^2$ is 1.96. From the data in Example 5.11, $\tilde{p} = 0.3590$. Substituting these values, along with $w = 0.08$ into Equation 5.11, we obtain $n \approx 135$.

Example 5.13

In Example 5.11, how large a sample is needed to guarantee that the width of the 95% confidence interval will be no greater than ± 0.08, if no preliminary sample has been taken?

Solution

Since the desired level is 95%, the value of $z_{\alpha/2}^2$ is 1.96. We have no preliminary value for $\tilde{p}$, so we substitute $z_{\alpha/2}^2 = 1.96$ and $w = 0.08$ into Equation 5.12, obtaining $n \approx 147$. Note that this estimate is somewhat larger than the one obtained in Example 5.12.

The Traditional Method

The method we recommend was developed quite recently (although it was created by simplifying a much older method). Many people still use a more traditional method, which we described earlier in this section. The traditional method uses the actual sample size n in place of $\tilde{n}$, and the actual sample proportion $\hat{p}$ in place of $\tilde{p}$. Although this method is still widely used, it fails to achieve its stated coverage probability even for some fairly large values of n. This means that $100(1 - \alpha)\%$ confidence intervals computed by the traditional method will cover the true proportion less than $100(1 - \alpha)\%$ of the time. The traditional method does not work at all for small samples; one rule of thumb regarding the sample size is that both $n\hat{p}$ (the number of successes) and $n(1 - \hat{p})$ (the number of failures) should be greater than 10.

Since the traditional method is still widely used, we summarize it in the following box. For very large sample sizes, the results of the traditional method are almost identical to those of the modern method. For small or moderately large sample sizes, the modern approach is better.

> ## Summary
>
> **The Traditional Method for Computing Confidence Intervals for a Proportion** (widely used but not recommended)
>
> Let $\hat{p}$ be the proportion of successes in a *large* number n of independent Bernoulli trials with success probability p. Then the traditional level $100(1-\alpha)\%$ confidence interval for p is
>
> $$\hat{p} \pm z_{\alpha/2}\sqrt{\frac{\hat{p}(1-\hat{p})}{n}} \qquad (5.13)$$
>
> The method must not be used unless the sample contains at least 10 successes and 10 failures.

Exercises for Section 5.3

1. In a simple random sample of 70 automobiles registered in a certain state, 28 of them were found to have emission levels that exceed a state standard.

   a. What proportion of the automobiles in the sample had emission levels that exceed the standard?

   b. Find a 95% confidence interval for the proportion of automobiles in the state whose emission levels exceed the standard.

   c. Find a 98% confidence interval for the proportion of automobiles whose emission levels exceed the standard.

   d. How many automobiles must be sampled to specify the proportion that exceed the standard to within ± 0.10 with 95% confidence?

   e. How many automobiles must be sampled to specify the proportion that exceed the standard to within ± 0.10 with 98% confidence?

2. During a recent drought, a water utility in a certain town sampled 100 residential water bills and found that 73 of the residences had reduced their water consumption over that of the previous year.

   a. Find a 95% confidence interval for the proportion of residences that reduced their water consumption.

   b. Find a 99% confidence interval for the proportion of residences that reduced their water consumption.

   c. Find the sample size needed for a 95% confidence interval to specify the proportion to within ± 0.05.

   d. Find the sample size needed for a 99% confidence interval to specify the proportion to within ± 0.05.

3. Leakage from underground fuel tanks has been a source of water pollution. In a random sample of 87 gasoline stations, 13 were found to have at least one leaking underground tank.

   a. Find a 95% confidence interval for the proportion of gasoline stations with at least one leaking underground tank.

   b. Find a 90% confidence interval for the proportion of gasoline stations with at least one leaking underground tank.

   c. How many stations must be sampled so that a 95% confidence interval specifies the proportion to within ± 0.04?

   d. How many stations must be sampled so that a 90% confidence interval specifies the proportion to within ± 0.04?

4. In a random sample of 150 households with an Internet connection, 32 said that they had changed their Internet service provider within the past six months.

   a. Find a 95% confidence interval for the proportion of customers who changed their Internet service provider within the past six months.

   b. Find a 99% confidence interval for the proportion of customers who changed their Internet service provider within the past six months.

   c. Find the sample size needed for a 95% confidence interval to specify the proportion to within ± 0.05.

d. Find the sample size needed for a 99% confidence interval to specify the proportion to within ± 0.05.

5. The article "The Functional Outcomes of Total Knee Arthroplasty" (R. Kane, K. Saleh, et al., *Journal of Bone and Joint Surgery,* 2005:1719–1724) reports that out of 10,501 surgeries, 859 resulted in complications within six months of surgery.

   a. Find a 95% confidence interval for the proportion of surgeries that result in complications within six months.

   b. Find a 99% confidence interval for the proportion of surgeries that result in complications within six months.

   c. A surgeon claims that the rate of complications is less than 8.5%. With what level of confidence can this claim be made?

6. Refer to Exercise 1. Find a 95% lower confidence bound for the proportion of automobiles whose emissions exceed the standard.

7. Refer to Exercise 2. Find a 98% upper confidence bound for the proportion of residences that reduced their water consumption.

8. Refer to Exercise 3. Find a 99% lower confidence bound for the proportion of gasoline stations with at least one leaking underground tank.

9. The article "Leachate from Land Disposed Residential Construction Waste" (W. Weber, Y. Jang, et al., *Journal of Environmental Engineering*, 2002:237–245) presents a study of contamination at landfills containing construction and demolition waste. Specimens of leachate were taken from a test site. Out of 42 specimens, 26 contained detectable levels of lead, 41 contained detectable levels of arsenic, and 32 contained detectable levels of chromium.

   a. Find a 90% confidence interval for the probability that a specimen will contain a detectable level of lead.

   b. Find a 95% confidence interval for the probability that a specimen will contain a detectable level of arsenic.

   c. Find a 99% confidence interval for the probability that a specimen will contain a detectable level of chromium.

10. A voltmeter is used to record 100 independent measurements of a known standard voltage. Of the 100 measurements, 85 are within 0.01 V of the true voltage.

    a. Find a 95% confidence interval for the probability that a measurement is within 0.01 V of the true voltage.

    b. Find a 98% confidence interval for the probability that a measurement is within 0.01 V of the true voltage.

    c. Find the sample size needed for a 95% confidence interval to specify the probability to within ± 0.05.

    d. Find the sample size needed for a 98% confidence interval to specify the probability to within ± 0.05.

11. A sociologist is interested in surveying workers in computer-related jobs to estimate the proportion of such workers who have changed jobs within the past year.

    a. In the absence of preliminary data, how large a sample must be taken to ensure that a 95% confidence interval will specify the proportion to within ± 0.05?

    b. In a sample of 100 workers, 20 of them had changed jobs within the past year. Find a 95% confidence interval for the proportion of workers who have changed jobs within the past year.

    c. Based on the data in part (b), estimate the sample size needed so that the 95% confidence interval will specify the proportion to within ± 0.05.

12. Stainless steels can be susceptible to stress corrosion cracking under certain conditions. A materials engineer is interested in determining the proportion of steel alloy failures that are due to stress corrosion cracking.

    a. In the absence of preliminary data, how large a sample must be taken so as to be sure that a 98% confidence interval will specify the proportion to within ± 0.05?

    b. In a sample of 200 failures, 30 of them were caused by stress corrosion cracking. Find a 98% confidence interval for the proportion of failures caused by stress corrosion cracking.

    c. Based on the data in part (b), estimate the sample size needed so that the 98% confidence interval will specify the proportion to within ± 0.05.

13. For major environmental remediation projects to be successful, they must have public support. The article "Modelling the Non-Market Environmental Costs and Benefits of Biodiversity Using Contingent Value Data" (D. Macmillan, E. Duff, and D. Elston, *Environmental and Resource Economics*, 2001:391–410) reported the results of a survey in which Scottish voters were asked whether they would be willing to pay additional taxes in order to restore the Affric forest. Out of 189 who responded, 61 said they would be willing to pay.

    a. Assuming that the 189 voters who responded constitute a random sample, find a 90% confidence interval for the proportion of voters who would be willing to pay to restore the Affric forest.

    b. How many voters should be sampled to specify the proportion to within ±0.03 with 90% confidence?

    c. Another survey is planned, in which voters will be asked whether they would be willing to pay in order to restore the Strathspey forest. At this point, no estimate of this proportion is available. Find a conservative estimate of the sample size needed so that the proportion will be specified to within ±0.03 with 90% confidence.

14. A stock market analyst notices that in a certain year, the price of IBM stock increased on 131 out of 252 trading days. Can these data be used to find a 95% confidence interval for the proportion of days that IBM stock increases? Explain.

5.4 Small-Sample Confidence Intervals for a Population Mean

The methods described in Section 5.2 for computing confidence intervals for a population mean require that the sample size be large. When the sample size is small, there are no good general methods for finding confidence intervals. However, when the population is approximately normal, a probability distribution called the Student's t distribution can be used to compute confidence intervals for a population mean. In this section, we describe this distribution and show how to use it.

The Student's t Distribution

If $\overline{X}$ is the mean of a large sample of size n from a population with mean μ and variance σ^2, then the Central Limit Theorem specifies that $\overline{X} \sim N(\mu, \sigma^2/n)$. The quantity $(\overline{X} - \mu)/(\sigma/\sqrt{n})$ then has a normal distribution with mean 0 and variance 1. In addition, the sample standard deviation s will almost certainly be close to the population standard deviation σ. For this reason, the quantity $(\overline{X} - \mu)/(s/\sqrt{n})$ is approximately normal with mean 0 and variance 1, so we can look up probabilities pertaining to this quantity in the standard normal table (z table). This enables us to compute confidence intervals of various levels for the population mean μ, as discussed in Section 5.2.

What can we do if $\overline{X}$ is the mean of a *small* sample? If the sample size is small, s may not be close to σ, and $\overline{X}$ may not be approximately normal. However, if the population is approximately normal, then $\overline{X}$ will be approximately normal even when the sample size is small. It turns out that we can still use the quantity $(\overline{X} - \mu)/(s/\sqrt{n})$, but since s is not necessarily close to σ, this quantity will not have a normal distribution. Instead, it has the **Student's t distribution** with $n - 1$ degrees of freedom, which we denote t_{n-1}. The number of degrees of freedom for the t distribution is one less than the sample size.

The Student's t distribution was discovered in 1908 by William Sealy Gosset, a statistician who worked for the Guinness Brewing Company in Dublin, Ireland. The management at Guinness considered the discovery to be proprietary information and forbade Gosset to publish it under his own name, so that their competitors wouldn't realize how useful the results could be. Gosset did publish it, using the pseudonym "Student." He had done this before; see Section 4.2.

Summary

Let $X_1, \ldots, X_n$ be a *small* (e.g., $n < 30$) sample from a *normal* population with mean μ. Then the quantity

$$\frac{\overline{X} - \mu}{s/\sqrt{n}}$$

has a Student's t distribution with $n - 1$ degrees of freedom, denoted t_{n-1}.

   When n is large, the distribution of the quantity $(\overline{X} - \mu)/(s/\sqrt{n})$ is very close to normal, so the normal curve can be used, rather than the Student's t.

The probability density function of the Student's t distribution is different for different degrees of freedom. Figure 5.7 presents plots of the probability density function for several choices of degrees of freedom. The curves all have a shape similar to that of the normal, or z, curve with mean 0 and standard deviation 1. The t curves are more spread out, however. When there are only one or two degrees of freedom, the t curve is much more spread out than the normal. When there are more degrees of freedom, the sample size is larger, and s tends to be closer to σ. Then the difference between the t curve and the normal curve is not great.

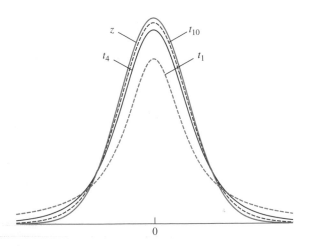

FIGURE 5.7 Plots of the probability density function of the Student's t curve for various degrees of freedom. The normal curve with mean 0 and variance 1 (z curve) is plotted for comparison. The t curves are more spread out than the normal, but the amount of extra spread decreases as the number of degrees of freedom increases.

Table A.3 (in Appendix A) called a *t* **table**, provides probabilities associated with the Student's *t* distribution. We present some examples to show how to use the table.

Example 5.14

A random sample of size 10 is to be drawn from a normal distribution with mean 4. The Student's t statistic $t = (\overline{X} - 4)/(s/\sqrt{10})$ is to be computed. What is the probability that $t > 1.833$?

Solution

This t statistic has $10 - 1 = 9$ degrees of freedom. From the t table, $P(t > 1.833) = 0.05$. See Figure 5.8.

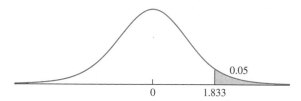

FIGURE 5.8 Solution to Example 5.14.

Example 5.15

Refer to Example 5.14. Find $P(t > 1.5)$.

Solution

Looking across the row corresponding to 9 degrees of freedom, we see that the t table does not list the value 1.5. We find that $P(t > 1.383) = 0.10$ and $P(t > 1.833) = 0.05$. We conclude that $0.05 < P(t > 1.5) < 0.10$. See Figure 5.9. A computer package gives the answer correct to three significant digits as 0.0839.

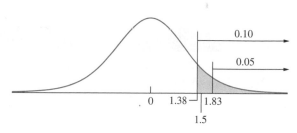

FIGURE 5.9 Solution to Example 5.15.

Example 5.16

Find the value for the t_{12} distribution whose upper-tail probability is 0.025.

Solution

Look down the column headed "0.025" to the row corresponding to 12 degrees of freedom. The value for t_{12} is 2.179.

xample
5.17

Find the value for the t_{14} distribution whose lower-tail probability is 0.01.

Solution
Look down the column headed "0.01" to the row corresponding to 14 degrees of freedom. The value for t_{14} is 2.624. This value cuts off an area, or probability, of 1% in the upper tail. The value whose lower-tail probability is 1% is -2.624.

Don't Use the Student's *t* Distribution If the Sample Contains Outliers

For the Student's t distribution to be valid, the sample must come from a population that is approximately normal. Such samples rarely contain outliers. Therefore, methods involving the Student's t distribution should *not* be used for samples that contain outliers.

Confidence Intervals Using the Student's *t* Distribution

When the sample size is small, and the population is approximately normal, we can use the Student's t distribution to compute confidence intervals. The confidence interval in this situation is constructed much like the ones in Section 5.2, except that the z-score is replaced with a value from the Student's t distribution.

To be specific, let $X_1, \ldots, X_n$ be a small random sample from an approximately normal population. To construct a level $100(1-\alpha)\%$ confidence interval or the population mean μ, we use the point estimate $\overline{X}$ and approximate the standard error $\sigma_{\overline{X}} = \sigma/\sqrt{n}$ with $s/\sqrt{n}$. The critical value is the $1 - \alpha/2$ quantile of the Student's t distribution with $n - 1$ degrees of freedom—that is, the value that cuts off an area of $\alpha/2$ in the upper tail. We denote this critical value by $t_{n-1,\alpha/2}$. For example, in Example 5.16 we found that $t_{12,.025} = 2.179$. A level $100(1 - \alpha)\%$ confidence interval for the population mean μ is $\overline{X} - t_{n-1,\alpha/2}(s/\sqrt{n}) < \mu < \overline{X} + t_{n-1,\alpha/2}(s/\sqrt{n})$, or $\overline{X} \pm t_{n-1,\alpha/2}(s/\sqrt{n})$.

Summary

Let $X_1, \ldots, X_n$ be a *small* random sample from a *normal* population with mean μ. Then a level $100(1 - \alpha)\%$ confidence interval for μ is

$$\overline{X} \pm t_{n-1,\alpha/2}\frac{s}{\sqrt{n}} \tag{5.14}$$

How Do I Determine Whether the Student's *t* Distribution Is Appropriate?

The Student's t distribution is appropriate whenever the sample comes from a population that is approximately normal. Sometimes one knows from past experience whether a process produces data that are approximately normally distributed. In many cases, however, one must decide whether a population is approximately normal by examining the sample. Unfortunately, when the sample size is small, departures from normality may be hard to detect. A reasonable way to proceed is to construct a boxplot or dotplot of the sample. If these plots do not reveal a strong asymmetry or any outliers, then in most cases the Student's t distribution will be reliable. In principle, one can also determine whether a population is approximately normal by constructing a probability plot. With small samples, however, boxplots and dotplots are easier to draw, especially by hand.

Example
5.18

The article "Direct Strut-and-Tie Model for Prestressed Deep Beams" (K. Tan, K. Tong, and C. Tang, *Journal of Structural Engineering*, 2001:1076–1084) presents measurements of the nominal shear strength (in kN) for a sample of 15 prestressed concrete beams. The results are

580	400	428	825	850	875	920	550
575	750	636	360	590	735	950	

Is it appropriate to use the Student's t distribution to construct a 99% confidence interval for the mean shear strength? If so, construct the confidence interval. If not, explain why not.

Solution

To determine whether the Student's t distribution is appropriate, we will make a boxplot and a dotplot of the sample. These are shown in the following figure.

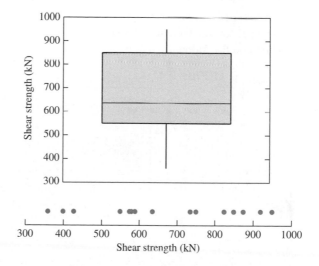

There is no evidence of a major departure from normality; in particular, the plots are not strongly asymmetric, and there are no outliers. The Student's t method is appropriate. We therefore compute $\overline{X} = 668.27$ and $s = 192.089$. We use expression (5.14) with $n = 15$ and $\alpha/2 = 0.005$. From the t table with 14 degrees of freedom, we find $t_{14,.005} = 2.977$. The 99% confidence interval is $668.27 \pm (2.977)(192.089)/\sqrt{15}$, or $(520.62, 815.92)$.

The following computer output (from MINITAB) presents the confidence interval calculated in Example 5.18.

```
One-Sample T: Strength

Test of mu = 0 vs not = 0

Variable    N      Mean      StDev   SE Mean          99% CI
Strength   15   668.2667   192.0891  49.59718   (520.6159, 815.9175)
```

Most of the output is self-explanatory. The quantity labeled "SE Mean" is the standard error $s/\sqrt{n}$.

In the article referred to in Example 5.18, cylindrical compressive strength (in MPa) was measured for 11 beams. The results were

38.43   38.43   38.39   38.83   38.45   38.35   38.43   38.31   38.32   38.48   38.50

Is it appropriate to use the Student's t distribution to construct a 95% confidence interval for the mean cylindrical compressive strength? If so, construct the confidence interval. If not, explain why not.

Solution

As in Example 5.18, we will make a boxplot and a dotplot of the sample. These are shown in the following figure.

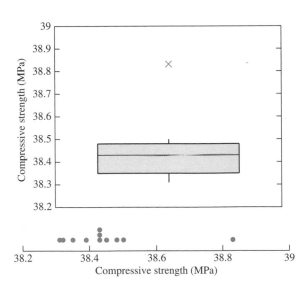

There is an outlier in this sample. The Student's t distribution should not be used.

The Student's t distribution can be used to compute one-sided confidence intervals. The formulas are analogous to those used with large samples.

Let $X_1, \ldots, X_n$ be a *small* random sample from a *normal* population with mean μ. Then a level $100(1 - \alpha)\%$ upper confidence bound for μ is

$$\overline{X} + t_{n-1,\alpha} \frac{s}{\sqrt{n}} \tag{5.15}$$

and a level $100(1 - \alpha)\%$ lower confidence bound for μ is

$$\overline{X} - t_{n-1,\alpha} \frac{s}{\sqrt{n}} \tag{5.16}$$

Use z, Not t, If σ Is Known

Occasionally a small sample may be taken from a normal population whose standard deviation σ is known. In these cases, we do not use the Student's t curve, because we are not approximating σ with s. Instead, we use the z table. Example 5.20 illustrates the method.

Example 5.20

Refer to Example 5.18. Assume that on the basis of a very large number of previous measurements of other beams, the population of shear strengths is known to be approximately normal, with standard deviation $\sigma = 180.0$ kN. Find a 99% confidence interval for the mean shear strength.

Solution
We compute $\overline{X} = 668.27$. We do not need to compute s, because we know the population standard deviation σ. Since we want a 99% confidence interval, $\alpha/2 = 0.005$. Because we know σ, we use $z_{\alpha/2} = z_{.005}$, rather than a Student's t value, to compute the confidence interval. From the z table, we obtain $z_{.005} = 2.58$. The confidence interval is $668.27 \pm (2.58)(180.0)/\sqrt{15}$, or $(548.36, 788.18)$.

It is important to remember that when the sample size is small, the population must be approximately normal, whether or not the standard deviation is known.

Summary

Let $X_1, \ldots, X_n$ be a random sample (of any size) from a *normal* population with mean μ. If the standard deviation σ is known, then a level $100(1 - \alpha)\%$ confidence interval for μ is

$$\overline{X} \pm z_{\alpha/2} \frac{\sigma}{\sqrt{n}} \tag{5.17}$$

Occasionally one has a single value that is sampled from a normal population with known standard deviation. In these cases a confidence interval for μ can be derived as a special case of expression (5.17) by setting $n = 1$.

Summary

Let X be a single value sampled from a *normal* population with mean μ. If the standard deviation σ is known, then a level $100(1 - \alpha)\%$ confidence interval for μ is

$$X \pm z_{\alpha/2}\sigma \qquad (5.18)$$

Exercises for Section 5.4

1. Find the value of $t_{n-1,\alpha/2}$ needed to construct a two-sided confidence interval of the given level with the given sample size:

a. Level 90%, sample size 9.

b. Level 95%, sample size 5.

c. Level 99%, sample size 29.

d. Level 95%, sample size 2.

2. Find the value of $t_{n-1,\alpha}$ needed to construct an upper or lower confidence bound in each of the situations in Exercise 1.

3. Find the level of a two-sided confidence interval that is based on the given value of $t_{n-1,\alpha/2}$ and the given sample size.

a. $t = 2.179$, sample size 13.

b. $t = 3.365$, sample size 6.

c. $t = 1.729$, sample size 20.

d. $t = 3.707$, sample size 7.

e. $t = 3.707$, sample size 27.

4. Five measurements are taken of the octane rating for a particular type of gasoline. The results (in %) are 87.0, 86.0, 86.5, 88.0, 85.3. Find a 99% confidence interval for the mean octane rating for this type of gasoline.

5. A model of heat transfer from a cylinder immersed in a liquid predicts that the heat transfer coefficient for the cylinder will become constant at very low flow rates of the fluid. A sample of 10 measurements is taken. The results, in W/m^2 K, are

13.7	12.0	13.1	14.1	13.1
14.1	14.4	12.2	11.9	11.8

Find a 95% confidence interval for the heat transfer coefficient.

6. A chemist made eight independent measurements of the melting point of tungsten. She obtained a sample mean of 3410.14°C and a sample standard deviation of 1.018°C.

a. Use the Student's t distribution to find a 95% confidence interval for the melting point of tungsten.

b. Use the Student's t distribution to find a 98% confidence interval for the melting point of tungsten.

c. If the eight measurements had been 3409.76, 3409.80, 3412.66, 3409.79, 3409.76, 3409.77, 3409.80, 3409.78, would the confidence intervals above be valid? Explain.

7. Eight independent measurements are taken of the diameter of a piston. The measurements (in inches) are 3.236, 3.223, 3.242, 3.244, 3.228, 3.253, 3.253, 3.230.

a. Make a dotplot of the eight values.

b. Should the Student's t distribution be used to find a 99% confidence interval for the diameter of this piston? If so, find the confidence interval. If not, explain why not.

c. Eight independent measurements are taken of the diameter of another piston. The measurements this time are 3.295, 3.232, 3.261, 3.248, 3.289, 3.245, 3.576, 3.201. Make a dotplot of these values.

d. Should the Student's t distribution be used to find a 95% confidence interval for the diameter of this piston? If so, find the confidence interval. If not, explain why not.

8. True or false: The Student's *t* distribution may be used to construct a confidence interval for the mean of any population, so long as the sample size is small.

9. The article "Ozone for Removal of Acute Toxicity from Logyard Run-off" (M. Zenaitis and S. Duff, *Ozone Science and Engineering*, 2002:83–90) presents chemical analyses of runoff water from sawmills in British Columbia. Included were measurements of pH for six water specimens: 5.9, 5.0, 6.5, 5.6, 5.9, 6.5. Assuming these to be a random sample of water specimens from an approximately normal population, find a 95% confidence interval for the mean pH.

10. The following are summary statistics for a data set. Would it be appropriate to use the Student's *t* distribution to construct a confidence interval from these data? Explain.

N	Mean	Median	StDev
10	8.905	6.105	9.690

Minimum	Maximum	Q1	Q3
0.512	39.920	1.967	8.103

11. The article "An Automatic Visual System for Marble Tile Classification" (L. Carrino, W. Polini, and S. Turchetta, *Journal of Engineering Manufacture*, 2002:1095–1108) describes a measure for the shade of marble tile in which the amount of light reflected by the tile is measured on a scale of 0–255. A perfectly black tile would reflect no light and measure 0, and a perfectly white tile would measure 255. A sample of nine Mezza Perla tiles were measured, with the following results:

    204.999 206.149 202.102 207.048 203.496

    206.343 203.496 206.676 205.831

    Is it appropriate to use the Student's *t* distribution to construct a 95% confidence interval for the mean shade of Mezza Perla tile? If so, construct the confidence interval. If not, explain why not.

12. Surfactants are chemical agents, such as detergents, that lower the surface tension of a liquid. Surfactants play an important role in the cleaning of contaminated soils. In an experiment to determine the effectiveness of a certain method for removing toluene from sand, the sand was washed with a surfactant and then rinsed with deionized water. Of interest was the amount of toluene that came out in the rinse. In five such experiments, the amounts of toluene removed in the rinse cycle, expressed as a percentage of the total amount originally present, were 5.0, 4.8, 9.0, 10.0, and 7.3. Find a 95% confidence interval for the percentage of toluene removed in the rinse. (This exercise is based on the article "Laboratory Evaluation of the Use of Surfactants for Ground Water Remediation and the Potential for Recycling Them," D. Lee, R. Cody, and B. Hoyle, *Ground Water Monitoring and Remediation*, 2001:49–57.)

13. In an experiment to measure the rate of absorption of pesticides through skin, 500 μg of uniconazole was applied to the skin of four rats. After 10 hours, the amounts absorbed (in μg) were 0.5, 2.0, 1.4, and 1.1. Find a 90% confidence interval for the mean amount absorbed.

14. The following MINITAB output presents a confidence interval for a population mean.

```
One-Sample T: X

Variable    N     Mean     StDev   SE Mean         95% CI
X          10   6.59635   0.11213   0.03546   (6.51613, 6.67656)
```

a. How many degrees of freedom does the Student's *t* distribution have?

b. Use the information in the output, along with the *t* table, to compute a 99% confidence interval.

15. The following MINITAB output presents a confidence interval for a population mean, but some of the numbers got smudged and are now illegible. Fill in the missing numbers.

```
One-Sample T: X

Variable    N     Mean     StDev   SE Mean        99% CI
X          20   2.39374    (a)    0.52640    (  (b), (c)  )
```

16. The concentration of carbon monoxide (CO) in a gas sample is measured by a spectrophotometer and found to be 85 ppm. Through long experience with this instrument, it is believed that its measurements are unbiased and normally distributed, with an standard deviation of 8 ppm. Find a 95% confidence interval for the concentration of CO in this sample.

17. The article "Filtration Rates of the Zebra Mussel (*Dreissena polymorpha*) on Natural Seston from Saginaw Bay, Lake Huron" (D. Fanslow, T. Nalepa, and G. Lang, *Journal of Great Lakes Research*, 1995:489–500) reports measurements of the rates (in mL/mg/h) at which mussels filter seston (particulate matter suspended in seawater).

    a. In the year 1992, 5 measurements were made in the Outer Bay; these averaged 21.7 with a standard deviation of 9.4. Use the Student's t distribution to find a 95% confidence interval for the mean filtration rate in the Outer Bay.

    b. In the year 1992, 7 measurements were made in the Inner Bay; these averaged 8.6 with a standard deviation of 4.5. Should the Student's t distribution be used to find a 95% confidence interval for the mean filtration rate for the Inner Bay? If so, find the confidence interval. If not, explain why not.

5.5   Prediction Intervals and Tolerance Intervals

A confidence interval for a parameter such as a population mean is an interval that is likely to contain the true value of the parameter. In contrast, prediction and tolerance intervals are concerned with the population itself and with values that may be sampled from it in the future. Prediction intervals and tolerance intervals are useful only when the shape of the population is known. The methods we present here, which are the most commonly used, are valid only when the population is known to be normal.

Prediction Intervals

A **prediction interval** is an interval that is likely to contain the value of an item that will be sampled from a population at a future time. In other words, we "predict" that a value that is yet to be sampled from the population will fall within the prediction interval. We illustrate with an example.

Assume that the silicon content (in %) has been measured for a sample of five steel beams, and that the sample mean is $\overline{X} = 0.26$ with a sample standard deviation of $s = 0.05$. Assume further that the silicon content in this type of beam is normally distributed. At some future time we will observe the silicon content Y of some other beam, and we wish to construct an interval that will contain the value of Y with probability 0.95. To see how this is done, let μ and σ denote the mean and standard deviation of the normal population of silicon contents. Then $Y \sim N(\mu, \sigma^2)$ and, since the sample size is

$n = 5$, $\overline{X} \sim N(\mu, \sigma^2/5)$. The difference $Y - \overline{X}$ is therefore normally distributed with mean 0 and variance $\sigma^2(1 + 1/5)$. It follows that

$$\frac{Y - \overline{X}}{\sigma\sqrt{1 + 1/5}} \sim N(0, 1)$$

Approximating σ with s, we find that the quantity $(Y - \overline{X}) / (s\sqrt{1 + 1/5})$ has a Student's t distribution with $5 - 1 = 4$ degrees of freedom (the number of degrees of freedom is based on the sample size used to compute s). From the Student's t table (Table A.3), we find that 95% of the area under the t curve with 4 degrees of freedom is contained between the values -2.776 and 2.776. It follows that

$$P\left(-2.776 < \frac{Y - \overline{X}}{s\sqrt{1 + 1/5}} < 2.776\right) = 0.95$$

Performing some algebra, we obtain

$$P\left(\overline{X} - 2.776s\sqrt{1 + 1/5} < Y < \overline{X} + 2.776s\sqrt{1 + 1/5}\right) = 0.95$$

The interval $\overline{X} \pm 2.776s\sqrt{1 + 1/5}$ is a 95% prediction interval for Y. In this example, $\overline{X} = 0.26$ and $s = 0.05$, so the 95% prediction interval is 0.26 ± 0.15, or $(0.11, 0.41)$.

Generalizing the procedure just described, a $100(1 - \alpha)\%$ prediction interval based on a sample of size n is given by $\overline{X} \pm t_{n-1,\alpha/2}\left(s\sqrt{1 + 1/n}\right)$.

Summary

Let $X_1, \ldots, X_n$ be a sample from a *normal* population. Let Y be another item to be sampled from this population, whose value has not yet been observed. A $100(1 - \alpha)\%$ prediction interval for Y is

$$\overline{X} \pm t_{n-1,\alpha/2}s\sqrt{1 + \frac{1}{n}} \tag{5.19}$$

The probability is $1 - \alpha$ that the value of Y will be contained in this interval.

Example

5.21

A sample of 10 concrete blocks manufactured by a certain process had a mean compressive strength of $\overline{X} = 1312$ MPa, with standard deviation $s = 25$ MPa. Find a 95% prediction interval for the strength of a block that has not yet been measured.

Solution

For a 95% prediction interval, $\alpha = 0.025$. We have a sample size of $n = 10$, so we consult the Student's t table (Table A.3) to find $t_{9,.025} = 2.262$. Using expression (5.19) with $\overline{X} = 1312$ and $s = 25$, the 95% prediction interval is $1312 \pm 2.262(25)\sqrt{1 + 1/10}$, or $(1253, 1371)$.

Comparison of Prediction Intervals and Confidence Intervals

The formula for a prediction interval is similar to that for a confidence interval for the mean of a normal population; in fact, the prediction interval can be obtained from the confidence interval by replacing the expression $s\sqrt{1/n}$ with $s\sqrt{1 + 1/n}$. The quantity $1/n$ under the square root reflects the random variation in the sample mean as an estimator of the population mean, and it is present in both the confidence interval and the prediction interval. The quantity 1 under the square root in the prediction interval reflects the random variation in the value of the sampled item that is to be predicted. Note that since 1 is larger than $1/n$, most of the margin of error for the prediction interval is due to the variability in the value to be predicted, and the prediction interval is much wider the confidence interval. Increasing the sample size affects the confidence interval and prediction interval in different ways as well. As the sample size gets larger, the margin of error for a level $100(1 - \alpha)\%$ confidence interval, which is $t_{n-1,\alpha/2}s\sqrt{1/n}$, shrinks to 0. In contrast, the margin of error for the level $100(1 - \alpha)\%$ prediction interval is $t_{n-1,\alpha/2}s\sqrt{1 + 1/n}$. As n becomes large, $t_{n-1,\alpha/2}$ becomes close to $z_{\alpha/2}$, s becomes close to σ, and $1 + 1/n$ becomes close to 1. The margin of error for the prediction interval, therefore, becomes close to $z_{\alpha/2}\sigma$. This reflects the fact that there is always random variation in the value of an item to be sampled from a population.

One-Sided Prediction Intervals

One-sided prediction intervals can be computed by a method analogous to that for computing one-sided confidence intervals.

Let $X_1, \ldots, X_n$ be a sample from a *normal* population. Let Y be another item to be sampled from this population, whose value has not been observed. A $100(1 - \alpha)\%$ upper prediction bound for Y is

$$\overline{X} + t_{n-1,\alpha}s\sqrt{1 + \frac{1}{n}} \tag{5.20}$$

and a level $100(1 - \alpha)\%$ lower prediction bound for Y is

$$\overline{X} - t_{n-1,\alpha}s\sqrt{1 + \frac{1}{n}} \tag{5.21}$$

Prediction Intervals Are Sensitive to Departures from Normality

The method presented here for computing prediction intervals is sensitive to the assumption that the population is normal. If the shape of the population differs much from the normal curve, the prediction interval may be misleading. For this reason, prediction intervals must be interpreted with caution. Large samples do not help. No matter how large the sample is, the prediction interval will not be valid unless the population is normal.

Tolerance Intervals for a Normal Population

A **tolerance interval** is an interval that is likely to contain a specified proportion of the population. The method we present here, which is the one most commonly used, requires that the population be normal. To illustrate the idea, first assume that we have a normal population whose mean μ and standard deviation σ are known. If we wish to find an interval that contains 90% of this population, we can do so exactly. The interval $\mu \pm 1.645\sigma$ contains 90% of the population. In general, the interval $\mu \pm z_{\gamma/2}\sigma$ will contain $100(1 - \gamma)\%$ of the population.

In practice, we do not know μ and σ. Instead, we have a sample of size n, and we estimate μ with the sample mean $\overline{X}$ and σ with the sample standard deviation s. This estimation has two consequences. First, we must make the interval wider than it would be if μ and σ were known. Second, we cannot be 100% confident that the interval actually contains the required proportion of the population.

To construct a tolerance interval, therefore, we must specify the proportion $100(1 - \gamma)\%$ of the population that we wish the interval to contain, along with a level of confidence $100(1 - \alpha)\%$ that the interval actually contains the specified proportion. It is then possible, using advanced methods, to find a number $k_{n,\alpha,\gamma}$ such that the interval

$$\overline{X} \pm k_{n,\alpha,\gamma} s$$

will contain at least $100(1 - \gamma)\%$ of the population with confidence $100(1 - \alpha)\%$. Values of $k_{n,\alpha,\gamma}$ are presented in Table A.4 for various values of α, γ, and the sample size n.

Summary

Let $X_1, \ldots, X_n$ be a sample from a *normal* population. A tolerance interval for containing at least $100(1 - \gamma)\%$ of the population with confidence $100(1 - \alpha)\%$ is

$$\overline{X} \pm k_{n,\alpha,\gamma} s \qquad (5.22)$$

Of all the tolerance intervals that are computed by this method, $100(1 - \alpha)\%$ will actually contain at least $100(1 - \gamma)\%$ of the population.

Example 5.22

The lengths of bolts manufactured by a certain process are known to be normally distributed. In a sample of 30 bolts, the average length was 10.25 cm, with a standard deviation of 0.20 cm. Find a tolerance interval that includes 90% of the lengths of the bolts with 95% confidence.

Solution
We have $\overline{X} = 10.25$ and $s = 0.20$. The value of γ is 0.10 and the value of α is 0.05. The sample size is $n = 30$. From Table A.4, we find that $k_{n,\alpha,\gamma} = 2.1398$. The tolerance interval is therefore $10.25 \pm 2.1398(0.20)$, or $(9.82, 10.68)$.

Exercises for Section 5.5

1. A sample of 25 resistors, each labeled 100 Ω, had an average resistance of 101.4 Ω with a standard deviation of 2.3 Ω. Assume the resistances are normally distributed.

   a. Find a 95% prediction interval for the resistance of a single resistor.

   b. Find a tolerance interval for the resistance that includes 90% of the resistors with 95% confidence.

2. In a sample of 20 bolts, the average breaking torque was 89.7 J with a standard deviation of 8.2 J. Assume that the breaking torques are normally distributed.

   a. Find a 99% prediction interval for the breaking torque of a single bolt.

   b. Find a tolerance interval for the breaking torque that includes 95% of the bolts with 99% confidence.

3. The article "Ozone for Removal of Acute Toxicity from Logyard Run-off" (M. Zenaitis and S. Duff, *Ozone Science and Engineering*, 2002:83–90) presents chemical analyses of runoff water from sawmills in British Columbia. Included were measurements of pH for six water specimens: 5.9, 5.0, 6.5, 5.6, 5.9, 6.5. Assume that these are a random sample

of water specimens from a normal population. These data also appear in Exercise 9 in Section 5.4.

   a. Find a 98% prediction interval for a pH of a single specimen.

   b. Find a tolerance interval for the pH that includes 95% of the specimens with 95% confidence.

4. Six measurements were made of the concentration (in percent) of ash in a certain variety of spinach. The sample mean was 19.35, and the sample standard deviation was 0.577. Assume that the concentrations are normally distributed.

   a. Find a 90% prediction interval for a single measurement.

   b. Find a tolerance interval for the pH that includes 99% of the measurements with 95% confidence.

5. Five measurements are taken of the octane rating for a particular type of gasoline. The results (in %) are 87.0, 86.0, 86.5, 88.0, 85.3. These data also appear in Exercise 4 in Section 5.4.

   a. Find a 95% prediction interval for a single measurement.

   b. Find a tolerance interval for the pH that includes 90% of the measurements with 99% confidence.

Supplementary Exercises for Chapter 5

1. Concentrations of atmospheric pollutants such as carbon monoxide (CO) can be measured with a spectrophotometer. In a calibration test, 50 measurements were taken of a laboratory gas sample that is known to have a CO concentration of 70 parts per million (ppm). A measurement is considered to be satisfactory if it is within 5 ppm of the true concentration. Of the 50 measurements, 37 were satisfactory.

   a. What proportion of the sample measurements were satisfactory?

   b. Find a 95% confidence interval for the proportion of measurements made by this instrument that will be satisfactory.

   c. How many measurements must be taken to specify the proportion of satisfactory measurements to within ±0.10 with 95% confidence?

   d. Find a 99% confidence interval for the proportion of measurements made by this instrument that will be satisfactory.

   e. How many measurements must be taken to specify the proportion of satisfactory measurements to within ±0.10 with 99% confidence?

2. In a sample of 150 boards of a certain grade, the average modulus of elasticity was 1.57 psi, and the standard deviation was 0.23 psi.

   a. Find a 95% confidence interval for the mean modulus of elasticity.

   b. Find a 99.5% confidence interval for the mean modulus of elasticity.

   c. What is the confidence level of the interval (1.54, 1.60)?

d. How many boards must be sampled so that a 95% confidence interval specifies the mean to within ± 0.02 psi?

e. How many boards must be sampled so that a 99.5% confidence interval specifies the mean to within ± 0.02 psi?

3. Three confidence intervals for the mean shear strength (in ksi) of anchor bolts of a certain type are computed, all from the same sample. The intervals are (4.01, 6.02), (4.20, 5.83), and (3.57, 6.46). The levels of the intervals are 90%, 95%, and 99%. Which interval has which level?

4. A new catalyst is being investigated for use in the production of a plastic chemical. Ten batches of the chemical are produced. The mean yield of the 10 batches is 72.5%, and the standard deviation is 5.8%. Assume the yields are independent and approximately normally distributed. Find a 99% confidence interval for the mean yield when the new catalyst is used.

5. A survey is to be conducted in which a random sample of residents in a certain city will be asked whether they favor or oppose a modernization project for the civic center. How many residents should be polled to be sure that a 98% confidence interval for the proportion who are in favor specifies that proportion to within ± 0.04?

6. In a random sample of 53 concrete specimens, the average porosity (in %) was 21.6, and the standard deviation was 3.2.

a. Find a 90% confidence interval for the mean porosity of specimens of this type of concrete.

b. Find a 95% confidence interval for the mean porosity of specimens of this type of concrete.

c. What is the confidence level of the interval (21.0, 22.2)?

d. How many specimens must be sampled so that a 90% confidence interval specifies the mean to within ± 0.3?

e. How many specimens must be sampled so that a 95% confidence interval specifies the mean to within ± 0.3?

7. A 90% confidence interval for a population mean based on 144 observations is computed to be (2.7, 3.4). How many observations must be made so

that a 90% confidence interval will specify the mean to within ± 0.2?

8. A sample of 100 components is drawn, and a 95% confidence interval for the proportion defective specifies this proportion to within ± 0.06. To get a more precise estimate of the number defective, the sample size will be increased to 400, and the confidence interval will be recomputed. What will be the approximate width of the new confidence interval? Choose the best answer:

   i. ± 0.015

   ii. ± 0.03

   iii. ± 0.06

   iv. ± 0.12

   v. ± 0.24

9. A random sample of 400 electronic components manufactured by a certain process are tested, and 30 are found to be defective.

a. Let p represent the proportion of components manufactured by this process that are defective. Find a 95% confidence interval for p.

b. How many components must be sampled so that the 95% confidence interval will specify the proportion defective to within ± 0.02?

c. (Hard) The company ships the components in lots of 200. Lots containing more than 20 defective components may be returned. Find a 95% confidence interval for the proportion of lots that will be returned.

10. Refer to Exercise 9. A device will be manufactured in which two of the components in Exercise 9 will be connected in series. The components function independently, and the device will function only if both components function. Let q be the probability that a device functions. Find a 95% confidence interval for q. (*Hint:* Express q in terms of p, then use the result of Exercise 9a.)

11. A metallurgist makes several measurements of the melting temperature of a certain alloy and computes a 95% confidence interval to be $2038 \pm 2°C$. Assume the measuring process for temperature is unbiased. True or false:

a. There is 95% probability that the true melting temperature is in the interval $2038 \pm 2°C$.

b. If the experiment were repeated, the probability is 95% that the mean measurement from that experiment would be in the interval $2038 \pm 2°C$.

c. If the experiment were repeated, and a 95% confidence interval computed, there is 95% probability that the confidence interval would cover the true melting point.

d. If one more measurement were made, the probability is 95% that it would be in the interval $2038 \pm 2°C$.

12. In a study of the lifetimes of electronic components, a random sample of 400 components are tested until they fail to function. The sample mean lifetime was 370 hours and the standard deviation was 650 hours. True or false:

a. An approximate 95% confidence interval for the mean lifetime of this type of component is from 306.3 to 433.7 hours.

b. About 95% of the sample components had lifetimes between 306.3 and 433.7 hours.

c. If someone takes a random sample of 400 components, divides the sample standard deviation of their lifetimes by 20, and then adds and subtracts that quantity from the sample mean, there is about a 68% chance that the interval so constructed will cover the mean lifetime of this type of component.

d. The z table can't be used to construct confidence intervals here, because the lifetimes of the components don't follow the normal curve.

e. About 68% of the components had lifetimes in the interval 370 ± 650 hours.

13. An investigator computes a 95% confidence interval for a population mean on the basis of a sample of size 70. If she wishes to compute a 95% confidence interval that is half as wide, how large a sample does she need?

14. A 95% confidence interval for a population mean is computed from a sample of size 400. Another 95% confidence interval will be computed from a sample of size 100 drawn from the same population. Choose the best answer to fill in the blank: The interval from the sample of size 400 will be approximately _____ as the interval from the sample of size 100.

   i. one-eighth as wide

   ii. one-fourth as wide

   iii. one-half as wide

   iv. the same width

   v. twice as wide

   vi. four times as wide

   vii. eight times as wide

15. Based on a large sample of capacitors of a certain type, a 95% confidence interval for the mean capacitance, in μF, was computed to be $(0.213, 0.241)$. Find a 90% confidence interval for the mean capacitance of this type of capacitor.

16. Sixty-four independent measurements were made of the speed of light. They averaged 299,795 km/s and had a standard deviation of 8 km/s. True or false:

a. A 95% confidence interval for the speed of light is $299,795 \pm 1.96$ km/s.

b. The probability is 95% that the speed of light is in the interval $299,795 \pm 1.96$.

c. If a 65th measurement is made, the probability is 95% that it would fall in the interval $299,795 \pm 1.96$.

17. A large box contains 10,000 ball bearings. A random sample of 120 is chosen. The sample mean diameter is 10 mm, and the standard deviation is 0.24 mm. True or false:

a. A 95% confidence interval for the mean diameter of the 120 bearings in the sample is $10 \pm (1.96)(0.24)/\sqrt{120}$.

b. A 95% confidence interval for the mean diameter of the 10,000 bearings in the box is $10 \pm (1.96)(0.24)/\sqrt{120}$.

c. A 95% confidence interval for the mean diameter of the 10,000 bearings in the box is $10 \pm (1.96)(0.24)/\sqrt{10,000}$.

18. A meteorologist measures the temperature in downtown Denver at noon on each day for one year. The 365 readings average 57°F and have standard deviation 20°F. The meteorologist computes a 95% confidence interval for the mean temperature at noon to be $57° \pm (1.96)(20)/\sqrt{365}$. Is this correct? Why or why not?

19. The temperature of a certain solution is estimated by taking a large number of independent measurements and averaging them. The estimate is 37°C, and the standard error is 0.1°C.

a. Find a 95% confidence interval for the temperature.

b. What is the confidence level of the interval $37 \pm 0.1°C$?

c. If only a small number of independent measurements had been made, what additional assumption would be necessary in order to compute a confidence interval?

d. Making the additional assumption, compute a 95% confidence interval for the temperature if 10 measurements were made.

20. Boxes of nails contain 100 nails each. A sample of 10 boxes is drawn, and each of the boxes is weighed. The average weight is 1500 g, and the standard deviation is 5 g. Assume the weight of the box itself is negligible, so that all the weight is due to the nails in the box.

a. Let μ_{box} denote the mean weight of a box of nails. Find a 95% confidence interval for μ_{box}.

b. Let μ_{nail} denote the mean weight of a nail. Express μ_{nail} in terms of μ_{box}.

c. Find a 95% confidence interval for μ_{nail}.

21. Let X represent the number of events that are observed to occur in n units of time or space, and assume $X \sim \text{Poisson}(n\lambda)$, where λ is the mean number of events that occur in one unit of time or space. Assume X is large, so that $X \sim N(n\lambda, n\lambda)$. Follow steps

(a) through (d) to derive a level $100(1 - \alpha)\%$ confidence interval for λ. Then in part (e), you are asked to apply the result found in part (d).

a. Show that for a proportion $1 - \alpha$ of all possible samples, $X - z_{\alpha/2}\sigma_X < n\lambda < X + z_{\alpha/2}\sigma_X$.

b. Let $\hat{\lambda} = X/n$. Show that $\sigma_{\hat{\lambda}} = \sigma_X/n$.

c. Conclude that for a proportion $1 - \alpha$ of all possible samples, $\hat{\lambda} - z_{\alpha/2}\sigma_{\hat{\lambda}} < \lambda < \hat{\lambda} + z_{\alpha/2}\sigma_{\hat{\lambda}}$.

d. Use the fact that $\sigma_{\hat{\lambda}} \approx \sqrt{\hat{\lambda}/n}$ to derive an expression for the level $100(1 - \alpha)\%$ confidence interval for λ.

e. A 5 mL sample of a certain suspension is found to contain 300 particles. Let λ represent the mean number of particles per mL in the suspension. Find a 95% confidence interval for λ.

22. The answer to Exercise 21 part (d) is needed for this exercise. A geologist counts 64 emitted particles in one minute from a certain radioactive rock.

a. Find a 95% confidence interval for the rate of emissions in units of particles per minute.

b. After four minutes, 256 particles are counted. Find a 95% confidence interval for the rate of emissions in units of particles per minute.

c. For how many minutes should errors be counted in order that the 95% confidence interval specifies the rate to within ± 1 particle per minute?

6

Hypothesis Tests for a Single Sample

Introduction

In Example 5.4 (in Section 5.2), a sample of 50 microdrills had an average lifetime of $\overline{X} = 12.68$ holes drilled and a standard deviation of $s = 6.83$. Let us assume that the question of interest is whether the population mean lifetime μ is greater than 11. We address this question by examining the value of the sample mean $\overline{X}$. We see that $\overline{X} > 11$, but because of the random variation in $\overline{X}$, this does not guarantee that $\mu > 11$. We would like to know just how certain we can be that $\mu > 11$. A confidence interval is not quite what we need. In Example 5.4, a 95% confidence interval for the population mean μ was computed to be (10.79, 14.57). This tells us that we are 95% confident that μ is between 10.79 and 14.57, but it does not directly tell us how confident we can be that $\mu > 11$.

The statement "$\mu > 11$" is a **hypothesis** about the population mean μ. To determine just how certain we can be that a hypothesis such as this is true, we must perform a **hypothesis test**. A hypothesis test produces a number between 0 and 1 that measures the degree of certainty we may have in the truth of a hypothesis. It turns out that hypothesis tests are closely related to confidence intervals. In general, whenever a confidence interval can be computed, a hypothesis test can also be performed, and vice versa.

6.1 Large-Sample Tests for a Population Mean

We begin with an example. A new coating has become available that is supposed to reduce the wear on a certain type of rotary gear. The mean wear on uncoated gears is known from long experience to be 80 μm per month. Engineers perform an experiment to determine whether the coating will reduce the wear. They apply the coating to a simple random sample of 60 gears and measure the wear on each gear after one month of use. The sample mean wear is 74 μm, and the sample standard deviation is $s = 18$ μm.

The population in this case consists of the amounts of wear on all the gears that would be used if the coating is applied. If there were no random variation in the sample mean, then we could conclude that the coating would reduce wear—from 80 to 74 μm. Of course, there is random variation in the sample mean. The population mean will actually be somewhat higher or lower than 74.

The engineers are concerned that applying the coating might not reduce the wear—that is, that the population mean for the coated gears might be 80 or more. They want to know whether this concern is justified. The question, therefore, is this: Is it plausible that this sample, with its mean of 74, could have come from a population whose mean is 80 or more?

This is the sort of question that hypothesis tests are designed to address, and we will now construct a hypothesis test to address this question. We have observed a sample with mean 74. There are two possible interpretations of this observation:

1. The population mean is actually greater than or equal to 80, and the sample mean is lower than this only because of random variation from the population mean. Thus the mean wear will actually be 80 or more if the coating is applied, and the sample is misleading.

2. The population mean is actually less than 80, and the sample mean reflects this fact. Thus the sample represents a real difference that can be expected if the coating is applied.

These two explanations have standard names. The first is called the **null hypothesis**. In most situations, the null hypothesis says that the effect indicated by the sample is due only to random variation between the sample and the population. The second explanation is called the **alternate hypothesis**. The alternate hypothesis says that the effect indicated by the sample is real, in that it accurately represents the whole population.

In our example, the engineers are concerned that the null hypothesis might be true. A hypothesis test assigns a quantitative measure to the plausibility of the null hypothesis. After performing a hypothesis test, we will be able to tell the engineers, in numerical terms, precisely how valid their concern is.

To make things more precise, we express everything in symbols. The null hypothesis is denoted H_0. The alternate hypothesis is denoted H_1. As usual, the population mean is denoted μ. We have, therefore,

$$H_0 : \mu \geq 80 \quad \text{versus} \quad H_1 : \mu < 80$$

In performing a hypothesis test, we essentially put the null hypothesis on trial. We begin by assuming that H_0 is true, just as we begin a trial by assuming a defendant to be innocent. The random sample provides the evidence. The hypothesis test involves measuring the strength of the disagreement between the sample and H_0 to produce a number between 0 and 1, called a **P-value**. The P-value measures the plausibility of H_0. The smaller the P-value, the stronger the evidence is against H_0. If the P-value is sufficiently small, we may be willing to abandon our assumption that H_0 is true and believe H_1 instead. This is referred to as **rejecting** the null hypothesis.

In this example, let $X_1, \ldots, X_{60}$ be the amounts of wear on the 60 sample gears. The observed value of the sample mean is $\overline{X} = 74$. We will also need to know the sample standard deviation, which is $s = 18$. We must assess the plausibility of H_0, which says that the population mean is 80 or more, given that we have observed a sample from this population whose mean is only 74. We will do this in two steps, as follows:

1. We will compute the distribution of $\overline{X}$ under the assumption that H_0 is true. This distribution is called the **null distribution** of $\overline{X}$.

2. We will compute the P-value. This is the probability, under the assumption that H_0 is true, of observing a value of $\overline{X}$ whose disagreement with H_0 is as least as great as that of the observed value of 74.

To perform step 1, note that $\overline{X}$ is the mean of a large sample, so the Central Limit Theorem specifies that it comes from a normal distribution whose mean is μ and whose variance is $\sigma^2/60$, where σ^2 is the population variance and 60 is the sample size. We must specify values for μ and for σ in order to determine the null distribution. Since we are assuming that H_0 is true, we assume that $\mu \geq 80$. This does not provide a specific value for μ. We take as the assumed value for μ the value closest to the alternate hypothesis H_1, for reasons that will be explained later in this section. Thus we assume $\mu = 80$. We do not know the population standard deviation σ. However, since the sample is large, we may approximate σ with the sample standard deviation $s = 18$. Therefore we have determined that under H_0, $\overline{X}$ has a normal distribution with mean 80 and standard deviation $18/\sqrt{60} = 2.324$. The null distribution is $\overline{X} \sim N(80, \ 2.324^2)$.

We are now ready for step 2. Figure 6.1 illustrates the null distribution. The number 74 indicates the point on the distribution corresponding to the observed value of $\overline{X}$. How plausible is it that a number sampled from this distribution would be as small as 74? This is measured by the P-value. The P-value is the probability that a number drawn from the null distribution would disagree with H_0 at least as strongly as the observed value of $\overline{X}$, which is 74. Since H_0 specifies that the mean of $\overline{X}$ is greater than or equal to 80, values less than 74 are in greater disagreement with H_0. The P-value, therefore, is the probability that a number drawn from an $N(80, 2.324^2)$ distribution is less than or equal to 74. This probability is determined by computing the z-score:

$$z = \frac{74 - 80}{2.324} = -2.58$$

From the z table, the probability that a standard normal random variable z is less than or equal to -2.58 is 0.0049. The P-value for this test is 0.0049.

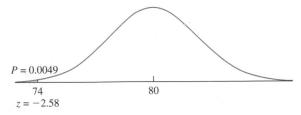

FIGURE 6.1 The null distribution of $\overline{X}$ is $N(80, \ 2.324^2)$. Thus if H_0 is true, the probability that $\overline{X}$ takes on a value as extreme as or more extreme than the observed value of 74 is 0.0049. This is the P-value.

The P-value, as promised, provides a quantitative measure of the plausibility of H_0. But how do we interpret this quantity? The proper interpretation is rather subtle. The P-value tells us that if H_0 were true, the probability of drawing a sample whose mean was as far from H_0 as the observed value of 74 is only 0.0049. Therefore, one of the following two conclusions is possible:

- H_0 is false.
- H_0 is true, which implies that of all the samples that might have been drawn, only 0.49% of them have a mean as small as or smaller than that of the sample actually drawn. In other words, our sample mean lies in the most extreme 0.49% of its distribution.

In practice, events in the most extreme 0.49% of their distributions very seldom occur. Therefore we reject H_0 and conclude that the coating will reduce the mean wear to a value less than 80μm.

The null hypothesis in this case specified only that $\mu \geq 80$. In assuming H_0 to be true, why did we choose the value $\mu = 80$, which is closest to H_1? To give H_0 a fair test, we must test it in its most plausible form. The most plausible value for μ is the value closest to $\overline{X}$. Now $\overline{X} = 74$, so among the values $\mu \geq 80$, the closest to $\overline{X}$ is $\mu = 80$. This value is also the one closest to H_1. This is typical. In practice, when it is of interest to perform a hypothesis test, the most plausible value for H_0 will be the value closest to H_1.

It is natural to ask how small the P-value should be in order to reject H_0. Some people use the "5% rule"; they reject H_0 if $P \leq 0.05$. However, there is no scientific justification for this or any other rule. We discuss this issue in more detail in Section 6.2.

Note that the method we have just described uses the Central Limit Theorem. It follows that for this method to be valid, the sample size must be reasonably large—say, 30 or more. Hypothesis tests that are sometimes valid for small samples are presented in Section 6.4.

Finally, note that the calculation of the P-value was done by computing a z-score. For this reason, the z-score is called a **test statistic**. A test that uses a z-score as a test statistic is called a z test.

There are many kinds of hypothesis tests. All of them follow a basic series of steps, which are outlined in the following box.

Steps in Performing a Hypothesis Test

1. Define the null hypothesis, H_0, and the alternate hypothesis, H_1.
2. Assume H_0 to be true.
3. Compute a **test statistic**. A test statistic is a statistic that is used to assess the strength of the evidence against H_0.
4. Compute the P-value of the test statistic. The P-value is the probability, assuming H_0 to be true, that the test statistic would have a value whose disagreement with H_0 is as great as or greater than that actually observed. The P-value is also called the **observed significance level**.

Another Way to Express H_0

We have mentioned that when assuming H_0 to be true, we use the value closest to H_1. Some authors consider this single value to be H_0, so that, in the previous example, they would write $H_0 : \mu = 80$ instead of $H_0 : \mu \geq 80$. There is an advantage to this notation, which is that it makes it clear which value is being used when H_0 is assumed to be true. But there is a disadvantage when it comes to interpretation. Generally, the value closest to H_1 is of no special interest. For example, in the gear wear example just discussed, we are not specifically concerned with the possibility $\mu = 80$ but, rather, with the possibility $\mu \geq 80$. The importance of rejecting H_0 is not that we reject the single value $\mu = 80$, but that we reject all values $\mu \geq 80$. For this reason, we choose to write $H_0 : \mu \geq 80$.

Example

6.1

The article "Wear in Boundary Lubrication" (S. Hsu, R. Munro, and M. Shen, *Journal of Engineering Tribology*, 2002:427–441) discusses several experiments involving various lubricants. In one experiment, 45 steel balls lubricated with purified paraffin were subjected to a 40 kg load at 600 rpm for 60 minutes. The average reduction in diameter was 673.2 μm, and the standard deviation was 14.9 μm. Assume that the specification for a lubricant is that the mean reduction be less than 675 μm. Find the P-value for testing $H_0 : \mu \geq 675$ versus $H_1 : \mu < 675$.

Solution

First let's translate the problem into statistical language. We have a simple random sample $X_1, \ldots, X_{45}$ of diameter reductions. The sample mean and standard deviation are $\overline{X} = 673.2$ and $s = 14.9$. The population mean is unknown and denoted by μ. Before getting into the construction of the test, we'll point out again that the basic issue is the random variation in the sample mean. If there were no random variation in the sample mean, we could conclude that the lubricant would meet the specification, since $673.2 < 675$. The question is whether the random variation in the sample mean is large enough so that the population mean could plausibly be as high as 675.

To perform the hypothesis test, we follow the steps given earlier. The null hypothesis is that the lubricant does not meet the specification and that the difference between the sample mean of 673.2 and 675 is due to chance. The alternate hypothesis is that the lubricant does indeed meet the specification. We assume H_0 is true, so that the sample was drawn from a population with mean $\mu = 675$ (the value closest to H_1). We estimate the population standard deviation σ with the sample standard deviation $s = 14.9$. The test is based on $\overline{X}$. Under H_0, $\overline{X}$ comes from a normal population with mean 675 and standard deviation $14.9/\sqrt{45} = 2.22$. The P-value is the probability of observing a sample mean less than or equal to 673.2. The test statistic is the z-score, which is

$$z = \frac{673.2 - 675}{2.22} = -0.81$$

The P-value is 0.209 (see Figure 6.2). Therefore if H_0 is true, there is a 20.9% chance to observe a sample whose disagreement with H_0 is as least as great as that

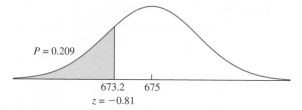

FIGURE 6.2 The null distribution of $\overline{X}$ is $N(675, 2.22^2)$. Thus if H_0 is true, the probability that $\overline{X}$ takes on a value as extreme as or more extreme than the observed value of 673.2 is 0.209. This is the P-value.

which was actually observed. Since 0.209 is not a very small probability, we do not reject H_0. Instead, we conclude that H_0 is plausible. The data do not show conclusively that the lubricant meets the specification. Note that we are *not* concluding that H_0 is *true*, only that it is *plausible*. We will discuss this distinction further in Section 6.2.

The following computer output (from MINITAB) presents the results of Example 6.1.

```
One-Sample Z: Wear

Test of mu = 675 vs < 675
The assumed standard deviation = 14.9

                                          95%
                                        Upper
Variable    N      Mean   StDev  SE Mean   Bound        Z       P
Wear       45   673.200    14.9    2.221  676.853   -0.81   0.209
```

The output states the null hypothesis as $\mu = 675$ rather than $\mu \geq 675$. This reflects the fact that the value $\mu = 675$ is used to construct the null distribution. The quantity "SE Mean" is the standard deviation of $\overline{X}$, estimated by $s/\sqrt{n}$. The output also provides a 95% upper confidence bound for μ.

In the examples shown so far, the null hypothesis specified that the population mean was less than or equal to something, or greater than or equal to something. In some cases, a null hypothesis specifies that the population mean is equal to a specific value. Example 6.2 provides an illustration.

Example
6.2

A scale is to be calibrated by weighing a 1000 g test weight 60 times. The 60 scale readings have mean 1000.6 g and standard deviation 2 g. Find the P-value for testing $H_0: \mu = 1000$ versus $H_1: \mu \neq 1000$.

Solution

Let μ denote the population mean reading. The null hypothesis says that the scale is in calibration, so that the population mean μ is equal to the true weight of 1000 g, and the difference between the sample mean reading and the true weight is due entirely to chance. The alternate hypothesis says that the scale is out of calibration.

In this example, the null hypothesis specifies that μ is *equal* to a specific value, rather than greater than or equal to or less than or equal to. For this reason, values of $\overline{X}$ that are *either* much larger *or* much smaller than μ will provide evidence against H_0. In the previous examples, only the values of $\overline{X}$ on one side of μ provided evidence against H_0.

We assume that H_0 is true, and therefore that the sample readings were drawn from a population with mean $\mu = 1000$. We approximate the population standard deviation σ with $s = 2$. The null distribution of $\overline{X}$ is normal with mean 1000 and standard deviation $2/\sqrt{60} = 0.258$. The z-score of the observed value $\overline{X} = 1000.6$ is

$$z = \frac{1000.6 - 1000}{0.258} = 2.32$$

Since H_0 specifies $\mu = 1000$, regions in both tails of the curve are in greater disagreement with H_0 than the observed value of 1000.6. The P-value is the sum of the areas in both of these tails, which is 0.0204 (see Figure 6.3). Therefore, if H_0 is true, the probability of a result as extreme as or more extreme than that observed is only 0.0204. The evidence against H_0 is pretty strong. It would be prudent to reject H_0 and to recalibrate the scale.

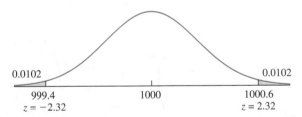

FIGURE 6.3 The null distribution of $\overline{X}$ is $N(1000,\ 0.258^2)$. Thus if H_0 is true, the probability that $\overline{X}$ takes on a value as extreme as or more extreme than the observed value of 1000.6 is 0.0204. This is the P-value.

When H_0 specifies a single value for μ, as in Example 6.2, both tails contribute to the P-value, and the test is said to be a **two-sided** or **two-tailed** test. When H_0 specifies only that μ is greater than or equal to, or less than or equal to a value, only one tail contributes to the P-value, and the test is called a **one-sided** or **one-tailed** test.

We conclude this section by summarizing the procedure used to perform a large-sample hypothesis test for a population mean.

Summary

Let $X_1, \ldots, X_n$ be a *large* (e.g., $n > 30$) sample from a population with mean μ and standard deviation σ.

To test a null hypothesis of the form $H_0: \mu \le \mu_0$, $H_0: \mu \ge \mu_0$, or $H_0: \mu = \mu_0$:

■ Compute the z-score: $z = \dfrac{\overline{X} - \mu_0}{\sigma/\sqrt{n}}$.

If σ is unknown, it may be approximated with s.

■ Compute the P-value. The P-value is an area under the normal curve, which depends on the alternate hypothesis as follows:

Alternate Hypothesis	P-value
$H_1: \mu > \mu_0$	Area to the right of z
$H_1: \mu < \mu_0$	Area to the left of z
$H_1: \mu \ne \mu_0$	Sum of the areas in the tails cut off by z and $-z$

Exercises for Section 6.1

1. Recently many companies have been experimenting with telecommuting, allowing employees to work at home on their computers. Among other things, telecommuting is supposed to reduce the number of sick days taken. Suppose that at one firm, it is known that over the past few years employees have taken a mean of 5.4 sick days. This year, the firm introduces telecommuting. Management chooses a simple random sample of 80 employees to follow in detail, and, at the end of the year, these employees average 4.5 sick days with a standard deviation of 2.7 days. Let μ represent the mean number of sick days for all employees of the firm.

   a. Find the P-value for testing $H_0: \mu \ge 5.4$ versus $H_1: \mu < 5.4$.

   b. Either the mean number of sick days has declined since the introduction of telecommuting, or the sample is in the most extreme _____% of its distribution.

2. The pH of an acid solution used to etch aluminum varies somewhat from batch to batch. In a sample of 50 batches, the mean pH was 2.6, with a standard deviation of 0.3. Let μ represent the mean pH for batches of this solution.

   a. Find the P-value for testing $H_0: \mu \le 2.5$ versus $H_1: \mu > 2.5$.

   b. Either the mean pH is greater than 2.5 mm, or the sample is in the most extreme _____% of its distribution.

3. The article "Evaluation of Mobile Mapping Systems for Roadway Data Collection" (H. Karimi, A. Khattak, and J. Hummer, *Journal of Computing in Civil Engineering*, 2000:168–173) describes a system for remotely measuring roadway elements such as the width of lanes and the heights of traffic signs. For a sample of 160 such elements, the average error (in percent) in the measurements was 1.90, with a standard deviation of 21.20. Let μ represent the mean error in this type of measurement.

   a. Find the P-value for testing $H_0: \mu = 0$ versus $H_1: \mu \ne 0$.

   b. Either the mean error for this type of measurement is nonzero, or the sample is in the most extreme _____% of its distribution.

4. In a process that manufactures tungsten-coated silicon wafers, the target resistance for a wafer is 85 mΩ. In a simple random sample of 50 wafers, the sample mean resistance was 84.8 mΩ, and the standard deviation was 0.5 mΩ. Let μ represent the mean resistance of the wafers manufactured by this process. A quality engineer tests $H_0: \mu = 85$ versus $H_1: \mu \ne 85$.

   a. Find the P-value.

b. Do you believe it is plausible that the mean is on target, or are you convinced that the mean is not on target? Explain your reasoning.

5. There is concern that increased industrialization may be increasing the mineral content of river water. Ten years ago, the silicon content of the water in a certain river was 5 mg/L. Eighty-five water samples taken recently from the river have mean silicon content 5.4 mg/L and standard deviation 1.2 mg/L.

   a. Find the P-value.

   b. Do you believe it is plausible that the silicon content of the water is no greater than it was 10 years ago, or are you convinced that the level has increased? Explain your reasoning.

6. A certain type of stainless steel powder is supposed to have a mean particle diameter of $\mu = 15$ μm. A random sample of 87 particles had a mean diameter of 15.2 μm, with a standard deviation of 1.8 μm. A test is made of $H_0: \mu = 15$ versus $H_1: \mu \neq 15$.

   a. Find the P-value.

   b. Do you believe it is plausible that the mean diameter is 15 μm, or are you convinced that it differs from 15 μm? Explain your reasoning.

7. When it is operating properly, a chemical plant has a mean daily production of at least 740 tons. The output is measured on a simple random sample of 60 days. The sample had a mean of 715 tons/day and a standard deviation of 24 tons/day. Let μ represent the mean daily output of the plant. An engineer tests $H_0: \mu \geq 740$ versus $H_1: \mu < 740$.

   a. Find the P-value.

   b. Do you believe it is plausible that the plant is operating properly, or are you convinced that the plant is not operating properly? Explain your reasoning.

8. Lasers can provide highly accurate measurements of small movements. To determine the accuracy of such a laser, it was used to take 100 measurements of a known quantity. The sample mean error was 25 μm with a standard deviation of 60 μm. The laser is properly calibrated if the mean error is $\mu = 0$. A test is made of $H_0: \mu = 0$ versus $H_1: \mu \neq 0$.

   a. Find the P-value.

   b. Do you believe it is plausible that the laser is properly calibrated, or are you convinced that it is out of calibration? Explain your reasoning.

9. In an experiment to measure the lifetimes of parts manufactured from a certain aluminum alloy, 67 parts were loaded cyclically until failure. The mean number of kilocycles to failure was 763, and the standard deviation was 120. Let μ represent the mean number of kilocycles to failure for parts of this type. A test is made of $H_0: \mu \leq 750$ versus $H_1: \mu > 750$.

   a. Find the P-value.

   b. Do you believe it is plausible that the mean lifetime is 750 kilocycles or less, or are you convinced that it is greater than 750? Explain your reasoning.

10. A new concrete mix is being designed to provide adequate compressive strength for concrete blocks. The specification for a particular application calls for the blocks to have a mean compressive strength μ greater than 1350 kPa. A sample of 100 blocks is produced and tested. Their mean compressive strength is 1356 kPa, and their standard deviation is 70 kPa. A test is made of $H_0: \mu \leq 1350$ versus $H_1: \mu > 1350$.

    a. Find the P-value.

    b. Do you believe it is plausible that the blocks do not meet the specification, or are you convinced that they do? Explain your reasoning.

11. Fill in the blank: If the null hypothesis is $H_0: \mu \leq 4$, then the mean of $\overline{X}$ under the null distribution is _____ .

    i. 0

    ii. 4

    iii. Any number less than or equal to 4.

    iv. We can't tell unless we know H_1.

12. Fill in the blank: In a test of $H_0: \mu \geq 10$ versus $H_1: \mu < 10$, the sample mean was $\overline{X} = 8$ and the P-value was 0.04. This means that if $\mu = 10$, and the experiment were repeated 100 times, we would expect to obtain a value of $\overline{X}$ of 8 or less approximately _____ times.

    i. 8

    ii. 0.8

    iii. 4

    iv. 0.04

    v. 80

13. Fill volumes, in oz, of a large number of beverage cans were measured, with $\overline{X} = 11.98$ and $\sigma_{\overline{X}} = 0.02$. Use this information to find the P-value for testing $H_0: \mu = 12.0$ versus $H_1: \mu \neq 12.0$.

14. The following MINITAB output presents the results of a hypothesis test for a population mean μ.

```
One-Sample Z: X

Test of mu = 73.5 vs not = 73.5
The assumed standard deviation = 2.3634
```

614 747 5936

```
Variable    N     Mean    StDev   SE Mean        95% CI              Z      P
X          145  73.2461  2.3634   0.1963   (72.8614, 73.6308)   -1.29  0.196
```

---

a. Is this a one-tailed or two-tailed test?
b. What is the null hypothesis?
c. What is the *P*-value?
d. Use the output and an appropriate table to compute the *P*-value for the test of $H_0 : \mu \geq 73.6$ versus $H_1 : \mu < 73.6$.
e. Use the output and an appropriate table to compute a 99% confidence interval for μ.

15. The following MINITAB output presents the results of a hypothesis test for a population mean μ. Some of the numbers are missing. Fill them in.

```
One-Sample Z: X

Test of mu = 3.5 vs > 3.5
The assumed standard deviation = 2.00819


                                                95%
                                               Lower
Variable    N     Mean    StDev   SE Mean      Bound     Z      P
X          87   4.07114  2.00819    (a)       3.71700   (b)    (c)
```

6.2 Drawing Conclusions from the Results of Hypothesis Tests

Let's take a closer look at the conclusions reached in Examples 6.1 and 6.2 (in Section 6.1). In Example 6.2, we rejected H_0; in other words, we concluded that H_0 was false. In Example 6.1, we did not reject H_0. However, we did not conclude that H_0 was true. We could only conclude that H_0 was plausible.

In fact, the only two conclusions that can be reached in a hypothesis test are that H_0 is false or that H_0 is plausible. In particular, one can never conclude that H_0 is true. To understand why, think of Example 6.1 again. The sample mean was $\overline{X} = 673.2$, and the null hypothesis was $\mu \geq 675$. The conclusion was that 673.2 is close enough to 675 so that the null hypothesis *might* be true. But a sample mean of 673.2 obviously could not lead us to conclude that $\mu \geq 675$ *is* true, since 673.2 is less than 675. This is typical of many situations of interest. The test statistic is consistent with the alternate hypothesis and disagrees somewhat with the null. The only issue is whether the level of disagreement, measured with the P-value, is great enough to render the null hypothesis implausible.

How do we know when to reject H_0? The smaller the P-value, the less plausible H_0 becomes. A common rule of thumb is to draw the line at 5%. According to this rule of thumb, if $P \leq 0.05$, H_0 is rejected; otherwise, H_0 is not rejected. In fact, there is no sharp dividing line between conclusive evidence against H_0 and inconclusive evidence, just as there is no sharp dividing line between hot and cold weather. So while this rule of thumb is convenient, it has no real scientific justification.

Summary

- The smaller the P-value, the more certain we can be that H_0 is false.
- The larger the P-value, the more plausible H_0 becomes, but we can never be certain that H_0 is true.
- A rule of thumb suggests to reject H_0 whenever $P \leq 0.05$. While this rule is convenient, it has no scientific basis.

Statistical Significance

Whenever the P-value is less than a particular threshold, the result is said to be "statistically significant" at that level. So, for example, if $P \leq 0.05$, the result is statistically significant at the 5% level; if $P \leq 0.01$, the result is statistically significant at the 1% level, and so on. If a result is statistically significant at the $100\alpha\%$ level, we can also say that the null hypothesis is "rejected at level $100\alpha\%$."

A hypothesis test is performed of the null hypothesis $H_0 : \mu = 0$. The P-value turns out to be 0.02. Is the result statistically significant at the 10% level? The 5% level? The 1% level? Is the null hypothesis rejected at the 10% level? The 5% level? The 1% level?

Solution
The result is statistically significant at any level greater than or equal to 2%. Thus it is statistically significant at the 10% and 5% levels, but not at the 1% level. Similarly, we can reject the null hypothesis at any level greater than or equal to 2%, so H_0 is rejected at the 10% and 5% levels, but not at the 1% level.

Sometimes people report only that a test result was statistically significant at a certain level, without giving the P-value. It is common, for example, to read that a result was "statistically significant at the 5% level" or "statistically significant ($P < 0.05$)." This is a poor practice, for three reasons. First, it provides no way to tell whether the P-value was just barely less than 0.05 or whether it was a lot less. Second, reporting that a result was statistically significant at the 5% level implies that there is a big difference between a P-value just under 0.05 and one just above 0.05, when in fact there is little difference. Third, a report like this does not allow readers to decide for themselves whether the P-value is small enough to reject the null hypothesis. If a reader believes that the null hypothesis should not be rejected unless $P < 0.01$, then reporting only that $P < 0.05$ does not allow that reader to decide whether to reject H_0.

Reporting the P-value gives more information about the strength of the evidence against the null hypothesis and allows each reader to decide for himself or herself whether to reject. Software packages always output P-values; these should be included whenever the results of a hypothesis test are reported.

Summary

Let α be any value between 0 and 1. Then, if $P \leq \alpha$,

■ The result of the test is said to be statistically significant at the $100\alpha\%$ level.
■ The null hypothesis is rejected at the $100\alpha\%$ level.
■ When reporting the result of a hypothesis test, report the P-value, rather than just comparing it to 0.05 or 0.01.

The *P*-value Is Not the Probability That H_0 Is True

Since the P-value is a probability, and since small P-values indicate that H_0 is unlikely to be true, it is tempting to think that the P-value represents the probability that H_0 is true. This is emphatically not the case. The concept of probability discussed here is useful only when applied to outcomes that can turn out in different ways when experiments are repeated. It makes sense to define the P-value as the probability of observing an extreme value of a statistic such as $\overline{X}$, since the value of $\overline{X}$ could come out differently if the experiment were repeated. The null hypothesis, on the other hand, either is true or is not true. The truth or falsehood of H_0 cannot be changed by repeating the experiment. It is therefore not correct to discuss the "probability" that H_0 is true.

At this point we must mention that there is a notion of probability, different from that which we discuss in this book, in which one can compute a probability that a statement such as a null hypothesis is true. This kind of probability, called **subjective** probability, plays an important role in the theory of **Bayesian statistics**. The kind of probability we discuss in this book is called **frequentist** probability. A good reference for Bayesian statistics is Lee (1997).

Choose H_0 to Answer the Right Question

When performing a hypothesis test, it is important to choose H_0 and H_1 appropriately so that the result of the test can be useful in forming a conclusion. Examples 6.4 and 6.5 illustrate this.

Specifications for steel plate to be used in the construction of a certain bridge call for the minimum yield (F_y) to be greater than 345 MPa. Engineers will perform a hypothesis test to decide whether to use a certain type of steel. They will select a random sample of steel plates, measure their breaking strengths, and perform a hypothesis test. The steel will not be used unless the engineers can conclude that $\mu > 345$. Assume they test $H_0: \mu \leq 345$ versus $H_1: \mu > 345$. Will the engineers decide to use the steel if H_0 is rejected? What if H_0 is not rejected?

Solution
If H_0 is rejected, the engineers will conclude that $\mu > 345$, and they will use the steel. If H_0 is not rejected, the engineers will conclude that μ *might* be less than or equal to 345, and they will not use the steel.

In Example 6.4, the engineers' action with regard to using the steel will differ depending on whether H_0 is rejected or not rejected. This is therefore a useful test to perform, and H_0 and H_1 have been specified correctly.

In Example 6.4, assume the engineers test $H_0: \mu \geq 345$ versus $H_1: \mu < 345$. Will the engineers decide to use the steel if H_0 is rejected? What if H_0 is not rejected?

Solution
If H_0 is rejected, the engineers will conclude that $\mu < 345$, and they will not use the steel. If H_0 is not rejected, the engineers will conclude that μ *might* be greater than or equal to 345 but that it also might not be. So again, they won't use the steel.

In Example 6.5, the engineers' action with regard to using the steel will be the same—they won't use it—whether or not H_0 is rejected. There is no point in performing this test. The hypotheses H_0 and H_1 have not been specified correctly.

Final note: In a one-tailed test, the equality always goes with the null hypothesis. Thus if μ_0 is the point that divides H_0 from H_1, we may have $H_0: \mu \leq \mu_0$ or $H_0: \mu \geq \mu_0$, but never $H_0: \mu < \mu_0$ or $H_0: \mu > \mu_0$. The reason for this is that when defining the null distribution, we represent H_0 with the value of μ closest to H_1. Without the equality, there is no value specified by H_0 that is the closest to H_1. Therefore the equality must go with H_0.

Statistical Significance Is Not the Same as Practical Significance

When a result has a small P-value, we say that it is "statistically significant." In common usage, the word *significant* means "important." It is therefore tempting to think that statistically significant results must always be important. This is not the case. Sometimes statistically significant results do not have any scientific or practical importance. We will illustrate this with an example. Assume that a process used to manufacture synthetic fibers is known to produce fibers with a mean breaking strength of 50 N. A new process, which would require considerable retooling to implement, has been developed. In a sample of 1000 fibers produced by this new method, the average breaking strength was 50.1 N, and the standard deviation was 1 N. Can we conclude that the new process produces fibers with greater mean breaking strength?

To answer this question, let μ be the mean breaking strength of fibers produced by the new process. We need to test $H_0: \mu \leq 50$ versus $H_1: \mu > 50$. In this way, if we reject H_0, we will conclude that the new process is better. Under H_0, the sample mean $\overline{X}$ has a normal distribution with mean 50 and standard deviation $1/\sqrt{1000} = 0.0316$. The z-score is

$$z = \frac{50.1 - 50}{0.0316} = 3.16$$

The P-value is 0.0008. This is very strong evidence against H_0. The new process produces fibers with a greater mean breaking strength.

What practical conclusion should be drawn from this result? On the basis of the hypothesis test, we are quite sure that the new process is better. Would it be worthwhile to implement the new process? Probably not. The reason is that the difference between the old and new processes, although highly statistically significant, amounts to only 0.1 N. It is unlikely that this is difference is large enough to matter.

The lesson here is that a result can be statistically significant without being large enough to be of practical importance. How can this happen? A difference is statistically significant when it is large compared to its standard deviation. In the example, a difference of 0.1 N was statistically significant because the standard deviation was only 0.0316 N. When the standard deviation is very small, even a small difference can be statistically significant.

The P-value does not measure practical significance. What it does measure is the degree of confidence we can have that the true value is really different from the value specified by the null hypothesis. When the P-value is small, then we can be confident that the true value is really different. This does not necessarily imply that the difference is large enough to be of practical importance.

The Relationship Between Hypothesis Tests and Confidence Intervals

Both confidence intervals and hypothesis tests are concerned with determining plausible values for a quantity such as a population mean μ. In a hypothesis test for a population mean μ, we specify a particular value of μ (the null hypothesis) and determine whether that value is plausible. In contrast, a confidence interval for a population mean μ can be thought of as the collection of all values for μ that meet a certain criterion of plausibility, specified by the confidence level $100(1 - \alpha)\%$. In fact, the relationship between confidence intervals and hypothesis tests is very close.

To be specific, the values contained within a two-sided level $100(1-\alpha)\%$ confidence interval for a population mean μ are precisely those values for which the P-value of a two-tailed hypothesis test will be greater than α. To illustrate this, consider the following example (presented as Example 5.4 in Section 5.2). The sample mean lifetime of 50 microdrills was $\overline{X} = 12.68$ holes drilled, and the standard deviation was $s = 6.83$. Setting α to 0.05 (5%), the 95% confidence interval for the population mean lifetime μ was computed to be (10.79, 14.57). Suppose we wanted to test the hypothesis that μ was equal to one of the endpoints of the confidence interval. For example, consider testing $H_0: \mu = 10.79$ versus $H_1: \mu \neq 10.79$. Under H_0, the observed value $\overline{X} = 12.68$ comes from a normal distribution with mean 10.79 and standard deviation $6.83/\sqrt{50} = 0.9659$. The z-score is $(12.68 - 10.79)/0.9659 = 1.96$. Since H_0 specifies that μ is *equal* to 10.79, both tails contribute to the P-value, which is 0.05 and thus equal to α (see Figure 6.4).

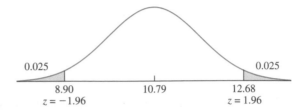

| 0.025 | | 0.025 |

8.90   10.79   12.68
$z = -1.96$ $z = 1.96$

FIGURE 6.4 The sample mean $\overline{X}$ is equal to 12.68. Since 10.79 is an endpoint of a 95% confidence interval based on $\overline{X} = 12.68$, the P-value for testing $H_0: \mu = 10.79$ is equal to 0.05.

Now consider testing the hypothesis $H_0: \mu = 14.57$ versus $H_1: \mu \neq 14.57$, where 14.57 is the other endpoint of the confidence interval. This time we will obtain $z = (12.68 - 14.57)/0.9659 = -1.96$, and again the P-value is 0.05. It is easy to check that if we choose any value μ_0 in the interval (10.79, 14.57) and test $H_0: \mu = \mu_0$ versus $H_1: \mu \neq \mu_0$, the P-value will be greater than 0.05. On the other hand, if we choose $\mu_0 < 10.79$ or $\mu_0 > 14.57$, the P-value will be less than 0.05. Thus the 95% confidence interval consists of precisely those values of μ whose P-values are greater than 0.05 in a hypothesis test. In this sense, the confidence interval contains all the values that are plausible for the population mean μ.

It is easy to check that a one-sided level $100(1 - \alpha)\%$ confidence interval consists of all the values for which the P-value in a one-tailed test would be greater than α. For example, with $\overline{X} = 12.68$, $s = 6.83$, and $n = 50$, the 95% lower confidence bound for the lifetime of the drills is 11.09. If $\mu_0 > 11.09$, then the P-value for testing $H_0: \mu \leq \mu_0$ will be greater than 0.05. Similarly, the 95% upper confidence bound for the lifetimes of the drills is 14.27. If $\mu_0 < 14.27$, then the P-value for testing $H_0: \mu \geq \mu_0$ will be greater than 0.05.

Exercises for Section 6.2

1. For which P-value is the null hypothesis more plausible: $P = 0.10$ or $P = 0.01$?

2. True or false:

a. If we reject H_0, then we conclude that H_0 is false.

b. If we do not reject H_0, then we conclude that H_0 is true.

c. If we reject H_0, then we conclude that H_1 is true.

d. If we do not reject H_0, then we conclude that H_1 is false.

3. If $P = 0.01$, which is the best conclusion?

   i. H_0 is definitely false.

   ii. H_0 is definitely true.

   iii. There is a 1% probability that H_0 is true.

   iv. H_0 might be true, but it's unlikely.

   v. H_0 might be false, but it's unlikely.

   vi. H_0 is plausible.

4. If $P = 0.50$, which is the best conclusion?

   i. H_0 is definitely false.

   ii. H_0 is definitely true.

   iii. There is a 50% probability that H_0 is true.

   iv. H_0 is plausible, and H_1 is false.

   v. Both H_0 and H_1 are plausible.

5. H_0 is rejected at the 5% level. True or false:

   a. The result is statistically significant at the 10% level.

   b. The result is statistically significant at the 5% level.

   c. The result is statistically significant at the 1% level.

6. George performed a hypothesis test. Luis checked George's work by redoing the calculations. Both George and Luis agree that the result was statistically significant the 5% level, but they got different P-values. George got a P-value of 0.20, and Luis got a P-value of 0.02.

   a. Is is possible that George's work is correct? Explain.

   b. Is is possible that Luis's work is correct? Explain.

7. The following MINITAB output presents the results of a hypothesis test for a population mean μ.

```
One-Sample Z: X

Test of mu = 37 vs not = 37
The assumed standard deviation = 3.2614


Variable   N     Mean    StDev   SE Mean        95% CI          Z       P
X          87   36.5280  3.2614   0.3497   (35.8247, 37.2133)  −1.35   0.177
```

   a. Can H_0 be rejected at the 5% level? How can you tell?

   b. Someone asks you whether the null hypothesis $H_0 : \mu = 36$ versus $H_1 : \mu \neq 36$ can be rejected at the 5% level. Can you answer without doing any calculations? How?

8. Let μ be the radiation level to which a radiation worker is exposed during the course of a year. The Environmental Protection Agency has set the maximum safe level of exposure at 5 rem per year. If a hypothesis test is to be performed to determine whether a workplace is safe, which is the most appropriate null hypothesis: $H_0 : \mu \leq 5$, $H_0 : \mu \geq 5$, or $H_0 : \mu = 5$? Explain.

9. In each of the following situations, state the most appropriate null hypothesis regarding the population mean μ.

   a. A new type of epoxy will be used to bond wood pieces if it can be shown to have a mean shear stress greater than 10 MPa.

   b. A quality control inspector will recalibrate a flowmeter if the mean flow rate differs from 20 mL/s.

   c. A new type of battery will be installed in heart pacemakers if it can be shown to have a mean lifetime greater than eight years.

10. The installation of a radon abatement device is recommended in any home where the mean radon concentration is 4.0 picocuries per liter (pCi/L) or more, because it is thought that long-term exposure to sufficiently high doses of radon can increase the risk of cancer. Seventy-five measurements are made in a particular home. The mean concentration was 3.72 pCi/L, and the standard deviation was 1.93 pCi/L.

   a. The home inspector who performed the test says that since the mean measurement is less than 4.0, radon abatement is not necessary. Explain why this reasoning is incorrect.

b. Because of health concerns, radon abatement is recommended whenever it is plausible that the mean radon concentration may be 4.0 pCi/L or more. State the appropriate null and alternate hypotheses for determining whether radon abatement is appropriate.

c. Compute the P-value. Would you recommend radon abatement? Explain.

11. It is desired to check the calibration of a scale by weighing a standard 10 g weight 100 times. Let μ be the population mean reading on the scale, so that the scale is in calibration if $\mu = 10$. A test is made of the hypotheses $H_0 : \mu = 10$ versus $H_1 : \mu \neq 10$. Consider three possible conclusions: (i) The scale is in calibration. (ii) The scale is out of calibration. (iii) The scale might be in calibration.

a. Which of the three conclusions is best if H_0 is rejected?

b. Which of the three conclusions is best if H_0 is not rejected?

c. Is it possible to perform a hypothesis test in a way that makes it possible to demonstrate conclusively that the scale is in calibration? Explain.

12. A machine that fills cereal boxes is supposed to be calibrated so that the mean fill weight is 12 oz. Let μ denote the true mean fill weight. Assume that in a test of the hypotheses $H_0 : \mu = 12$ versus $H_1 : \mu \neq 12$, the P-value is 0.30.

a. Should H_0 be rejected on the basis of this test? Explain.

b. Can you conclude that the machine is calibrated to provide a mean fill weight of 12 oz? Explain.

13. A method of applying zinc plating to steel is supposed to produce a coating whose mean thickness is no greater than 7 microns. A quality inspector measures the thickness of 36 coated specimens and tests $H_0 : \mu \leq 7$ versus $H_1 : \mu > 7$. She obtains a P-value of 0.40. Since $P > 0.05$, she concludes that the mean thickness is within the specification. Is this conclusion correct? Explain.

14. Fill in the blank: A 95% confidence interval for μ is (1.2, 2.0). Based on the data from which the confidence interval was constructed, someone wants to test $H_0 : \mu = 1.4$ versus $H_1 : \mu \neq 1.4$. The P-value will be _____.

i. greater than 0.05
ii. less than 0.05
iii. equal to 0.05

15. Refer to Exercise 14. For which null hypothesis will $P = 0.05$?

i. $H_0 : \mu = 1.2$
ii. $H_0 : \mu \leq 1.2$
iii. $H_0 : \mu \geq 1.2$

16. A scientist computes a 90% confidence interval to be (4.38, 6.02). Using the same data, she also computes a 95% confidence interval to be (4.22, 6.18), and a 99% confidence interval to be (3.91, 6.49). Now she wants to test $H_0 : \mu = 4$ versus $H_1 : \mu \neq 4$. Regarding the P-value, which one of the following statements is true?

i. $P > 0.10$
ii. $0.05 < P < 0.10$
iii. $0.01 < P < 0.05$
iv. $P < 0.01$

17. The strength of a certain type of rubber is tested by subjecting pieces of the rubber to an abrasion test. For the rubber to be acceptable, the mean weight loss μ must be less than 3.5 mg. A large number of pieces of rubber that were cured in a certain way were subject to the abrasion test. A 95% upper confidence bound for the mean weight loss was computed from these data to be 3.45 mg. Someone suggests using these data to test $H_0 : \mu \geq 3.5$ versus $H_1 : \mu < 3.5$.

a. Is it possible to determine from the confidence bound whether $P < 0.05$? Explain.

b. Is it possible to determine from the confidence bound whether $P < 0.01$? Explain.

18. A shipment of fibers is not acceptable if the mean breaking strength of the fibers is less than 50 N. A large sample of fibers from this shipment was tested, and a 98% lower confidence bound for the mean breaking strength was computed to be 50.1 N. Someone suggests using these data to test the hypotheses $H_0 : \mu \leq 50$ versus $H_1 : \mu > 50$.

a. Is it possible to determine from the confidence bound whether $P < 0.01$? Explain.

b. Is it possible to determine from the confidence bound whether $P < 0.05$? Explain.

19. Refer to Exercise 17. It is discovered that the mean of the sample used to compute the confidence bound is $\overline{X} = 3.40$. Is it possible to determine whether $P < 0.01$? Explain.

20. Refer to Exercise 18. It is discovered that the standard deviation of the sample used to compute the confidence interval is 5 N. Is it possible to determine whether $P < 0.01$? Explain.

6.3 Tests for a Population Proportion

Hypothesis tests for proportions are similar to those discussed in Section 6.1 for population means. Here is an example.

A supplier of semiconductor wafers claims that of all the wafers he supplies, no more than 10% are defective. A sample of 400 wafers is tested, and 50 of them, or 12.5%, are defective. Can we conclude that the claim is false?

The hypothesis test here proceeds much like those in Section 6.1. What makes this problem distinct is that the sample consists of successes and failures, with "success" indicating a defective wafer. If the population proportion of defective wafers is denoted by p, then the supplier's claim is that $p \leq 0.1$. Now if we let X represent the number of wafers in the sample that are defective, then $X \sim \text{Bin}(n, p)$ where $n = 400$ is the sample size. In this example, we have observed $X = 50$. Since our hypothesis concerns a population proportion, it is natural to base the test on the sample proportion $\hat{p} = X/n$. In this example, we have observed $\hat{p} = 50/400 = 0.125$. Making the reasonable assumption that the wafers are sampled independently, then since the sample size is large, it follows from the Central Limit Theorem (Equation 4.35 in Section 4.8) that

$$X \sim N(np,\ np(1 - p)) \tag{6.1}$$

Since $\hat{p} = X/n$, it follows that

$$\hat{p} \sim N\left(p,\ \frac{p(1 - p)}{n}\right) \tag{6.2}$$

We must define the null hypothesis. The question asked is whether the data allow us to conclude that the supplier's claim is false. Therefore, the supplier's claim, which is that $p \leq 0.1$, must be H_0. Otherwise, it would be impossible to prove the claim false, no matter what the data showed.

The null and alternate hypotheses are

$$H_0: p \leq 0.1 \quad \text{versus} \quad H_1: p > 0.1$$

To perform the hypothesis test, we assume H_0 to be true and take $p = 0.1$. Substituting $p = 0.1$ and $n = 400$ in expression (6.2) yields the null distribution of $\hat{p}$:

$$\hat{p} \sim N(0.1,\ 2.25 \times 10^{-4})$$

The standard deviation of $\hat{p}$ is $\sigma_{\hat{p}} = \sqrt{2.25 \times 10^{-4}} = 0.015$. The observed value of $\hat{p}$ is $50/400 = 0.125$. The z-score of $\hat{p}$ is

$$z = \frac{0.125 - 0.100}{0.015} = 1.67$$

The z table indicates that the probability that a standard normal random variable has a value greater than 1.67 is approximately 0.0475. The P-value is therefore 0.0475 (see Figure 6.5).

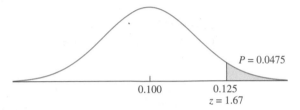

$P = 0.0475$

0.100   0.125
              $z = 1.67$

FIGURE 6.5 The null distribution of $\hat{p}$ is $N(0.1, \ 0.015^2)$. Therefore if H_0 is true, the probability that $\hat{p}$ takes on a value as extreme as or more extreme than the observed value of 0.125 is 0.0475. This is the P-value.

What do we conclude about H_0? Either the supplier's claim is false, or we have observed a sample that is as extreme as all but 4.75% of the samples we might have drawn. Such a sample would be unusual but not fantastically unlikely. There is every reason to be quite skeptical of the claim, but we probably shouldn't convict the supplier quite yet. If possible, it would be a good idea to sample more wafers.

Note that under the commonly used rule of thumb, we would reject H_0 and condemn the supplier, because P is less than 0.05. This example illustrates the weakness of this rule. If you do the calculations, you will find that if only 49 of the sample wafers had been defective rather than 50, the P-value would have risen to 0.0668, and the supplier would be off the hook. Thus the fate of the supplier hangs on the outcome of one single wafer out of 400. It doesn't make sense to draw such a sharp line. It's better just to report the P-value and wait for more evidence before reaching a firm conclusion.

The Sample Size Must Be Large

The test just described requires that the sample proportion be approximately normally distributed. This assumption will be justified whenever both $np_0 > 10$ and $n(1 - p_0) > 10$, where p_0 is the population proportion specified in the null distribution. Then the z-score can be used as the test statistic, making this a z test.

Example
6.6

The article "Refinement of Gravimetric Geoid Using GPS and Leveling Data" (W. Thurston, *Journal of Surveying Engineering*, 2000:27–56) presents a method for measuring orthometric heights above sea level. For a sample of 1225 baselines, 926 gave results that were within the class C spirit leveling tolerance limits. Can we conclude that this method produces results within the tolerance limits more than 75% of the time?

Solution

Let p denote the probability that the method produces a result within the tolerance limits. The null and alternate hypotheses are

$$H_0: p \le 0.75 \quad \text{versus} \quad H_1: p > 0.75$$

The sample proportion is $\hat{p} = 926/1225 = 0.7559$. Under the null hypothesis, $\hat{p}$ is normally distributed with mean 0.75 and standard deviation $\sqrt{(0.75)(1 - 0.75)/1225} = 0.0124$. The z-score is

$$z = \frac{0.7559 - 0.7500}{0.0124} = 0.48$$

The P-value is 0.3156 (see Figure 6.6). We cannot conclude that the method produces good results more than 75% of the time.

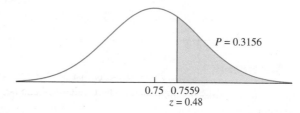

FIGURE 6.6 The null distribution of $\hat{p}$ is $N(0.75,\ 0.0124^2)$. Thus if H_0 is true, the probability that $\hat{p}$ takes on a value as extreme as or more extreme than the observed value of 0.7559 is 0.3156. This is the P-value.

The following computer output (from MINITAB) presents the results from Example 6.6.

```
Test and CI for One Proportion: GPS

Test of p = 0.75 vs p > 0.75
                                    95%
                                  Lower
Variable    X     N   Sample p    Bound   Z-Value  P-Value
GPS        926  1225  0.755918  0.735732    0.48     0.316
```

The output contains a 95% lower confidence bound as well as the P-value.

Relationship with Confidence Intervals for a Proportion

A level $100(1 - \alpha)\%$ confidence interval for a population mean μ contains those values for a parameter for which the P-value of a hypothesis test will be greater than α. For the confidence intervals for a proportion presented in Section 5.3 and the hypothesis test presented here, this statement is only approximately true. The reason for this is that the methods presented in Section 5.3 are slight modifications (which are much easier to compute) of a more complicated confidence interval method for which the statement is exactly true.

Summary

Let X be the number of successes in n independent Bernoulli trials, each with success probability p; in other words, let $X \sim \text{Bin}(n, p)$.

To test a null hypothesis of the form $H_0: p \leq p_0$, $H_0: p \geq p_0$, or $H_0: p = p_0$, assuming that both np_0 and $n(1 - p_0)$ are greater than 10:

- Compute the z-score: $z = \dfrac{\hat{p} - p_0}{\sqrt{p_0(1 - p_0)/n}}$.

- Compute the P-value. The P-value is an area under the normal curve, which depends on the alternate hypothesis as follows:

Alternate Hypothesis	P-value
$H_1: p > p_0$	Area to the right of z
$H_1: p < p_0$	Area to the left of z
$H_1: p \neq p_0$	Sum of the areas in the tails cut off by z and $-z$

Exercises for Section 6.3

1. Gravel pieces are classified as small, medium, or large. A vendor claims that at least 10% of the gravel pieces from her plant are large. In a random sample of 1600 pieces, 130 pieces were classified as large. Is this enough evidence to reject the claim?

2. Do patients value interpersonal skills more than technical ability when choosing a primary care physician? The article "Patients' Preferences for Technical Versus Interpersonal Quality When Selecting a Primary Care Physician" (C. Fung, M. Elliot, et al., *Health Services Research*, 2005:957–977) reports the results of a study in which 304 people were asked to choose a physician based on two hypothetical descriptions. One physician was described as having high technical skills and average interpersonal skills, and the other was described as having average technical skills and high interpersonal skills. Sixty-two percent of the people chose the physician with high technical skills. Can you conclude that more than half of patients prefer a physician with high technical skills?

3. Do bathroom scales tend to underestimate a person's true weight? A 150 lb test weight was placed on each of 50 bathroom scales. The readings on 29 of the scales were too light, and the readings on the other 21 were too heavy. Can you conclude that more than half of bathroom scales underestimate weight?

4. Incinerators can be a source of hazardous emissions into the atmosphere. Stack gas samples were collected from a sample of 50 incinerators in a major city. Of the 50 samples, only 18 met an environmental standard for the concentration of a hazardous compound. Can it be concluded that fewer than half of the incinerators in the city meet the standard?

5. In a survey of 500 residents in a certain town, 274 said they were opposed to constructing a new shopping mall. Can you conclude that more than half of the residents in this town are opposed to constructing a new shopping mall?

6. A random sample of 80 bolts is sampled from a day's production, and 4 of them are found to have diameters below specification. It is claimed that the proportion of defective bolts among those manufactured that day is less than 0.10. Is it appropriate to use the methods of this section to determine whether we can reject this claim? If so, state the appropriate null and alternate hypotheses and compute the P-value. If not, explain why not.

7. In a sample of 150 households in a certain city, 110 had high-speed Internet access. Can you conclude that more than 70% of the households in this city have high-speed Internet access?

8. A grinding machine will be qualified for a particular task if it can be shown to produce less than 8% defective parts. In a random sample of 300 parts, 12 were defective. On the basis of these data, can the machine be qualified?

9. The manufacturer of a certain voltmeter claims that 95% or more of its readings are within 0.1% of the true value. In a sample of 500 readings, 470 were within 0.1% of the true value. Is there enough evidence to reject the claim?

10. Refer to Exercise 1 in Section 5.3. Can it be concluded that less than half of the automobiles in the state have pollution levels that exceed the standard?

11. Refer to Exercise 2 in Section 5.3. Can it be concluded that more than 60% of the residences in the town reduced their water consumption?

12. The following MINITAB output presents the results of a hypothesis test for a population proportion p.

```
Test and CI for One Proportion: X

Test of p = 0.4 vs p < 0.4
```

				95% Upper		
Variable	X	N	Sample p	Bound	Z-Value	P-Value
X	73	240	0.304167	0.353013	−3.03	0.001

a. Is this a one-tailed or two-tailed test?
b. What is the null hypothesis?
c. Can H_0 be rejected at the 2% level? How can you tell?
d. Someone asks you whether the null hypothesis $H_0 : p \geq 0.45$ versus $H_1 : p < 0.45$ can be rejected at the 2% level. Can you answer without doing any calculations? How?
e. Use the output and an appropriate table to compute the P-value for the test of $H_0 : p \leq 0.25$ versus $H_1 : p > 0.25$.
f. Use the output and an appropriate table to compute a 90% confidence interval for p.

13. The following MINITAB output presents the results of a hypothesis test for a population proportion p. Some of the numbers are missing. Fill them in.

```
Test and CI for One Proportion: X

Test of p = 0.7 vs p < 0.7
```

				95% Upper		
Variable	X	N	Sample p	Bound	Z-Value	P-Value
X	345	500	(a)	0.724021	(b)	(c)

6.4 Small-Sample Tests for a Population Mean

In Section 6.1, we described a method for testing a hypothesis about a population mean, based on a large sample. A key step in the method is to approximate the population standard deviation σ with the sample standard deviation s. The normal curve is then used to find the P-value. When the sample size is small, s may not be close to σ, which invalidates this large-sample method. However, when the population is approximately normal, the Student's t distribution can be used. We illustrate with an example.

Spacer collars for a transmission countershaft have a thickness specification of 38.98–39.02 mm. The process that manufactures the collars is supposed to be calibrated so that the mean thickness is 39.00 mm, which is in the center of the specification window. A sample of six collars is drawn and measured for thickness. The six thicknesses are 39.030, 38.997, 39.012, 39.008, 39.019, and 39.002. Assume that the population of thicknesses of the collars is approximately normal. Can we conclude that the process needs recalibration?

Denoting the population mean by μ, the null and alternate hypotheses are

$$H_0: \mu = 39.00 \quad \text{versus} \quad H_1: \mu \neq 39.00$$

Note that H_0 specifies a single value for μ, since calibration requires that the mean be equal to the correct value. To construct the test statistic, note that since the population is assumed to follow a normal distribution, the quantity

$$t = \frac{\overline{X} - \mu}{s/\sqrt{n}}$$

has a Student's t distribution with $n - 1 = 5$ degrees of freedom. This is the test statistic.

In this example, the observed values of the sample mean and standard deviation are $\overline{X} = 39.01133$ and $s = 0.011928$. The sample size is $n = 6$. The null hypothesis specifies that $\mu = 39$. The value of the test statistic is therefore

$$t = \frac{39.01133 - 39.00}{0.011928/\sqrt{6}} = 2.327$$

The P-value is the probability of observing a value of the test statistic whose disagreement with H_0 is as great as or greater than that actually observed. Since H_0 specifies that $\mu = 39.00$, this is a two-tailed test, so values both above and below 39.00 disagree with H_0. Therefore the P-value is the sum of the areas under the curve corresponding to $t > 2.327$ and $t < -2.327$.

Figure 6.7 illustrates the null distribution and indicates the location of the test statistic. From the t table (Table A.3 in Appendix A), the row corresponding to 5 degrees

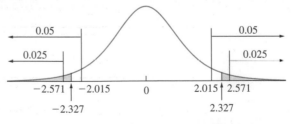

FIGURE 6.7 The null distribution of $t = (\overline{X} - 39.00)/(s/\sqrt{6})$ is Student's t with five degrees of freedom. The observed value of t, corresponding to the observed values $\overline{X} = 39.01133$ and $s = 0.011928$, is 2.327. If H_0 is true, the probability that t takes on a value as extreme as or more extreme than that observed is between 0.05 and 0.10. Because H_0 specified that μ was *equal* to a specific value, both tails of the curve contribute to the P-value.

of freedom indicates that the value $t = \pm 2.015$ cuts off an area of 0.05 in each tail, for a total of 0.10, and that the value $t = \pm 2.571$ cuts off an area of 0.025 in each tail, for a total of 0.05. Thus the P-value is between 0.05 and 0.10. While we cannot conclusively state that the process is out of calibration, it doesn't look too good. It would be prudent to recalibrate.

In this example, the test statistic was a t statistic rather than a z-score. For this reason, this test is referred to as a t test.

Example

6.7

Before a substance can be deemed safe for landfilling, its chemical properties must be characterized. The article "Landfilling Ash/Sludge Mixtures" (J. Benoît, T. Eighmy, and B. Crannell, *Journal of Geotechnical and Geoenvironmental Engineering*, 1999: 877–888) reports that in a sample of six replicates of sludge from a New Hampshire wastewater treatment plant, the mean pH was 6.68 with a standard deviation of 0.20. Can we conclude that the mean pH is less than 7.0?

Solution

Let μ denote the mean pH for this type of sludge. The null and alternate hypotheses are

$$H_0 : \mu \geq 7.0 \quad \text{versus} \quad H_1 : \mu < 7.0$$

Under H_0, the test statistic

$$t = \frac{\overline{X} - 7.0}{s/\sqrt{n}}$$

has a Student's t distribution with five degrees of freedom. Substituting $\overline{X} = 6.68$, $s = 0.20$, and $n = 6$, the value of the test statistic is

$$t = \frac{6.68 - 7.00}{0.20/\sqrt{6}} = -3.919$$

Consulting the t table, we find that the value $t = -3.365$ cuts off an area of 0.01 in the left-hand tail, and the value $t = -4.033$ cuts off an area of 0.005 (see Figure 6.8). We conclude that the P-value is between 0.005 and 0.01. There is strong evidence that the mean pH is less than 7.0.

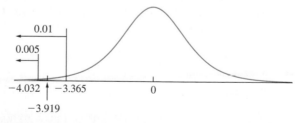

FIGURE 6.8 Solution to Example 6.7. The null distribution is Student's t with five degrees of freedom. The observed value of t is -3.919. If H_0 is true, the probability that t takes on a value as extreme as or more extreme than that observed is between 0.005 and 0.01.

```
One-Sample T: pH

Test of mu = 7 vs < 7
                                                    95%
                                                  Upper
Variable    N    Mean    StDev    SE Mean        Bound        T        P
pH          6    6.680   0.200    0.081665      6.84453    -3.92    0.006
```

Note that the upper 95% confidence bound provided in the output is consistent with the alternate hypothesis. This indicates that the P-value is less than 5%.

Use z, Not t, If σ Is Known

Occasionally a small sample may be taken from a normal population whose standard deviation σ is known. In these cases, we do not use the Student's t curve because we are not approximating σ with s. Instead, we use the z table and perform a z test. Example 6.8 illustrates the method.

At the beginning of this section, we described a sample of six spacer collars, whose thicknesses (in mm) were 39.030, 38.997, 39.012, 39.008, 39.019, and 39.002. We denoted the population mean thickness by μ and tested the hypotheses

$$H_0: \mu = 39.00 \quad \text{versus} \quad H_1: \mu \neq 39.00$$

Now assume that these six spacer collars were manufactured just after the machine that produces them had been moved to a new location. Assume that on the basis of a very large number of collars manufactured before the move, the population of collar thicknesses is known to be very close to normal, with standard deviation $\sigma = 0.010$ mm, and that it is reasonable to assume that the move has not changed this. On the basis of the given data, can we reject H_0? We compute $\overline{X} = 39.01133$. We do not need the value of s, because we know that $\sigma = 0.010$. Since the population is normal, $\overline{X}$ is normal even though the sample size is small. The null distribution is therefore

$$\overline{X} \sim N(39.00, \ 0.010^2)$$

The z-score is

$$z = \frac{39.01133 - 39.000}{0.010/\sqrt{6}} = 2.78$$

The P-value is 0.0054, so H_0 can be rejected.

Summary

Let $X_1, \ldots, X_n$ be a sample from a *normal* population with mean μ and standard deviation σ, where σ is unknown.

To test a null hypothesis of the form $H_0 : \mu \leq \mu_0$, $H_0 : \mu \geq \mu_0$, or $H_0 : \mu = \mu_0$:

- Compute the test statistic $t = \dfrac{\overline{X} - \mu_0}{s/\sqrt{n}}$.

- Compute the P-value. The P-value is an area under the Student's t curve with $n - 1$ degrees of freedom, which depends on the alternate hypothesis as follows:

Alternate Hypothesis	**P-value**
$H_1 : \mu > \mu_0$	Area to the right of t
$H_1 : \mu < \mu_0$	Area to the left of t
$H_1 : \mu \neq \mu_0$	Sum of the areas in the tails cut off by t and $-t$

- If σ is known, the test statistic is $z = \dfrac{\overline{X} - \mu_0}{\sigma/\sqrt{n}}$, and a z test should be performed.

Exercises for Section 6.4

1. Each of the following hypothetical data sets represents some repeated measurements on the concentration of carbon monoxide (CO) in a gas sample whose CO concentration is known to be 60 ppm. Assume that the readings are a random sample from a population that follows the normal curve. Perform a t test to see whether the measuring instrument is properly calibrated, if possible. If impossible, explain why.

   a. 60.02, 59.98, 60.03

   b. 60.01

2. A geologist is making repeated measurements (in grams) on the mass of a rock. It is not known whether the measurements are a random sample from an approximately normal population. Following are three sets of replicate measurements, listed in the order they were made. For each set of readings, state whether the assumptions necessary for the validity of the t test appear to be met. If the assumptions are not met, explain why.

   a. 213.03 212.95 213.04 213.00 212.99
      213.01 221.03 213.05

   b. 213.05 213.00 212.94 213.09 212.98
      213.02 213.06 212.99

   c. 212.92 212.95 212.97 213.00 213.01 213.04
      213.05 213.06

3. Suppose you have purchased a filling machine for candy bags that is supposed to fill each bag with 16 oz of candy. Assume that the weights of filled bags are approximately normally distributed. A random sample of 10 bags yields the following data (in oz):

15.87	16.02	15.78	15.83	15.69	15.81
16.04	15.81	15.92	16.10		

   On the basis of these data, can you conclude that the mean fill weight is actually less than 16 oz?

   a. State the appropriate null and alternate hypotheses.

   b. Compute the value of the test statistic.

   c. Find the P-value and state your conclusion.

4. A certain manufactured product is supposed to contain 23% potassium by weight. A sample of 10 specimens of this product had an average percentage of 23.2 with a standard deviation of 0.2. If the mean percentage is found to differ from 23, the manufacturing process will be recalibrated.

   a. State the appropriate null and alternate hypotheses.

b. Compute the P-value.

c. Should the process be recalibrated? Explain.

5. Measurements were made of total solids for seven wastewater sludge specimens. The results, in grams, were

$$15 \quad 25 \quad 21 \quad 28 \quad 23 \quad 17 \quad 29$$

a. Can you conclude that the mean amount of total solids is greater than 20 g?

b. Can you conclude that the mean amount of total solids is less than 30 g?

c. An environmental scientists claims that the mean amount of total solids is 25 g. Does the sample provide evidence to reject this claim?

6. The thicknesses of six pads designed for use in aircraft engine mounts were measured. The results, in mm, were 40.93, 41.11, 41.47, 40.96, 40.80, and 41.32.

a. Can you conclude that the mean thickness is greater than 41 mm?

b. Can you conclude that the mean thickness is less than 41.4 mm?

c. The target thickness is 41.2 mm. Can you conclude that the mean thickness differs from the target value?

7. Specifications call for the wall thickness of two-liter polycarbonate bottles to average 4.0 mils. A quality control engineer samples 7 two-liter polycarbonate bottles from a large batch and measures the wall thickness (in mils) in each. The results are 3.999, 4.037, 4.116, 4.063, 3.969, 3.955, and 4.091. It is desired to test $H_0: \mu = 4.0$ versus $H_1: \mu \neq 4.0$.

a. Make a dotplot of the seven values.

b. Should a Student's t test be used to test H_0? If so, perform the test. If not, explain why not.

c. Measurements are taken of the wall thicknesses of seven bottles of a different type. The measurements this time are: 4.065, 3.967, 4.028, 4.008, 4.195, 4.057, and 4.010. Make a dotplot of these values.

d. Should a Student's t test be used to test $H_0: \mu = 4.0$ versus $H_1: \mu \neq 4.0$? If so, perform the test. If not, explain why not.

8. The article "Solid-Phase Chemical Fractionation of Selected Trace Metals in Some Northern Kentucky Soils" (A. Karathanasis and J. Pils, *Soil and Sediment Contamination*, 2005:293–308) reports that in a sample of 26 soil specimens taken in a region of northern Kentucky, the average concentration of chromium (Cr) in mg/kg was 20.75 with a standard deviation of 3.93.

a. Can you conclude that the mean concentration of Cr is greater than 20 mg/kg?

b. Can you conclude that the mean concentration of Cr is less than 25 mg/kg?

9. Benzene conversions (in mole percent) were measured for 16 different benzenehydroxylation reactions. The sample mean was 45.2 with a standard deviation of 11.3.

a. Can you conclude that the mean conversion is greater than 35?

b. Can you conclude that the mean conversion differs from 50?

10. Refer to Exercise 12 in Section 5.4. Can you conclude that the mean amount of toluene removed in the rinse is less than 8%?

11. Refer to Exercise 13 in Section 5.4. Can you conclude that the mean amount of uniconazole absorbed is less than 2.5 μg?

---

12. The following MINITAB output presents the results of a hypothesis test for a population mean μ.

```
One-Sample T: X

Test of mu = 5.5 vs > 5.5
```

Variable	N	Mean	StDev	SE Mean	95% Lower Bound	T	P
X	5	5.92563	0.15755	0.07046	5.77542	6.04	0.002

a. Is this a one-tailed or two-tailed test?
b. What is the null hypothesis?
c. Can H_0 be rejected at the 1% level? How can you tell?
d. Use the output and an appropriate table to compute the P-value for the test of $H_0: \mu \geq 6.5$ versus $H_1: \mu < 6.5$.
e. Use the output and an appropriate table to compute a 99% confidence interval for μ.

13. The following MINITAB output presents the results of a hypothesis test for a population mean μ. Some of the numbers are missing. Fill them in.

```
One-Sample T: X

Test of mu = 16 vs not = 16

Variable   N     Mean   StDev   SE Mean    95% CI      T      P
X         11   13.2874   (a)    1.8389   ((b), (c))  (d)   0.171
```

6.5 The Chi-Square Test

In Section 6.3, we learned how to test a null hypothesis about a success probability p. The data involve a number of trials, each of which results in one of two outcomes: success or failure. A generalization of this concept is the **multinomial trial**, which is an experiment that can result in any one of k outcomes, where $k \geq 2$. The probabilities of the k outcomes are denoted $p_1, \ldots, p_k$. For example, the roll of a fair die is a multinomial trial with six outcomes $1, 2, 3, 4, 5, 6$; and probabilities $p_1 = p_2 = p_3 = p_4 = p_5 = p_6 = 1/6$. In this section, we generalize the tests for a success probability to multinomial trials. We begin with an example in which we test the null hypothesis that the multinomial probabilities $p_1, p_2, \ldots, p_k$ are equal to a prespecified set of values $p_{01}, p_{02}, \ldots, p_{0k}$, so that the null hypothesis has the form $H_0: p_1 = p_{01}, p_2 = p_{02}, \ldots, p_k = p_{0k}$.

Imagine that a gambler wants to test a die to see whether it deviates from fairness. Let p_i be the probability that the number i comes up. The null hypothesis will state that the die is fair, so the null hypothesis is $H_0: p_1 = \cdots = p_6 = 1/6$. The gambler rolls the die 600 times and obtains the results shown in Table 6.1, in the column labeled "Observed." The

TABLE 6.1 Observed and expected values for 600 rolls of a die

Category	Observed	Expected
1	115	100
2	97	100
3	91	100
4	101	100
5	110	100
6	86	100
Total	600	600

results obtained are called the **observed values**. To test the null hypothesis, we construct a second column, labeled "Expected." This column contains the **expected values**. The expected value for a given outcome is the mean number of trials that would result in that outcome if H_0 were true. To compute the expected values, let N be the total number of trials. (In the die example, $N = 600$.) When H_0 is true, the probability that a trial results in outcome i is p_{0i}, so the expected number of trials resulting in outcome i is Np_{0i}. In the die example, the expected number of trials for each outcome is 100.

The idea behind the hypothesis test is that if H_0 is true, then the observed and expected values are likely to be close to each other. Therefore, we will construct a test statistic that measures the closeness of the observed to the expected values. The statistic is called the **chi-square statistic**. To define it, let k be the number of outcomes ($k = 6$ in the die example), and let O_i and E_i be the observed and expected numbers of trials, respectively, that result in outcome i. The chi-square statistic is

$$\chi^2 = \sum_{i=1}^{k} \frac{(O_i - E_i)^2}{E_i} \tag{6.3}$$

The larger the value of χ^2, the stronger the evidence against H_0. To determine the P-value for the test, we must know the null distribution of this test statistic. In general, we cannot determine the null distribution exactly. However, when the expected values are all sufficiently large, a good approximation is available. It is called the **chi-square distribution** with $k-1$ degrees of freedom, denoted χ^2_{k-1}. Note that the number of degrees of freedom is one less than the number of categories. Use of the chi-square distribution is appropriate whenever all the expected values are greater than or equal to 5.

A table for the chi-square distribution (Table A.5) is provided in Appendix A. The table provides values for certain quantiles, or upper percentage points, for a large number of choices of degrees of freedom. As an example, Figure 6.9 presents the probability density function of the χ^2_{10} distribution. The upper 5% of the distribution is shaded. To find the upper 5% point in the table, look under $\alpha = 0.05$ and degrees of freedom $\nu = 10$. The value is 18.307.

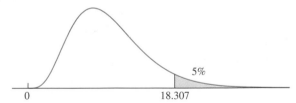

FIGURE 6.9 Probability density function of the χ^2_{10} distribution. The upper 5% point is 18.307. (See the chi-square table, Table A.5, in Appendix A.)

We now compute the value of the chi-square statistic for the data in Table 6.1. The number of degrees of freedom is 5 (one less than the number of outcomes). Using Equation (6.3), we find that the value of the statistic is

$$\chi^2 = \frac{(115 - 100)^2}{100} + \cdots + \frac{(86 - 100)^2}{100}$$
$$= 2.25 + \cdots + 1.96$$
$$= 6.12$$

To determine the P-value for the test statistic, we first note that all the expected values are greater than or equal to 5, so use of the chi-square distribution is appropriate. We consult the chi-square table under five degrees of freedom. The upper 10% point is 9.236. We conclude that $P > 0.10$. (See Figure 6.10.) There is no evidence to suggest that the die is not fair.

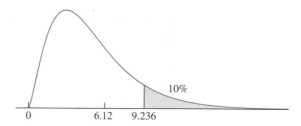

FIGURE 6.10 Probability density function of the χ_5^2 distribution. The observed value of the test statistic is 6.12. The upper 10% point is 9.236. Therefore the P-value is greater than 0.10.

The Chi-Square Test for Homogeneity

In the previous example, we tested the null hypothesis that the probabilities of the outcomes for a multinomial trial were equal to a prespecified set of values. Sometimes several multinomial trials are conducted, each with the same set of possible outcomes. The null hypothesis is that the probabilities of the outcomes are the same for each experiment. We present an example.

Four machines manufacture cylindrical steel pins. The pins are subject to a diameter specification. A pin may meet the specification, or it may be too thin or too thick. Pins are sampled from each machine, and the number of pins in each category is counted. Table 6.2 presents the results.

TABLE 6.2 Observed numbers of pins in various categories with regard to a diameter specification

	Too Thin	OK	Too Thick	Total
Machine 1	10	102	8	120
Machine 2	34	161	5	200
Machine 3	12	79	9	100
Machine 4	10	60	10	80
Total	66	402	32	500

Table 6.2 is an example of a **contingency table**. Each row specifies a category regarding one criterion (machine, in this case), and each column specifies a category regarding another criterion (thickness, in this case). Each intersection of row and column is called a **cell**, so there are 12 cells in Table 6.2.

The number in the cell at the intersection of row i and column j is the number of trials whose outcome was observed to fall into row category i and into column category j. This number is called the **observed value** for cell ij. Note that we have included the totals of the observed values for each row and column. These are called the **marginal totals**.

The null hypothesis is that the proportion of pins that are too thin, OK, or too thick is the same for all machines. More generally, the null hypothesis says that no matter which row is chosen, the probabilities of the outcomes associated with the columns are the same. We will develop some notation with which to express H_0 and to define the test statistic.

Let I denote the number of rows in the table, and let J denote the number of columns. Let p_{ij} denote the probability that the outcome of a trial falls into column j given that it is in row i. Then the null hypothesis is

$$H_0: \text{For each column } j, \ p_{1j} = \cdots = p_{Ij} \tag{6.4}$$

Let O_{ij} denote the observed value in cell ij. Let $O_{i.}$ denote the sum of the observed values in row i, let $O_{.j}$ denote the sum of the observed values in column j, and let $O_{..}$ denote the sum of the observed values in all the cells (see Table 6.3).

TABLE 6.3 Notation for observed values

	Column 1	Column 2	$\cdots$	Column J	Total
Row 1	O_{11}	O_{12}	$\cdots$	O_{1J}	$O_{1.}$
Row 2	O_{21}	O_{22}	$\cdots$	O_{2J}	$O_{2.}$
$\vdots$	$\vdots$	$\vdots$	$\vdots$	$\vdots$	$\vdots$
Row I	O_{I1}	O_{I2}	$\cdots$	O_{IJ}	$O_{I.}$
Total	$O_{.1}$	$O_{.2}$	$\cdots$	$O_{.J}$	$O_{..}$

To define a test statistic, we must compute an expected value for each cell in the table. Under H_0, the probability that the outcome of a trial falls into column j is the same for each row i. The best estimate of this probability is the proportion of trials whose outcome falls into column j. This proportion is $O_{.j}/O_{..}$. We need to compute the expected *number* of trials whose outcome falls into cell ij. We denote this expected value by E_{ij}. It is equal to the proportion of trials whose outcome falls into column j, multiplied by the number $O_{i.}$ of trials in row i. That is,

$$E_{ij} = \frac{O_{i.}O_{.j}}{O_{..}} \tag{6.5}$$

The test statistic is based on the differences between the observed and expected values:

$$\chi^2 = \sum_{i=1}^{I}\sum_{j=1}^{J}\frac{(O_{ij} - E_{ij})^2}{E_{ij}} \tag{6.6}$$

Under H_0, this test statistic has a chi-square distribution with $(I-1)(J-1)$ degrees of freedom. Use of the chi-square distribution is appropriate whenever the expected values are all greater than or equal to 5.

Example

6.9

Use the data in Table 6.2 to test the null hypothesis that the proportions of pins that are too thin, OK, or too thick are the same for all the machines.

Solution

We begin by using Equation (6.5) to compute the expected values E_{ij}. We show the calculations of E_{11} and E_{23} in detail:

$$E_{11} = \frac{(120)(66)}{500} = 15.84$$

$$E_{23} = \frac{(200)(32)}{500} = 12.80$$

The complete table of expected values is as follows:

Expected values for Table 6.2   *homogenity*

	Too Thin	OK	Too Thick	Total
Machine 1	15.84	96.48	7.68	120.00
Machine 2	26.40	160.80	12.80	200.00
Machine 3	13.20	80.40	6.40	100.00
Machine 4	10.56	64.32	5.12	80.00
Total	66.00	402.00	32.00	500.00

Machines work same, figure out if thickness same

We note that all the expected values are greater than 5. Therefore, the chi-square test is appropriate. We use Equation (6.6) to compute the value of the chi-square statistic:

$$\chi^2 = \frac{(10-15.84)^2}{15.84} + \cdots + \frac{(10-5.12)^2}{5.12}$$

$$= \frac{34.1056}{15.84} + \cdots + \frac{23.8144}{5.12}$$

$$= 15.5844$$

Since there are four rows and three columns, the number of degrees of freedom is $(4-1)(3-1) = 6$. To obtain the P-value, we consult the chi-square table (Table A.5). Looking under six degrees of freedom, we find that the upper 2.5% point is 14.449, and the upper 1% point is 16.812. Therefore $0.01 < P < 0.025$. It is reasonable to conclude that the machines differ in the proportions of pins that are too thin, OK, or too thick.

Note that the observed row and column totals are identical to the expected row and column, totals. This is always the case.

The following computer output (from MINITAB) presents the results of this hypothesis test.

```
Chi-Square Test: Thin, OK, Thick

Expected counts are printed below observed counts
Chi-Square contributions are printed below expected counts

         Thin        OK     Thick   Total
   1       10       102         8     120
         15.84     96.48      7.68
          2.153     0.316     0.013

   2       34       161         5     200
         26.40    160.80     12.80
          2.188     0.000     4.753

   3       12        79         9     100
         13.20     80.40      6.40
          0.109     0.024     1.056

   4       10        60        10      80
         10.56     64.32      5.12
          0.030     0.290     4.651

Total      66       402        32     500

Chi-Sq = 15.584, DF = 6, P-Value = 0.016
```

In the output, each cell (intersection of row and column) contains three numbers. The top number is the observed value, the middle number is the expected value, and the bottom number is the contribution $(O_{ij} - E_{ij})^2/E_{ij}$ made to the chi-square statistic from that cell.

The Chi-Square Test for Independence

In Example 6.9, the column totals were random, while the row totals were presumably fixed in advance, since they represented numbers of items sampled from various machines. In some cases, both row and column totals are random. In either case, we can test the null hypothesis that the probabilities of the column outcomes are the same for each row outcome, and the test is exactly the same in both cases. When both row and column totals are random, this is a test of for **independence** of the row and column categories.

Exercises for Section 6.5

1. Fasteners are manufactured for an application involving aircraft. Each fastener is categorized as conforming (suitable for its intended use), downgraded (unsuitable for its intended use but usable for another purpose), or scrap (not usable). It is thought that 85% of the fasteners are conforming, while 10% are downgraded and 5% are scrap. In a sample of 500 fasteners, 405 were conforming, 55 were downgraded, and 40 were scrap. Can you conclude that the true percentages differ from 85%, 10%, and 5%?

   a. State the appropriate null hypothesis.
   b. Compute the expected values under the null hypothesis.
   c. Compute the value of the chi-square statistic.
   d. Find the P-value. What do you conclude?

2. At an assembly plant for light trucks, routine monitoring of the quality of welds yields the following data:

   relationship between shift & quality Independence
   And if they produce the same

   | | **Number of Welds** | | | |
	High Quality	**Moderate Quality**	**Low Quality**	**Total**
Day Shift	467	191	42	700
Evening Shift	445	171	34	650
Night Shift	254	129	17	400
Total	1166	491	93	1750

   Can you conclude that the quality varies among shifts?

   a. State the appropriate null hypothesis.
   b. Compute the expected values under the null hypothesis. $H_0 : P_{1j} = P_{2j} = P_{3j} \ | \ j = 1, 2, 3$
   c. Compute the value of the chi-square statistic.
   d. Find the P-value. What do you conclude?

3. The article "An Investment Tax Credit for Investing in New Technology: A Survey of California

Firms" (R. Pope, *The Engineering Economist*, 1997: 269–287) examines the potential impact of a tax credit on capital investment. A number of firms were categorized by size (> 100 employees vs. ≤ 100 employees) and net excess capacity. The numbers of firms in each of the categories are presented in the following table:

homogenity

Net Excess Capacity	Small	Large
$< 0\%$	66	115
0–10%	52	47
·11–20%	13	18
21–30%	6	5
$> 30\%$	36	25

   Can you conclude that the distribution of net excess capacity differs between small and large firms? Compute the relevant test statistic and P-value.

4. The article "Analysis of Time Headways on Urban Roads: Case Study from Riyadh" (A. Al-Ghamdi, *Journal of Transportation Engineering*, 2001: 289–294) presents a model for the time elapsed between the arrival of consecutive vehicles on urban roads. Following are 137 arrival times (in seconds) along with the values expected from a theoretical model.

   test against expected
   goodness of fit value

Time	Observed	Expected
0–2	18	23
2–4	28	18
4–6	14	16
6–8	7	13
8–10	11	11
10–12	11	9
12–18	10	20
18–22	8	8
> 22	30	19

Can you conclude that the theoretical model does not explain the observed values well?

5. The article "Chronic Beryllium Disease and Sensitization at a Beryllium Processing Facility" (K. Rosenman, V. Hertzberg, et al., *Environmental Health Perspectives*, 2005:1366–1372) discusses the effects of exposure to beryllium in a cohort of workers. Workers were categorized by their duration of exposure (in years) and by their disease status (chronic beryllium disease, sensitization to beryllium, or no disease). The results were as follows:

	Duration of Exposure		
	< 1	1 to < 5	≥ 5
Diseased	10	8	23
Sensitized	9	19	11
Normal	70	136	206

Can you conclude that the proportions of workers in the various disease categories differ among exposure levels?

6. The article referred to in Exercise 3 categorized firms by size and percentage of full-operating-capacity labor force currently employed. The numbers of firms in each of the categories are presented in the following table.

Percent of Full-Operating-Capacity Labor Force Currently Employed	Small	Large
> 100%	6	8
95–100%	29	45
90–94%	12	28
85–89%	20	21
80–84%	17	22
75–79%	15	21
70–74%	33	29
< 70%	39	34

Can you conclude that the distribution of labor force currently employed differs between small and large firms? Compute the relevant test statistic and *P*-value.

7. For the given table of observed values:
   a. Construct the corresponding table of expected values.
   b. If appropriate, perform the chi-square test for the null hypothesis that the row and column outcomes are independent. If not appropriate, explain why.

	Observed Values		
	1	2	3
A	15	10	12
B	3	11	11
C	9	14	12

8. For the given table of observed values:
   a. Construct the corresponding table of expected values.
   b. If appropriate, perform the chi-square test for the null hypothesis that the row and column outcomes are independent. If not appropriate, explain why.

	Observed Values		
	1	2	3
A	25	4	11
B	3	3	4
C	42	3	5

9. Fill in the blank: For observed and expected values, _____

   i. the row totals in the observed table must be the same as the row totals in the expected table, but the column totals need not be the same.

   ii. the column totals in the observed table must be the same as the column totals in the expected table, but the row totals need not be the same.

   iii. both the row and the column totals in the observed table must be the same as the row and the column totals, respectively, in the expected table.

iv. neither the row nor the column totals in the observed table need be the same as the row or the column totals in the expected table.

10. Because of printer failure, none of the observed values in the following table were printed, but some of the marginal totals were. Is it possible to construct the corresponding table of expected values from the information given? If so, construct it. If not, describe the additional information you would need.

	Observed Values			
	1	**2**	**3**	**Total**
A	—	—	—	25
B	—	—	—	—
C	—	—	—	40
D	—	—	—	75
Total	50	20	—	150

11. Plates are evaluated according to their surface finish and placed into four categories: Premium, Conforming, Downgraded, and Unacceptable. A quality engineer claims that the proportions of plates in the four categories are 10%, 70%, 15%, and 5%, respectively. In a sample of 200 plates, 19 were classified as premium, 133 were classified as conforming, 35 were classified as downgraded, and 13 were classified as unacceptable. Can you conclude that the engineer's claim is incorrect?

12. The article "Determination of Carboxyhemoglobin Levels and Health Effects on Officers Working at the Istanbul Bosphorus Bridge" (G. Kocasoy and H. Yalin, *Journal of Environmental Science and Health*, 2004:1129–1139) presents assessments of health outcomes of people working in an environment with high levels of carbon monoxide (CO). Following are the numbers of workers reporting various symptoms, categorized by work shift. The numbers were read from a graph.

	Shift		
	Morning	**Evening**	**Night**
Influenza	16	13	18
Headache	24	33	46
Weakness	11	16	5
Shortness of Breath	7	9	9

Can you conclude that the proportions of workers with the various symptoms differ among the shifts?

13. The article "Analysis of Unwanted Fire Alarm: Case Study" (W. Chow, N. Fong, and C. Ho, *Journal of Architectural Engineering*, 1999:62–65) presents a count of the number of false alarms at several sites. The numbers of false alarms each month, divided into those with known causes and those with unknown causes, are given in the following table. Can you conclude that the proportion of false alarms whose cause is known differs from month to month?

	Month											
	1	**2**	**3**	**4**	**5**	**6**	**7**	**8**	**9**	**10**	**11**	**12**
Known	20	13	21	26	23	18	14	10	20	20	18	14
Unknown	12	2	16	12	22	30	32	32	14	16	10	12

14. At a certain genetic locus on a chromosome, each individual has one of three different DNA sequences (alleles). The three alleles are denoted A, B, C. At another genetic locus on the same chromosome, each organism has one of three alleles, denoted 1, 2, 3. Each individual therefore has one of nine possible allele pairs: A1, A2, A3, B1, B2, B3, C1, C2, or C3. These allele pairs are called *haplotypes*. The loci are said to be in *linkage equilibrium* if the two alleles in an individual's haplotype are independent. Haplotypes were determined for 316 individuals. The following MINITAB output presents the results of a chi-square test for independence.

```
Chi-Square Test: A, B, C

Expected counts are printed below observed counts
Chi-Square contributions are printed below expected counts

            A       B       C    Total
   1       66      44      34     144
        61.06   47.39   35.54
        0.399   0.243   0.067

   2       36      38      20      94
        39.86   30.94   23.20
        0.374   1.613   0.442

   3       32      22      24      78
        33.08   25.67   19.25
        0.035   0.525   1.170

Total     134     104      78     316

Chi-Sq = 4.868, DF = 4, P-Value = 0.301
```

a. How many individuals were observed to have the haplotype B3?
b. What is the expected number of individuals with the haplotype A2?
c. Which of the nine haplotypes was least frequently observed?
d. Which of the nine haplotypes has the smallest expected count?
e. Can you conclude that the loci are not in linkage equilibrium (i.e., not independent)? Explain.
f. Can you conclude that the loci are in linkage equilibrium (i.e., independent)? Explain.

6.6 Fixed-Level Testing

Critical Points and Rejection Regions

A hypothesis test measures the plausibility of the null hypothesis by producing a P-value. The smaller the P-value, the less plausible the null. We have pointed out that there is no scientifically valid dividing line between plausibility and implausibility, so it is impossible to specify a "correct" P-value below which we should reject H_0. When possible, it is best simply to report the P-value, and not to make a firm decision whether to reject. Sometimes, however, a decision has to be made. For example, if items are sampled from an assembly line to test whether the mean diameter is within tolerance, a decision must be made whether to recalibrate the process. If a sample of parts is drawn from a shipment and checked for defects, a decision must be made whether to accept or to return the shipment. If a decision is going to be made on the basis of a hypothesis test, there is no choice but to pick a cutoff point for the P-value. When this is done, the test is referred to as a **fixed-level** test.

Fixed-level testing is just like the hypothesis testing we have been discussing so far, except that a firm rule is set ahead of time for rejecting the null hypothesis. A value α, where $0 < \alpha < 1$, is chosen. Then the P-value is computed. If $P \leq \alpha$, the null hypothesis is rejected, and the alternate hypothesis is taken as truth. If $P > \alpha$, then the null hypothesis is considered to be plausible. The value of α is called the **significance level** or, more simply, the **level**, of the test. Recall from Section 6.2 that if a test results in a P-value less than or equal to α, we say that the null hypothesis is rejected at level α (or $100\alpha\%$), or that the result is statistically significant at level α (or $100\alpha\%$). As we have mentioned, a common choice for α is 0.05.

Summary

To conduct a fixed-level test:

- Choose a number α, where $0 < \alpha < 1$. This is called the significance level, or the level, of the test.
- Compute the P-value in the usual way.
- If $P \leq \alpha$, reject H_0. If $P > \alpha$, do not reject H_0.

Example

6.10

Refer to Example 6.1 in Section 6.1. The mean wear in a sample of 45 steel balls was $\overline{X} = 673.2\,\mu\mathrm{m}$, and the standard deviation was $s = 14.9\,\mu\mathrm{m}$. Let μ denote the population mean wear. A test of $H_0: \mu \geq 675$ versus $H_1: \mu < 675$ yielded a P-value of 0.209. Can we reject H_0 at the 25% level? Can we reject H_0 at the 5% level?

Solution

The P-value of 0.209 is less than 0.25, so if we had chosen a significance level of $\alpha = 0.25$, we would reject H_0. Thus we reject H_0 at the 25% level. Since $0.209 > 0.05$, we do not reject H_0 at the 5% level.

In a fixed-level test, a **critical point** is a value of the test statistic that produces a P-value exactly equal to α. A critical point is a dividing line for the test statistic just as the significance level is a dividing line for the P-value. If the test statistic is on one side of the critical point, the P-value will be less than α, and H_0 will be rejected. If the test statistic is on the other side of the critical point, the P-value will be greater than α, and H_0 will not be rejected. The region on the side of the critical point that leads to rejection is called the **rejection region**. The critical point itself is also in the rejection region.

Example

6.11

A new concrete mix is being evaluated. The plan is to sample 100 concrete blocks made with the new mix, compute the sample mean compressive strength $\overline{X}$, and then test $H_0: \mu \leq 1350$ versus $H_1: \mu > 1350$, where the units are MPa. It is assumed from previous tests of this sort that the population standard deviation σ will be close to 70 MPa. Find the critical point and the rejection region if the test will be conducted at a significance level of 5%.

Solution

We will reject H_0 if the P-value is less than or equal to 0.05. The P-value for this test will be the area to the right of the value of $\overline{X}$. Therefore the P-value will be less than 0.05, and H_0 will be rejected, if the value of $\overline{X}$ is in the upper 5% of the null distribution (see Figure 6.11). The rejection region therefore consists of the upper 5% of the null distribution. The critical point is the boundary of the upper 5%. The null distribution is normal, and from the z table we find that the z-score of the point that cuts off the upper 5% of the normal curve is $z_{.05} = 1.645$. Therefore we can express the critical point as $z = 1.645$ and the rejection region as $z \geq 1.645$. It is often more convenient to express the critical point and rejection region in terms of $\overline{X}$, by converting the z-score to the original units. The null distribution has mean $\mu = 1350$ and standard deviation $\sigma_{\overline{X}} = \sigma/\sqrt{n} \approx 70/\sqrt{100} = 7$. Therefore the critical point can be expressed as $\overline{X} = 1350 + (1.645)(7) = 1361.5$. The rejection region consists of all values of $\overline{X}$ greater than or equal to 1361.5.

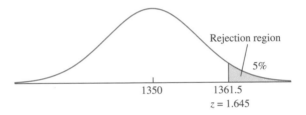

FIGURE 6.11 The rejection region for this one-tailed test consists of the upper 5% of the null distribution. The critical point is 1361.5, on the boundary of the rejection region.

Type I and Type II Errors

Since a fixed-level test results in a firm decision, there is a chance that the decision could be the wrong one. There are exactly two ways in which the decision can be wrong. One can reject H_0 when it is in fact true. This is known as a type I error. Or, one can fail to reject H_0 when it is false. This is known as a type II error.

When designing experiments whose data will be analyzed with a fixed-level test, it is important to try to make the probabilities of type I and type II errors reasonably small. There is no use in conducting an experiment that has a large probability of leading to an incorrect decision. It turns out that it is easy to control the probability of a type I error, as shown by the following result.

> If α is the significance level that has been chosen for the test, then the probability of a type I error is never greater than α.

We illustrate this fact with the following example. Let $X_1, \ldots, X_n$ be a large random sample from a population with mean μ and variance σ^2. Then $\overline{X}$ is normally distributed with mean μ and variance σ^2/n. Assume that we are to test $H_0: \mu \leq 0$ versus $H_1: \mu > 0$

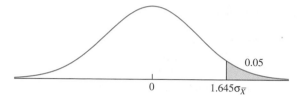

FIGURE 6.12 The null distribution with the rejection region for $H_0: \mu \leq 0$.

at the fixed level $\alpha = 0.05$. That is, we will reject H_0 if $P \leq 0.05$. The null distribution, shown in Figure 6.12, is normal with mean 0 and variance $\sigma_{\overline{X}}^2 = \sigma^2/n$. Assume the null hypothesis is true. We will compute the probability of a type I error and show that it is no greater than 0.05.

A type I error will occur if we reject H_0, which will occur if $P \leq 0.05$, which in turn will occur if $\overline{X} \geq 1.645\sigma_{\overline{X}}$. Therefore the rejection region is the region $\overline{X} \geq 1.645\sigma_{\overline{X}}$. Now since H_0 is true, $\mu \leq 0$. First, we'll consider the case where $\mu = 0$. Then the distribution of $\overline{X}$ is given by Figure 6.12. In this case, $P(\overline{X} \geq 1.645\sigma_{\overline{X}}) = 0.05$, so the probability of rejecting H_0 and making a type I error is equal to 0.05. Next, consider the case where $\mu < 0$. Then the distribution of $\overline{X}$ is obtained by shifting the curve in Figure 6.12 to the left, so $P(\overline{X} \geq 1.645\sigma_{\overline{X}}) < 0.05$, and the probability of a type I error is less than 0.05. We could repeat this illustration using any number α in place of 0.05. We conclude that if H_0 is true, the probability of a type I error is never greater than α. Furthermore, note that if μ is on the boundary of H_0 ($\mu = 0$ in this case), then the probability of a type I error is equal to α.

We can therefore make the probability of a type I error as small as we please, because it is never greater than the significance level α that we choose. Unfortunately, as we will see in Section 6.7, the smaller we make the probability of a type I error, the larger the probability of a type II error becomes. The usual strategy is to begin by choosing a value for α so that the probability of a type I error will be reasonably small. As we have mentioned, a conventional choice for α is 0.05. Then one computes the probability of a type II error and hopes that it is not too large. If it is large, it can be reduced only by redesigning the experiment—for example, by increasing the sample size. Calculating and controlling the size of the type II error is somewhat more difficult than calculating and controlling the size of the type I error. We will discuss this in Section 6.7.

Summary

When a fixed-level test is conducted at significance level α, two types of errors can be made. These are

- Type I error: Reject H_0 when it is true.
- Type II error: Fail to reject H_0 when it is false.

The probability of a type I error is never greater than α.

Exercises for Section 6.6

1. A hypothesis test is performed, and the P-value is 0.07. True or false:

   a. H_0 is rejected at the 5% level.

   b. H_0 is not rejected at the 2% level.

   c. H_0 is rejected at the 10% level.

2. A test is made of the hypotheses $H_0: \mu \geq 10$ versus $H_1: \mu < 10$. For each of the following situations, determine whether the decision was correct, a type I error occurred, or a type II error occurred.

   a. $\mu = 12$, H_0 is rejected.

   b. $\mu = 10$, H_0 is not rejected.

   c. $\mu = 6$, H_0 is not rejected.

   d. $\mu = 8$, H_0 is rejected.

3. A new coal liquefaction process is supposed to increase the mean yield μ of distillate fuel. The new process is very expensive, so it would be a costly error to put it into production unless $\mu > 20$. A test of $H_0: \mu \leq 20$ versus $H_1: \mu > 20$ will be performed, and the new process will be put into production if H_0 is rejected. Which procedure provides a smaller probability for this costly error, to test at the 5% level or to test at the 1% level?

4. A hypothesis test is to be performed, and the null hypothesis will be rejected if $P \leq 0.05$. If H_0 is in fact true, what is the maximum probability that it will be rejected?

5. The manufacturer of a heavy-duty cooling fan claims that the mean lifetime of these fans under severe conditions is greater than 6 months. Let μ represent the actual mean lifetime of these fans. A test was made of the hypotheses $H_0: \mu \geq 6$ versus $H_1: \mu < 6$. For each of the following situations, determine whether the decision was correct, a type I error occurred, or a type II error occurred.

   a. The claim is true, and H_0 is rejected.

   b. The claim is false, and H_0 is rejected.

   c. The claim is true, and H_0 is not rejected.

   d. The claim is false, and H_0 is not rejected.

6. A wastewater treatment program is designed to produce treated water with a pH of 7. Let μ represent the mean pH of water treated by this process. The pH of 60 water specimens will be measured, and a test of the hypotheses $H_0: \mu = 7$ versus $H_1: \mu \neq 7$ will be made. Assume it is known from previous experiments that the standard deviation of the pH of water specimens is approximately 0.5.

   a. If the test is made at the 5% level, what is the rejection region?

   b. If the sample mean pH is 6.87, will H_0 be rejected at the 10% level?

   c. If the sample mean pH is 6.87, will H_0 be rejected at the 1% level?

   d. If the value 7.2 is a critical point, what is the level of the test?

7. A machine that grinds valves is set to produce valves whose lengths have mean 100 mm and standard deviation 0.1 mm. The machine is moved to a new location. It is thought that the move may have upset the calibration for the mean length but that it is unlikely to have changed the standard deviation. Let μ represent the mean length of valves produced after the move. To test the calibration, a sample of 100 valves will be ground, their lengths will be measured, and a test will be made of the hypotheses $H_0: \mu = 100$ versus $H_1: \mu \neq 100$.

   a. Find the rejection region if the test is made at the 5% level.

   b. Find the rejection region if the test is made at the 10% level.

   c. If the sample mean length is 99.97 mm, will H_0 be rejected at the 5% level?

   d. If the sample mean length is 100.01 mm, will H_0 be rejected at the 10% level?

   e. A critical point is 100.015 mm. What is the level of the test?

6.7 Power

A hypothesis test results in a type II error if H_0 is not rejected when it is false. The **power** of a test is the probability of *rejecting* H_0 when it is false. Therefore

$$\text{Power} = 1 - P(\text{type II error})$$

To be useful, a test must have reasonably small probabilities of both type I and type II errors. The type I error is kept small by choosing a small value of α as the significance level. Then the power of the test is calculated. If the power is large, then the probability of a type II error is small as well, and the test is a useful one. Note that power calculations are generally done before data are collected. The purpose of a power calculation is to determine whether a hypothesis test, when performed, is likely to reject H_0 in the event that H_0 is false.

As an example of a power calculation, assume that a new chemical process has been developed that may increase the yield over that of the current process. The current process is known to have a mean yield of 80 and a standard deviation of 5, where the units are the percentage of a theoretical maximum. If the mean yield of the new process is shown to be greater than 80, the new process will be put into production. Let μ denote the mean yield of the new process. It is proposed to run the new process 50 times and then to test the hypothesis

$$H_0 : \mu \leq 80 \quad \text{versus} \quad H_1 : \mu > 80$$

at a significance level of 5%. If H_0 is rejected, it will be concluded that $\mu > 80$, and the new process will be put into production. Let us assume that if the new process had a mean yield of 81, then it would be a substantial benefit to put this process into production. If it is in fact the case that $\mu = 81$, what is the power of the test—that is, the probability that H_0 will be rejected?

Before presenting the solution, we note that in order to compute the power, it is necessary to specify a particular value of μ, in this case $\mu = 81$, for the alternate hypothesis. The reason for this is that the power is different for different values of μ. We will see that if μ is close to H_0, the power will be small, while if μ is far from H_0, the power will be large.

Computing the power involves two steps:

1. Compute the rejection region.
2. Compute the probability that the test statistic falls in the rejection region if the alternate hypothesis is true. This is the power.

We'll begin to find the power of the test by computing the rejection region, using the method illustrated in Example 6.11 in Section 6.6. We must first find the null distribution. We know that the statistic $\overline{X}$ has a normal distribution with mean μ and standard deviation $\sigma_{\overline{X}} = \sigma/\sqrt{n}$, where $n = 50$ is the sample size. Under H_0, we take $\mu = 80$. We must now find an approximation for σ. In practice this can be a difficult problem, because the sample has not yet been drawn, so there is no sample standard deviation s. There are several ways in which it may be possible to approximate σ. Sometimes a small preliminary sample has been drawn—for example, in a feasibility study—and the standard deviation of this sample may be a satisfactory approximation for σ. In other cases, a sample from a similar population may exist, whose standard deviation may be used. In this example, there is a long history of a currently used process, whose standard deviation is 5. Let's say that it is reasonable to assume that the standard deviation of the new process will be similar to that of the current process. We will therefore assume that the population standard deviation for the new process is $\sigma = 5$ and that $\sigma_{\overline{X}} = 5/\sqrt{50} = 0.707$.

Figure 6.13 presents the null distribution of $\overline{X}$. Since H_0 specifies that $\mu \leq 80$, large values of $\overline{X}$ disagree with H_0, so the P-value will be the area to the right of the observed value of $\overline{X}$. The P-value will be less than or equal to 0.05 if $\overline{X}$ falls into the upper 5% of the null distribution. This upper 5% is the rejection region. The critical point has a z-score of 1.645, so its value is $80 + (1.645)(0.707) = 81.16$. We will reject H_0 if $\overline{X} \geq 81.16$. This is the rejection region.

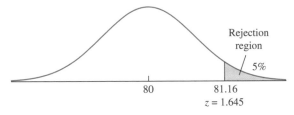

FIGURE 6.13 The hypothesis test will be conducted at a significance level of 5%. The rejection region for this test is the region where the P-value will be less than 0.05.

We are now ready to compute the power, which is the probability that $\overline{X}$ will fall into the rejection region if the alternate hypothesis $\mu = 81$ is true. Under this alternate hypothesis, the distribution of $\overline{X}$ is normal with mean 81 and standard deviation 0.707. Figure 6.14 presents the alternate distribution and the null distribution on the same plot. Note that the alternate distribution is obtained by shifting the null distribution so that

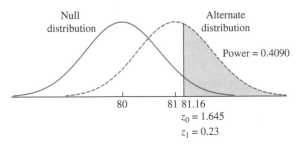

FIGURE 6.14 The rejection region, consisting of the upper 5% of the null distribution, is shaded. The z-score of the critical point is $z_0 = 1.645$ under the null distribution and $z_1 = 0.23$ under the alternate. The power is the area of the rejection region under the alternate distribution, which is 0.4090.

the mean becomes the alternate mean of 81 rather than the null mean of 80. Because the alternate distribution is shifted over, the probability that the test statistic falls into the rejection region is greater than it is under H_0. To be specific, the z-score under H_1 for the critical point 81.16 is $z = (81.16 - 81)/0.707 = 0.23$. The area to the right of $z = 0.23$ is 0.4090. This is the power of the test.

A power of 0.4090 is very low. It means that if the mean yield of new process is actually equal to 81, there is only a 41% chance that the proposed experiment will detect the improvement over the old process and allow the new process to be put into production. It would be unwise to invest time and money to run this experiment, since it has a large chance to fail.

It is natural to wonder how large the power must be for a test to be worthwhile to perform. As with P-values, there is no scientifically valid dividing line between sufficient and insufficient power. In general, tests with power greater than 0.80 or perhaps 0.90 are considered acceptable, but there are no well-established rules of thumb.

We have mentioned that the power depends on the value of μ chosen to represent the alternate hypothesis and is larger when the value is far from the null mean. Example 6.12 illustrates this.

Example

6.12

Find the power of the 5% level test of $H_0 : \mu \leq 80$ versus $H_1 : \mu > 80$ for the mean yield of the new process under the alternative $\mu = 82$, assuming $n = 50$ and $\sigma = 5$.

Solution
We have already completed the first step of the solution, which is to compute the rejection region. We will reject H_0 if $\overline{X} \geq 81.16$. Figure 6.15 presents the alternate and null distributions on the same plot. The z-score for the critical point of 81.16 under the alternate hypothesis is $z = (81.16 - 82)/0.707 = -1.19$. The area to the right of $z = -1.19$ is 0.8830. This is the power.

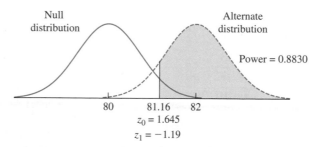

FIGURE 6.15 The rejection region, consisting of the upper 5% of the null distribution, is shaded. The z-score of the critical point is $z_0 = 1.645$ under the null distribution and $z_1 = -1.19$ under the alternate. The power is the area of the rejection region under the alternate distribution, which is 0.8830.

Since the alternate distribution is obtained by shifting the null distribution, the power depends on which alternate value is chosen for μ, and it can range from barely greater than the significance level α all the way up to 1. If the alternate mean is chosen very close to the null mean, the alternate curve will be almost identical to the null, and the power will be very close to α. If the alternate mean is far from the null, almost all the area under the alternate curve will lie in the rejection region, and the power will be close to 1.

When power is not large enough, it can be increased by increasing the sample size. When planning an experiment, one can determine the sample size necessary to achieve a desired power. Example 6.13 illustrates this.

Example

6.13

In testing the hypothesis $H_0 : \mu \leq 80$ versus $H_1 : \mu > 80$ regarding the mean yield of the new process, how many times must the new process be run so that a test conducted at a significance level of 5% will have power 0.90 against the alternative $\mu = 81$, if it is assumed that $\sigma = 5$?

Solution

Let n represent the necessary sample size. We first use the null distribution to express the critical point for the test in terms of n. The null distribution of $\overline{X}$ is normal with mean 80 and standard deviation $5/\sqrt{n}$. Therefore the critical point is $80 + 1.645(5/\sqrt{n})$. Now, we use the alternate distribution to obtain a different expression for the critical point in terms of n. Refer to Figure 6.16. The power of the test is the area of the rejection region under the alternate curve. This area must be 0.90. Therefore, the z-score for the critical point, under the alternate hypothesis, is $z = -1.28$. The critical point is thus $81 - 1.28(5/\sqrt{n})$. We now have two different expressions for the critical point. Since there is only one critical point, these two expressions are equal. We therefore set them equal and solve for n.

$$80 + 1.645 \left(\frac{5}{\sqrt{n}} \right) = 81 - 1.28 \left(\frac{5}{\sqrt{n}} \right)$$

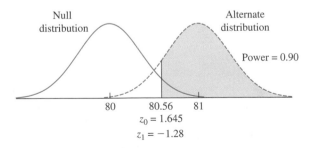

FIGURE 6.16 To achieve power of 0.90 with a significance level of 0.05, the z-score for the critical point must be $z_0 = 1.645$ under the null distribution and $z_1 = -1.28$ under the alternate distribution.

Solving for n yields $n \approx 214$. The critical point can be computed by substituting this value for n into either side of the previous equation. The critical point is 80.56.

Using a Computer to Calculate Power

We have presented a method for calculating the power, and the sample size needed to attain a specified power, for a one-tailed large-sample test of a population mean. It is reasonably straightforward to extend this method to compute power and needed sample sizes for two-tailed tests and for tests for proportions. It is more difficult to compute power for a t test, F test, or chi-square test. Computer packages, however, can compute power and needed sample sizes for all these tests. We present some examples.

E*xample*

6.14

A pollster will conduct a survey of a random sample of voters in a community to estimate the proportion who support a measure on school bonds. Let p be the proportion of the population who support the measure. The pollster will test $H_0: p = 0.50$ versus $H_1: p \neq 0.50$ at the 5% level. If 200 voters are sampled, what is the power of the test if the true value of p is 0.55?

Solution
The following computer output (from MINITAB) presents the solution:

```
Power and Sample Size

Test for One Proportion

Testing proportion = 0.5 (versus not = 0.5)
Alpha = 0.05

Alternative   Sample
 Proportion    Size      Power
       0.55     200   0.292022
```

The first two lines of output state that this is a power calculation for a test for a single population proportion p. The next two lines state the null and alternate hypotheses, and the significance level of the test. Note that we have specified a two-tailed test with significance level $\alpha = 0.05$. Next is the alternative proportion, which is the value of p (0.55) that we are assuming to be true when the power is calculated. The sample size has been specified to be 200, and the power is computed to be 0.292.

E*xample*

6.15

Refer to Example 6.14. How many voters must be sampled so that the power will be 0.8 when the true value of $p = 0.55$?

Solution
The following computer output (from MINITAB) presents the solution:

```
Power and Sample Size

Test for One Proportion

Testing proportion = 0.5 (versus not = 0.5)
Alpha = 0.05

Alternative  Sample  Target
 Proportion   Size    Power   Actual Power
       0.55    783     0.8        0.800239
```

The needed sample size is 783. Note that the actual power is slightly higher than 0.80. Because the sample size is discrete, it is not possible to find a sample size that provides exactly the power requested (the target power). So MINITAB calculates the smallest sample size for which the power is greater than that requested.

Example
6.16

Shipments of coffee beans are checked for moisture content. High moisture content indicates possible water contamination, leading to rejection of the shipment. Let μ represent the mean moisture content (in percent by weight) in a shipment. Five moisture measurements will be made on beans chosen at random from the shipment. A test of the hypothesis $H_0 : \mu \leq 10$ versus $H_1 : \mu > 10$ will be made at the 5% level, using the Student's t test. What is the power of the test if the true moisture content is 12% and the standard deviation is $\sigma = 1.5\%$?

Solution
The following computer output (from MINITAB) presents the solution:

```
Power and Sample Size

1-Sample t Test

Testing mean = null (versus > null)
Calculating power for mean = null + difference
Alpha = 0.05   Assumed standard deviation = 1.5

                  Sample
Difference         Size      Power
         2            5   0.786485
```

The power depends only on the difference between the true mean and the null mean, which is $12 - 10 = 2$, and not on the means themselves. The power is 0.786. Note that the output specifies that this is the power for a one-tailed test.

Example

6.17

Refer to Example 6.16. Find the sample size needed so that the power will be at least 0.9.

Solution

The following computer output (from MINITAB) presents the solution:

```
Power and Sample Size

1-Sample t Test

Testing mean = null (versus > null)
Calculating power for mean = null + difference
Alpha = 0.05   Assumed standard deviation = 1.5

                  Sample  Target
Difference         Size   Power   Actual Power
        2            7    0.9        0.926750
```

The smallest sample size for which the power is 0.9 or more is 7. The actual power is 0.927.

To summarize, power calculations are important to ensure that experiments have the potential to provide useful conclusions. Many agencies that provide funding for scientific research require that power calculations be provided with every proposal in which hypothesis tests are to be performed.

Exercises for Section 6.7

1. A test has power 0.85 when $\mu = 10$. True or false:

   a. The probability of rejecting H_0 when $\mu = 10$ is 0.85.

   b. The probability of making a correct decision when $\mu = 10$ is 0.85.

   c. The probability of making a correct decision when $\mu = 10$ is 0.15.

   d. The probability that H_0 is true when $\mu = 10$ is 0.15.

2. A test has power 0.90 when $\mu = 7.5$. True or false:

   a. The probability of rejecting H_0 when $\mu = 7.5$ is 0.90.

   b. The probability of making a type I error when $\mu = 7.5$ is 0.90.

   c. The probability of making a type I error when $\mu = 7.5$ is 0.10.

   d. The probability of making a type II error when $\mu = 7.5$ is 0.90.

   e. The probability of making a type II error when $\mu = 7.5$ is 0.10.

   f. The probability that H_0 is false when $\mu = 7.5$ is 0.90.

3. If the sample size remains the same, and the level α increases, then the power will _____. *Options: increase, decrease.*

4. If the level α remains the same, and the sample size increases, then the power will _____. *Options: increase, decrease.*

5. A power calculation has shown that if $\mu = 10$, the power of a test of $H_0 : \mu \leq 8$ versus $H_1 : \mu > 8$ is 0.80. If instead $\mu = 12$, which one of the following statements is true?

   i. The power of the test will be less than 0.80.

   ii. The power of the test will be greater than 0.80.

   iii. We cannot determine the power of the test without specifying the population standard deviation σ.

6. A process that manufactures glass sheets is supposed to be calibrated so that the mean thickness μ of the sheets is more than 4 mm. The standard deviation of the sheet thicknesses is known to be well approximated by $\sigma = 0.20$ mm. Thicknesses of each sheet in a sample of sheets will be measured, and a test of the hypothesis $H_0 : \mu \leq 4$ versus $H_1 : \mu > 4$ will be performed. Assume that, in fact, the true mean thickness is 4.04 mm.

   a. If 100 sheets are sampled, what is the power of a test made at the 5% level?

   b. How many sheets must be sampled so that a 5% level test has power 0.95?

   c. If 100 sheets are sampled, at what level must the test be made so that the power is 0.90?

   d. If 100 sheets are sampled, and the rejection region is $\overline{X} \geq 4.02$, what is the power of the test?

7. A tire company claims that the lifetimes of its tires average 50,000 miles. The standard deviation of tire lifetimes is known to be 5000 miles. You sample 100 tires and will test the hypothesis that the mean tire lifetime is at least 50,000 miles against the alternative that it is less. Assume, in fact, that the true mean lifetime is 49,500 miles.

   a. State the null and alternate hypotheses. Which hypothesis is true?

   b. It is decided to reject H_0 if the sample mean is less than 49,400. Find the level and power of this test.

   c. If the test is made at the 5% level, what is the power?

   d. At what level should the test be conducted so that the power is 0.80?

   e. You are given the opportunity to sample more tires. How many tires should be sampled in total so that the power is 0.80 if the test is made at the 5% level?

8. Water quality in a large estuary is being monitored in order to measure the PCB concentration (in parts per billion).

   a. If the population mean is 1.6 ppb and the population standard deviation is 0.33 ppb, what is the probability that the null hypothesis $H_0 : \mu \leq 1.50$ is rejected at the 5% level, if the sample size is 80?

   b. If the population mean is 1.6 ppb and the population standard deviation is 0.33 ppb, what sample size is needed so that the probability is 0.99 that $H_0 : \mu \leq 1.50$ is rejected at the 5% level?

9. The following MINITAB output presents the results of a power calculation for a test concerning a population proportion p.

```
Power and Sample Size

Test for One Proportion

Testing proportion = 0.5
(versus not = 0.5)
Alpha = 0.05

Alternative   Sample
 Proportion    Size      Power
        0.4     150   0.691332
```

---

   a. Is the power calculated for a one-tailed or two-tailed test?

   b. What is the null hypothesis for which the power is calculated?

   c. For what alternative value of p is the power calculated?

   d. If the sample size were 100, would the power be less than 0.7, greater than 0.7, or is it impossible to tell from the output? Explain.

   e. If the sample size were 200, would the power be less than 0.6, greater than 0.6, or is it impossible to tell from the output? Explain.

   f. For a sample size of 150, is the power against the alternative $p = 0.3$ less than 0.65, greater than 0.65, or is it impossible to tell from the output? Explain.

   g. For a sample size of 150, is the power against the alternative $p = 0.45$ less than 0.65, greater than 0.65, or is it impossible to tell from the output? Explain.

10. The following MINITAB output presents the results of a power calculation for a test concerning a population mean μ.

```
Power and Sample Size

1-Sample t Test

Testing mean = null (versus > null)
Calculating power for mean = null + difference
Alpha = 0.05  Assumed standard deviation = 1.5

             Sample  Target
Difference    Size    Power  Actual Power
         1     18     0.85     0.857299
```

---

a. Is the power calculated for a one-tailed or two-tailed test?
b. Assume that the value of μ used for the null hypothesis is $\mu = 3$. For what alternate value of μ is the power calculated?
c. If the sample size were 25, would the power be less than 0.85, greater than 0.85, or is it impossible to tell from the output? Explain.
d. If the difference were 0.5, would the power be less than 0.90, greater than 0.90, or is it impossible to tell from the output? Explain.
e. If the sample size were 17, would the power be less than 0.85, greater than 0.85, or is it impossible to tell from the output? Explain.

11. The following MINITAB output presents the results of a power calculation for a test of the difference between two means $\mu_1 - \mu_2$.

```
Power and Sample Size

2-Sample t Test

Testing mean 1 = mean 2 (versus not =)
Calculating power for mean 1 = mean 2 + difference
Alpha = 0.05  Assumed standard deviation = 5

             Sample  Target
Difference    Size    Power  Actual Power
         3     60     0.9      0.903115

The sample size is for each group.
```

---

a. Is the power calculated for a one-tailed or two-tailed test?
b. If the sample sizes were 50 in each group, would the power be less than 0.9, greater than 0.9, or is it impossible to tell from the output? Explain.
c. If the difference were 4, would the power be less than 0.9, greater than 0.9, or is it impossible to tell from the output? Explain.

6.8 Multiple Tests

Sometimes a situation occurs in which it is necessary to perform many hypothesis tests. The basic rule governing this situation is that as more tests are performed, the confidence that we can place in our results decreases. In this section, we present an example to illustrate this point.

It is thought that applying a hard coating containing very small particles of tungsten carbide may reduce the wear on cam gears in a certain industrial application. There are many possible formulations for the coating, varying in the size and concentration of the tungsten carbide particles. Twenty different formulations were manufactured. Each one was tested by applying it to a large number of gears, and then measuring the wear on the gears after a certain period of time had elapsed. It is known on the basis of long experience that the mean wear for uncoated gears over this period of time is 100 μm. For each formulation, a test was made of the null hypothesis $H_0: \mu \geq 100$ μm. H_0 says that the formulation does not reduce wear. For 19 of the 20 formulations, the P-value was greater than 0.05, so H_0 was not rejected. For one formulation, H_0 was rejected. It might seem natural to conclude that this formulation really does reduce wear. Examples 6.18 through 6.21 will show that this conclusion is premature.

Example 6.18

If only one formulation were tested, and it in fact had no effect on wear, what is the probability that H_0 would be rejected, leading to a wrong conclusion?

Solution

If the formulation has no effect on wear, then $\mu = 100$ μm, so H_0 is true. Rejecting H_0 is then a type I error. The question is therefore asking for the probability of a type I error. In general, this probability is always less than or equal to the significance level of the test, which in this case is 5%. Since $\mu = 100$ is on the boundary of H_0, the probability of a type I error is equal to the significance level. The probability is 0.05 that H_0 will be rejected.

Example 6.19

Given that H_0 was rejected for one of the 20 formulations, is it plausible that this formulation actually has no effect on wear?

Solution

Yes. It is plausible that none of the formulations, including the one for which H_0 was rejected, have any effect on wear. Twenty hypothesis tests were made. For each test there was a 5% chance (i.e., 1 chance in 20) of a type I error. We therefore expect on the average that out of every 20 true null hypotheses, one will be rejected. So rejecting H_0 in one out of the 20 tests is exactly what one would expect in the case that none of the formulations made any difference.

Example 6.20

If in fact none of the 20 formulations have any effect on wear, what is the probability that H_0 will be rejected for one or more of them?

Solution

We first find the probability that the right conclusion (not rejecting H_0) is made for all the formulations. For each formulation, the probability that H_0 is not rejected is $1 - 0.05 = 0.95$, so the probability that H_0 is not rejected for any of the 20 formulations is $(0.95)^{20} = 0.36$. The probability is therefore $1 - 0.36 = 0.64$ that we incorrectly reject H_0 for one or more of the formulations.

The experiment is repeated. This time, the operator forgets to apply the coatings, so each of the 20 wear measurements is actually made on uncoated gears. Is it likely that one or more of the formulations will appear to reduce wear, in that H_0 will be rejected?

Solution

Yes. Example 6.20 shows that the probability is 0.64 that one or more of the coatings will appear to reduce wear, even if they are not actually applied.

Examples 6.18 through 6.21 illustrate a phenomenon known as the **multiple testing problem**. Put simply, the multiple testing problem is this: When H_0 is rejected, we have strong evidence that it is false. But strong evidence is not certainty. Occasionally a true null hypothesis will be rejected. When many tests are performed, it is more likely that some true null hypotheses will be rejected. Thus when many tests are performed, it is difficult to tell which of the rejected null hypotheses are really false and which correspond to type I errors.

The Bonferroni Method

The Bonferroni method provides a way to adjust P-values upward when several hypothesis tests are performed. If a P-value remains small after the adjustment, the null hypothesis may be rejected. To make the Bonferroni adjustment, simply multiply the P-value by the number of tests performed. Here are two examples.

Four different coating formulations are tested to see if they reduce the wear on cam gears to a value below 100 μm. The null hypothesis $H_0 : \mu \geq 100$ μm is tested for each formulation, and the results are

$$
\begin{array}{ll}
\text{Formulation A:} & P = 0.37 \\
\text{Formulation B:} & P = 0.41 \\
\text{Formulation C:} & P = 0.005 \\
\text{Formulation D:} & P = 0.21
\end{array}
$$

The operator suspects that formulation C may be effective, but he knows that the P-value of 0.005 is unreliable, because several tests have been performed. Use the Bonferroni adjustment to produce a reliable P-value.

Solution

Four tests were performed, so the Bonferroni adjustment yields $P = (4)(0.005) = 0.02$ for formulation C. So the evidence is reasonably strong that formulation C is in fact effective.

Example

6.23

In Example 6.22, assume the P-value for formulation C had been 0.03 instead of 0.005. What conclusion would you reach then?

Solution

The Bonferroni adjustment would yield $P = (4)(0.03) = 0.12$. This is probably not strong enough evidence to conclude that formulation C is in fact effective. Since the original P-value was small, however, it is likely that one would not want to give up on formulation C quite yet.

The Bonferroni adjustment is conservative; in other words, the P-value it produces is never smaller than the true P-value. So when the Bonferroni-adjusted P-value is small, the null hypothesis can be rejected conclusively. Unfortunately, as Example 6.23 shows, there are many occasions in which the original P-value is small enough to arouse a strong suspicion that a null hypothesis may be false, but the Bonferroni adjustment does not allow the hypothesis to be rejected.

When the Bonferroni-adjusted P-value is too large to reject a null hypothesis, yet the original P-value is small enough to lead one to suspect that the hypothesis is in fact false, often the best thing to do is to retest the hypothesis that appears to be false, using data from a new experiment. If the P-value is again small, this time without multiple tests, this provides real evidence against the null hypothesis.

Real industrial processes are monitored frequently by sampling and testing process output to see whether it meets specifications. Every so often, the output appears to be outside the specifications. But in these cases, how do we know whether the process is really malfunctioning (out of control) or whether the result is a type I error? This is a version of the multiple testing problem that has received much attention. The subject of statistical quality control (see Chapter 10) is dedicated in large part to finding ways to overcome the multiple testing problem.

Exercises for Section 6.8

1. Six different settings are tried on a machine to see if any of them will reduce the proportion of defective parts. For each setting, an appropriate null hypothesis is tested to see if the proportion of defective parts has been reduced. The six P-values are 0.34, 0.27, 0.002, 0.45, 0.03, and 0.19.

   a. Find the Bonferroni-adjusted P-value for the setting whose P-value is 0.002. Can you conclude that this setting reduces the proportion of defective parts? Explain.

   b. Find the Bonferroni-adjusted P-value for the setting whose P-value is 0.03. Can you conclude that this setting reduces the proportion of defective parts? Explain.

2. Five different variations of a bolt-making process are run to see if any of them can increase the mean breaking strength of the bolts over that of the current process. The P-values are 0.13, 0.34, 0.03, 0.28, and 0.38. Of the following choices, which is the best thing to do next?

i. Implement the process whose P-value was 0.03, since it performed the best.

ii. Since none of the processes had Bonferroni-adjusted P-values less than 0.05, we should stick with the current process.

iii. Rerun the process whose P-value was 0.03 to see if it remains small in the absence of multiple testing.

iv. Rerun all the five variations again, to see if any of them produce a small P-value the second time around.

3. Twenty formulations of a coating are being tested to see if any of them reduce gear wear. For the Bonferroni-adjusted P-value for a formulation to be 0.05, what must the original P-value be?

4. Five new paint additives have been tested to see if any of them can reduce the mean drying time from the current value of 12 minutes. Ten specimens have been painted with each of the new types of paint, and the drying times (in minutes) have been measured. The results are as follows:

			Additive		
	A	**B**	**C**	**D**	**E**
1	14.573	10.393	15.497	10.350	11.263
2	12.012	10.435	9.162	7.324	10.848
3	13.449	11.440	11.394	10.338	11.499
4	13.928	9.719	10.766	11.600	10.493
5	13.123	11.045	11.025	10.725	13.409
6	13.254	11.707	10.636	12.240	10.219
7	12.772	11.141	15.066	10.249	10.997
8	10.948	9.852	11.991	9.326	13.196
9	13.702	13.694	13.395	10.774	12.259
10	11.616	9.474	8.276	11.803	11.056

For each additive, perform a hypothesis test of the null hypothesis $H_0: \mu \geq 12$ against the alternate $H_1: \mu < 12$. You may assume that each population is approximately normal.

a. What are the P-values for the five tests?

b. On the basis of the results, which of the three following conclusions seems most appropriate? Explain your answer.

i. At least one of the new additives results in an improvement.

ii. None of the new additives result in an improvement.

iii. Some of the new additives may result in improvement, but the evidence is inconclusive.

5. Each day for 200 days, a quality engineer samples 144 fuses rated at 15 A and measures the amperage at which they burn out. He performs a hypothesis test of $H_0: \mu = 15$ versus $H_1: \mu \neq 15$, where μ is the mean burnout amperage of the fuses manufactured that day.

a. On 10 of the 200 days, H_0 is rejected at the 5% level. Does this provide conclusive evidence that the mean burnout amperage was different from 15 A on at least one of the 200 days? Explain.

b. Would the answer to part (a) be different if H_0 had been rejected on 20 of the 200 days? Explain.

Supplementary Exercises for Chapter 6

1. Specifications call for the mean tensile strength μ of paper used in a certain packaging application to be greater than 50 psi. A new type of paper is being considered for this application. The tensile strength is measured for a sample of 110 specimens of this paper. The mean strength was 51.2 psi and the standard deviation was 4.0 psi. Can you conclude that the mean strength is greater than 50 psi?

2. Are answer keys to multiple-choice tests generated randomly, or are they constructed to make it less likely for the same answer to occur twice in a row? This question was addressed in the article "Seek Whence: Answer Sequences and Their Consequences in Key-Balanced Multiple-Choice Tests" (M. Bar-Hillel and Y. Attali, *The American Statistician*, 2002:299–303). They studied 1280 questions on 10 real Scholastic Assessment

Tests (SATs). Assume that all the questions had five choices (in fact, 150 of them had only four choices). They found that for 192 of the questions, the correct choice (A, B, C, D, or E) was the same as the correct choice for the question immediately preceding. If the choices were generated at random, then the probability that a question would have the same correct choice as the one immediately preceding would be 0.20. Can you conclude that the choices for the SAT are not generated at random?

a. State the appropriate null and alternate hypotheses.

b. Compute the value of the test statistic.

c. Find the P-value and state your conclusion.

3. A new braking system is being evaluated for a certain type of car. The braking system will be installed if it can be conclusively demonstrated that the stopping distance under certain controlled conditions at a speed of 30 mph is less than 90 ft. It is known that under these conditions the standard deviation of stopping distance is approximately 5 ft. A sample of 150 stops will be made from a speed of 30 mph. Let μ represent the mean stopping distance for the new braking system.

a. State the appropriate null and alternate hypotheses.

b. Find the rejection region if the test is to be conducted at the 5% level.

c. Someone suggests rejecting H_0 if $\overline{X} \geq 89.4$ ft. Is this an appropriate rejection region, or is something wrong? If this is an appropriate rejection region, find the level of the test. Otherwise explain what is wrong.

d. Someone suggests rejecting H_0 if $\overline{X} \leq 89.4$ ft. Is this an appropriate rejection region, or is something wrong? If this is an appropriate rejection region, find the level of the test. Otherwise, explain what is wrong.

e. Someone suggests rejecting H_0 if $\overline{X} \leq 89.4$ ft or if $\overline{X} \geq 90.6$ ft. Is this an appropriate rejection region, or is something wrong? If this is an appropriate rejection region, find the level of the test. Otherwise, explain what is wrong.

4. The mean drying time of a certain paint in a certain application is 12 minutes. A new additive will be tested to see if it reduces the drying time. One hundred specimens will be painted, and the sample mean drying time $\overline{X}$ will be computed. Assume the population standard deviation of drying times is $\sigma = 2$ minutes. Let μ be

the mean drying time for the new paint. The null hypothesis $H_0 : \mu \geq 12$ will be tested against the alternate $H_1 : \mu < 12$. Assume that unknown to the investigators, the true mean drying time of the new paint is 11.5 minutes.

a. It is decided to reject H_0 if $\overline{X} \leq 11.7$. Find the level and power of this test.

b. For what values of $\overline{X}$ should H_0 be rejected so that the power of the test will be 0.90? What will the level then be?

c. For what values of $\overline{X}$ should H_0 be rejected so that the level of the test will be 5%? What will the power then be?

d. How large a sample is needed so that a 5% level test has power 0.90?

5. A machine manufactures bolts that are supposed to be 3 inches in length. Each day a quality engineer selects a random sample of 50 bolts from the day's production, measures their lengths, and performs a hypothesis test of $H_0 : \mu = 3$ versus $H_1 : \mu \neq 3$, where μ is the mean length of all the bolts manufactured that day. Assume that the population standard deviation for bolt lengths is 0.1 in. If H_0 is rejected at the 5% level, the machine is shut down and recalibrated.

a. Assume that on a given day, the true mean length of bolts is 3 in. What is the probability that the machine will be shut down? (This is called the **false alarm rate**.)

b. If the true mean bolt length on a given day is 3.01 in., find the probability that the equipment will be recalibrated.

6. Electric motors are assembled on four different production lines. Random samples of motors are taken from each line and inspected. The number that pass and that fail the inspection are counted for each line, with the following results:

		Line		
	1	2	3	4
Pass	482	467	458	404
Fail	57	59	37	47

Can you conclude that the failure rates differ among the four lines?

7. Refer to Exercise 6. The process engineer notices that the sample from line 3 has the lowest proportion of failures. Use the Bonferroni adjustment to determine whether she can conclude that the population proportion of failures on line 3 is less than 0.10.

8. The article "Valuing Watershed Quality Improvements Using Conjoint Analysis" (S. Farber and B. Griner, *Ecological Economics*, 2000:63–76) presents the results of a mail survey designed to assess opinions on the value of improvement efforts in an acid-mine degraded watershed in Western Pennsylvania. Of the 510 respondents to the survey, 347 were male. Census data show that 48% of the target population is male. Can you conclude that the survey method employed in this study tends to oversample males? Explain.

9. Anthropologists can estimate the birthrate of an ancient society by studying the age distribution of skeletons found in ancient cemeteries. The numbers of skeletons found at two such sites, as reported in the article "Paleoanthropological Traces of a Neolithic Demographic Transition" (J. Bocquet-Appel, *Current Anthropology*, 2002:637–650) are given in the following table:

	Ages of Skeletons		
Site	**0−4 years**	**5−19 years**	**20 years or more**
Casa da Moura	27	61	126
Wandersleben	38	60	118

Do these data provide convincing evidence that the age distributions differ between the two sites?

10. Deforestation is a serious problem throughout much of India. The article "Factors Influencing People's Participation in Forest Management in India" (W. Lise, *Ecological Economics*, 2000:379–392) discusses the social forces that influence forest management policies in three Indian states: Haryana, Bihar, and Uttar Pradesh. The forest quality in Haryana is somewhat degraded, in Bihar it is very degraded, and in Uttar Pradesh it is well stocked. In order to study the relationship between educational levels and attitudes toward forest management, researchers surveyed random samples of adults in each of these states and ascertained their educational levels. The numbers of adults at each of several educational levels were recorded. The data are presented in the following table.

	Years of Education					
State	**0**	**1−4**	**5−6**	**7−9**	**10−11**	**12 or more**
Haryana	48	6	16	26	24	7
Bihar	34	24	7	32	16	10
Uttar Pradesh	20	9	25	30	17	34

Can you conclude that the educational levels differ among the three states? Explain.

Chapter

7

Chapter

Inferences for Two Samples

Introduction

In Chapters 5 and 6, we saw how to construct confidence intervals and perform hypothesis tests concerning a single mean or proportion. There are cases in which we have two populations, and we wish to study the difference between their means, proportions, or variances. For example, suppose that a metallurgist is interested in estimating the difference in strength between two types of welds. She conducts an experiment in which a sample of 6 welds of one type has an average ultimate testing strength (in ksi) of 83.2 with a standard deviation of 5.2, and a sample of 8 welds of the other type has an average strength of 71.3 with a standard deviation of 3.1. It is easy to compute a point estimate for the difference in strengths. The difference between the sample means is $83.2 - 71.3 = 11.9$. To construct a confidence interval, however, we will need to know how to find a standard error and a critical value for this point estimate. To perform a hypothesis test to determine whether we can conclude that the mean strengths differ, we will need to know how to construct a test statistic. This chapter presents methods for constructing confidence intervals and performing hypothesis tests in situations like this.

7.1 Large-Sample Inferences on the Difference Between Two Population Means

Confidence Intervals on the Difference Between Two Means

We now investigate examples in which we wish to estimate the difference between the means of two populations. The data will consist of two samples, one from each population.

The method we describe is based on the result concerning the difference of two independent normal random variables that was presented in Section 4.3. We review this result here:

Let X and Y be independent, with $X \sim N(\mu_X, \sigma_X^2)$ and $Y \sim N(\mu_Y, \sigma_Y^2)$. Then

$$X - Y \sim N(\mu_X - \mu_Y, \sigma_X^2 + \sigma_Y^2) \tag{7.1}$$

If $\overline{X}$ is the mean of large sample of size n_X, and $\overline{Y}$ is the mean of large sample of size n_Y, then $\overline{X} \sim N(\mu_X, \sigma_X^2/n_X)$ and $\overline{Y} \sim N(\mu_Y, \sigma_Y^2/n_Y)$. If $\overline{X}$ and $\overline{Y}$ are independent, it follows that $\overline{X} - \overline{Y}$ has a normal distribution with mean $\mu_X - \mu_Y$ and standard deviation $\sigma_{\overline{X}-\overline{Y}} = \sqrt{\sigma_X^2/n_X + \sigma_Y^2/n_Y}$. Figure 7.1 illustrates the distribution of $\overline{X} - \overline{Y}$ and indicates that the middle 95% of the curve has width $\pm 1.96\sigma_{\overline{X}-\overline{Y}}$.

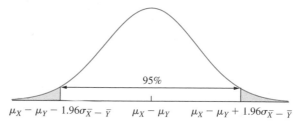

$$\mu_X - \mu_Y - 1.96\sigma_{\overline{X}-\overline{Y}} \qquad \mu_X - \mu_Y \qquad \mu_X - \mu_Y + 1.96\sigma_{\overline{X}-\overline{Y}}$$

FIGURE 7.1 The observed difference $\overline{X} - \overline{Y}$ is drawn from a normal distribution with mean $\mu_X - \mu_Y$ and standard deviation $\sigma_{\overline{X}-\overline{Y}} = \sqrt{\sigma_X^2/n_X + \sigma_Y^2/n_Y}$.

The point estimate of $\mu_X - \mu_Y$ is $\overline{X} - \overline{Y}$, and the standard error is $\sqrt{\sigma_X^2/n_X + \sigma_Y^2/n_Y}$. It follows that a 95% confidence interval for $\mu_X - \mu_Y$ is

$$\overline{X} - \overline{Y} \pm 1.96\sqrt{\frac{\sigma_X^2}{n_X} + \frac{\sigma_Y^2}{n_Y}}$$

In general, to obtain a $100(1-\alpha)\%$ confidence interval for $\mu_X - \mu_Y$, replace the critical value of 1.96 in the preceding expression with $z_{\alpha/2}$.

Summary

Let $X_1, \ldots, X_{n_X}$ be a *large* random sample of size n_X from a population with mean μ_X and standard deviation σ_X, and let $Y_1, \ldots, Y_{n_Y}$ be a *large* random sample of size n_Y from a population with mean μ_Y and standard deviation σ_Y. If the two samples are independent, then a level $100(1-\alpha)\%$ confidence interval for $\mu_X - \mu_Y$ is

$$\overline{X} - \overline{Y} \pm z_{\alpha/2}\sqrt{\frac{\sigma_X^2}{n_X} + \frac{\sigma_Y^2}{n_Y}} \tag{7.2}$$

When the values of σ_X and σ_Y are unknown, they can be replaced with the sample standard deviations s_X and s_Y.

Example 7.1

The chemical composition of soil varies with depth. The article "Sampling Soil Water in Sandy Soils: Comparative Analysis of Some Common Methods" (M. Ahmed, M. Sharma, et al., *Communications in Soil Science and Plant Analysis*, 2001:1677–1686) describes chemical analyses of soil taken from a farm in western Australia. Fifty specimens were taken at each of the depths 50 and 250 cm. At a depth of 50 cm, the average NO_3 concentration (in mg/L) was 88.5 with a standard deviation of 49.4. At a depth of 250 cm, the average concentration was 110.6 with a standard deviation of 51.5. Find a 95% confidence interval for the difference between the NO_3 concentrations at the two depths.

Solution

Let $X_1, \ldots, X_{50}$ represent the concentrations of the 50 specimens taken at 50 cm, and let $Y_1, \ldots, Y_{50}$ represent the concentrations of the 50 specimens taken at 250 cm. Then $\overline{X} = 88.5$, $\overline{Y} = 110.6$, $s_X = 49.4$, and $s_Y = 51.5$. The sample sizes are $n_X = n_Y = 50$. Both samples are large, so we can use expression (7.2). Since we want a 95% confidence interval, $z_{\alpha/2} = 1.96$. The 95% confidence interval for the difference $\mu_Y - \mu_X$ is $110.6 - 88.5 \pm 1.96\sqrt{49.4^2/50 + 51.5^2/50}$, or 22.1 ± 19.8.

Hypothesis Tests on the Difference Between Two Means

Sometimes we have large samples and we wish to determine whether the difference between two means might be equal to some specified value. In these cases we perform a hypothesis test on the difference $\mu_X - \mu_Y$. Tests on this difference are based on $\overline{X} - \overline{Y}$.

The null hypothesis will specify a value for $\mu_X - \mu_Y$. Denote this value Δ_0. The null distribution of $\overline{X} - \overline{Y}$ is

$$\overline{X} - \overline{Y} \sim N\left(\Delta_0, \frac{\sigma_X^2}{n_X} + \frac{\sigma_Y^2}{n_Y}\right) \tag{7.3}$$

The test statistic is

$$z = \frac{\overline{X} - \overline{Y} - \Delta_0}{\sqrt{\sigma_X^2/n_X + \sigma_Y^2/n_Y}} \tag{7.4}$$

If σ_X and σ_Y are unknown, they may be replaced with s_X and s_Y, respectively. The P-value is the area in one or both tails of the normal curve, depending on whether the test is one- or two-tailed.

Example 7.2

The article "Effect of Welding Procedure on Flux Cored Steel Wire Deposits" (N. Ramini de Rissone, I. de S. Bott, et al., *Science and Technology of Welding and Joining*, 2003:113–122) compares properties of welds made using carbon dioxide as a shielding gas with those of welds made using a mixture of argon and carbon dioxide. One property studied was the diameter of inclusions, which are particles embedded in the weld. A sample of 544 inclusions in welds made using argon shielding averaged 0.37 μm in diameter, with a standard deviation of 0.25 μm. A sample of 581 inclusions in welds made using carbon dioxide shielding averaged 0.40 μm in diameter, with a standard deviation of 0.26 μm. (Standard deviations were estimated from a graph.) Can you conclude that the mean diameters of inclusions differ between the two shielding gases?

Solution

Let $\overline{X} = 0.37$ denote the sample mean diameter for argon welds. Then $s_X = 0.25$ and the sample size is $n_X = 544$. Let $\overline{Y} = 0.40$ denote the sample mean diameter for carbon dioxide welds. Then $s_Y = 0.26$ and the sample size is $n_Y = 581$. Let μ_X denote the population mean diameter for argon welds, and let μ_Y denote the population mean diameter for carbon dioxide welds. The null and alternate hypotheses are

$$H_0: \mu_X - \mu_Y = 0 \quad \text{versus} \quad H_1: \mu_X - \mu_Y \neq 0$$

The value of the test statistic is

$$z = \frac{0.37 - 0.40 - 0}{\sqrt{0.25^2/544 + 0.26^2/581}} = -1.97$$

This is a two-tailed test, and the P-value is 0.0488 (see Figure 7.2). A follower of the 5% rule would reject the null hypothesis. It is certainly reasonable to be skeptical about the truth of H_0.

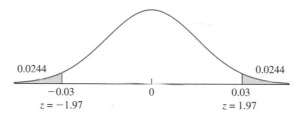

FIGURE 7.2 Solution to Example 7.2.

The following computer output (from MINITAB) presents the results of Example 7.2.

```
Two-sample T for Argon vs CO2

         N    Mean   StDev    SE Mean
Argon   544   0.37   0.25     0.010719
CO2     581   0.40   0.26     0.010787


Difference = mu (Argon) - mu (CO2)
Estimate for difference: 0.030000
95% confidence bound for difference: (-0.0598366, -0.000163)
T-Test of difference = 0 (vs not = 0): T-Value = -1.97 P-Value = 0.049 DF = 1122
```

Note that the computer uses the t statistic rather than the z statistic for this test. Many computer packages use the t statistic whenever a sample standard deviation is used to estimate a population standard deviation. When the sample size is large, the difference between t and z is negligible for practical purposes. When using tables rather than a computer, the z-score has the advantage that the P-value can be determined with greater precision with a z table than with a t table.

Example 7.3 presents a situation in which the null hypothesis specifies that the two population means differ by a constant.

Example
7.3

Refer to Example 7.2. Can you conclude that the mean diameter for carbon dioxide welds (μ_Y) exceeds that for argon welds (μ_X) by more than 0.015 μm?

Solution
The null and alternate hypotheses are

$$H_0 : \mu_X - \mu_Y \geq -0.015 \quad \text{versus} \quad H_1 : \mu_X - \mu_Y < -0.015$$

We observe $\overline{X} = 0.37$, $\overline{Y} = 0.40$, $s_X = 0.25$, $s_Y = 0.26$, $n_X = 544$, and $n_Y = 581$. Under H_0, we take $\mu_X - \mu_Y = -0.015$. The null distribution of $\overline{X} - \overline{Y}$ is given by expression (7.3) to be

$$\overline{X} - \overline{Y} \sim N(-0.015,\ 0.01521^2)$$

We observe $\overline{X} - \overline{Y} = 0.37 - 0.40 = -0.03$. The z-score is

$$z = \frac{-0.03 - (-0.015)}{0.01521} = -0.99$$

This is a one-tailed test. The P-value is 0.1611. We cannot conclude that the mean diameter of inclusions from carbon dioxide welds exceeds that of argon welds by more than 0.015 μm.

Summary

Let $X_1, \ldots, X_{n_X}$ and $Y_1, \ldots, Y_{n_Y}$ be *large* (e.g., $n_X > 30$ and $n_Y > 30$) samples from populations with means μ_X and μ_Y and standard deviations σ_X and σ_Y, respectively. Assume the samples are drawn independently of each other.

To test a null hypothesis of the form $H_0 : \mu_X - \mu_Y \leq \Delta_0$, $H_0 : \mu_X - \mu_Y \geq \Delta_0$, or $H_0 : \mu_X - \mu_Y = \Delta_0$:

- Compute the z-score: $z = \dfrac{(\overline{X} - \overline{Y}) - \Delta_0}{\sqrt{\sigma_X^2/n_X + \sigma_Y^2/n_Y}}$. If σ_X and σ_Y are unknown

  they may be approximated with s_X and s_Y, respectively.

- Compute the P-value. The P-value is an area under the normal curve, which depends on the alternate hypothesis as follows:

Alternate Hypothesis	P-value
$H_1 : \mu_X - \mu_Y > \Delta_0$	Area to the right of z
$H_1 : \mu_X - \mu_Y < \Delta_0$	Area to the left of z
$H_1 : \mu_X - \mu_Y \neq \Delta_0$	Sum of the areas in the tails cut off by z and $-z$

Exercises for Section 7.1

1. The article "Vehicle-Arrival Characteristics at Urban Uncontrolled Intersections" (V. Rengaraju and V. Rao, *Journal of Transportation Engineering*, 1995:317–323) presents data on traffic characteristics at 10 intersections in Madras, India. At one particular intersection, the average speed for a sample of 39 cars was 26.50 km/h, with a standard deviation of 2.37 km/h. The average speed for a sample of 142 motorcycles was 37.14 km/h, with a standard deviation of 3.66 km/h. Find a 95% confidence interval for the difference between the mean speeds of motorcycles and cars.

2. The article "Some Parameters of the Population Biology of Spotted Flounder (*Ciutharus linguatula* Linnaeus, 1758) in Edremit Bay (North Aegean Sea)" (D. Türker, B. Bayhan, et al., *Turkish Journal of Veterinary and Animal Science,* 2005:1013–1018) reports that a sample of 87 one-year-old spotted flounder had an average length of 126.31 mm with a standard deviation of 18.10 mm, and a sample of 132 two-year-old spotted flounder had an average length of 162.41 mm with a standard deviation of 28.49 mm. Find a 95% confidence interval for the mean length increase between one- and two-year-old spotted flounder.

3. The melting points of two alloys are being compared. Thirty-five specimens of alloy 1 were melted. The average melting temperature was 517.0°F and the standard deviation was 2.4°F. Forty-seven specimens of alloy 2 were melted. The average melting temperature was 510.1°F and the standard deviation was 2.1°F. Find a 99% confidence interval for the difference between the melting points.

4. A stress analysis was conducted on random samples of epoxy-bonded joints from two species of wood. A random sample of 120 joints from species A had a mean shear stress of 1250 psi and a standard deviation of 350 psi, and a random sample of 90 joints from species B had a mean shear stress of 1400 psi and a standard deviation of 250 psi. Find a 98% confidence interval for the difference in mean shear stress between the two species.

5. The article "Capillary Leak Syndrome in Children with C4A-Deficiency Undergoing Cardiac Surgery with Cardiopulmonary Bypass: A Double-Blind, Randomised Controlled Study" (S. Zhang, S. Wang, et al., *Lancet*, 2005:556–562) presents the results of a study of the effectiveness of giving blood plasma containing complement component C4A to pediatric cardiopulmonary bypass patients. Of 58 patients receiving C4A-rich plasma, the average length of hospital stay was 8.5 days and the standard deviation was 1.9 days. Of 58 patients receiving C4A-free plasma, the average length of hospital stay was 11.9 days and the standard deviation was 3.6 days. Find a 99% confidence interval for the reduction in mean hospital stay for patients receiving C4A-rich plasma rather than C4A-free plasma.

6. A sample of 45 room air conditioners of a certain model had a mean sound pressure of 52 decibels (dB) and a standard deviation of 5 dB, and a sample of 60 air conditioners of a different model had a mean sound pressure of 46 dB and a standard deviation of 2 dB. Find a 98% confidence interval for the difference in mean sound pressure between the two models.

7. The article "The Prevalence of Daytime Napping and Its Relationship to Nighttime Sleep" (J. Pilcher, K. Michalkowski, and R. Carrigan), *Behavioral Medicine*, 2001:71–76) presents results of a study of sleep habits in a large number of subjects. In a sample of 87 young adults, the average time per day spent in bed (either awake or asleep) was 7.70 hours, with a standard deviation of 1.02 hours, and the average time spent in bed asleep was 7.06 hours, with a standard deviation of 1.11 hours. The mean time spent in bed awake was estimated to be $7.70 - 7.06 = 0.64$ hours. Is it possible to compute a 95% confidence interval for the mean time spent in bed awake? If so, construct the confidence interval. If not possible, explain why not.

8. The article "Occurrence and Distribution of Ammonium in Iowa Groundwater" (K. Schilling, *Water Environment Research*, 2002:177–186) describes measurements of ammonium concentrations (in mg/L) at a large number of wells in the state of Iowa. These included 349 alluvial wells and 143 quaternary wells. The concentrations at the alluvial wells averaged 0.27 with a standard deviation of 0.40, and those at the quaternary wells averaged 1.62 with a standard deviation of 1.70. Find a 95% confidence interval for the

difference in mean concentrations between alluvial and quaternary wells.

9. The article "Measurement of Complex Permittivity of Asphalt Paving Materials" (J. Shang, J. Umana, et al., *Journal of Transportation Engineering*, 1999: 347–356) compared the dielectric constants between two types of asphalt, HL3 and HL8, commonly used in pavements. For 42 specimens of HL3 asphalt the average dielectric constant was 5.92 with a standard deviation of 0.15, and for 37 specimens of HL8 asphalt the average dielectric constant was 6.05 with a standard deviation of 0.16. Find a 95% confidence interval for the difference between the mean dialectric constants for the two types of asphalt.

10. The article referred to in Exercise 2 reported that a sample of 482 female spotted flounder had an average weight of 20.95 g with a standard deviation of 14.5 g, and a sample of 614 male spotted flounder had an average weight of 22.79 g with a standard deviation of 15.6 g. Find a 90% confidence interval for the difference between the mean weights of male and female spotted flounder.

11. Two corrosion inhibitors, one more expensive than the other, are being considered for treating stainless steel. To compare them, specimens of stainless steel were immersed for four hours in a solution containing sulfuric acid and a corrosion inhibitor. Thirty-eight specimens immersed in the less expensive product had an average weight loss of 242 mg and a standard deviation of 20 mg, and 42 specimens immersed in the more expensive product had an average weight loss of 180 mg and a standard deviation of 31 mg. It is determined that the more expensive product will be used if it can be shown that its additional mean weight loss over that of the less expensive method is greater than 50 mg. Perform an appropriate hypothesis test, and on the basis of the results, determine which inhibitor to use.

12. In an experiment involving high-temperature performance of two types of transistors, a sample of 60 transistors of type A were tested and were found to have a mean lifetime of 1427 hours and a standard deviation of 183 hours. A sample of 180 transistors of type B were tested and were found to have a mean lifetime of 1358 hours and a standard deviation of 240 hours. Can you conclude that the mean lifetimes differ between the two types of transistors?

13. In a study of the effect of cooling rate on the hardness of welded joints, 70 welds cooled at a rate of 10°C/s had an average Rockwell (B) hardness of 92.3 and a standard deviation of 6.2, and 60 welds cooled at a rate of 40°C/s had an average hardness of 90.2 and a standard deviation of 4.4. Can you conclude that the mean hardness of welds cooled at a rate of 10°C/s is greater than that of welds cooled at a rate of 40°C/s?

14. A crayon manufacturer is comparing the effects of two kinds of yellow dye on the brittleness of crayons. Dye B is more expensive than dye A, but it is thought that it might produce a stronger crayon. Forty crayons are tested with each kind of dye, and the impact strength (in joules) is measured for each. For the crayons made with dye A, the strength averaged 2.6 with a standard deviation of 1.4. For the crayons made with dye B, the strength averaged 3.8 with a standard deviation of 1.2.

   a. Can you conclude that the mean strength of crayons made with dye B is greater than that of crayons made with dye A?

   b. Can you conclude that the mean strength of crayons made with dye B exceeds that of crayons made with dye A by more than 1 J?

15. An engineer is investigating the effect of air flow speed on the recovery of heat lost due to exhaust gases. Eighty measurements were made of the proportion of heat recovered in a furnace with a flow speed of 2 m/s. The average was 0.67 and the standard deviation was 0.46. The flow speed was then set to 4 m/s, and 60 measurements were taken. The average was 0.59 and the standard deviation was 0.38. Can you conclude that the mean proportion of heat recovered is greater at the lower flow speed?

16. In a study of the relationship of the shape of a tablet to its dissolution time, 36 disk-shaped ibuprofen tablets and 48 oval-shaped ibuprofen tablets were dissolved in water. The dissolution times for the disk-shaped tablets averaged 258 seconds with a standard deviation of 12 seconds, and the times for the oval-shaped tablets averaged 262 seconds with a standard deviation of 15 seconds. Can you conclude that the mean dissolve times differ between the two shapes?

17. A statistics instructor who teaches a lecture section of 160 students wants to determine whether students have more difficulty with one-tailed hypothesis tests

or with two-tailed hypothesis tests. On the next exam, 80 of the students, chosen at random, get a version of the exam with a 10-point question that requires a one-tailed test. The other 80 students get a question that is identical except that it requires a two-tailed test. The one-tailed students average 7.79 points, and their standard deviation is 1.06 points. The two-tailed students average 7.64 points, and their standard deviation is 1.31 points.

a. Can you conclude that the mean score μ_1 on one-tailed hypothesis test questions is higher than the mean score μ_2 on two-tailed hypothesis test questions? State the appropriate null and alternate hypotheses, and then compute the P-value.

b. Can you conclude that the mean score μ_1 on one-tailed hypothesis test questions differs from the mean score μ_2 on two-tailed hypothesis test questions? State the appropriate null and alternate hypotheses, and then compute the P-value.

18. Fifty specimens of a new computer chip were tested for speed in a certain application, along with 50 specimens of chips with the old design. The average speed, in MHz, for the new chips was 495.6, and the standard deviation was 19.4. The average speed for the old chips was 481.2, and the standard deviation was 14.3.

a. Can you conclude that the mean speed for the new chips is greater than that of the old chips? State the appropriate null and alternate hypotheses, and then find the P-value.

b. A sample of 60 even older chips had an average speed of 391.2 MHz with a standard deviation of 17.2 MHz. Someone claims that the new chips average more than 100 MHz faster than these very old ones. Do the data provide convincing evidence for this claim? State the appropriate null and alternate hypotheses, and then find the P-value.

19. Two methods are being considered for a paint manufacturing process, in an attempt to increase production. In a random sample of 100 days, the mean daily production using the first method was 625 tons with a standard deviation of 40 tons. In a random sample of 64 days, the mean daily production using the second method was 645 tons with a standard deviation of 50 tons. Assume the samples are independent.

a. Can you conclude that the second method yields the greater mean daily production?

b. Can you conclude that the mean daily production for the second method exceeds that of the first method by more than 5 tons?

20. The following MINITAB output presents the results of a hypothesis test for the difference $\mu_X - \mu_Y$ between two population means:

```
Two-sample T for X vs Y

      N   Mean   StDev   SE Mean
X    135  3.94    2.65    0.23
Y    180  4.43    2.38    0.18

Difference = mu (X) — mu (Y)
Estimate for difference: −0.484442
95% upper bound for difference: −0.007380
T-Test of difference = 0 (vs <): T-Value = −1.68 P-Value = 0.047 DF = 270
```

a. Is this a one-tailed or two-tailed test?
b. What is the null hypothesis?
c. Can H_0 be rejected at the 5% level? How can you tell?
d. The output presents a Student's t test. Compute the P-value using a z test. Are the two results similar?
e. Use the output and an appropriate table to compute a 99% confidence interval for $\mu_X - \mu_Y$ based on the z statistic.

21. The following MINITAB output presents the results of a hypothesis test for the difference $\mu_X - \mu_Y$ between two population means. Some of the numbers are missing.

```
Two-sample T for X vs Y

      N    Mean   StDev   SE Mean
X    78   23.3    (i)      1.26
Y    63   20.63   3.02     (ii)

Difference = mu (X) − mu (Y)
Estimate for difference: 2.670
95% CI for difference: (0.05472, 5.2853)
T-Test of difference = 0 (vs not =): T-Value = 2.03 P-Value = 0.045 DF = 90
```

---

a. Fill in the missing numbers for (i) and (ii).
b. The output presents a Student's t test. Compute the P-value using a z test. Are the two results similar?
c. Use the output and an appropriate table to compute a 98% confidence interval for $\mu_X - \mu_Y$ based on the z statistic.

7.2   Inferences on the Difference Between Two Proportions

Confidence Intervals on the Difference Between Two Proportions

In Section 7.1, we presented a method for constructing a confidence interval for the difference between two population means. In this section, we present a similar method that is designed to find a confidence interval for the difference between two population proportions.

Let X be the number of successes in n_X trials where the success probability is p_X. Let Y be the number of successes in n_Y trials where the success probability is p_Y. The sample proportions are $\hat{p}_X = X/n_X$ and $\hat{p}_Y = Y/n_Y$. If n_X and n_Y are sufficiently large, then X and Y are both approximately normally distributed. It follows that $\hat{p}_X$, $\hat{p}_Y$, and the difference $\hat{p}_X - \hat{p}_Y$ are also approximately normally distributed. In particular, if X and Y are independent, the difference $\hat{p}_X - \hat{p}_Y$ has the following normal distribution:

$$\hat{p}_X - \hat{p}_Y \sim N\left(p_X - p_Y, \frac{p_X(1 - p_X)}{n_X} + \frac{p_Y(1 - p_Y)}{n_Y}\right)$$

The standard error of $\hat{p}_X - \hat{p}_Y$ is $\sqrt{p_X(1 - p_X)/n_X + p_Y(1 - p_Y)/n_Y}$. We can't use this value in a confidence interval because it contains the unknowns p_X and p_Y. The traditional approach is to replace p_X and p_Y with $\hat{p}_X$ and $\hat{p}_Y$, obtaining $\sqrt{\hat{p}_X(1 - \hat{p}_X)/n_X + \hat{p}_Y(1 - \hat{p}_Y)/n_Y}$. Since the point estimate $\hat{p}_X - \hat{p}_Y$ is approximately normal, the critical value for a $100(1 - \alpha)\%$ confidence interval is $z_{\alpha/2}$.

Recent research has shown that this interval can be improved by slightly modifying the point estimate and the standard error. Simply add 1 to each of the numbers of successes X and Y, and add 2 to each of the numbers of trials n_X and n_Y. Thus we define $\tilde{n}_X = n_X + 2$, $\tilde{n}_Y = n_Y + 2$, $\tilde{p}_X = (X + 1)/\tilde{n}_X$, and $\tilde{p}_Y = (Y + 1)/\tilde{n}_Y$.

The level $100(1 - \alpha)\%$ confidence interval for the difference $p_X - p_Y$ is

$$\tilde{p}_X - \tilde{p}_Y \pm z_{\alpha/2}\sqrt{\tilde{p}_X(1 - \tilde{p}_X)/\tilde{n}_X + \tilde{p}_Y(1 - \tilde{p}_Y)/\tilde{n}_Y}$$

This method has been found to give good results for almost all sample sizes.

Summary

Let X be the number of successes in n_X independent Bernoulli trials with success probability p_X, and let Y be the number of successes in n_Y independent Bernoulli trials with success probability p_Y, so that $X \sim \text{Bin}(n_X, p_X)$ and $Y \sim \text{Bin}(n_Y, p_Y)$. Define $\tilde{n}_X = n_X + 2$, $\tilde{n}_Y = n_Y + 2$, $\tilde{p}_X = (X + 1)/\tilde{n}_X$, and $\tilde{p}_Y = (Y + 1)/\tilde{n}_Y$.

Then a level $100(1 - \alpha)\%$ confidence interval for the difference $p_X - p_Y$ is

$$\tilde{p}_X - \tilde{p}_Y \pm z_{\alpha/2}\sqrt{\frac{\tilde{p}_X(1 - \tilde{p}_X)}{\tilde{n}_X} + \frac{\tilde{p}_Y(1 - \tilde{p}_Y)}{\tilde{n}_Y}} \tag{7.5}$$

If the lower limit of the confidence interval is less than -1, replace it with -1. If the upper limit of the confidence interval is greater than 1, replace it with 1.

The adjustment described here for the two-sample confidence interval is similar to the one described in Section 5.3 for the one-sample confidence interval. In both cases, a total of two successes and four trials are added. In the two-sample case, these are divided between the samples, so that one success and two trials are added to each sample. In the one-sample case, two successes and four trials are added to the one sample. The confidence interval given by expression (7.5) can be called the *Agresti–Caffo* interval, after Alan Agresti and Brian Caffo, who developed it. For more information about this confidence interval, consult the article "Simple and Effective Confidence Intervals for Proportions and Differences of Proportions Result from Adding Two Successes and Two Failures" (A. Agresti and B. Caffo, *The American Statistician*, 2000:280–288).

Example
7.4

Methods for estimating strength and stiffness requirements should be conservative, in that they should overestimate rather than underestimate. The success rate of such a method can be measured by the probability of an overestimate. The article "Discrete Bracing Analysis for Light-Frame Wood-Truss Compression Webs" (M. Waltz, T. McLain, et al., *Journal of Structural Engineering*, 2000:1086–1093) presents the results of an experiment that evaluated a standard method (Plaut's method) for estimating the brace force for a compression web brace. In a sample of 380 short test columns (4 to 6 ft in length), the method overestimated the force for 304 of them, and in a sample of 394 long test columns (8 to 10 ft in length), the method overestimated the force for 360 of them. Find a 95% confidence interval for the difference between the success rates for long columns and short columns.

Solution

The number of successes in the sample of short columns is $X = 304$, and the number of successes in the sample of long columns is $Y = 360$. The numbers of trials are $n_X = 380$ and $n_Y = 394$. We compute $\tilde{n}_X = 382$, $\tilde{n}_Y = 396$, $\tilde{p}_X = (304 + 1)/382 = 0.7984$, and $\tilde{p}_Y = (360 + 1)/396 = 0.9116$. The value of $z_{\alpha/2}$ is 1.96. The 95% confidence interval is $0.9116 - 0.7984 \pm 1.96\sqrt{(0.7984)(0.2016)/382 + (0.9116)(0.0884)/396}$, or 0.1132 ± 0.0490.

The Traditional Method

Many people use a traditional method for computing confidence intervals for the difference between proportions. This method uses the sample proportions $\hat{p}_X$ and $\hat{p}_Y$ and the actual sample sizes n_X and n_Y. The traditional method gives results very similar to those of the modern method previously described for large or moderately large sample sizes. For small sample sizes, however, the traditional confidence interval fails to achieve its coverage probability; in other words, level $100(1 - \alpha)\%$ confidence intervals computed by the traditional method cover the true value less than $100(1 - \alpha)\%$ of the time.

Summary

The Traditional Method for Computing Confidence Intervals for the Difference Between Proportions (widely used but not recommended)

Let $\hat{p}_X$ be the proportion of successes in a *large* number n_X of independent Bernoulli trials with success probability p_X, and let $\hat{p}_Y$ be the proportion of successes in a *large* number n_Y of independent Bernoulli trials with success probability p_Y. Then the traditional level $100(1 - \alpha)\%$ confidence interval for $p_X - p_Y$ is

$$\hat{p}_X - \hat{p}_Y \pm z_{\alpha/2}\sqrt{\frac{\hat{p}_X(1 - \hat{p}_X)}{n_X} + \frac{\hat{p}_Y(1 - \hat{p}_Y)}{n_Y}} \tag{7.6}$$

This method must not be used unless both samples contain at least 10 successes and 10 failures.

Hypothesis Tests on the Difference Between Two Proportions

We can perform hypothesis tests on the difference $p_X - p_Y$ between two proportions whenever the sample sizes are large enough so that the sample proportions $\hat{p}_X$ and $\hat{p}_Y$ are approximately normal. One rule of thumb is that the samples may be considered to be large enough if each sample contains at least 10 successes and 10 failures.

When performing a hypothesis test, we do not adjust the values of X, Y, n_X and n_Y as we do for a confidence interval. We use the sample proportions $\hat{p}_X = X/n_X$ and $\hat{p}_Y = Y/n_Y$. The test is based on the statistic $\hat{p}_X - \hat{p}_Y$. We must determine the null distribution of this statistic. Since n_X and n_Y are both large,

$$\hat{p}_X - \hat{p}_Y \sim N\left(p_X - p_Y, \; \frac{p_X(1-p_X)}{n_X} + \frac{p_Y(1-p_Y)}{n_Y}\right) \tag{7.7}$$

To obtain the null distribution, we must substitute values for the mean $p_X - p_Y$ and the variance $p_X(1-p_X)/n_X + p_Y(1-p_Y)/n_Y$. The mean is easy. In problems we will consider, the null hypothesis will specify that $p_X - p_Y = 0$. The variance is a bit trickier. At first glance, it might seem reasonable to approximate the standard deviation by substituting the sample proportions $\hat{p}_X$ and $\hat{p}_Y$ for the population proportions p_X and p_Y. However, the null hypothesis H_0 specifies that the population proportions are equal. Therefore we must estimate them both with a common value. The appropriate value is the **pooled proportion**, obtained by dividing the total number of successes in both samples by the total sample size. This value is

$$\hat{p} = \frac{X + Y}{n_X + n_Y}$$

The null distribution of $\hat{p}_X - \hat{p}_Y$ is therefore estimated by substituting the pooled proportion $\hat{p}$ for both p_X and p_Y into expression (7.7). This yields

$$\hat{p}_X - \hat{p}_Y \sim N\left(0, \; \hat{p}(1-\hat{p})\left(\frac{1}{n_X} + \frac{1}{n_Y}\right)\right) \tag{7.8}$$

The test statistic is

$$z = \frac{\hat{p}_X - \hat{p}_Y}{\hat{p}(1-\hat{p})(1/n_X + 1/n_Y)} \tag{7.9}$$

Example

7.5

Industrial firms often employ methods of "risk transfer," such as insurance or indemnity clauses in contracts, as a technique of risk management. The article "Survey of Risk Management in Major U.K. Companies" (S. Baker, K. Ponniah, and S. Smith, *Journal of Professional Issues in Engineering Education and Practice*, 1999:94–102) reports the results of a survey in which managers were asked which methods played a major role in the risk management strategy of their firms. In a sample of 43 oil companies, 22 indicated that risk transfer played a major role, while in a sample of 93 construction companies, 55 reported that risk transfer played a major role. (These figures were read from a graph.) Can we conclude that the proportion of oil companies that employ the method of risk transfer is less than the proportion of construction companies that do?

Solution

Let $\hat{p}_X = 22/43 = 0.5116$ be the sample proportion of oil companies employing risk transfer methods, and let $\hat{p}_Y = 55/93 = 0.5914$ be the corresponding sample proportion of industrial firms. The sample sizes are $n_X = 43$ and $n_Y = 93$. Let p_X and p_Y denote the population proportions for oil and industrial companies, respectively. The null and alternate hypotheses are

$$H_0: p_X - p_Y \geq 0 \quad \text{versus} \quad H_1: p_X - p_Y < 0$$

The pooled proportion is

$$\hat{p} = \frac{22 + 55}{43 + 93} = 0.5662$$

The null distribution is normal with mean 0 and standard deviation

$$\sqrt{0.5662(1 - 0.5662)(1/43 + 1/93)} = 0.0914$$

The observed value of $\hat{p}_X - \hat{p}_Y$ is $0.5116 - 0.5914 = -0.0798$. The z-score is

$$z = \frac{-0.0798 - 0}{0.0914} = -0.87$$

The P-value is 0.1922 (see Figure 7.3). We cannot conclude that the proportion of oil companies employing risk transfer methods is less than the proportion of industrial firms that do.

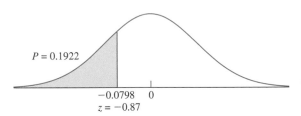

$P = 0.1922$

$-0.0798 \quad 0$

$z = -0.87$

FIGURE 7.3 Solution to Example 7.5.

The following computer output (from MINITAB) presents the results of Example 7.5.

```
Test and CI for Two Proportions: Oil, Indus.

Variable   X    N  Sample p
Oil       22   43  0.511628
Indus.    55   93  0.591398

Difference = p (Oil) - p (Indus.)
Estimate for difference: -0.079770
95% Upper Bound for difference: 0.071079
Test for difference = 0 (vs < 0): Z = -0.87 P-Value = 0.192
```

Summary

Let $X \sim \text{Bin}(n_X, p_X)$ and let $Y \sim \text{Bin}(n_Y, p_Y)$. Assume that there are at least 10 successes and 10 failures in each sample, and that X and Y are independent.

To test a null hypothesis of the form $H_0: p_X - p_Y \leq 0$, $H_0: p_X - p_Y \geq 0$, or $H_0: p_X - p_Y = 0$:

■ Compute $\hat{p}_X = \dfrac{X}{n_X}$, $\hat{p}_Y = \dfrac{Y}{n_Y}$, and $\hat{p} = \dfrac{X + Y}{n_X + n_Y}$.

■ Compute the z-score: $z = \dfrac{\hat{p}_X - \hat{p}_Y}{\sqrt{\hat{p}(1 - \hat{p})(1/n_X + 1/n_Y)}}$.

■ Compute the P-value. The P-value is an area under the normal curve, which depends on the alternate hypothesis as follows:

Alternate Hypothesis	P-value
$H_1: p_X - p_Y > 0$	Area to the right of z
$H_1: p_X - p_Y < 0$	Area to the left of z
$H_1: p_X - p_Y \neq 0$	Sum of the areas in the tails cut off by z and $-z$

Exercises for Section 7.2

1. The article "Genetically Based Tolerance to Endosulfan, Chromium (VI) and Fluoranthene in the Grass Shrimp *Palaemonetes pugio*" (R. Harper-Arabie, Ph.D. thesis, Colorado School of Mines, 2002) reported that out of 1985 eggs produced by shrimp at the Diesel Creek site in Charleston, South Carolina, 1919 hatched, and at the Shipyard Creed site, also in Charleston, 4561 out of 4988 eggs hatched. Find a 99% confidence interval for the difference between the proportions of eggs that hatch at the two sites.

2. The article "Accidents on Suburban Highways—Tennessee's Experience" (R. Margiotta and A. Chatterjee, *Journal of Transportation Engineering*, 1995:255–261) compares rates of traffic accidents at intersections with raised medians with rates at intersections with two-way left-turn lanes. Out of 4644 accidents at intersections with raised medians, 2280 were rear-end accidents, and out of 4584 accidents at two-way left-turn lanes, 1982 were rear-end accidents. Assuming these to be random samples of accidents from two types of intersections, find a 90% confidence interval for the difference between the proportions of accidents that are of the rear-end type at the two types of intersections.

3. A mail order company is investigating the potential advantages of using a new heavy-duty packaging ma-terial. They shipped 1000 orders in the new packaging and another 1000 orders in the regular light packaging. Of the orders shipped in the regular packaging, 32 were damaged in shipment, and of the orders shipped in the new packaging, only 15 were damaged in shipment. Find a 95% confidence interval for the difference between the proportions of damaged shipments.

4. The article "Occurrence and Distribution of Ammonium in Iowa Groundwater" (K. Schilling, *Water Environment Research*, 2002:177–186) describes measurements of ammonium concentrations (in mg/L) at a large number of wells in the state of Iowa. These included 349 alluvial wells and 143 quaternary wells. Of the alluvial wells, 182 had concentrations above 0.1, and 112 of the quaternary wells had concentrations above 0.1. Find a 95% confidence interval for the difference between the proportions of the two types of wells with concentrations above 0.1.

5. To judge the effectiveness of an advertising campaign for a certain brand of potato chip, a random sample of 500 supermarket receipts was taken during the week before the ad campaign began, and 65 of them showed a purchase of the potato chips. Another random sample of 500 receipts was taken during the week after the campaign ended, and 92 of them showed a potato chip purchase. Find a 99% confidence interval for the

difference in the proportions of customers purchasing potato chips before and after the ad campaign.

6. The article "Case Study Based Instruction of DOE and SPC" (J. Brady and T. Allen, *The American Statistician*, 2002:312–315) describes an effort by an engineering team to reduce the defect rate in the manufacture of a certain printed circuit board. The team decided to reconfigure the transistor heat sink. A total of 1500 boards were produced the week before the reconfiguration was implemented, and 345 of these were defective. A total of 1500 boards were produced the week after reconfiguration, and 195 of these were defective. Find a 95% confidence interval for the decrease in the defective rate after the reconfiguration.

7. The article referred to in Exercise 9 in Section 5.3 describes an experiment in which a total of 42 leachate specimens were tested for the presence of several contaminants. Of the 42 specimens, 26 contained detectable levels of lead, and 32 contained detectable levels of chromium. Is it possible, using the methods of this section, to find a 95% confidence interval for the difference between the probability that a specimen will contain a detectable amount of lead and the probability that it will contain a detectable amount of chromium? If so, find the confidence interval. If not, explain why not.

8. In a clinical trial to study the effectiveness of a new pain reliever, a sample of 250 patients was given the new drug, and an independent sample of 100 patients was given a placebo. Of the patients given the new drug, 186 reported substantial relief, and of the patients given the placebo, 56 reported substantial relief. Find a 99% confidence interval for the difference between the proportions of patients experiencing substantial relief.

9. Two extrusion machines that manufacture steel rods are being compared. In a sample of 1000 rods taken from machine 1, 960 met specifications regarding length and diameter. In a sample of 600 rods taken from machine 2, 582 met the specifications. Machine 2 is more expensive to run, so it is decided that machine 1 will be used unless it can be convincingly shown that machine 2 produces a larger proportion of rods meeting specifications.

    a. State the appropriate null and alternate hypotheses for making the decision as to which machine to use.

    b. Compute the *P*-value.

    c. Which machine should be used?

10. Resistors labeled as 100 Ω are purchased from two different vendors. The specification for this type of resistor is that its actual resistance be within 5% of its labeled resistance. In a sample of 180 resistors from vendor A, 150 of them met the specification. In a sample of 270 resistors purchased from vendor B, 233 of them met the specification. Vendor A is the current supplier, but if the data demonstrate convincingly that a greater proportion of the resistors from vendor B meet the specification, a change will be made.

    a. State the appropriate null and alternate hypotheses.

    b. Find the *P*-value.

    c. Should a change be made?

11. The article "Strategic Management in Engineering Organizations" (P. Chinowsky, *Journal of Management in Engineering*, 2001:60–68) presents results of a survey on management styles that was administered both to private construction firms and to public agencies. Out of a total of 400 private firms contacted, 133 responded by completing the survey, while out of a total of 100 public agencies contacted, 50 responded. Can you conclude that the response rate differs between private firms and public agencies?

12. The article "Training Artificial Neural Networks with the Aid of Fuzzy Sets" (C. Juang, S. Ni, and C. Lu, *Computer-Aided Civil and Infrastructure Engineering*, 1999:407–415) describes the development of artificial neural networks designed to predict the collapsibility of soils. A model with one hidden layer made a successful prediction in 48 out of 60 cases, while a model with two hidden layers made a successful prediction in 44 out of 60 cases. Assuming these samples to be independent, can you conclude that the model with one hidden layer has a greater success rate?

13. In a certain state, a referendum is being held to determine whether the transportation authority should issue additional highway bonds. A sample of 500 voters is taken in county A, and 285 say that they favor the bond proposal. A sample of 600 voters is taken in county B, and 305 say that they favor the bond issue. Can you conclude that the proportion of voters favoring the proposal is greater in county A than in county B?

14. Out of 1200 pieces of gravel from one plant, 110 pieces are classified as "large." Out of 900 pieces from another plant, 95 are classified as large. Can you conclude that there is a difference between the proportions of large gravel pieces produced at the two plants?

15. In a random sample of 280 cars driven at sea level, 38 of them produced more than 10 g of particulate pollution per gallon of fuel consumed. In a random sample of 77 cars driven at a higher elevation, 18 of them produced more than 10 g of particulate pollution per gallon of fuel consumed. Can you conclude that the proportion of cars whose emissions exceed 10 g per gallon is greater at the higher elevation?

16. The article "Modeling the Inactivation of Particle-Associated Coliform Bacteria" (R. Emerick, F. Loge, et al., *Water Environment Research*, 2000:432–438) presents counts of the numbers of particles of various sizes in wastewater samples that contained coliform bacteria. Out of 161 particles 75–80 μm in diameter, 19 contained coliform bacteria, and out of 95 particles 90–95 μm in diameter, 22 contained coliform bacteria. Can you conclude that the larger particles are more likely to contain coliform bacteria?

17. In the article "Nitrate Contamination of Alluvial Groundwaters in the Nakdong River Basin, Korea" (J. Min, S. Yun, et al., *Geosciences Journal*, 2002: 35–46), 41 water samples taken from wells in the Daesan area were described, and 22 of these were found to meet quality standards for drinking water. Thirty-one samples were taken from the Yongdang area, and 18 of them were found to meet the standards.

Can you conclude that the proportion of wells that meet the standards differs between the two areas?

18. The article "Using Conditional Probability to Find Driver Age Effect in Crashes" (M. Abdel-Aty, C. Chen, and A. Radwan, *Journal of Transportation Engineering*, 1999:502–507) presents a tabulation of types of car crashes by the age of the driver over a three-year period in Florida. Out of a total of 82,486 accidents involving drivers aged 15–24 years, 4243, or 5.1%, occurred in a driveway. Out of a total of 219,170 accidents involving drivers aged 25–64 years, 10,701 of them, or 4.9%, occurred in a driveway.

   a. Can you conclude that accidents involving drivers aged 15–24 are more likely to occur in driveways than accidents involving drivers aged 25–64?

   b. Someone suggests that since accidents involving younger drivers are more likely to occur in driveways, young drivers should be required to take a special course in driveway driving. Is this justified by the data? Explain.

19. In a certain year, there were 80 days with measurable snowfall in Denver and 63 days with measurable snowfall in Chicago. A meteorologist computes $(80 + 1)/(365 + 2) = 0.22$, $(63 + 1)/(365 + 2) = 0.17$, and he proposes to compute a 95% confidence interval for the difference between the proportions of snowy days in the two cities as follows:

$$0.22 - 0.17 \pm 1.96\sqrt{\frac{(0.22)(0.78)}{367} + \frac{(0.17)(0.83)}{367}}$$

Is this a valid confidence interval? Explain.

20. The following MINITAB output presents the results of a hypothesis test for the difference $p_1 - p_2$ between two population proportions.

```
Test and CI for Two Proportions

Sample   X   N   Sample p
1       41  97   0.422680
2       37  61   0.606557

Difference = p (1) − p (2)
Estimate for difference: −0.183877
95% CI for difference: (−0.341016, −0.026738)
Test for difference = 0 (vs not = 0): Z = −2.25 P-Value = 0.024
```

a. Is this a one-tailed or two-tailed test?
b. What is the null hypothesis?
c. Can H_0 be rejected at the 5% level? How can you tell?

21. The following MINITAB output presents the results of a hypothesis test for the difference $p_1 - p_2$ between two population proportions. Some of the numbers are missing. Fill them in.

```
Test and CI for Two Proportions

Sample   X    N   Sample p
1       101  153     (a)
2       (b)   90  0.544444


Difference = p (1) − p (2)
Estimate for difference: 0.115686
95% CI for difference: (−0.0116695, 0.243042)
Test for difference = 0 (vs not = 0): Z = (c)   P-Value = (d)
```

7.3 Small-Sample Inferences on the Difference Between Two Means

Small-Sample Confidence Intervals on the Difference Between Two Means

The Student's t distribution can be used in some cases where samples are small, and thus, where the Central Limit Theorem does not apply. Confidence intervals for the difference between two means in these cases are based on the following result:

If $X_1, \ldots, X_{n_X}$ is a sample of size n_X from a *normal* population with mean μ_X and $Y_1, \ldots, Y_{n_Y}$ is a sample of size n_Y from a *normal* population with mean μ_Y, then the quantity

$$\frac{(\overline{X} - \overline{Y}) - (\mu_X - \mu_Y)}{\sqrt{s_X^2/n_X + s_Y^2/n_Y}}$$

has an approximate Student's t distribution.

The number of degrees of freedom to use for this distribution is given by

$$\nu = \frac{\left(\dfrac{s_X^2}{n_X} + \dfrac{s_Y^2}{n_Y}\right)^2}{\dfrac{(s_X^2/n_X)^2}{n_X - 1} + \dfrac{(s_Y^2/n_Y)^2}{n_Y - 1}} \qquad \text{rounded down to the nearest integer.} \qquad (7.10)$$

By the reasoning used in Section 5.4, a $100(1 - \alpha)\%$ confidence interval for the difference $\mu_X - \mu_Y$ is

$$\overline{X} - \overline{Y} \pm t_{\nu,\alpha/2}\sqrt{s_X^2/n_X + s_Y^2/n_Y}$$

Summary

Let $X_1, \ldots, X_{n_X}$ be a random sample of size n_X from a *normal* population with mean μ_X, and let $Y_1, \ldots, Y_{n_Y}$ be a random sample of size n_Y from a *normal* population with mean μ_Y. Assume the two samples are independent.

A level $100(1 - \alpha)\%$ confidence interval for $\mu_X - \mu_Y$ is

$$\overline{X} - \overline{Y} \pm t_{\nu,\alpha/2}\sqrt{\frac{s_X^2}{n_X} + \frac{s_Y^2}{n_Y}} \tag{7.11}$$

The number of degrees of freedom, ν, is given by

$$\nu = \frac{\left(\dfrac{s_X^2}{n_X} + \dfrac{s_Y^2}{n_Y}\right)^2}{\dfrac{(s_X^2/n_X)^2}{n_X - 1} + \dfrac{(s_Y^2/n_Y)^2}{n_Y - 1}} \qquad \text{rounded down to the nearest integer.}$$

 *Example*
7.6

Resin-based composites are used in restorative dentistry. The article "Reduction of Polymerization Shrinkage Stress and Marginal Leakage Using Soft-Start Polymerization" (C. Ernst, N. Brand, et al., *Journal of Esthetic and Restorative Dentistry*, 2003: 93–104) presents a comparison of the surface hardness of specimens cured for 40 seconds with constant power with that of specimens cured for 40 seconds with exponentially increasing power. Fifteen specimens were cured with each method. Those cured with constant power had an average surface hardness (in N/mm^2) of 400.9 with a standard deviation of 10.6. Those cured with exponentially increasing power had an average surface hardness of 367.2 with a standard deviation of 6.1. Find a 98% confidence interval for the difference in mean hardness between specimens cured by the two methods.

Solution
We have $\overline{X} = 400.9$, $s_X = 10.6$, $n_X = 15$, $\overline{Y} = 367.2$, $s_Y = 6.1$, and $n_Y = 15$. The number of degrees of freedom is given by Equation (7.10) to be

$$\nu = \frac{\left(\dfrac{10.6^2}{15} + \dfrac{6.1^2}{15}\right)^2}{\dfrac{(10.6^2/15)^2}{15 - 1} + \dfrac{(6.1^2/15)^2}{15 - 1}} = 22.36 \approx 22$$

From the t table (Table A.3 in Appendix A), we find that $t_{22, .01} = 2.508$. We use expression (7.11) to find that the 98% confidence interval is

$$400.9 - 367.2 \pm 2.508\sqrt{10.6^2/15 + 6.1^2/15}, \text{ or } 33.7 \pm 7.9.$$

An Alternate Method When the Population Variances Are Equal

When the two population variances, σ_X and σ_Y, are known to be equal, there is an alternate method for computing a confidence interval. This alternate method was widely used in the past, and is still an option in many computer packages. We will describe the method here, because it is still sometimes used. In practice, use of this method is rarely advisable.

A method for constructing a confidence interval when $\sigma_X = \sigma_Y$
(Rarely advisable)

Step 1: Compute the **pooled standard deviation**, s_p, as follows:

$$s_p = \sqrt{\frac{(n_X - 1)s_X^2 + (n_Y - 1)s_Y^2}{n_X + n_Y - 2}}$$

Step 2: Compute the degrees of freedom:

$$\text{Degrees of freedom} = n_X + n_Y - 2$$

A level C confidence interval is

$$\overline{X} - \overline{Y} - t_C s_p \sqrt{\frac{1}{n_X} + \frac{1}{n_Y}} < \mu_X - \mu_Y < \overline{X} - \overline{Y} + t_C s_p \sqrt{\frac{1}{n_X} + \frac{1}{n_Y}}$$

The major problem with this method is that the assumption that the population variances are equal is very strict. The method can be quite unreliable if it is used when the population variances are not equal. Now in practice, the population variances σ_X and σ_Y are almost always unknown. Therefore it almost always impossible to be sure that they are equal. Computer packages often offer a choice of assuming variances to be equal or unequal. The best practice is to assume the variances to be unequal unless it is quite certain that they are equal.

Small-Sample Hypothesis Tests on the Difference Between Two Means

Since the quantity

$$\frac{(\overline{X} - \overline{Y}) - (\mu_X - \mu_Y)}{\sqrt{s_X^2/n_X + s_Y^2/n_Y}} \tag{7.12}$$

has an approximate Student's t distribution, it can used to derive a test statistic for testing hypotheses about the difference $\mu_X - \mu_Y$.

The null hypothesis will specify a value for $\mu_X - \mu_Y$. Denote this value Δ_0. The test statistic is

$$t = \frac{(\overline{X} - \overline{Y}) - \Delta_0}{\sqrt{s_X^2/n_X + s_Y^2/n_Y}} \tag{7.13}$$

The null distribution is Student's t with ν degrees of freedom, where ν is calculated using Equation (7.10).

Example 7.7

Good website design can make Web navigation easier. The article "The Implications of Visualization Ability and Structure Preview Design for Web Information Search Tasks" (H. Zhang and G. Salvendy, *International Journal of Human-Computer Interaction*, 2001:75–95) presents a comparison of item recognition between two designs. A sample of 10 users using a conventional Web design averaged 32.3 items identified, with a standard deviation of 8.56. A sample of 10 users using a new structured Web design averaged 44.1 items identified, with a standard deviation of 10.09. Can we conclude that the mean number of items identified is greater with the new structured design?

Solution

Let $\overline{X} = 44.1$ be the sample mean for the structured Web design. Then $s_X = 10.09$ and $n_X = 10$. Let $\overline{Y} = 32.3$ be the sample mean for the conventional Web design. Then $s_Y = 8.56$ and $n_Y = 10$. Let μ_X and μ_Y denote the population mean measurements made by the structured and conventional methods, respectively. The null and alternate hypotheses are

$$H_0: \mu_X - \mu_Y \le 0 \quad \text{versus} \quad H_1: \mu_X - \mu_Y > 0$$

The test statistic is

$$t = \frac{(\overline{X} - \overline{Y}) - 0}{\sqrt{s_X^2/n_X + s_Y^2/n_Y}}$$

Substituting values for $\overline{X}$, $\overline{Y}$, s_X, s_Y, n_X, and n_Y, we compute the value of the test statistic to be $t = 2.820$. Under H_0, this statistic has an approximate Student's t distribution, with the number of degrees of freedom given by

$$\nu = \frac{\left(\dfrac{10.09^2}{10} + \dfrac{8.56^2}{10}\right)^2}{\dfrac{(10.09^2/10)^2}{9} + \dfrac{(8.56^2/10)^2}{9}} = 17.53 \approx 17$$

Consulting the t table with 17 degrees of freedom, we find that the value cutting off 1% in the right-hand tail is 2.567, and the value cutting off 0.5% in the right-hand tail is 2.898. Therefore the area in the right-hand tail corresponding to values as extreme as or more extreme than the observed value of 2.820 is between 0.005 and 0.010. Therefore $0.005 < P < 0.01$ (see Figure 7.4, page 288). There is strong evidence that the mean number of items identified is greater for the new design.

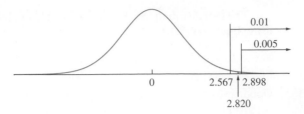

FIGURE 7.4 Solution to Example 7.7. The *P*-value is the area in the right-hand tail, which is between 0.005 and 0.01.

The following computer output (from MINITAB) presents the results from Example 7.7.

```
Two-Sample T-Test and CI: Struct, Conven

Two-sample T for C1 vs C2

           N    Mean   StDev   SE Mean
Struct    10   44.10   10.09   3.19074
Conven    10   32.30    8.56   2.70691

Difference = mu (Struct) - mu (Conven)
Estimate for difference:   11.8000
95% lower bound for difference:   4.52100
T-Test of difference = 0 (vs >):
T-Value = 2.82   P-Value = 0.006   DF = 17
```

Note that the 95% lower confidence bound is consistent with the alternate hypothesis. This indicates that the *P*-value is less than 5%.

Example 7.8

Refer to Example 7.7. Can you conclude that the mean number of items identified with the new structured design exceeds that of the conventional design by more than 2?

Solution
The null and alternate hypotheses are

$$H_0: \mu_X - \mu_Y \leq 2 \quad \text{versus} \quad H_1: \mu_X - \mu_Y > 2$$

We observe $\overline{X} = 44.1$, $\overline{Y} = 32.3$, $s_X = 10.09$, $s_Y = 8.56$, $n_X = 10$, and $n_Y = 10$. Under H_0, we take $\mu_X - \mu_Y = 2$. The test statistic given by expression (7.13) is

$$t = \frac{(\overline{X} - \overline{Y}) - 2}{\sqrt{s_X^2/n_X + s_Y^2/n_Y}}$$

The number of degrees of freedom is calculated in the same way as in Example 7.7, so there are 17 degrees of freedom. The value of the test statistic is $t = 2.342$. This is a one-tailed test. The P-value is between 0.01 and 0.025. We conclude that the mean number of items identified with the new structured design exceeds that of the conventional design by more than 2.

Summary

Let $X_1, \ldots, X_{n_X}$ and $Y_1, \ldots, Y_{n_Y}$ be samples from *normal* populations with means μ_X and μ_Y and standard deviations σ_X and σ_Y, respectively. Assume the samples are drawn independently of each other.

If σ_X and σ_Y are not known to be equal, then, to test a null hypothesis of the form $H_0: \mu_X - \mu_Y \leq \Delta_0$, $H_0: \mu_X - \mu_Y \geq \Delta_0$, or $H_0: \mu_X - \mu_Y = \Delta_0$:

■ Compute $\nu = \dfrac{[(s_X^2/n_X) + (s_Y^2/n_Y)]^2}{[(s_X^2/n_X)^2/(n_X - 1)] + [(s_Y^2/n_Y)^2/(n_Y - 1)]}$, rounded down to the nearest integer.

■ Compute the test statistic $t = \dfrac{(\overline{X} - \overline{Y}) - \Delta_0}{\sqrt{s_X^2/n_X + s_Y^2/n_Y}}$.

■ Compute the P-value. The P-value is an area under the Student's t curve with ν degrees of freedom, which depends on the alternate hypothesis as follows:

Alternate Hypothesis	**P-value**
$H_1: \mu_X - \mu_Y > \Delta_0$	Area to the right of t
$H_1: \mu_X - \mu_Y < \Delta_0$	Area to the left of t
$H_1: \mu_X - \mu_Y \neq \Delta_0$	Sum of the areas in the tails cut off by t and $-t$

An Alternate Method When the Population Variances Are Equal

When the two population variances, σ_X^2 and σ_Y^2, are known to be equal, there is an alternate method for testing hypotheses about $\mu_1 - \mu_2$. It is similar to the alternate method for finding confidence intervals that we described earlier in this section. This alternate method was widely used in the past, and is still an option in many computer packages. We will describe the method here, because it is still sometimes used. However, as with the alternate method for finding confidence intervals, use of this method is rarely advisable, because it is very sensitive to the assumption that the population variances are equal, which is usually difficult or impossible to verify in practice. Computer packages often offer a choice of assuming variances to be equal or unequal. The best practice is always to assume the variances to be unequal unless it is quite certain that they are equal.

A method for testing a hypothesis about $\mu_X - \mu_Y$ when $\sigma_X = \sigma_Y$
(Rarely advisable)

Step 1: Compute the **pooled standard deviation**, s_p, as follows:

$$s_p = \sqrt{\frac{(n_X - 1)s_X^2 + (n_Y - 1)s_Y^2}{n_X + n_Y - 2}}$$

Step 2: Let Δ_0 denote the value of $\mu_X - \mu_Y$ specified by H_0. Compute the test statistic

$$t = \frac{(\overline{X} - \overline{Y}) - \Delta_0}{s_p \sqrt{\dfrac{1}{n_X} + \dfrac{1}{n_Y}}}$$

Step 3: Compute the degrees of freedom:

Degrees of freedom $= n_X + n_Y - 2$

Step 4: Compute the P-value using a t distribution with $n_X + n_Y - 2$ degrees of freedom.

Exercises for Section 7.3

1. A new postsurgical treatment is being compared with a standard treatment. Seven subjects receive the new treatment, while seven others (the controls) receive the standard treatment. The recovery times, in days, are given below.

Treatment	12	13	15	17	19	20	21
Control	15	18	21	25	29	33	36

   Find a 95% confidence interval for the reduction in mean recovery time for patients receiving the new treatment.

2. In a study of the rate at which the pesticide hexaconazole is absorbed through skin, skin specimens were exposed to 24 μg of hexaconazole. Four specimens were exposed for 30 minutes and four others were exposed for 60 minutes. The amounts (in μg) that were absorbed were

30 minutes:	3.1	3.3	3.4	3.0
60 minutes:	3.7	3.6	3.7	3.4

   Find a 95% confidence interval for the mean amount absorbed in the time interval between 30 and 60 minutes after exposure.

3. The article "Differences in Susceptibilities of Different Cell Lines to Bilirubin Damage" (K. Ngai, C. Yeung, and C. Leung, *Journal of Paediatric Child Health*, 2000:36–45) reports an investigation into the toxicity of bilirubin on several cell lines. Ten sets of human liver cells and 10 sets of mouse fibroblast cells were placed into solutions of bilirubin in albumin with a 1.4 bilirubin/albumin molar ratio for 24 hours. In the 10 sets of human liver cells, the average percentage of cells surviving was 53.9 with a standard deviation of 10.7. In the 10 sets of mouse fibroblast cells, the average percentage of cells surviving was 73.1 with a standard deviation of 9.1. Find a 98% confidence interval for the difference in survival percentages between the two cell lines.

4. An experiment was performed in a manufacturing plant by making 5 batches of a chemical using the standard method (A) and 5 batches using a new method (B). The yields, expressed as a percent of a theoretical maximum, were as follows:

Method A:	77.0	69.1	71.5	73.0	73.7
Method B:	78.5	79.6	76.1	76.0	78.5

   Find a 99% confidence interval for the difference in the mean yield between the two methods.

5. During the spring of 1999, many fuel storage facilities in Serbia were destroyed by bombing. As a

result, significant quantities of oil products were spilled and burned, resulting in soil pollution. The article "Mobility of Heavy Metals Originating from Bombing of Industrial Sites" (B. Škrbić, J. Novaković, and N. Miljević, *Journal of Environmental Science and Health*, 2002:7–16) reports measurements of heavy metal concentrations at several industrial sites in June 1999, just after the bombing, and again in March of 2000. At the Smederevo site, on the banks of the Danube River, eight soil specimens taken in 1999 had an average lead concentration (in mg/kg) of 10.7 with a standard deviation of 3.3. Four specimens taken in 2000 had an average lead concentration of 33.8 with a standard deviation of 0.50. Find a 95% confidence interval for the increase in lead concentration between June 1999 and March 2000.

6. The article "Quality of the Fire Clay Coal Bed, Southeastern Kentucky" (J. Hower, W. Andrews, et al., *Journal of Coal Quality*, 1994:13–26) contains measurements on samples of coal from several counties in Kentucky. In units of percent ash, five samples from Knott County had an average aluminum dioxide (AlO_2) content of 32.17 and a standard deviation of 2.23. Six samples from Leslic County had an average AlO_2 content of 26.48 and a standard deviation of 2.02. Find a 98% confidence interval for the difference in AlO_2 content between coal samples from the two counties.

7. The article "The Frequency Distribution of Daily Global Irradiation at Kumasi" (F. Akuffo and A. Brew-Hammond, *Solar Energy*, 1993:145–154) defines the daily clearness index for a location to be the ratio of daily global irradiation to extraterrestrial irradiation. Measurements were taken in the city of Ibadan, Nigeria, over a five-year period. For five months of May, the clearness index averaged 0.498 and had standard deviation 0.036. For five months of July, the average was 0.389 and the standard deviation was 0.049. Find a 95% confidence interval for the difference in mean clearness index between May and July.

8. In the article "Bactericidal Properties of Flat Surfaces and Nanoparticles Derivatized with Alkylated Polyethylenimines" (J. Lin, S. Qiu, et al., *Biotechnology Progress*, 2002:1082–1086), experiments were described in which alkylated polyethylenimines were attached to surfaces and to nanoparticles to make them bactericidal. In one series of experiments, the bacteri-

cidal efficiency against the bacterium *E. coli* was compared for a methylated versus a nonmethylated polymer. The mean percentage of bacterial cells killed with the methylated polymer was 95 with a standard deviation of 1, and the mean percentage of bacterial cells killed with the nonmethylated polymer was 70 with a standard deviation of 6. Assume that five independent measurements were made on each type of polymer. Find a 95% confidence interval for the increase in bactericidal efficiency of the methylated polymer.

9. A computer scientist is studying the tendency for a computers running a certain operating system to run more slowly as the operating system ages. She measures the time (in seconds) for a certain application to load for nine computers one month after installation and for seven computers six months after installation. The results are as follows:

One month after install:	84.3	53.2	127.3
	201.3	174.2	246.2
	149.4	156.4	103.3

Six months after install:	207.4	233.1	215.9
	235.1	225.6	244.4
	245.3		

Find a 95% confidence interval for the mean difference in time to load between the first month and the sixth.

10. In a comparison of the effectiveness of distance learning with traditional classroom instruction, 12 students took a business administration course online, while 14 students took it in a classroom. The final exam scores were as follows.

| Online | 64 | 66 | 74 | 69 | 75 | 72 | 77 | 83 | 77 |
| | 91 | 85 | 88 | | | | | | |

| Classroom | 80 | 77 | 74 | 64 | 71 | 80 | 68 | 85 | 83 |
| | 59 | 55 | 75 | 81 | 81 | | | | |

Can you conclude that the mean score differs between the two types of course?

11. The breaking strength of hockey stick shafts made of two different graphite-Kevlar composites yield the following results (in newtons):

| Composite A: | 487.3 | 444.5 | 467.7 | 456.3 | 449.7 |
| | 459.2 | 478.9 | 461.5 | 477.2 | |

Composite B:	488.5	501.2	475.3	467.2	462.5
	499.7	470.0	469.5	481.5	485.2
	509.3	479.3	478.3	491.5	

Can you conclude that the mean breaking strength is greater for hockey sticks made from composite B?

12. Eight independent measurements were taken of the dissolution rate of a certain chemical at a temperature of 0°C, and seven independent measurements were taken of the rate at a temperature of 10°C. The results are as follows:

0°C: 2.28 1.66 2.56 2.64 1.92 3.09 3.09 2.48
10°C: 4.63 4.56 4.42 4.79 4.26 4.37 4.44

Can you conclude that the dissolution rates differ between the two temperatures?

13. The article "Permeability, Diffusion and Solubility of Gases" (B. Flaconnèche, et al., *Oil and Gas Science and Technology*, 2001:262–278) reported on a study of the effect of temperature and other factors on gas transport coefficients in semicrystalline polymers. The permeability coefficient (in 10^{-6} cm^3 (STP)/cm · s · MPa) of CO_2 was measured for extruded medium-density polyethelene at both 60°C and 61°C. The results are as follows:

60°C: 54 51 61 67 57 69 60
        60 63 62

61°C: 58 60 66 66 68 61 60

Can you conclude that the mean permeability coefficient at 60°C differs from that at 61°C?

14. In an experiment to determine the effect of curing time on compressive strength of concrete blocks, two samples of 14 blocks each were prepared identically except for curing time. The blocks in one sample were cured for 2 days, while the blocks in the other were cured for 6 days. The compressive strengths of the blocks, in MPa, are presented below.

Cured 2 days 1287 1326 1270 1255 1314 1329 1318
              1306 1310 1302 1291 1328 1255 1296

Cured 6 days 1301 1364 1332 1372 1321 1397 1349
              1378 1341 1376 1396 1343 1399 1387

Can you conclude that the mean strength is greater for blocks cured 6 days?

15. The article "Modeling Resilient Modulus and Temperature Correction for Saudi Roads" (H. Wahhab, I. Asi, and R. Ramadhan, *Journal of Materials in Civil Engineering*, 2001:298–305) describes a study designed to predict the resilient modulus of pavement from physical properties. One of the questions addressed was whether the modulus differs between rutted and nonrutted pavement. Measurements of the resilient modulus at 40°C (in 10^6 kPa) are presented below for 5 sections of rutted pavement and 9 sections of nonrutted pavement.

Rutted: 1.48 1.88 1.90 1.29 1.00
Nonrutted: 3.06 2.58 1.70 2.44 2.03
            1.76 2.86 2.82 1.04

Perform a hypothesis test to determine whether it is plausible that the mean resilient modulus is the same for rutted and nonrutted pavement. Compute the *P*-value. What do you conclude?

16. The article "Time Series Analysis for Construction Productivity Experiments" (T. Abdelhamid and J. Everett, *Journal of Construction Engineering and Management*, 1999:87–95) presents a study comparing the effectiveness of a video system that allows a crane operator to see the lifting point while operating the crane with the old system in which the operator relies on hand signals from a tagman. A lift of moderate difficulty was performed several times, both with the new video system and with the old tagman system. The time (in seconds) required to perform each lift was recorded. The following table presents the means, standard deviations, and sample sizes.

	Mean	Standard Deviation	Sample Size
Tagman	69.33	6.26	12
Video	58.50	5.59	24

Can you conclude that the mean time to perform a lift is less when using the video system than when using the tagman system? Explain.

17. The article "Calibration of an FTIR Spectrometer" (P. Pankratz, *Statistical Case Studies for Industrial and Process Improvement*, SIAM-ASA, 1997:19–38) describes the use of a spectrometer to make five measurements of the carbon content (in ppm) of a certain

silicon wafer on each of two successive days. The results were as follows:

Day 1:   2.1321 2.1385 2.0985 2.0941 2.0680

Day 2:   2.0853 2.1476 2.0733 2.1194 2.0717

Can you conclude that the calibration of the spectrometer has changed from the first day to the second day?

18. The article "Effects of Aerosol Species on Atmospheric Visibility in Kaohsiung City, Taiwan" (C. Lee, C. Yuan, and J. Chang, *Journal of Air and Waste Management*, 2005:1031–1041) reported that for a sample of 20 days in the winter, the mass ratio of fine to coarse particles averaged 0.51 with a standard deviation of 0.09, and for a sample of 14 days in the spring the mass ratio averaged 0.62 with a standard deviation of 0.09. Can you conclude that the mean mass ratio differs between winter and spring?

19. The article "Mechanical Grading of Oak Timbers" (D. Kretschmann and D. Green, *Journal of Materials in Civil Engineering*, 1999:91–97) presents measurements of the ultimate compressive stress, in MPa, for green mixed oak 7 by 9 timbers from West Virginia and Pennsylvania. For 11 specimens of no. 1 grade lumber, the average compressive stress was 22.1 with a standard deviation of 4.09. For 7 specimens of no. 2 grade lumber, the average compressive stress was 20.4 with a standard deviation of 3.08. Can you conclude that the mean compressive stress is greater for no. 1 grade lumber than for no. 2 grade?

20. A real estate agent in a certain city claims that apartment rents are rising faster on the east side of town than on the west. To test this claim, rents in a sample of 12 apartments on the east side of town were found to have increased by a mean of $50 per month since last year, with a standard deviation of $18 per month. A sample of 15 apartments on the west side of town had a mean increase of $35 with a standard deviation of $12. Does this provide sufficient evidence to prove the agent's claim?

21. Refer to Exercise 4.

   a. Can you conclude that the mean yield for method B is greater than that of method A?

   b. Can you conclude that the mean yield for method B exceeds that of method A by more than 3?

22. The following MINITAB output presents the results of a hypothesis test for the difference $\mu_X - \mu_Y$ between two population means.

```
Two-sample T for X vs Y

       N    Mean   StDev   SE Mean
X     10   39.31   8.71       2.8
Y     10   29.12   4.79       1.5


Difference = mu (X) − mu (Y)
Estimate for difference: 10.1974
95% lower bound for difference: 4.6333
T-Test of difference = 0 (vs >): T-Value = 3.25 P-Value = 0.003 DF = 13
```

   a. Is this a one-tailed or two-tailed test?
   b. What is the null hypothesis?
   c. Can H_0 be rejected at the 1% level? How can you tell?

23. The following MINITAB output presents the results of a hypothesis test for the difference $\mu_X - \mu_Y$ between two population means. Some of the numbers are missing. Fill them in.

```
Two-sample T for X vs Y

     N    Mean   StDev   SE Mean
X    6    1.755  0.482    (a)
Y    13   3.239  (b)     0.094


Difference = mu (X) − mu (Y)
Estimate for difference: (c)
95% CI for difference: (−1.99996, −0.96791)
T-Test of difference = 0 (vs not =): T-Value = (d) P-Value = 0.000 DF = 7
```

7.4 Inferences Using Paired Data

Confidence Intervals Using Paired Data

The methods discussed so far for finding confidence intervals on the basis of two samples have required that the samples be independent. In some cases, it is better to design an experiment so that each item in one sample is paired with an item in the other. Following is an example.

A tire manufacturer wishes to compare the tread wear of tires made of a new material with that of tires made of a conventional material. One tire of each type is placed on each front wheel of each of 10 front-wheel-drive automobiles. The choice as to which type of tire goes on the right wheel and which goes on the left is made with the flip of a coin. Each car is driven for 40,000 miles, then the depth of the tread on each tire is measured. The results are presented in Figure 7.5.

The column on the right-hand side of Figure 7.5 presents the results for all 20 tires. There is considerable overlap in tread wear for the two samples. It is difficult to tell from the column whether there is a difference between the old and the new types of tire. However, when the data are examined in pairs, it is clear that the tires of the new type generally have more tread than those of the old type. The reason that analyzing the pairs presents a clearer picture of the result is that the cars vary greatly in the amount of wear they produce. Heavier cars, and those whose driving patterns involve many starts and stops, will generally produce more wear than others. The aggregated data in the column on the right-hand side of the figure includes this variability between cars as well as the variability in wear between tires. When the data are considered in pairs, the variability between the cars disappears, because both tires in a pair come from the same car.

Table 7.1 presents, for each car, the depths of tread for both the tires as well as the difference between them. We wish to find a 95% confidence interval for the mean

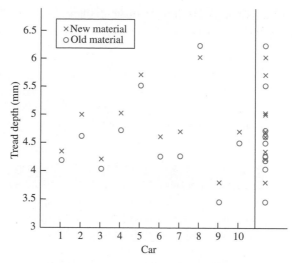

FIGURE 7.5 Tread depth for ten pairs of tires.

TABLE 7.1 Depths of tread, in mm, for tires made of new and old material

	Car									
	1	2	3	4	5	6	7	8	9	10
New material	4.35	5.00	4.21	5.03	5.71	4.61	4.70	6.03	3.80	4.70
Old material	4.19	4.62	4.04	4.72	5.52	4.26	4.27	6.24	3.46	4.50
Difference	0.16	0.38	0.17	0.31	0.19	0.35	0.43	−0.21	0.34	0.20

difference in tread wear between old and new materials in a way that takes advantage of the reduced variability produced by the paired design. The way to do this is to think of a population of *pairs* of values, in which each pair consists of measurements from an old type tire and a new type tire on the same car. For each pair in the population, there is a difference (New − Old); thus there is a population of differences. The data are then a random sample from the population of pairs, and their differences are a random sample from the population of differences.

To put this into statistical notation, let $(X_1, Y_1), \ldots, (X_{10}, Y_{10})$ be the 10 observed pairs, with X_i representing the tread on the tire made from the new material on the ith car and Y_i representing the tread on the tire made from the old material on the ith car. Let $D_i = X_i - Y_i$ represent the difference between the treads for the tires on the ith car. Let μ_X and μ_Y represent the population means for X and Y, respectively. We wish to find a 95% confidence interval for the difference $\mu_X - \mu_Y$. Let μ_D represent the population mean of the differences. Then $\mu_D = \mu_X - \mu_Y$. It follows that a confidence interval for μ_D will also be a confidence interval for $\mu_X - \mu_Y$.

Since the sample $D_1, \ldots, D_{10}$ is a random sample from a population with mean μ_D, we can use one-sample methods to find confidence intervals for μ_D. In this example, since the sample size is small, we use the Student's t method of Section 5.4. The observed values of the sample mean and sample standard deviation are

$$\overline{D} = 0.232 \qquad s_D = 0.183$$

The sample size is 10, so there are nine degrees of freedom. The appropriate t value is $t_{9,.025} = 2.262$. The confidence interval using expression (5.14) (in Section 5.4) is therefore $0.232 \pm (2.262)(0.183)/\sqrt{10}$, or (0.101, 0.363). When the number of pairs is large, the large-sample methods of Section 5.2, specifically expression (5.4), can be used.

Summary

Let $D_1, \ldots, D_n$ be a *small* random sample ($n \leq 30$) of differences of pairs. If the population of differences is approximately normal, then a level $100(1 - \alpha)\%$ confidence interval for the mean difference μ_D is given by

$$\overline{D} \pm t_{n-1,\alpha/2} \frac{s_D}{\sqrt{n}} \tag{7.14}$$

where s_D is the sample standard deviation of $D_1, \ldots, D_n$. Note that this interval is the same as that given by expression (5.14).

If the sample size is large, a level $100(1 - \alpha)\%$ confidence interval for the mean difference μ_D is given by

$$\overline{D} \pm z_{\alpha/2} \sigma_{\overline{D}} \tag{7.15}$$

In practice $\sigma_{\overline{D}}$ is approximated with $s_D/\sqrt{n}$. Note that this interval is the same as that given by expression (5.4) in Section 5.2.

Hypothesis Tests Using Paired Data

When we have paired data, we can use the sample mean difference $\overline{D}$ and the sample standard deviation s_D to test hypotheses about the mean difference μ_D. We begin with an example.

Particulate matter (PM) emissions from automobiles are a serious environmental concern. Eight vehicles were chosen at random from a fleet, and their emissions were measured under both highway driving and stop-and-go driving conditions. The differences (stop-and-go emission − highway emission) were computed as well. The results, in milligrams of particulates per gallon of fuel, were as follows:

	Vehicle							
	1	2	3	4	5	6	7	8
Stop-and-go	1500	870	1120	1250	3460	1110	1120	880
Highway	941	456	893	1060	3107	1339	1346	644
Difference	559	414	227	190	353	−229	−226	236

Can we conclude that the mean level of emissions is less for highway driving than for stop-and-go driving?

There are only eight differences, which is a small sample. If we assume that the population of differences is approximately normal, we can use the Student's t test, as presented in Section 6.4. The observed value of the sample mean of the differences is $\overline{D} = 190.5$. The sample standard deviation is $s_D = 284.1$. The null and alternate hypotheses are

$$H_0 : \mu_D \le 0 \quad \text{versus} \quad H_1 : \mu_D > 0$$

The test statistic is

$$t = \frac{\overline{D} - 0}{s_D/\sqrt{n}} = \frac{190.5 - 0}{284.1/\sqrt{8}} = 1.897$$

The null distribution of the test statistic is Student's t with seven degrees of freedom. Figure 7.6 presents the null distribution and indicates the location of the test statistic. This is a one-tailed test. The t table indicates that 5% of the area in the tail is cut off by a t value of 1.895, very close to the observed value of 1.897. The P-value is approximately 0.05. The following computer output (from MINITAB) presents this result.

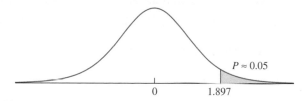

FIGURE 7.6 The null distribution of $t = (\overline{D} - 0)/(s_D/\sqrt{8})$ is t_7. The observed value of t, corresponding to the observed values $\overline{D} = 190.5$ and $s_p = 284.1$, is 1.897. If H_0 is true, the probability that t takes on a value as extreme as or more extreme than that observed is very close to 0.05.

```
Paired T-Test and CI: StopGo, Highway

Paired T for StopGo − Highway

             N      Mean     StDev    SE Mean
StopGo       8   1413.75   850.780    300.796
Highway      8   1223.25   820.850    290.214
Difference   8    190.50   284.104    100.446

95% lower bound for mean difference: 0.197215
T-Test of mean difference = 0 (vs > 0):
T-Value = 1.90   P-Value = 0.050
```

Note that the 95% lower bound is just barely consistent with the alternate hypothesis. This indicates that the P-value is just barely less than 0.05 (although it is given by 0.050 to two significant digits).

Summary

Let $(X_1, Y_1), \ldots, (X_n, Y_n)$ be a sample of ordered pairs whose differences $D_1, \ldots, D_n$ are a sample from a *normal* population with mean μ_D. Let s_D be the sample standard deviation of $D_1, \ldots, D_n$.

To test a null hypothesis of the form $H_0 : \mu_D \leq \mu_0$, $H_0 : \mu_D \geq \mu_0$, or $H_0 : \mu_D = \mu_0$:

■ Compute the test statistic $t = \dfrac{\overline{D} - \mu_0}{s_D/\sqrt{n}}$.

■ Compute the *P*-value. The *P*-value is an area under the Student's t curve with $n - 1$ degrees of freedom, which depends on the alternate hypothesis as follows:

Alternate Hypothesis	P-value
$H_1 : \mu_D > \mu_0$	Area to the right of t
$H_1 : \mu_D < \mu_0$	Area to the left of t
$H_1 : \mu_D \neq \mu_0$	Sum of the areas in the tails cut off by t and $-t$

■ If the sample is large, the D_i need not be normally distributed, the test statistic is $z = \dfrac{\overline{D} - \mu_0}{s_D/\sqrt{n}}$, and a z test should be performed.

Exercises for Section 7.4

1. In a study to compare the absorption rates of two antifungal ointments (labeled "A" and "B"), equal amounts of the two drugs were applied to the skin of 14 volunteers. After 6 hours, the amounts absorbed into the skin in (μg/cm²) were measured. The results are presented in the following table.

Subject	A	B	Difference
1	2.56	2.34	0.22
2	2.08	1.49	0.59
3	2.46	1.86	0.60
4	2.61	2.17	0.44
5	1.94	1.29	0.65
6	4.50	4.07	0.43
7	2.27	1.83	0.44
8	3.19	2.93	0.26
9	3.16	2.58	0.58
10	3.02	2.73	0.29
11	3.46	2.84	0.62
12	2.57	2.37	0.20
13	2.64	2.42	0.22
14	2.63	2.44	0.19

Find a 95% confidence interval for the mean difference between the amounts absorbed.

2. Two gauges that measure tire tread depth were compared by measuring ten different locations on a tire with each gauge. The results, in mm, are presented in the following table.

Location	Gauge 1	Gauge 2	Difference
1	3.90	3.80	0.10
2	3.28	3.30	−0.02
3	3.65	3.59	0.06
4	3.63	3.61	0.02
5	3.94	3.88	0.06
6	3.81	3.73	0.08
7	3.60	3.57	0.03
8	3.06	3.02	0.04
9	3.87	3.77	0.10
10	3.48	3.49	−0.01

Find a 99% confidence interval for the mean difference between the readings of the two gauges.

3. The water content (in percent) for 7 bricks was measured shortly after manufacture, then again after the bricks were dried in a furnace. The results were as follows.

Specimen	Before Drying	After Drying
1	6.56	3.60
2	5.33	2.00
3	7.12	3.95
4	7.28	4.47
5	5.85	2.12
6	8.19	5.77
7	8.18	3.00

Find a 90% confidence interval for the reduction in percent water content after drying.

4. A sample of 10 diesel trucks were run both hot and cold to estimate the difference in fuel economy. The results, in mpg, are presented in the following table. (From "In-use Emissions from Heavy-Duty Diesel Vehicles," J. Yanowitz, Ph.D. thesis, Colorado School of Mines, 2001.)

Truck	Hot	Cold
1	4.56	4.26
2	4.46	4.08
3	6.49	5.83
4	5.37	4.96
5	6.25	5.87
6	5.90	5.32
7	4.12	3.92
8	3.85	3.69
9	4.15	3.74
10	4.69	4.19

Find a 98% confidence interval for the difference in mean fuel mileage between hot and cold engines.

5. The article "Simulation of the Hot Carbonate Process for Removal of CO_2 and H_2S from Medium Btu Gas" (K. Park and T. Edgar, *Energy Progress*, 1984:174–180) presents an equation used to estimate the equilibrium vapor pressure of CO_2 in a potassium carbonate solution. The actual equilibrium pressure (in kPa) was measured in nine different reactions and compared with the value estimated from the equation. The results are presented in the following table:

Reaction	Estimated	Experimental	Difference
1	45.10	42.95	2.15
2	85.77	79.98	5.79
3	151.84	146.17	5.67
4	244.30	228.22	16.08
5	257.67	240.63	17.04
6	44.32	41.99	2.33
7	84.41	82.05	2.36
8	150.47	149.62	0.85
9	253.81	245.45	8.36

Find a 95% confidence interval for the mean difference between the estimated and actual pressures.

6. The article "Effect of Refrigeration on the Potassium Bitartrate Stability and Composition of Italian Wines" (A. Versari, D. Barbanti, et al., *Italian Journal of Food Science,* 2002:45–52) reports a study in which eight types of white wine had their tartaric acid concentration (in g/L) measured both before and after a cold stabilization process. The results are presented in the following table:

Wine Type	Before	After	Difference
1	2.86	2.59	0.27
2	2.85	2.47	0.38
3	1.84	1.58	0.26
4	1.60	1.56	0.04
5	0.80	0.78	0.02
6	0.89	0.66	0.23
7	2.03	1.87	0.16
8	1.90	1.71	0.19

Find a 95% confidence interval for the mean difference between the tartaric acid concentrations before and after the cold stabilization process.

7. A tire manufacturer is interested in testing the fuel economy for two different tread patterns. Tires of each tread type are driven for 1000 miles on each of 18 different cars. The mileages, in mpg, are presented in the following table.

Car	Tread A	Tread B
1	24.1	20.3
2	22.3	19.7
3	24.5	22.5
4	26.1	23.2
5	22.6	20.4
6	23.3	23.5
7	22.4	21.9
8	19.9	18.6
9	27.1	25.8
10	23.5	21.4
11	25.4	20.6
12	24.9	23.4
13	23.7	20.3
14	23.9	22.5
15	24.6	23.5
16	26.4	24.5
17	21.5	22.4
18	24.6	24.9

a. Find a 99% confidence interval for the mean difference in fuel economy.

b. A confidence interval based on the data in the table has width ± 0.5 mpg. Is the level of this confidence interval closest to 80%, 90%, or 95%?

8. Refer to Exercise 7. In a separate experiment, 18 cars were outfitted with tires with tread type A, and another 18 were outfitted with tires with tread type B. Each car was driven 1000 miles. The cars with tread type A averaged 23.93 mpg, with a standard deviation of 1.79 mpg. The cars with tread type B averaged 22.19 mpg, with a standard deviation of 1.95 mpg.

a. Which method should be used to find a confidence interval for the difference between the mean mileages of the two tread types: expression (7.14) (in this section) or expression (7.11) (in Section 7.3)?

b. Using the appropriate method, find a 99% confidence interval for the difference between the mean mileages of the two tread types.

c. Is the confidence interval found in part (b) wider than the one found in Exercise 7a? Why is this so?

9. An automobile manufacturer wishes to compare the lifetimes of two brands of tire. She obtains samples of six tires of each brand. On each of six cars, she mounts one tire of each brand on each front wheel. The cars are driven until only 20% of the original tread remains. The distances, in miles, for each tire are presented in the following table. Can you conclude that there is a difference between the mean lifetimes of the two brands of tire?

Car	Brand 1	Brand 2
1	36,925	34,318
2	45,300	42,280
3	36,240	35,500
4	32,100	31,950
5	37,210	38,015
6	48,360	47,800
7	38,200	33,215

a. State the appropriate null and alternate hypotheses.

b. Compute the value of the test statistic.

c. Find the P-value and state your conclusion.

10. The article "Modeling of Urban Area Stop-and-Go Traffic Noise" (P. Pamanikabud and C. Tharasawatipipat, *Journal of Transportation Engineering*, 1999:152–159) presents measurements of traffic noise, in dBA, from 10 locations in Bangkok, Thailand. Measurements, presented in the following table, were made at each location, in both the acceleration and deceleration lanes.

Location	Acceleration	Deceleration
1	78.1	78.6
2	78.1	80.0
3	79.6	79.3
4	81.0	79.1
5	78.7	78.2
6	78.1	78.0
7	78.6	78.6
8	78.5	78.8
9	78.4	78.0
10	79.6	78.4

Can you conclude that there is a difference in the mean noise levels between acceleration and deceleration lanes?

11. Six bean plants had their carbohydrate concentrations (in percent by weight) measured both in the shoot and in the root. The following results were obtained:

Plant	Shoot	Root
1	4.42	3.66
2	5.81	5.51
3	4.65	3.91
4	4.77	4.47
5	5.25	4.69
6	4.75	3.93

Can you conclude that there is a difference in mean concentration between the shoot and the root?

12. In an experiment to determine whether there is a systematic difference between the weights obtained with two different scales, 10 rock specimens were weighed, in grams, on each scale. The following data were obtained:

Specimen	Weight on Scale 1	Weight on Scale 2
1	11.23	11.27
2	14.36	14.41
3	8.33	8.35
4	10.50	10.52
5	23.42	23.41
6	9.15	9.17
7	13.47	13.52
8	6.47	6.46
9	12.40	12.45
10	19.38	19.35

Assume that the difference between the scales, if any, does not depend on the object being weighed. Can you conclude that scale 2 reads heavier, on the average, than scale 1?

13. Muscles flex when stimulated through electric impulses either to motor points (points on the muscle) or to nerves. The article "Force Assessment of the Stimulated Arm Flexors: Quantification of Contractile Properties" (J. Hong and P. Iaizzo, *Journal of Medical Engineering and Technology*, 2002:28–35) reports a study in which both methods were applied to the upper-arm regions of each of several subjects. The latency time (delay between stimulus and contraction) was measured (in milliseconds) for each subject. The results for seven subjects are presented in the following table (one outlier has been deleted).

	Subject						
	1	**2**	**3**	**4**	**5**	**6**	**7**
Nerve	59	57	58	38	53	47	51
Motor point	56	52	56	32	47	42	48
Difference	3	5	2	6	6	5	3

Can you conclude that there is a difference in latency period between motor point and nerve stimulation?

14. For a sample of nine automobiles, the mileage (in 1000s of miles) at which the original front brake pads were worn to 10% of their original thickness was measured, as was the mileage at which the original rear brake pads were worn to 10% of their original thickness. The results are given in the following table.

Automobile	Front	Rear
1	32.8	41.2
2	26.6	35.2
3	35.6	46.1
4	36.4	46.0
5	29.2	39.9
6	40.9	51.7
7	40.9	51.6
8	34.8	46.1
9	36.6	47.3

a. Can you conclude that the mean lifetime of rear brake pads is greater than that of front brake pads?

b. Can you conclude that the mean lifetime of rear brake pads exceeds that of front brake pads by more than 10,000 miles?

15. Two rubber specimens from each of eight stocks were subject to an abrasion test. One specimen in each stock was cured at 90°C and the other was cured at 140°C. The weight loss (in mg) for each specimen was measured. The results are presented in the following table. Can you conclude that there is a difference in mean weight loss between specimens cured at the two temperatures?

Specimen	90°C	140°C
1	3.33	2.75
2	2.15	2.59
3	4.76	4.70
4	2.88	2.68
5	2.32	2.64
6	4.16	3.35
7	3.89	3.17
8	2.13	2.33

16. The Valsalva maneuver involves blowing into a closed tube in order to create pressure in respiratory airways. Impedance cardiography is used during this maneuver to assess cardiac function. The article "Impedance Cardiographic Measurement of the Physiological Response to the Valsalva Manoeuvre" (R. Patterson and J. Zhang, *Medical and Biological Engineering and Computing*, 2003:40–43) presents a study in which the impedance ratio was measured for each of 11 subjects in both a standing and a reclining position. The results at an airway pressure of 10 mmHg are presented in the following table.

Subject	Standing	Reclining	Difference
1	1.45	0.98	0.47
2	1.71	1.42	0.29
3	1.81	0.70	1.11
4	1.01	1.10	−0.09
5	0.96	0.78	0.18
6	0.83	0.54	0.29
7	1.23	1.34	−0.11
8	1.00	0.72	0.28
9	0.80	0.75	0.05
10	1.03	0.82	0.21
11	1.39	0.60	0.79

Can you conclude that there is a difference between the mean impedance ratio measured in the standing position and that measured in the reclining position?

17. The management of a taxi cab company is trying to decide if they should switch from bias tires to radial tires to improve fuel economy. Each of 10 taxis was equipped with one of the two tire types and driven on a test course. Without changing drivers, tires were then switched to the other tire type and the test course was repeated. The fuel economy (in mpg) for the 10 cars is as follows:

Car	Radial	Bias
1	32.1	27.1
2	36.1	31.5
3	32.3	30.4
4	29.5	26.9
5	34.3	29.9
6	31.9	28.7
7	33.4	30.2
8	34.6	31.8
9	35.2	33.6
10	32.7	29.9

a. Because switching tires on the taxi fleet is expensive, management does not want to switch unless a hypothesis test provides strong evidence that the mileage will be improved. State the appropriate null and alternate hypotheses, and find the P-value.

b. A cost-benefit analysis shows that it will be profitable to switch to radial tires if the mean mileage improvement is greater than 2 mpg. State the appropriate null and alternate hypotheses, and find the P-value, for a hypothesis test that is designed to form the basis for the decision whether to switch.

18. The following MINITAB output presents the results of a paired hypothesis test for the difference $\mu_X - \mu_Y$ between two population means.

```
Paired T for X - Y

                N      Mean    StDev   SE Mean
X              12   134.233   68.376    19.739
Y              12   100.601   94.583    27.304
Difference     12   33.6316  59.5113   17.1794


95% lower bound for mean difference: 2.7793
T-Test of mean difference = 0 (vs > 0): T-Value = 1.96   P-Value = 0.038
```

a. Is this a one-tailed or two-tailed test?
b. What is the null hypothesis?
c. Can H_0 be rejected at the 1% level? How can you tell?
d. Use the output and an appropriate table to compute a 98% confidence interval for $\mu_X - \mu_Y$.

19. The following MINITAB output presents the results of a paired hypothesis test for the difference $\mu_X - \mu_Y$ between two population means. Some of the numbers are missing. Fill them in.

```
Paired T for X - Y

             N     Mean    StDev   SE Mean
X            7   12.4141   2.9235    (a)
Y            7    8.3476    (b)    1.0764
Difference   7    (c)     3.16758  1.19723

95% lower bound for mean difference: 1.74006
T-Test of mean difference = 0 (vs > 0): T-Value = (d) P-Value = 0.007
```

7.5 The *F* Test for Equality of Variance

The tests we have studied so far have involved means or proportions. Sometimes it is desirable to test a null hypothesis that two populations have equal variances. In general there is no good way to do this. In the special case where both populations are normal, however, a method is available.

Let $X_1, \ldots, X_m$ be a simple random sample from a $N(\mu_1, \sigma_1^2)$ population, and let $Y_1, \ldots, Y_n$ be a simple random sample from a $N(\mu_2, \sigma_2^2)$ population. Assume that the samples are chosen independently. The values of the means, μ_1 and μ_2, are irrelevant here; we are concerned only with the variances σ_1^2 and σ_2^2. Note that the sample sizes, m and n, may be different. Let s_1^2 and s_2^2 be the sample variances. That is,

$$s_1^2 = \frac{1}{m-1} \sum_{i=1}^{m} (X_i - \overline{X})^2 \qquad s_2^2 = \frac{1}{n-1} \sum_{i=1}^{n} (Y_i - \overline{Y})^2$$

Any of three null hypotheses may be tested. They are

$$H_0 : \frac{\sigma_1^2}{\sigma_2^2} \leq 1 \quad \text{or equivalently,} \quad \sigma_1^2 \leq \sigma_2^2$$

$$H_0 : \frac{\sigma_1^2}{\sigma_2^2} \geq 1 \quad \text{or equivalently,} \quad \sigma_1^2 \geq \sigma_2^2$$

$$H_0 : \frac{\sigma_1^2}{\sigma_2^2} = 1 \quad \text{or equivalently,} \quad \sigma_1^2 = \sigma_2^2$$

The procedures for testing these hypotheses are similar but not identical. We will describe the procedure for testing the null hypothesis $H_0 : \sigma_1^2/\sigma_2^2 \leq 1$ versus $H_1 : \sigma_1^2/\sigma_2^2 > 1$, and then discuss how the procedure may be modified to test the other two hypotheses.

The test statistic is the ratio of the two sample variances:

$$F = \frac{s_1^2}{s_2^2} \tag{7.16}$$

When H_0 is true, we assume that $\sigma_1^2/\sigma_2^2 = 1$ (the value closest to H_1) or, equivalently, that $\sigma_1^2 = \sigma_2^2$. When H_0 is true, s_1^2 and s_2^2 are, on average, the same size, so F is likely to be near 1. When H_0 is false, $\sigma_1^2 > \sigma_2^2$, so s_1^2 is likely to be larger than s_2^2, and F is likely to be greater than 1. In order to use F as a test statistic, we must know its null distribution. The null distribution is called an F distribution, which we now describe.

The *F* Distribution

Statistics that have an F distribution are ratios of quantities, such as the ratio of the two sample variances in Equation (7.16). The F distribution therefore has two values for the degrees of freedom: one associated with the numerator and one associated with the denominator. The degrees of freedom are indicated with subscripts under the letter F. For example, the symbol $F_{3,16}$ denotes the F distribution with 3 degrees of freedom for the numerator and 16 degrees of freedom for the denominator. Note that the degrees of freedom for the numerator are always listed first.

A table for the F distribution is provided (Table A.6 in Appendix A). The table provides values for certain quantiles, or upper percentage points, for a large number of choices for the degrees of freedom. As an example, Figure 7.7 presents the probability density function of the $F_{3,16}$ distribution. The upper 5% of the distribution is shaded. To find the upper 5% point in the table, look under $\alpha = 0.050$, and degrees of freedom $\nu_1 = 3$, $\nu_2 = 16$. The value is 3.24.

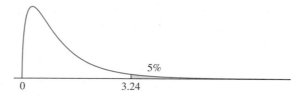

FIGURE 7.7 Probability density function of the $F_{3,16}$ distribution. The upper 5% point is 3.24. [See the *F* table (Table A.6) in Appendix A.]

The *F* Statistic for Testing Equality of Variance

The null distribution of the test statistic $F = s_1^2/s_2^2$ is $F_{m-1,\,n-1}$. The number of degrees of freedom for the numerator is one less than the sample size used to compute s_1^2, and the number of degrees of freedom for the denominator is one less than the sample size used to compute s_2^2. We illustrate the *F* test with an example.

Example

7.9

In a series of experiments to determine the absorption rate of certain pesticides into skin, measured amounts of two pesticides were applied to several skin specimens. After a time, the amounts absorbed (in μg) were measured. For pesticide A, the variance of the amounts absorbed in 6 specimens was 2.3, while for pesticide B, the variance of the amounts absorbed in 10 specimens was 0.6. Assume that for each pesticide, the amounts absorbed are a simple random sample from a normal population. Can we conclude that the variance in the amount absorbed is greater for pesticide A than for pesticide B?

Solution

Let σ_1^2 be the population variance for pesticide A, and let σ_2^2 be the population variance for pesticide B. The null hypothesis is

$$H_0 : \frac{\sigma_1^2}{\sigma_2^2} \le 1$$

The sample variances are $s_1^2 = 2.3$ and $s_2^2 = 0.6$. The value of the test statistic is

$$F = \frac{2.3}{0.6} = 3.83$$

The null distribution of the test statistic is $F_{5,9}$. If H_0 is true, then s_1^2 will on the average be smaller than s_2^2. It follows that the larger the value of F, the stronger the evidence against H_0. Consulting the *F* table with five and nine degrees of freedom, we find that the upper 5% point is 3.48, while the upper 1% point is 6.06. We conclude that $0.01 < P < 0.05$. There is reasonably strong evidence against the null hypothesis. See Figure 7.8 (page 306).

We now describe the modifications to the procedure shown in Example 7.9 that are necessary to test the other null hypotheses. To test

$$H_0 : \frac{\sigma_1^2}{\sigma_2^2} \ge 1$$

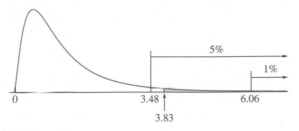

FIGURE 7.8 The observed value of the test statistic is 3.83. The upper 5% point of the $F_{5,9}$ distribution is 3.48; the upper 1% point is 6.06. Therefore the P-value is between 0.01 and 0.05.

one could in principle use the test statistic s_1^2/s_2^2, with *small* values of the statistic providing evidence against H_0. However, since the F table contains only large values (i.e., greater than 1) for the F statistic, it is easier to use the statistic s_2^2/s_1^2. Under H_0, the distribution of s_2^2/s_1^2 is $F_{n-1,\,m-1}$.

Finally, we describe the method for testing the two-tailed hypothesis

$$H_0: \frac{\sigma_1^2}{\sigma_2^2} = 1$$

For this hypothesis, both large and small values of the statistic s_1^2/s_2^2 provide evidence against H_0. The procedure is to use either s_1^2/s_2^2 or s_2^2/s_1^2, whichever is greater than 1. The P-value for the two-tailed test is twice the P-value for the one-tailed test. In other words, the P-value of the two-tailed test is twice the upper tail area of the F distribution. We illustrate with an example.

Example

7.10

In Example 7.9, $s_1^2 = 2.3$ with a sample size of 6, and $s_2^2 = 0.6$ with a sample size of 10. Test the null hypothesis

$$H_0: \sigma_1^2 = \sigma_2^2$$

Solution
The null hypothesis $\sigma_1^2 = \sigma_2^2$ is equivalent to $\sigma_1^2/\sigma_2^2 = 1$. Since $s_1^2 > s_2^2$, we use the test statistic s_1^2/s_2^2. In Example 7.9, we found that for the one-tailed test, $0.01 < P < 0.05$. Therefore for the two-tailed test, $0.02 < P < 0.10$.

The following computer output (from MINITAB) presents the solution to Example 7.10.

```
Test for Equal Variances

F-Test (normal distribution)
Test statistic = 3.83, p-value = 0.078
```

The *F* Test Is Sensitive to Departures from Normality

The *F* test, like the *t* test, requires that the samples come from normal populations. Unlike the *t* test, the *F* test for comparing variances is fairly sensitive to this assumption. If the shapes of the populations differ much from the normal curve, the *F* test may give misleading results. For this reason, the *F* test for comparing variances must be used with caution.

In Chapters 8 and 9, we will use the *F* distribution to perform certain hypothesis tests in the context of linear regression and analysis of variance. In these settings, the *F* test is less sensitive to violations of the normality assumption.

Exercises for Section 7.5

1. Find the upper 5% point of $F_{7,20}$.

2. Find the upper 1% point of $F_{2,5}$.

3. An *F* test with five degrees of freedom in the numerator and seven degrees of freedom in the denominator produced a test statistic whose value was 7.46.

   a. What is the *P*-value if the test is one-tailed?

   b. What is the *P*-value if the test is two-tailed?

4. A broth used to manufacture a pharmaceutical product has its sugar content, in mg/mL, measured several times on each of three successive days.

Day 1:	5.0	4.8	5.1	5.1	4.8	5.1	4.8
	4.8	5.0	5.2	4.9	4.9	5.0	
Day 2:	5.8	4.7	4.7	4.9	5.1	4.9	5.4
	5.3	5.3	4.8	5.7	5.1	5.7	
Day 3:	6.3	4.7	5.1	5.9	5.1	5.9	4.7
	6.0	5.3	4.9	5.7	5.3	5.6	

   a. Can you conclude that the variability of the process is greater on the second day than on the first day?

   b. Can you conclude that the variability of the process is greater on the third day than on the second day?

5. Refer to Exercise 11 in Section 7.3. Can you conclude that the variance of the breaking strengths differs between the two composites?

6. Refer to Exercise 9 in Section 7.3. Can you conclude that the time to load is more variable in the first month than in the sixth month after installation?

Supplementary Exercises for Chapter 7

1. In a test to compare the effectiveness of two drugs designed to lower cholesterol levels, 75 randomly selected patients were given drug A and 100 randomly selected patients were given drug B. Those given drug A reduced their cholesterol levels by an average of 40 with a standard deviation of 12, and those given drug B reduced their levels by an average of 42 with a standard deviation of 15. The units are milligrams of cholesterol per deciliter of blood serum. Can you conclude that the mean reduction using drug B is greater than that of drug A?

2. Two machines used to fill soft drink containers are being compared. The number of containers filled each minute is counted for 60 minutes for each machine. During the 60 minutes, machine 1 filled an average of 73.8 cans per minute with a standard deviation of 5.2 cans per minute, and machine 2 filled an average of 76.1 cans per minute with a standard deviation of 4.1 cans per minute.

   a. If the counts are made each minute for 60 consecutive minutes, what assumption necessary to the validity of a hypothesis test may be violated?

b. Assuming that all necessary assumptions are met, perform a hypothesis test. Can you conclude that machine 2 is faster than machine 1?

3. An engineer claims that a new type of power supply for home computers lasts longer than the old type. Independent random samples of 75 of each of the two types are chosen, and the sample means and standard deviations of their lifetimes (in hours) are computed:

New: $\overline{X}_1 = 4387$ $\quad s_1 = 252$
Old: $\overline{X}_2 = 4260$ $\quad s_2 = 231$

Can you conclude that the mean lifetime of new power supplies is greater than that of the old power supplies?

4. To determine the effect of fuel grade on fuel efficiency, 80 new cars of the same make, with identical engines, were each driven for 1000 miles. Forty of the cars ran on regular fuel and the other 40 received premium grade fuel. The cars with the regular fuel averaged 27.2 mpg, with a standard deviation of 1.2 mpg. The cars with the premium fuel averaged 28.1 mpg and had a standard deviation of 2.0 mpg. Can you conclude that this type of car gets better mileage with premium fuel?

5. In a test of the effect of dampness on electric connections, 100 electric connections were tested under damp conditions and 150 were tested under dry conditions. Twenty of the damp connections failed and only 10 of the dry ones failed. Find a 90% confidence interval for the difference between the proportions of connections that fail when damp as opposed to dry.

6. The specification for the pull strength of a wire that connects an integrated circuit to its frame is 10 g or more. In a sample of 85 units made with gold wire, 68 met the specification, and in a sample of 120 units made with aluminum wire, 105 met the specification. Find a 95% confidence interval for the difference in the proportions of units that meet the specification between units with gold wire and those with aluminum wire.

7. Two processes for manufacturing a certain microchip are being compared. A sample of 400 chips was selected from a less expensive process, and 62 chips were found to be defective. A sample of 100 chips

was selected from a more expensive process, and 12 were found to be defective.

a. Find a 95% confidence interval for the difference between the proportions of defective chips produced by the two processes.

b. In order to increase the precision of the confidence interval, additional chips will be sampled. Three sampling plans of equal cost are being considered. In the first plan, 100 additional chips from the less expensive process will be sampled. In the second plan 50 additional chips from the more expensive process will be sampled. In the third plan, 50 chips from the less expensive and 25 chips from the more expensive process will be sampled. Which plan is most likely to provide the greatest increase in the precision of the confidence interval? Explain.

8. A quality manager suspects that the quality of items that are manufactured on a Monday is less good than that of items manufactured on a Wednesday. In a sample of 300 items manufactured on Monday, 260 were rated as acceptable or better, and in a sample of 400 items manufactured on a Wednesday, 370 were rated acceptable or better. Can you conclude that the proportion of items rated acceptable or better is greater on Wednesday than on Monday?

9. In order to determine whether to pitch a new advertising campaign more toward men or women, an advertiser provided each couple in a random sample of 500 married couples with a new type of TV remote control that is supposed to be easier to find when needed. Of the 500 husbands, 62% said that the new remote was easier to find than their old one. Of the 500 wives, only 54% said the new remote was easier to find. Let p_1 be the population proportion of married men who think that the new remote is easier to find, and let p_2 be the corresponding proportion of married women. Can the statistic $\hat{p}_1 - \hat{p}_2 = 0.62 - 0.54$ be used to test $H_0: p_1 - p_2 = 0$ versus $H_1: p_1 - p_2 \neq 0$? If so, perform the test and compute the P-value. If not, explain why not.

10. Twenty-one independent measurements were taken of the hardness (on the Rockwell C scale) of HSLA-100 steel base metal, and another 21 independent measurements were made of the hardness of a weld produced on this base metal. The standard deviation of the measurements made on the base metal was 3.06, and the

standard deviation of the measurements made on the weld was 1.41. Assume that the measurements are independent random samples from normal populations. Can you conclude that measurements made on the base metal are more variable than measurements made on the weld?

11. In a survey of 100 randomly chosen holders of a certain credit card, 57 said that they were aware that use of the credit card could earn them frequent flier miles on a certain airline. After an advertising campaign to build awareness of this benefit, an independent survey of 200 credit card holders was made, and 135 said that they were aware of the benefit. Can you conclude that awareness of the benefit increased after the advertising campaign?

12. A new production process is being contemplated for the manufacture of stainless steel bearings. Measurements of the diameters of random samples of bearings from the old and the new processes produced the following data:

Old:	16.3	15.9	15.8	16.2	16.1	16.0
	15.7	15.8	15.9	16.1	16.3	16.1
	15.8	15.7	15.8	15.7		
New:	15.9	16.2	16.0	15.8	16.1	16.1
	15.8	16.0	16.2	15.9	15.7	16.2
	15.8	15.8	16.2	16.3		

a. Can you conclude that one process yields a different mean size bearing than the other?

b. Can you conclude that the variance of the size for the new procedure is lower than that of the older procedure?

13. A molecular biologist is studying the effectiveness of a particular enzyme to digest a certain sequence of DNA nucleotides. He divides six DNA samples into two parts, treats one part with the enzyme, and leaves the other part untreated. He then uses a polymerase chain reaction assay to count the number of DNA fragments that contain the given sequence. The results are as follows:

	Sample					
	1	**2**	**3**	**4**	**5**	**6**
Enzyme present	22	16	11	14	12	30
Enzyme absent	43	34	16	27	10	40

Find a 95% confidence interval for the difference between the mean numbers of fragments.

14. Refer to Exercise 13. Another molecular biologist repeats the study with a different design. She makes up 12 DNA samples, and then chooses 6 at random to be treated with the enzyme and 6 to remain untreated. The results are as follows:

Enzyme present:	12	15	14	22	22	20
Enzyme absent:	23	39	37	18	26	24

Find a 95% confidence interval for the difference between the mean numbers of fragments.

15. In the article "Groundwater Electromagnetic Imaging in Complex Geological and Topographical Regions: A Case Study of a Tectonic Boundary in the French Alps" (S. Houtot, P. Tarits, et al., *Geophysics*, 2002:1048–1060), the pH was measured for several water samples in various locations near Gittaz Lake in the French Alps. The results for 11 locations on the northern side of the lake and for 6 locations on the southern side are as follows:

Northern side:	8.1	8.2	8.1	8.2	8.2	7.4
	7.3	7.4	8.1	8.1	7.9	
Southern side:	7.8	8.2	7.9	7.9	8.1	8.1

Find a 98% confidence interval for the difference in pH between the northern and southern side.

16. Five specimens of untreated wastewater produced at a gas field had an average benzene concentration of 6.83 mg/L with a standard deviation of 1.72 mg/L. Seven specimens of treated wastewater had an average benzene concentration of 3.32 mg/L with a standard deviation of 1.17 mg/L. Find a 95% confidence interval for the reduction in benzene concentration after treatment.

Exercises 17 and 18 describe experiments that require a hypothesis test. For each experiment, describe the appropriate test. State the appropriate null and alternate hypotheses, describe the test statistic, and specify which table should be used to find the P-value. If relevant, state the number of degrees of freedom for the test statistic.

17. A fleet of 100 taxis is divided into two groups of 50 cars each to see whether premium gasoline reduces

maintenance costs. Premium unleaded fuel is used in group A, while regular unleaded fuel is used in group B. The total maintenance cost for each vehicle during a one-year period is recorded. Premium fuel will be used if it is shown to reduce maintenance costs.

18. A group of 15 swimmers is chosen to participate in an experiment to see if a new breathing style will improve their stamina. Each swimmer's pulse recovery rate is measured after a 20 minute workout using the old breathing style. The swimmers practice the new style for two weeks and then measure their pulse recovery rates after a 20 minute workout using the new style. They will continue to use the new breathing style if it is shown to reduce pulse recovery time.

19. In a study comparing various methods of gold plating, 7 printed circuit edge connectors were gold plated with control-immersion tip plating. The average gold thickness was 1.5 μm, with a standard deviation of 0.25 μm. Five connectors were masked, then plated with total immersion plating. The average gold thickness was 1.0 μm, with a standard deviation of 0.15 μm. Find a 99% confidence interval for the difference between the mean thicknesses produced by the two methods.

20. In an experiment to determine the effect of ambient temperature on the emissions of oxides of nitrogen (NO_x) of diesel trucks, 10 trucks were run at temperatures of 40°F and 80°F. The emissions, in ppm, are presented in the following table.

Truck	40°F	80°F
1	0.8347	0.8152
2	0.7532	0.7652
3	0.8557	0.8426
4	0.9012	0.7971
5	0.7854	0.7643
6	0.8629	0.8195
7	0.8827	0.7836
8	0.7403	0.6945
9	0.7480	0.7729
10	0.8486	0.7947

Can you conclude that the mean emissions differ between the two temperatures?

21. Two formulations of a certain coating, designed to inhibit corrosion, are being tested. For each of eight pipes, half the pipe is coated with formulation A, and the other half is coated with formulation B. Each pipe is exposed to a salt environment for 500 hours. Afterward, the corrosion loss (in μm) is measured for each formulation on each pipe.

Pipe	A	B
1	197	204
2	161	182
3	144	140
4	162	178
5	185	183
6	154	163
7	136	156
8	130	143

Can you conclude that the mean amount of corrosion differs between the two formulations?

22. Two microprocessors are compared on a sample of six benchmark codes to determine whether there is a difference in speed. The times (in seconds) used by each processor on each code are given in the following table.

	Code					
	1	2	3	4	5	6
Processor A	27.2	18.1	27.2	19.7	24.5	22.1
Processor B	24.1	19.3	26.8	20.1	27.6	29.8

Can you conclude that the mean speeds of the two processors differ?

23. Two different chemical formulations of rocket fuel are considered for the peak thrust they deliver in a particular design for a rocket engine. The thrust/weight ratios (in kilograms force per gram) for each of the two

fuels are measured several times. The results are as follows:

Fuel A: 54.3   52.9   57.9   58.2   53.4   51.4
         56.8   55.9   57.9   56.8   58.4   52.9
         55.5   51.3   51.8   53.3

Fuel B: 55.1   55.5   53.1   50.5   49.7   50.1
         52.4   54.4   54.1   55.6   56.1   54.8
         48.4   48.3   55.5   54.7

a. Assume the fuel processing plant is presently configured to produce fuel B and changeover costs are high. Since an increased thrust/weight ratio for rocket fuel is beneficial, how should the null and alternate hypotheses be stated for a test on which to base a decision whether to switch to fuel A?

b. Can you conclude that the switch to fuel A should be made?

Chapter 8

Inference in Linear Models

Introduction

Data that consists of a collection of ordered pairs $(x_1, y_1), \ldots, (x_n, y_n)$ are called bivariate data. In Chapter 2, we introduced the least-squares line as a way to summarize a set of bivariate data and to predict a value of y given a value of x. In many situations, it is reasonable to assume that x and y are linearly related by an equation $y = \beta_0 + \beta_1 x + \varepsilon$, where ε is a random variable. In these situations, the equation $y = \beta_0 + \beta_1 x$ represents the "true" regression line, and the least-squares line computed from the sample is an estimate of the true line. In Section 8.1 we will learn to compute confidence intervals and to perform hypothesis tests on the slope and intercept of the true regression line. For these confidence intervals and hypothesis tests to be valid, certain assumptions must be satisfied. We will learn some methods for checking these assumptions, and for correcting violations of them, in Section 8.2.

With bivariate data, the variable represented by x is called the independent variable, and the variable represented by y is called the dependent variable. A linear model $y = \beta_0 + \beta_1 x + \varepsilon$ that relates the value of a dependent variable y to the value of a single independent variable x is called a simple linear regression model. In many situations, however, a single independent variable is not enough. In these cases, there are several independent variables, $x_1, x_2, \ldots, x_p$, that are related to a dependent variable y. If the relationship between the dependent variable and the independent variables is linear, the technique of **multiple regression** can be used to include all the dependent variables in the model.

For example, assume that for a number of light trucks, we measure fuel efficiency, along with independent variables weight, engine displacement, and age. We could predict fuel efficiency with a multiple regression model $y = \beta_0 + \beta_1 x_1 + \beta_2 x_2 + \beta_3 x_3 + \varepsilon$, where y represents fuel efficiency and x_1, x_2 and x_3 represent weight, engine displacement, and age. Multiple regression will be discussed in Sections 8.3 and 8.4.

8.1 Inferences Using the Least-Squares Coefficients

When two variables have a linear relationship, the scatterplot tends to be clustered around a line known as the least-squares line (see Figure 2.2 in Section 2.1). In many cases, we think of the slope and intercept of the least-squares line as estimates of the slope and intercept of a true regression line. In this section we will learn how to use the least-squares line to construct confidence intervals and to test hypothesis on the slope and intercept of the true line.

We begin by describing a hypothetical experiment. Springs are used in applications for their ability to extend (stretch) under load. The stiffness of a spring is measured by the "spring constant," which is the length that the spring will be extended by one unit of force or load.[1] To make sure that a given spring functions appropriately, it is necessary to estimate its spring constant with good accuracy and precision.

In our hypothetical experiment, a spring is hung vertically with the top end fixed, and weights are hung one at a time from the other end. After each weight is hung, the length of the spring is measured. Let $x_1, \ldots, x_n$ represent the weights, and let l_i represent the length of the spring under the load x_i. Hooke's law states that

$$l_i = \beta_0 + \beta_1 x_i \tag{8.1}$$

where β_0 is the length of the spring when unloaded and β_1 is the spring constant.

Let y_i be the *measured* length of the spring under load x_i. Because of measurement error, y_i will differ from the true length l_i. We write

$$y_i = l_i + \varepsilon_i \tag{8.2}$$

where ε_i is the error in the ith measurement. Combining (8.1) and (8.2), we obtain

$$y_i = \beta_0 + \beta_1 x_i + \varepsilon_i \tag{8.3}$$

In Equation (8.3) y_i is called the **dependent variable**, x_i is called the **independent variable**, β_0 and β_1 are the **regression coefficients**, and ε_i is called the **error**. The line $y = \beta_0 + \beta_1 x$ is called the **true regression line**. Equation (8.3) is called a **linear model**.

Table 8.1 (page 314) presents the results of the hypothetical experiment. We wish to use these data to estimate the spring constant β_1 and the unloaded length β_0. If there were no measurement error, the points would lie on a straight line with slope β_1 and intercept β_0, and these quantities would be easy to determine. Because of measurement error, β_0 and β_1 cannot be determined exactly, but they can be estimated by calculating the least-squares line. We write the equation of the line as

$$y = \hat{\beta}_0 + \hat{\beta}_1 x \tag{8.4}$$

The quantities $\hat{\beta}_0$ and $\hat{\beta}_1$ are called the **least-squares coefficients**. The coefficient $\hat{\beta}_1$, the slope of the least-squares line, is an estimate of the true spring constant β_1, and the

---

[1] The more traditional definition of the spring constant is the reciprocal of this quantity—namely, the force required to extend the spring one unit of length.

TABLE 8.1 Measured lengths of a spring under various loads

Weight (lb) x	Measured Length (in.) y	Weight (lb) x	Measured Length (in.) y
0.0	5.06	2.0	5.40
0.2	5.01	2.2	5.57
0.4	5.12	2.4	5.47
0.6	5.13	2.6	5.53
0.8	5.14	2.8	5.61
1.0	5.16	3.0	5.59
1.2	5.25	3.2	5.61
1.4	5.19	3.4	5.75
1.6	5.24	3.6	5.68
1.8	5.46	3.8	5.80

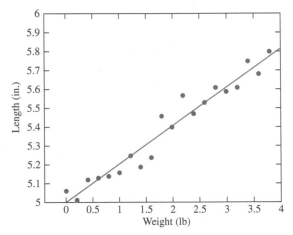

FIGURE 8.1 Plot of measured lengths of a spring versus load. The least-squares line is superimposed.

coefficient $\hat{\beta}_0$, the intercept of the least-squares line, is an estimate of the true unloaded length β_0. Figure 8.1 presents the scatterplot of y versus x with the least-squares line superimposed.

The formulas for computing the least-squares coefficients were presented as Equations (2.6) and (2.7) in Section 2.2. We repeat them here.

Summary

Given points $(x_1, y_1), \ldots, (x_n, y_n)$, the least-squares line is $\hat{y} = \hat{\beta}_0 + \hat{\beta}_1 x$, where

$$\hat{\beta}_1 = \frac{\sum_{i=1}^{n}(x_i - \overline{x})(y_i - \overline{y})}{\sum_{i=1}^{n}(x_i - \overline{x})^2} \qquad (8.5)$$

$$\hat{\beta}_0 = \overline{y} - \hat{\beta}_1 \overline{x} \qquad (8.6)$$

Example 8.1

Using the Hooke's law data in Table 8.1, compute the least-squares estimates of the spring constant and the unloaded length of the spring. Write the equation of the least-squares line.

Solution

The estimate of the spring constant is $\hat{\beta}_1$, and the estimate of the unloaded length is $\hat{\beta}_0$. From Table 8.1 we compute:

$$\overline{x} = 1.9000 \qquad \overline{y} = 5.3885$$

$$\sum_{i=1}^{n}(x_i - \overline{x})^2 = \sum_{i=1}^{n}x_i^2 - n\overline{x}^2 = 26.6000$$

$$\sum_{i=1}^{n}(x_i - \overline{x})(y_i - \overline{y}) = \sum_{i=1}^{n}x_i y_i - n\overline{x}\,\overline{y} = 5.4430$$

Using Equations (8.5) and (8.6), we compute

$$\hat{\beta}_1 = \frac{5.4430}{26.6000} = 0.2046$$

$$\hat{\beta}_0 = 5.3885 - (0.2046)(1.9000) = 4.9997$$

The equation of the least-squares line is $y = \hat{\beta}_0 + \hat{\beta}_1 x$. Substituting the computed values for $\hat{\beta}_0$ and $\hat{\beta}_1$, we obtain

$$y = 4.9997 + 0.2046x$$

Random Variation in the Least-Squares Estimates

It is important to understand the difference between the least-squares *estimates* $\hat{\beta}_0$ and $\hat{\beta}_1$, and the *true values* β_0 and β_1. The true values are constants whose values are unknown. The estimates are quantities that are computed from the data. We may use the estimates as approximations for the true values.

In principle, an experiment such as the Hooke's law experiment could be repeated many times. The true values β_0 and β_1 would remain constant over the replications of the experiment. But each replication would produce different data, and thus different values of the estimates $\hat{\beta}_0$ and $\hat{\beta}_1$. Therefore $\hat{\beta}_0$ and $\hat{\beta}_1$ are *random variables*, since their values vary from experiment to experiment. In order for the estimates $\hat{\beta}_1$ and $\hat{\beta}_0$ to be useful, we need to estimate their standard deviations to determine how much they are likely to vary. To estimate their standard deviations, we need to know something about the nature of the errors ε_i. We will begin by studying the simplest situation, in which four important assumptions are satisfied. These are as follows:

Assumptions for Errors in Linear Models

In the simplest situation, the following assumptions are satisfied:

1. The errors $\varepsilon_1, \ldots, \varepsilon_n$ are random and independent. In particular, the magnitude of any error ε_i does not influence the value of the next error ε_{i+1}.

2. The errors $\varepsilon_1, \ldots, \varepsilon_n$ all have mean 0.

3. The errors $\varepsilon_1, \ldots, \varepsilon_n$ all have the same variance, which we denote by σ^2.

4. The errors $\varepsilon_1, \ldots, \varepsilon_n$ are normally distributed.

These assumptions are restrictive, so it is worthwhile to discuss briefly the degree to which it is acceptable to violate them in practice. When the sample size is large, the normality assumption (4) becomes less important. Mild violations of the assumption of constant variance (3) do not matter too much, but severe violations should be corrected. We briefly describe some methods of correction in Section 8.2. See Draper and Smith (1998) for a more thorough treatment of the topic.

Under these assumptions, the effect of the ε_i is largely governed by the magnitude of the variance σ^2, since it is this variance that determines how large the errors are likely to be. Therefore, in order to estimate the standard deviations of $\hat{\beta}_0$ and $\hat{\beta}_1$, we must first estimate the error variance σ^2. This can be done by first computing the fitted values $\hat{y}_i = \hat{\beta}_0 + \hat{\beta}_1 x_i$ for each value x_i. The residuals are the values $e_i = y_i - \hat{y}_i$. The estimate of the error variance σ^2 is the quantity s^2 given by

$$s^2 = \frac{\sum_{i=1}^n e_i^2}{n-2} = \frac{\sum_{i=1}^n (y_i - \hat{y}_i)^2}{n-2} \tag{8.7}$$

The estimate of the error variance is thus the average of the squared residuals, except that we divide by $n - 2$ rather than n. Since the least-squares line minimizes the sum $\sum_{i=1}^n e_i^2$, the residuals tend to be a little smaller than the errors ε_i. It turns out that dividing by $n - 2$ rather than n appropriately compensates for this.

There is an equivalent formula for s^2, involving the correlation coefficient r, that is often easier to calculate:

$$s^2 = \frac{(1 - r^2) \sum_{i=1}^n (y_i - \overline{y})^2}{n-2} \tag{8.8}$$

Under assumptions 1 through 4, the observations y_i are also random variables. In fact, since $y_i = \beta_0 + \beta_1 x_i + \varepsilon_i$, it follows that y_i has a normal distribution with mean $\beta_0 + \beta_1 x_i$ and variance σ^2. In particular, β_1 represents the change in the mean of y associated with an increase of one unit in the value of x.

Summary

In the linear model $y_i = \beta_0 + \beta_1 x_i + \varepsilon_i$, under assumptions 1 through 4, the observations $y_1, \ldots, y_n$ are independent random variables that follow the normal distribution. The mean and variance of y_i are given by

$$\mu_{y_i} = \beta_0 + \beta_1 x_i$$

$$\sigma_{y_i}^2 = \sigma^2$$

The slope β_1 represents the change in the mean of y associated with an increase of one unit in the value of x.

It can be shown that the means and standard deviations of $\hat{\beta}_0$ and $\hat{\beta}_1$ are given by

$$\mu_{\hat{\beta}_0} = \beta_0 \qquad \mu_{\hat{\beta}_1} = \beta_1$$

$$\sigma_{\hat{\beta}_0} = \sigma \sqrt{\frac{1}{n} + \frac{\overline{x}^2}{\sum_{i=1}^{n}(x_i - \overline{x})^2}} \qquad \sigma_{\hat{\beta}_1} = \frac{\sigma}{\sqrt{\sum_{i=1}^{n}(x_i - \overline{x})^2}}$$

The estimators $\hat{\beta}_0$ and $\hat{\beta}_1$ are unbiased, since their means are equal to the true values. They are also normally distributed, because they are linear combinations of the independent normal random variables y_i. In practice, when computing the standard deviations, we usually don't know the value of σ, so we approximate it with s.

Summary

Under assumptions 1 through 4 (page 316),

■ The quantities $\hat{\beta}_0$ and $\hat{\beta}_1$ are normally distributed random variables.
■ The means of $\hat{\beta}_0$ and $\hat{\beta}_1$ are the true values β_0 and β_1, respectively.
■ The standard deviations of $\hat{\beta}_0$ and $\hat{\beta}_1$ are estimated with

$$s_{\hat{\beta}_0} = s \sqrt{\frac{1}{n} + \frac{\overline{x}^2}{\sum_{i=1}^{n}(x_i - \overline{x})^2}} \tag{8.9}$$

and

$$s_{\hat{\beta}_1} = \frac{s}{\sqrt{\sum_{i=1}^{n}(x_i - \overline{x})^2}} \tag{8.10}$$

where $s = \sqrt{\dfrac{(1 - r^2)\sum_{i=1}^{n}(y_i - \overline{y})^2}{n - 2}}$ is an estimate of the error standard deviation σ.

xample

8.2

For the Hooke's law data, compute s, $s_{\hat{\beta}_1}$, and $s_{\hat{\beta}_0}$. Estimate the spring constant and the unloaded length, and find their standard deviations.

Solution

In Example 8.1 we computed $\bar{x} = 1.9000$, $\bar{y} = 5.3885$, $\sum_{i=1}^{n}(x_i - \bar{x})^2 = 26.6000$, and $\sum_{i=1}^{n}(x_i - \bar{x})(y_i - \bar{y}) = 5.4430$. Now compute $\sum_{i=1}^{n}(y_i - \bar{y})^2 = 1.1733$. To compute s, we first compute the correlation coefficient r, which is given by

$$r = \frac{\sum_{i=1}^{n}(x_i - \bar{x})(y_i - \bar{y})}{\sqrt{\sum_{i=1}^{n}(x_i - \bar{x})^2}\sqrt{\sum_{i=1}^{n}(y_i - \bar{y})^2}} \qquad \text{(Equation 2.1 in Section 2.1)}$$

The correlation is $r = 5.4430/\sqrt{(26.6000)(1.1733)} = 0.9743$.

Using Equation (8.8), $s = \sqrt{\dfrac{(1 - 0.9743^2)(1.1733)}{18}} = 0.0575$.

Using Equation (8.9), $s_{\hat{\beta}_0} = 0.0575\sqrt{\dfrac{1}{20} + \dfrac{1.9000^2}{26.6000}} = 0.0248$.

Using Equation (8.10), $s_{\hat{\beta}_1} = \dfrac{0.0575}{\sqrt{26.6000}} = 0.0111$.

The More Spread in the *x* Values, the Better (Within Reason)

In the expressions for both of the standard deviations $s_{\hat{\beta}_0}$ and $s_{\hat{\beta}_1}$ in Equations (8.9) and (8.10), the quantity $\sum_{i=1}^{n}(x_i - \bar{x})^2$ appears in a denominator. This quantity measures the spread in the *x* values; when divided by the constant $n-1$, it is just the sample variance of the *x* values. It follows that, other things being equal, an experiment performed with more widely spread-out *x* values will result in smaller standard deviations for $\hat{\beta}_0$ and $\hat{\beta}_1$, and thus more precise estimation of the true values β_0 and β_1. Of course, it is important not to use *x* values so large or so small that they are outside the range for which the linear model holds.

Summary

When one is able to choose the *x* values, it is best to spread them out widely. The more spread out the *x* values, the smaller the standard deviations of $\hat{\beta}_0$ and $\hat{\beta}_1$.

  Specifically, the standard deviation $\sigma_{\hat{\beta}_1}$ of $\hat{\beta}_1$ is inversely proportional to $\sqrt{\sum_{i=1}^{n}(x_i - \bar{x})^2}$, or equivalently, to the sample standard deviation of $x_1, x_2, \ldots, x_n$.

  Caution: If the range of *x* values extends beyond the range where the linear model holds, the results will not be valid.

  There are two other ways to improve the accuracy of the estimated regression line. First, one can increase the size of the sum $\sum_{i=1}^{n}(x_i - \bar{x})^2$ by taking more observations, thus adding more terms to the sum. And second, one can decrease the size of the error variance σ^2, for example, by measuring more precisely. These two methods usually add to the cost of a project, however, while simply choosing more widely spread x values often does not.

Example

8.3

Two engineers are conducting independent experiments to estimate a spring constant for a particular spring. The first engineer suggests measuring the length of the spring with no load, and then applying loads of 1, 2, 3, and 4 lb. The second engineer suggests using loads of 0, 2, 4, 6, and 8 lb. Which result will be more precise? By what factor?

Solution
The sample standard deviation of the numbers 0, 2, 4, 6, 8 is twice as great as the sample standard deviation of the numbers 0, 1, 2, 3, 4. Therefore the standard deviation $\sigma_{\hat{\beta}_1}$ for the first engineer is twice as large as for the second engineer, so the second engineer's estimate is twice as precise.

We have made two assumptions in the solution to this example. First, we assumed that the error variance σ^2 is the same for both engineers. If they are both using the same apparatus and the same measurement procedure, this could be a safe assumption. But if one engineer is able to measure more precisely, this needs to be taken into account. Second, we have assumed that a load of 8 lb is within the elastic zone of the spring, so that the linear model applies throughout the range of the data.

Inferences on the Slope and Intercept

Given a scatterplot with points $(x_1, y_1), \ldots, (x_n, y_n)$, we can compute the slope $\hat{\beta}_1$ and intercept $\hat{\beta}_0$ of the least-squares line. We consider these to be estimates of a true slope β_1 and intercept β_0. We will now explain how to use these estimates to find confidence intervals for, and to test hypotheses about, the true values β_1 and β_0. It turns out that the methods for a population mean, based on the Student's t distribution, can be easily adapted for this purpose.

We have seen that under assumptions 1 through 4, $\hat{\beta}_0$ and $\hat{\beta}_1$ are normally distributed with means β_0 and β_1, and standard deviations that are estimated by $s_{\hat{\beta}_0}$ and $s_{\hat{\beta}_1}$. The quantities $(\hat{\beta}_0 - \beta_0)/s_{\hat{\beta}_0}$ and $(\hat{\beta}_1 - \beta_1)/s_{\hat{\beta}_1}$ have Student's t distributions with $n - 2$ degrees of freedom. The number of degrees of freedom is $n-2$ because in the computation of $s_{\hat{\beta}_0}$ and $s_{\hat{\beta}_1}$ we divide the sum of squared residuals by $n - 2$. When the sample size n is large enough, the normal distribution is nearly indistinguishable from the Student's t and may be used instead. However, most software packages use the Student's t distribution regardless of sample size.

Summary

Under assumptions 1 through 4, the quantities $\dfrac{\hat{\beta}_0 - \beta_0}{s_{\hat{\beta}_0}}$ and $\dfrac{\hat{\beta}_1 - \beta_1}{s_{\hat{\beta}_1}}$ have Student's t distributions with $n - 2$ degrees of freedom.

Confidence intervals for β_0 and β_1 can be derived in exactly the same way as the Student's t based confidence interval for a population mean. Let $t_{n-2,\alpha/2}$ denote the point on the Student's t curve with $n - 2$ degrees of freedom that cuts off an area of $\alpha/2$ in the right-hand tail. Then the point estimates for the confidence intervals are $\hat{\beta}_0$ and $\hat{\beta}_1$, the standard errors are $s_{\hat{\beta}_0}$ and $s_{\hat{\beta}_1}$, and the critical value is $t_{n-2,\alpha/2}$.

> Level $100(1 - \alpha)\%$ confidence intervals for β_0 and β_1 are given by
>
> $$\hat{\beta}_0 \pm t_{n-2,\alpha/2} \cdot s_{\hat{\beta}_0} \qquad \hat{\beta}_1 \pm t_{n-2,\alpha/2} \cdot s_{\hat{\beta}_1} \qquad (8.11)$$
>
> where
>
> $$s_{\hat{\beta}_0} = s\sqrt{\frac{1}{n} + \frac{\bar{x}^2}{\sum_{i=1}^{n}(x_i - \bar{x})^2}} \qquad s_{\hat{\beta}_1} = \frac{s}{\sqrt{\sum_{i=1}^{n}(x_i - \bar{x})^2}}$$

We illustrate the preceding method with some examples.

Example 8.4

Find a 95% confidence interval for the spring constant in the Hooke's law data.

Solution
The spring constant is β_1. We have previously computed $\hat{\beta}_1 = 0.2046$ (Example 8.1) and $s_{\hat{\beta}_1} = 0.0111$ (Example 8.2).

The number of degrees of freedom is $n - 2 = 20 - 2 = 18$, so the t value for a 95% confidence interval is $t_{18,.025} = 2.101$. The confidence interval for β_1 is therefore

$$0.2046 \pm (2.101)(0.0111) = 0.2046 \pm 0.0233 = (0.181, \ 0.228)$$

We are 95% confident that the increase in the length of the spring that will result from an increase of 1 lb in the load is between 0.181 and 0.228 in. Of course, this confidence interval is valid only within the range of the data (0 to 3.8 lb).

Example 8.5

In the Hooke's law data, find a 99% confidence interval for the unloaded length of the spring.

Solution
The unloaded length of the spring is β_0. We have previously computed $\hat{\beta}_0 = 4.9997$ (Example 8.1) and $s_{\hat{\beta}_0} = 0.0248$ (Example 8.2).

The number of degrees of freedom is $n - 2 = 20 - 2 = 18$, so the t value for a 99% confidence interval is $t_{18,.005} = 2.878$. The confidence interval for β_0 is therefore

$$4.9997 \pm (2.878)(0.0248) = 4.9997 \pm 0.0714 = (4.928, \ 5.071)$$

We are 99% confident that the unloaded length of the spring is between 4.928 and 5.071 in.

We can perform hypothesis tests on β_0 and β_1 as well. We present some examples.

Example

8.6

The manufacturer of the spring in the Hooke's law data claims that the spring constant β_1 is at least 0.215 in./lb. We have estimated the spring constant to be $\hat{\beta}_1 = 0.2046$ in./lb. Can we conclude that the manufacturer's claim is false?

Solution
This calls for a hypothesis test. The null and alternate hypotheses are

$$H_0: \beta_1 \geq 0.215 \quad \text{versus} \quad H_1: \beta_1 < 0.215$$

The quantity

$$\frac{\hat{\beta}_1 - \beta_1}{s_{\hat{\beta}_1}}$$

has a Student's t distribution with $n - 2 = 20 - 2 = 18$ degrees of freedom. Under H_0, we take $\beta_1 = 0.215$. The test statistic is therefore

$$\frac{\hat{\beta}_1 - 0.215}{s_{\hat{\beta}_1}}$$

We have previously computed $\hat{\beta}_1 = 0.2046$ and $s_{\hat{\beta}_1} = 0.0111$. The value of the test statistic is therefore

$$\frac{0.2046 - 0.215}{0.0111} = -0.937$$

Consulting the Student's t table, we find that the P-value is between 0.10 and 0.25. We cannot reject the manufacturer's claim on the basis of these data.

Example

8.7

Can we conclude from the Hooke's law data that the unloaded length of the spring is more than 4.9 in.?

Solution
This requires a hypothesis test. The null and alternate hypotheses are

$$H_0: \beta_0 \leq 4.9 \text{ versus } H_1: \beta_0 > 4.9$$

The quantity

$$\frac{\hat{\beta}_0 - \beta_0}{s_{\hat{\beta}_0}}$$

has a Student's t distribution with $n - 2 = 20 - 2 = 18$ degrees of freedom. Under H_0, we take $\beta_0 = 4.9$. The test statistic is therefore

$$\frac{\hat{\beta}_0 - 4.9}{s_{\hat{\beta}_0}}$$

We have previously computed $\hat{\beta}_0 = 4.9997$ and $s_{\hat{\beta}_0} = 0.0248$. The value of the test statistic is therefore

$$\frac{4.9997 - 4.9}{0.0248} = 4.020$$

Consulting the Student's t table, we find that the P-value is less than 0.0005. We can conclude that the unloaded length of the spring is more than 4.9 in.

The most commonly tested null hypothesis is $H_0 : \beta_1 = 0$. If this hypothesis is true, then there is no tendency for y either to increase or decrease as x increases. This implies that x and y have no linear relationship. In general, if the hypothesis that $\beta_1 = 0$ is not rejected, the linear model should not be used to predict y from x.

In an experiment to determine the effect of temperature on shrinkage of a synthetic fiber, 25 specimens were subjected to various temperatures. For each specimen, the temperature in °C (x) and shrinkage in % (y) were measured, and the following summary statistics were calculated:

$$\sum_{i=1}^{n}(x_i - \overline{x})^2 = 87.34 \qquad \sum_{i=1}^{n}(x_i - \overline{x})(y_i - \overline{y}) = 11.62 \qquad s = 0.951$$

Assuming that x and y follow a linear model, compute the estimated change in shrinkage due to an increase of 1°C in temperature. Should we use the linear model to predict shrinkage from temperature?

Solution

The linear model is $y = \beta_0 + \beta_1 x + \varepsilon$, and the change in shrinkage (y) due to a 1°C increase in temperature (x) is β_1. The null and alternate hypotheses are

$$H_0 : \beta_1 = 0 \quad \text{versus} \quad H_1 : \beta_1 \neq 0$$

The null hypothesis says that increasing the temperature does not affect the shrinkage, while the alternate hypothesis says that is does. The quantity

$$\frac{\hat{\beta}_1 - \beta_1}{s_{\hat{\beta}_1}}$$

has a Student's t distribution with $n - 2 = 25 - 2 = 23$ degrees of freedom. Under H_0, $\beta_1 = 0$. The test statistic is therefore

$$\frac{\hat{\beta}_1 - 0}{s_{\hat{\beta}_1}}$$

We compute $\hat{\beta}_1$ and $s_{\hat{\beta}_1}$:

$$\hat{\beta}_1 = \frac{\sum_{i=1}^{n}(x_i - \overline{x})(y_i - \overline{y})}{\sum_{i=1}^{n}(x_i - \overline{x})^2} = \frac{11.62}{87.34} = 0.13304$$

$$s_{\hat{\beta}_1} = \frac{s}{\sqrt{\sum_{i=1}^{n}(x_i - \overline{x})^2}} = 0.10176$$

The value of the test statistic is

$$\frac{0.13304 - 0}{0.10176} = 1.307$$

The t table shows that the P-value is greater than 0.20. We cannot conclude that the linear model is useful for predicting elongation from carbon content.

Inferences on the Mean Response

We can use the Hooke's law data to estimate the length of the spring under a load of x lbs by computing the fitted value $\hat{y} = \hat{\beta}_0 + \hat{\beta}_1 x$. Since the values $\hat{\beta}_0$ and $\hat{\beta}_1$ are subject to random variation, the value $\hat{y}$ is subject to random variation as well. For the estimate $\hat{y}$ to be more useful, we should construct a confidence interval around it to reflect its random variation. We now describe how to do this.

If a measurement y were taken of the length of the spring under a load of x lb, the mean of y would be the true length (or "mean response") $\beta_0 + \beta_1 x$, where β_1 is the true spring constant and β_0 is the true unloaded length of the spring. We estimate this length with $\hat{y} = \hat{\beta}_0 + \hat{\beta}_1 x$. Since $\hat{\beta}_0$ and $\hat{\beta}_1$ are normally distributed with means β_0 and β_1, respectively, it follows that $\hat{y}$ is normally distributed with mean $\beta_0 + \beta_1 x$.

To use $\hat{y}$ to find a confidence interval, we must know its standard deviation. It can be shown that the standard deviation of $\hat{y}$ can be approximated by

$$s_{\hat{y}} = s\sqrt{\frac{1}{n} + \frac{(x - \overline{x})^2}{\sum_{i=1}^{n}(x_i - \overline{x})^2}} \tag{8.12}$$

The quantity $[\hat{y} - (\beta_0 + \beta_1 x)]/s_{\hat{y}}$ has a Student's t distribution with $n - 2$ degrees of freedom. We can now provide the expression for a confidence interval for the mean response.

A level $100(1 - \alpha)\%$ confidence interval for the quantity $\beta_0 + \beta_1 x$ is given by

$$\hat{\beta}_0 + \hat{\beta}_1 x \pm t_{n-2,\alpha/2} \cdot s_{\hat{y}} \tag{8.13}$$

where $s_{\hat{y}} = s\sqrt{\dfrac{1}{n} + \dfrac{(x - \overline{x})^2}{\sum_{i=1}^{n}(x_i - \overline{x})^2}}.$

Example

8.9

Using the Hooke's law data, compute a 95% confidence interval for the length of a spring under a load of 1.4 lb.

Solution

We will calculate $\hat{y}$, $s_{\hat{y}}$, $\hat{\beta}_0$, and $\hat{\beta}_1$ and use expression (8.13). The number of points is $n = 20$. In Example 8.2, we computed $s = 0.0575$. In Example 8.1 we computed $\overline{x} = 1.9$, $\sum_{i=1}^{n}(x_i - \overline{x})^2 = 26.6$, $\hat{\beta}_1 = 0.2046$, and $\hat{\beta}_0 = 4.9997$. Using $x = 1.4$, we now compute

$$\hat{y} = \hat{\beta}_0 + \hat{\beta}_1 x = 4.9997 + (0.2046)(1.4) = 5.286$$

Using Equation (8.12) with $x = 1.4$, we obtain

$$s_{\hat{y}} = 0.0575\sqrt{\frac{1}{20} + \frac{(1.4 - 1.9)^2}{26.6}} = 0.0140$$

The number of degrees of freedom is $n - 2 = 20 - 2 = 18$. We find that the t value is $t_{18,.025} = 2.101$. Substituting into expression (8.13) we determine the 95% confidence interval for the length $\beta_0 + \beta_1(1.4)$ to be

$$5.286 \pm (2.101)(0.0140) = 5.286 \pm 0.0294 = (5.26, 5.32)$$

Example 8.10

In a study of the relationship between oxygen content (x) and ultimate testing strength (y) of welds, the data presented in the following table were obtained for 29 welds. Here oxygen content is measured in parts per thousand, and strength is measured in ksi. Using a linear model, find a 95% confidence interval for the mean strength for welds with oxygen content 1.7 parts per thousand. (From the article "Advances in Oxygen Equivalence Equations for Predicting the Properties of Titanium Welds," D. Harwig, W. Ittiwattana, and H. Castner, *The Welding Journal*, 2001:126s–136s.)

Oxygen Content	Strength	Oxygen Content	Strength	Oxygen Content	Strength
1.08	63.00	1.16	68.00	1.17	73.00
1.19	76.00	1.32	79.67	1.40	81.00
1.57	66.33	1.61	71.00	1.69	75.00
1.72	79.67	1.70	81.00	1.71	75.33
1.80	72.50	1.69	68.65	1.63	73.70
1.65	78.40	1.78	84.40	1.70	91.20
1.50	72.00	1.50	75.05	1.60	79.55
1.60	83.20	1.70	84.45	1.60	73.95
1.20	71.85	1.30	70.25	1.30	66.05
1.80	87.15	1.40	68.05		

Solution

We calculate the following quantities:

$$\bar{x} = 1.51966 \quad \bar{y} = 75.4966 \quad \sum_{i=1}^{n}(x_i - \bar{x})^2 = 1.33770 \quad \sum_{i=1}^{n}(y_i - \bar{y})^2 = 1304.23$$

$$\sum_{i=1}^{n}(x_i - \bar{x})(y_i - \bar{y}) = 22.6377 \quad \hat{\beta}_0 = 49.7796 \quad \hat{\beta}_1 = 16.9229 \quad s = 5.84090$$

The estimate of the mean strength for welds with an oxygen content of 1.7 is

$$\hat{y} = \hat{\beta}_0 + \hat{\beta}_1(1.7) = 49.7796 + (16.9229)(1.7) = 78.5485$$

The standard deviation of $\hat{y}$ is estimated to be

$$s_{\hat{y}} = s\sqrt{\frac{1}{n} + \frac{(x - \overline{x})^2}{\sum_{i=1}^{n}(x_i - \overline{x})^2}}$$

$$= 5.84090\sqrt{\frac{1}{29} + \frac{(1.7 - 1.51966)^2}{1.33770}}$$

$$= 1.4163$$

There are $n - 2 = 29 - 2 = 27$ degrees of freedom. The t value is therefore $t_{27,.025} = 2.052$. The 95% confidence interval is

$$78.5485 \pm (2.052)(1.4163) = 78.5485 \pm 2.9062 = (75.64, 81.45)$$

Hypothesis tests on the mean response can be conducted, using a Student's t distribution. Following is an example.

Example 8.11

Refer to Example 8.9. Let μ_0 represent the true length of the spring under a load of 1.6 lb. Test the hypothesis $H_0: \mu_0 \leq 5.3$ versus $H_1: \mu_0 > 5.3$.

Solution
Since μ_0 is the true length of the spring under a load of 1.6 lb, $\mu_0 = \beta_0 + \beta_1(1.6)$. Now let $\hat{y} = \hat{\beta}_0 + \hat{\beta}_1(1.6)$. The quantity

$$\frac{\hat{y} - [\beta_0 + \beta_1(1.6)]}{s_{\hat{y}}} = \frac{\hat{y} - \mu_0}{s_{\hat{y}}}$$

has a Student's t distribution with $n - 2 = 18$ degrees of freedom. Under H_0, we take $\mu_0 = 5.3$. The test statistic is therefore

$$\frac{\hat{y} - 5.3}{s_{\hat{y}}}$$

We compute $\hat{y}$ and $s_{\hat{y}}$:

$$\hat{y} = \hat{\beta}_0 + \hat{\beta}_1(1.6) = 4.9997 + (0.2046)(1.6) = 5.3271$$

$$s_{\hat{y}} = 0.0575\sqrt{\frac{1}{20} + \frac{(1.6 - 1.9)^2}{26.6000}} = 0.0133$$

The value of the test statistic is

$$\frac{5.3271 - 5.3}{0.0133} = 2.04$$

The P-value is between 0.025 and 0.05. It is reasonable to conclude that the true length is greater than 5.3 in.

Prediction Intervals for Future Observations

In Example 8.10 we found a confidence interval for the mean strength of welds with an oxygen content of 1.7 parts per thousand. Here is a somewhat different scenario: Assume we wish to predict the strength of a particular weld whose oxygen content is 1.7, rather than the mean strength of all such welds.

Using values calculated in Example 8.10, we predict this weld's strength to be $\hat{y} = \hat{\beta}_0 + \hat{\beta}_1(1.7) = 49.7796 + (16.9229)(1.7) = 78.5485$. This prediction is the same as the estimate of the mean strength for all welds with an oxygen content of 1.7. Now we wish to put an interval around this prediction, to describe how far the prediction is likely to be off. To compute this **prediction interval**, we must determine the amount of random variation in the prediction.

The mean strength of welds with an oxygen content of 1.7 is $\beta_0 + \beta_1(1.7)$. The actual strength of a particular weld is equal to $\beta_0 + \beta_1(1.7) + \varepsilon$, where ε represents the random difference between the strength of the particular weld and the mean strength of all welds whose oxygen content is 1.7. The error in predicting the strength of the particular weld with $\hat{y}$ is the prediction error

$$\hat{y} - [\beta_0 + \beta_1(1.7)] - \varepsilon \tag{8.14}$$

Since the quantity $\beta_0 + \beta_1(1.7)$ is constant, the variance of this prediction error is the sum of the variances of $\hat{y}$ and of ε. It follows that the standard deviation of the prediction error (expression 8.14) is approximated by

$$s_{\text{pred}} = \sqrt{s_{\hat{y}}^2 + s^2}$$

Using Equation (8.12) to substitute for $s_{\hat{y}}$ yields

$$s_{\text{pred}} = s\sqrt{1 + \frac{1}{n} + \frac{(x - \overline{x})^2}{\sum_{i=1}^{n}(x_i - \overline{x})^2}} \tag{8.15}$$

The appropriate expression for the prediction interval can now be determined.

A level $100(1 - \alpha)\%$ prediction interval for the quantity $\beta_0 + \beta_1 x$ is given by

$$\hat{\beta}_0 + \hat{\beta}_1 x \pm t_{n-2,\alpha/2} \cdot s_{\text{pred}} \tag{8.16}$$

where $s_{\text{pred}} = s\sqrt{1 + \dfrac{1}{n} + \dfrac{(x - \overline{x})^2}{\sum_{i=1}^{n}(x_i - \overline{x})^2}}$.

Note that the prediction interval is wider than the confidence interval, because the value 1 is added to the quantity under the square root to account for the additional random variation.

Example

8.12

For the weld data in Example 8.10, find a 95% prediction interval for the strength of a particular weld whose oxygen content is 1.7 parts per thousand.

Solution

The predicted strength is $\hat{y} = \hat{\beta}_0 + \hat{\beta}_1(1.7)$, which we have calculated in Example 8.10 to be 78.5485.

Using the quantities presented in Example 8.10, we compute the value of s_{pred} to be

$$s_{pred} = 5.84090\sqrt{1 + \frac{1}{29} + \frac{(1.7 - 1.51966)^2}{1.33770}} = 6.0102$$

There are $n - 2 = 29 - 2 = 27$ degrees of freedom. The t value is therefore $t_{27,.025} = 2.052$. The 95% prediction interval is therefore

$$78.5485 \pm (2.052)(6.0102) = 78.5485 \pm 12.3329 = (66.22,\ 90.88)$$

Interpreting Computer Output

Nowadays, least-squares calculations are usually done on a computer. The following output (from MINITAB) is for the Hooke's law data.

```
Regression Analysis: Length versus Weight

The regression equation is
Length = 5.00 + 0.205 Weight (1)

Predictor          Coef (2)  SE Coef (3)        T (4)      P (5)
Constant        4.99971      0.02477       201.81      0.000
Weight          0.20462      0.01115        18.36      0.000

S = 0.05749 (6)    R-Sq = 94.9% (7)    R-Sq(adj) = 94.6%

Analysis of Variance (8)

Source             DF          SS          MS          F          P
Regression          1      1.1138      1.1138     337.02      0.000
Residual Error     18      0.0595      0.0033
Total              19      1.1733

Unusual Observations (9)
Obs     Weight    Length        Fit      SE Fit    Residual    St Resid
 12       2.20    5.5700     5.4499      0.0133      0.1201       2.15R

R denotes an observation with a large standardized residual

Predicted Values for New Observations (10)

New Obs      Fit      SE Fit           95.0% CI                95.0% PI
1         5.2453      0.0150   (  5.2137,  5.2769)   (  5.1204,  5.3701)

Values of Predictors for New Observations (11)

New Obs     Weight
1             1.20
```

We will now explain the labeled quantities in the output:

(1) This is the equation of the least-squares line.

(2) **Coef:** The coefficients $\hat{\beta}_0 = 4.99971$ and $\hat{\beta}_1 = 0.20462$.

(3) **SE Coef:** The standard deviations $s_{\hat{\beta}_0}$ and $s_{\hat{\beta}_1}$. ("SE" stands for standard error, another term for standard deviation.)

(4) **T:** The values of the Student's t statistics for testing the hypotheses $\beta_0 = 0$ and $\beta_1 = 0$. The t statistic is equal to the coefficient divided by its standard deviation.

(5) **P:** The P-values for the tests of the hypotheses $\beta_0 = 0$ and $\beta_1 = 0$. The more important P-value is that for β_1. If this P-value is not small enough to reject the hypothesis that $\beta_1 = 0$, the linear model is not useful for predicting y from x. In this example, the P-values are extremely small, indicating that neither β_0 nor β_1 is equal to 0.

(6) **S:** The estimate s of the error standard deviation.

(7) **R-Sq:** This is r^2, the square of the correlation coefficient r, also called the coefficient of determination.

(8) **Analysis of Variance:** This table is not so important in simple linear regression, where there is only one independent variable. It is more important in multiple regression, where there are several independent variables. However, it is worth noting that the three numbers in the column labeled "SS" are the regression sum of squares $\sum_{i=1}^{n}(\hat{y}_i - \overline{y})^2$, the error sum of squares $\sum_{i=1}^{n}(y_i - \hat{y}_i)^2$, and their sum, the total sum of squares $\sum_{i=1}^{n}(y_i - \overline{y})^2$.

(9) **Unusual Observations:** Here the output tries to alert you to data points that may violate some of the assumptions 1 through 4 previously discussed. MINITAB is conservative and will often list several such points even when the data are well described by a linear model. In Section 8.2, we will learn some graphical methods for checking the assumptions of the linear model.

(10) **Predicted Values for New Observations:** These are confidence intervals and prediction intervals for values of x that are specified by the user. Here we specified $x = 1.2$ for the weight. The "Fit" is the fitted value $\hat{y} = \hat{\beta}_0 + \hat{\beta}_1 x$, and "SE Fit" is the standard deviation $s_{\hat{y}}$. Then come the 95% confidence and prediction intervals, respectively.

(11) **Values of Predictors for New Observations:** This is simply a list of the x values for which confidence and prediction intervals have been calculated. It shows that these intervals refer to a weight of $x = 1.2$.

Inference on the Population Correlation

When the points (x_i, y_i) are a random sample from a population of ordered pairs, the correlation coefficient r is often called the **sample correlation**. There is a population correlation as well, denoted by ρ. To get an intuitive understanding of the population correlation, you may imagine the population to consist of a large finite collection of points, and the population correlation ρ to be the quantity computed using

Equation (2.1) on the whole population, with sample means replaced by population means.

If the population of ordered pairs has a certain distribution known as a **bivariate normal distribution**, then the sample correlation r can be used to construct confidence intervals and perform hypothesis tests on the population correlation ρ. In practice, if the x-coordinates and the y-coordinates of the points are both normally distributed, then it is a virtual certainty that the population is bivariate normal, so the confidence intervals and tests described subsequently will be valid.

The null hypotheses most frequently tested are those of the form $\rho = 0$, $\rho \leq 0$, and $\rho \geq 0$. The method of testing these hypotheses is based on the test statistic

$$U = \frac{r\sqrt{n-2}}{\sqrt{1-r^2}}$$

When $\rho = 0$, U has a Student's t distribution with $n - 2$ degrees of freedom. Example 8.13 shows how to use U as a test statistic.

Example
8.13

The article "Withdrawal Strength of Threaded Nails" (D. Rammer, S. Winistorfer, and D. Bender, *Journal of Structural Engineering*, 2001:442–449) describes an experiment to investigate the relationship between the diameter of a nail (x) and its ultimate withdrawal strength (y). Annularly threaded nails were driven into Douglas fir lumber, and then their withdrawal strengths were measured in N/mm. The following results for 10 different diameters (in mm) were obtained.

x	2.52	2.87	3.05	3.43	3.68	3.76	3.76	4.50	4.50	5.26
y	54.74	59.01	72.92	50.85	54.99	60.56	69.08	77.03	69.97	90.70

The sample correlation coefficient is computed to be $r = 0.7492$. Test the hypothesis $H_0 : \rho \leq 0$ versus $H_1 : \rho > 0$.

Solution
Under H_0 we take $\rho = 0$, so the test statistic U has a Student's t distribution with $n - 2 = 8$ degrees of freedom. Since the sample correlation is $r = 0.7492$, the value of U is

$$U = \frac{r\sqrt{n-2}}{\sqrt{1-r^2}}$$

$$= \frac{0.7492\sqrt{10-2}}{\sqrt{1-0.7492^2}}$$

$$= 3.199$$

Consulting the Student's t table with eight degrees of freedom, we find that the P-value is between 0.005 and 0.01. It is reasonable to conclude that $\rho > 0$.

Exercises for Section 8.1

1. A chemical reaction is run 12 times, and the temperature x_i (in °C) and the yield y_i (in percent of a theoretical maximum) is recorded each time. The following summary statistics are recorded:

$$\bar{x} = 65.0 \qquad \bar{y} = 29.05 \qquad \sum_{i=1}^{12}(x_i - \bar{x})^2 = 6032.0$$

$$\sum_{i=1}^{12}(y_i - \bar{y})^2 = 835.42 \qquad \sum_{i=1}^{12}(x_i - \bar{x})(y_i - \bar{y}) = 1988.4$$

Let β_0 represent the hypothetical yield at a temperature of 0°C, and let β_1 represent the increase in yield caused by an increase in temperature of 1°C. Assume that assumptions 1 through 4 on page 316 hold.

a. Compute the least-squares estimates $\hat{\beta}_0$ and $\hat{\beta}_1$.
b. Compute the error variance estimate s^2.
c. Find 95% confidence intervals for β_0 and β_1.
d. A chemical engineer claims that the yield increases by more than 0.5 for each 1°C increase in temperature. Do the data provide sufficient evidence for you to conclude that this claim is false?
e. Find a 95% confidence interval for the the mean yield at a temperature of 40°C.
f. Find a 95% prediction interval for the yield of a particular reaction at a temperature of 40°C.

2. The following output (from MINITAB) describes the fit of a linear model $y = \beta_0 + \beta_1 x + \varepsilon$ that expresses the length of a spring in cm (y) in terms of the load upon it in kg (x). There are $n = 15$ observations.

Predictor	Coef	StDev	T	P
Constant	6.6361	1.1455	5.79	0.000
Load	2.9349	0.086738	33.8	0.000

a. How many degrees of freedom are there for the Student's t statistics?
b. Find a 98% confidence interval for β_1.
c. Find a 98% confidence interval for β_0.
d. Someone claims that if the load is increased by 1 kg, that the length will increase by exactly 3 cm. Use the given output to perform a hypothesis test to determine whether this claim is plausible.
e. Someone claims that the unloaded (load $= 0$) length of the spring is more than 10 cm. Use the given output to perform a hypothesis test to determine whether this claim is plausible.

3. Ozone (O_3) is a major component of air pollution in many cities. Atmospheric ozone levels are influenced by many factors, including weather. In one study, the mean percent relative humidity (x) and the mean ozone levels (y) were measured for 120 days in a western city. Mean ozone levels were measured in ppb. The following output (from MINITAB) describes the fit of a linear model to these data. Assume that assumptions 1 through 4 on page 316 hold.

The regression equation is
Ozone = 88.8 − 0.752 Humidity

Predictor	Coef	SE Coef	T	P
Constant	88.761	7.288	12.18	0.000
Humidity	−0.7524	0.13024	−5.78	0.000

```
S = 11.43        R-Sq = 22.0%      R-Sq(adj) = 21.4%

Predicted Values for New Observations

New Obs     Fit     SE Fit        95.0% CI            95.0% PI
1          43.62     1.20    (  41.23   46.00)  (  20.86,  66.37)

Values of Predictors for New Observations

New Obs    Humidity
1            60.0
```

a. What are the slope and intercept of the least-squares line?
b. Is the linear model useful for predicting ozone levels from relative humidity? Explain.
c. Predict the ozone level for a day when the relative humidity is 50%.
d. What is the correlation between relative humidity and ozone level?
e. The output provides a 95% confidence interval for the mean ozone level for days where the relative humidity is 60%. There are $n = 120$ observations in this data set. Using the value "SE Fit," find a 90% confidence interval.
f. Upon learning that the relative humidity on a certain day is 60%, someone predicts that the ozone level that day will be 80 ppb. Is this a reasonable prediction? If so, explain why. If not, give a reasonable range of predicted values.

4. In an study similar to the one in Exercise 3, the relative humidity and ozone levels were measured for 120 days in another city. The MINITAB output follows. Assume that assumptions 1 through 4 on page 316 hold.

```
The regression equation is
Ozone = 29.7 - 0.135 Humidity

Predictor       Coef    SE Coef        |        P
Constant       29.703     2.066     14.38    0.000
Humidity      -0.13468    0.03798    -3.55    0.001

S = 6.26    R-Sq =  9.6%    R-Sq(adj) =  8.9%
```

a. What is the slope of the least-squares line?
b. Find a 95% confidence interval for the slope.
c. Perform a test of the null hypothesis that the slope is greater than or equal to -0.1. What is the P-value?

5. Refer to Exercises 3 and 4. An atmospheric scientist notices that the slope of the least-squares line in the study described in Exercise 4 differs from the one in the study described in Exercise 3. He wishes to test the hypothesis that the effect of humidity on ozone level differs between the two cities. Let β_A denote the change in ozone level associated with an increase of 1% relative humidity for the city in Exercise 3, and β_B denote the corresponding increase for the city in Exercise 4.

a. Express the null hypothesis to be tested in terms of β_A and β_B.
b. Let $\hat{\beta}_A$ and $\hat{\beta}_B$ denote the slopes of the least-squares lines. Assume these slopes are independent. There are 120 observations in each data set. Test the null hypothesis in part (a). Can you conclude that the effect of humidity differs between the two cities?

6. In a study of reaction times, the time to respond to a visual stimulus (x) and the time to respond to an auditory stimulus (y) were recorded for each of 10 subjects. Times were measured in ms. The results are presented in the following table.

x	161	203	235	176	201	188	228	211	191	178
y	159	206	241	163	197	193	209	189	169	201

a. Compute the least-squares line for predicting auditory response time to visual response time.
b. Compute the error standard deviation estimate s.
c. Compute a 95% confidence interval for the slope.
d. Find a 95% confidence interval for the mean auditory response time for subjects with a visual response time of 200 ms.
e. Can you conclude that the mean auditory response time for subjects with a visual response time of 200 ms is greater than 190 ms? Perform a hypothesis test and report the P-value.
f. Find a 95% prediction interval for the auditory response time for a particular subject whose visual response time is 200 ms.

7. The article "Evaluation of the Expansion Attained to Date by Concrete Affected by Alkali-Silica Reaction. Part III: Application to Existing Structures" (M. Bérubé, N. Smaoui, et al, *Canadian Journal of Civil Engineering*, 2005: 463–479) reports measurements of expansion for several concrete bridges in the area of Quebec City. Following are measurements of horizontal and vertical expansion (in parts per hundred thousand) for several locations on the Pére-Leliévre bridge.

Horizontal (x)	20	15	43	5	18	24	32	10	21
Vertical (y)	58	58	55	80	58	68	57	69	63

a. Compute the least-squares line for predicting vertical expansion from horizontal expansion.
b. Compute 95% confidence intervals for β_0 and β_1.
c. Compute a 95% prediction interval for the vertical expansion at a location where the horizontal expansion is 25.
d. It is desired to predict the vertical expansion in two locations. In one location the horizontal expansion is 15, and in the other it is 30. If 95% prediction intervals are computed for the vertical expansion in both locations, which one will be wider? Explain.

8. The article "Computation of Equilibrium Oxidation and Reduction Potentials for Reversible and Dissociative Electron-Transfer Reactions in Solution" (P. Winget, C. Cramer, and D. Truhlar, *Theoretical Chemistry Accounts*, 2004:217–227) presents several models for estimating aqueous onc-electron potentials. The following table presents true potentials, measured experimentally in volts relative to the normal hydrogen electrode, for phenol and 23 substituted phenols, along with the corresponding value from the Austin model. Although the model values are not close to the true values, it is thought that they follow a linear model $y = \beta_0 + \beta_1 x + \varepsilon$, where y is the true value and x is the model value.

Model	True	Model	True	Model	True	Model	True
4.483	1.002	4.107	0.819	4.070	0.820	3.669	0.674
4.165	0.884	4.394	1.113	4.274	0.921	4.034	0.811
4.074	0.887	3.792	0.724	4.038	0.835	5.173	1.192
4.204	0.875	4.109	0.834	4.041	0.835	3.777	0.802
5.014	1.123	5.040	1.114	4.549	1.059	3.984	0.846
4.418	0.893	3.806	0.831	4.731	0.984	4.291	0.901

a. Compute the least-squares line for predicting the true potential from the model value.
b. Compute 95% confidence intervals for β_0 and β_1.
c. Two molecules differ in their model value by 0.5. By how much do you estimate that their true potentials will differ?
d. A molecule has a model value of 4.3. Find a 95% prediction interval for its true potential.
e. Can you conclude that the mean potential of molecules whose model value is 4.5 is greater than 0.93? Explain.

9. In a study to determine the relationship between ambient outdoor temperature and the rate of evaporation of water from soil, measurements of average daytime temperature in °C and evaporation in mm/day were taken for 40 days. The results are shown in the following table.

Temp.	Evap.	Temp.	Evap.	Temp.	Evap.	Temp.	Evap.
11.8	2.4	11.8	3.8	18.6	3.5	14.0	1.1
21.5	4.4	24.2	5.0	25.4	5.5	13.6	3.5
16.5	5.0	15.8	2.6	22.1	4.8	25.4	5.1
23.6	4.1	26.8	8.0	25.4	4.8	17.7	2.0
19.1	6.0	24.8	5.4	22.6	3.2	24.7	5.7
21.6	5.9	26.2	4.2	24.4	5.1	24.3	4.7
31.0	4.8	14.2	4.4	15.8	3.3	25.8	5.8
18.9	3.0	14.1	2.2	22.3	4.9	28.3	5.8
24.2	7.1	30.3	5.7	23.2	7.4	29.8	7.8
19.1	1.6	15.2	1.2	19.7	3.3	26.5	5.1

a. Compute the least-squares line for predicting evaporation (y) from temperature (x).
b. Compute 95% confidence intervals for β_0 and β_1.
c. Predict the evaporation rate when the temperature is 20°C.
d. Find a 95% confidence interval for the mean evaporation rate for all days with a temperature of 20°C.
e. Find a 95% prediction interval for the evaporation rate on a given day with a temperature of 20°C.

10. Three engineers are independently estimating the spring constant of a spring, using the linear model specified by Hooke's law. Engineer A measures the length of the spring under loads of 0, 1, 3, 4, and 6 lb, for a total of five measurements. Engineer B uses the same loads but repeats the experiment twice, for a total of 10 independent measurements. Engineer C uses loads of 0, 2, 6, 8, and 12 lb, measuring once for each load. The engineers all use the same measurement apparatus and procedure. Each engineer computes a 95% confidence interval for the spring constant.

a. If the width of the interval of engineer A is divided by the width of the interval of engineer B, the quotient will be approximately _____ .
b. If the width of the interval of engineer A is divided by the width of the interval of engineer C, the quotient will be approximately _____ .
c. Each engineer computes a 95% confidence interval for the length of the spring under a load of 2.5 lb. Which interval is most likely to be the shortest? Which interval is most likely to be the longest?

11. In the weld data (Example 8.10) imagine that 95% confidence intervals are computed for the mean strength of welds with oxygen contents of 1.3, 1.5, and 1.8 parts per thousand. Which of the confidence intervals would be the shortest? Which would be the longest?

12. Refer to Exercise 1. If 95% confidence intervals are constructed for the yield of the reaction at temperatures of 45°C, 60°C, and 75°C, which confidence interval would be the shortest? Which would be the longest?

13. In the following MINITAB output, some of the numbers have been accidentally erased. Recompute them, using the numbers still available. There are $n = 25$ points in the data set.

```
The regression equation is
Y = 1.71 + 4.27 X

Predictor      Coef    SE Coef       T       P
Constant    1.71348    6.69327     (a)     (b)
X           4.27473       (c)    3.768     (d)

S = 0.05749      R-Sq = 38.2%
```

14. In the following MINITAB output, some of the numbers have been accidentally erased. Recompute them, using the numbers still available. There are $n = 20$ points in the data set.

```
Predictor     Coef    SE Coef       T      P
Constant       (a)    0.43309    0.688    (b)
X          0.18917   0.065729      (c)    (d)

S = 0.67580      R-Sq = 31.0%
```

15. In order to increase the production of gas wells, a procedure known as "fracture treatment" is often used. Fracture fluid, which consists of fluid mixed with sand, is pumped into the well. The following figure presents a scatterplot of the monthly production versus the volume of fracture fluid pumped for 255 gas wells. Both production and fluid are expressed in units of volume per foot of depth of the well. The least-squares line is superimposed. The equation of the least-squares line is $y = 106.11 + 0.1119x$.

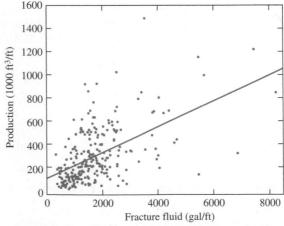

Production versus fracture fluid for 255 gas wells.

a. From the least-squares line, estimate the production for a well into which 4000 gal/ft are pumped.
b. From the least-squares line, estimate the production for a well into which 500 gal/ft are pumped.
c. A new well is dug, and 500 gal/ft of fracture fluid are pumped in. Based on the scatterplot, is it more likely that the production of this well will fall above or below the least-squares estimate?
d. What feature of the scatterplot indicates that assumption 3 on page 316 is violated?

16. In a sample of 400 ball bearings, the correlation coefficient between eccentricity and smoothness was $r = 0.10$.

a. Find the P-value for testing $H_0: \rho \leq 0$ versus $H_1: \rho > 0$. Can you conclude that $\rho > 0$?
b. Does the result in part (a) allow you to conclude that there is a strong correlation between eccentricity and smoothness? Explain.

17. The article " 'Little Ice Age' Proxy Glacier Mall Balance Records Reconstructed from Tree Rings in the Mt. Waddington Area, British Columbia Coast Mountains, Canada" (S. Larocque and D. Smith, *The Holocene*, 2005:748–757) evaluates the use of tree ring widths to estimate changes in the masses of glaciers. For the Sentinel glacier, the net mass balance (change in mass between the end of one summer and the end of the next summer) was measured for 23 years. During the same time period, the tree-ring index for white bark pine trees was measured, and the sample correlation between net mass balance and tree ring index was $r = -0.509$. Can you conclude that the population correlation ρ differs from 0?

8.2 Checking Assumptions

The methods discussed so far are valid under the assumption that the relationship between the variables x and y satisfies the linear model $y_i = \beta_0 + \beta_1 x_i + \varepsilon_i$, where the errors ε_i satisfy four assumptions. We repeat these assumptions here.

Assumptions for Errors in Linear Models

1. The errors $\varepsilon_1, \ldots, \varepsilon_n$ are random and independent. In particular, the magnitude of any error ε_i does not influence the value of the next error ε_{i+1}.
2. The errors $\varepsilon_1, \ldots, \varepsilon_n$ all have mean 0.
3. The errors $\varepsilon_1, \ldots, \varepsilon_n$ all have the same variance, which we denote by σ^2.
4. The errors $\varepsilon_1, \ldots, \varepsilon_n$ are normally distributed.

As mentioned earlier, the normality assumption (4) is less important when the sample size is large. While mild violations of the assumption of constant variance (3) do not matter too much, severe violations are a cause for concern.

We need ways to check these assumptions to assure ourselves that our methods are appropriate. Innumerable diagnostic tools have been proposed for this purpose. Many books have been written on the topic. We will restrict ourselves here to a few of the most basic procedures.

The Plot of Residuals versus Fitted Values

The single best diagnostic for least-squares regression is a plot of residuals e_i versus fitted values $\hat{y}_i$, sometimes called a **residual plot**. Figure 8.2 (page 336) presents such a plot for the Hooke's law data (see Figure 8.1 in Section 8.1 for a scatterplot of the

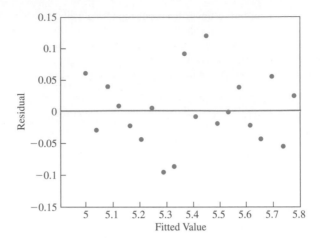

FIGURE 8.2 Plot of residuals (e_i) versus fitted values ($\hat{y}_i$) for the Hooke's law data. There is no substantial pattern to the plot, and the vertical spread does not vary too much, except perhaps near the edges. This is consistent with the assumptions of the linear model.

original data). By mathematical necessity, the residuals have mean 0, and the correlation between the residuals and fitted values is 0 as well. The least-squares line is therefore horizontal, passing through 0 on the vertical axis. When the linear model is valid, and assumptions 1 through 4 are satisfied, the plot will show no substantial pattern. There should be no curve to the plot, and the vertical spread of the points should not vary too much over the horizontal range of the plot, except perhaps near the edges. These conditions are reasonably well satisfied for the Hooke's law data. The plot confirms the validity of the linear model.

A bit of terminology: When the vertical spread in a scatterplot doesn't vary too much, the scatterplot is said to be **homoscedastic**. The opposite of homoscedastic is **heteroscedastic**.

A good-looking residual plot does not by itself prove that the linear model is appropriate, because the assumptions of the linear model can fail in other ways. On the other hand, a residual plot with a serious defect does clearly indicate that the linear model is inappropriate.

Summary

If the plot of residuals versus fitted values

■ Shows no substantial trend or curve, and

■ Is **homoscedastic**, that is, the vertical spread does not vary too much along the horizontal length of plot, except perhaps near the edges,

then it is *likely*, but not *certain*, that the assumptions of the linear model hold. However, if the residual plot *does* show a substantial trend or curve, or is **heteroscedastic**, it is certain that the assumptions of the linear model do *not* hold.

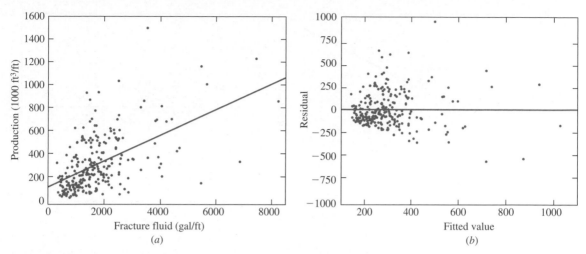

FIGURE 8.3 *(a)* Plot of monthly production versus volume of fracture fluid for 255 gas wells. *(b)* Plot of residuals (e_i) versus fitted values ($\hat{y}_i$) for the gas well data. The vertical spread clearly increases with the fitted value. This indicates a violation of the assumption of constant error variance.

Transforming the Variables

In many cases, the residual plot will exhibit curvature or heteroscedasticity, which reveal violations of assumptions. As an example, Figure 8.3 presents a scatterplot and a residual plot for a group of 255 gas wells. These data were presented in Exercise 15 in Section 8.1. The monthly production per foot of depth of the well is plotted against the volume of fracture fluid pumped into the well. The residual plot is strongly heteroscedastic, indicating that the error variance is larger for gas wells whose estimated production is larger. These of course are the wells into which more fracture fluid has been pumped. We conclude that we should not use this model to predict well production from the amount of fracture fluid pumped.

The model $y = \beta_0 + \beta_1 x + \varepsilon$ does not fit the data. In this case, we can fix the problem replacing x with $\ln x$ and y with $\ln y$. We fit the model $\ln y = \beta_0 + \beta_1 \ln x + \varepsilon$. Figure 8.4 (page 338) presents the results. We can see that an approximately linear relationship holds between the logarithm of production and the logarithm of the volume of fracture fluid.

We obtained a linear relationship in the gas well data by replacing the original variables x and y with functions of those variables. In general, replacing a variable with a function of itself is called **transforming** the variable. So for the gas well data, we applied a log transform to both x and y. In some cases, it works better to transform only one of the variables, either x or y. Functions other than the logarithm can be used as well. The most commonly used functions, other than the logarithm, are power transformations, in which x, y, or both are raised to a power.

Determining Which Transformation to Apply

It is possible with experience to look at a scatterplot or residual plot, and make an educated guess as to how to transform the variables. Mathematical methods are also

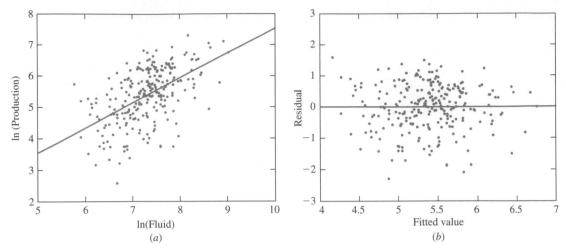

FIGURE 8.4 *(a)* Plot of the log of production versus the log of the volume of fracture fluid for 255 gas wells, with the least-squares line superimposed. *(b)* Plot of residuals versus fitted values. There is no substantial pattern to the residuals. The linear model looks good.

available to determine a good transformation. However, it is perfectly satisfactory to proceed by trial and error. Try various powers on both x and y (including $\ln x$ and $\ln y$), look at the residual plots, and hope to find one that is homoscedastic, with no discernible pattern. It is important to remember that transformations don't always work. Sometimes, none of the residual plots look good, no matter what transformations are tried. In these cases, other methods should be used. One of these is multiple regression, discussed in Section 8.3. A more advanced discussion of transformation selection can be found in Draper and Smith (1998).

Residual Plots with Only a Few Points Can Be Hard to Interpret

When there are only a few points in a residual plot, it can be hard to determine whether the assumptions of the linear model are met. Sometimes such a plot will at first glance appear to be heteroscedastic, or to exhibit a pattern, but upon closer inspection it turns out that this visual impression is caused by the placement of just one or two points. It is sometimes even difficult to determine whether such a plot contains an outlier. When one is faced with a sparse residual plot that is hard to interpret, a reasonable thing to do is to fit a linear model but to consider the results tentative, with the understanding that the appropriateness of the model has not been established. If and when more data become available, a more informed decision can be made. Of course, not all sparse residual plots are hard to interpret. Sometimes there is a clear pattern, which cannot be changed just by shifting one or two points. In these cases, the linear model should not be used.

To summarize, we present some generic examples of residual plots in Figure 8.5. For each one, we present a diagnosis and a prescription.

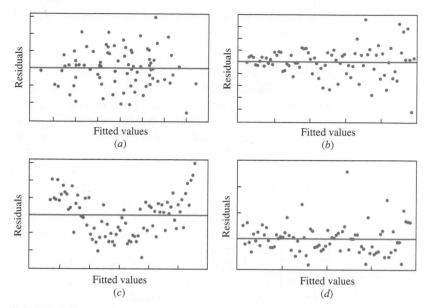

FIGURE 8.5 *(a)* No substantial pattern, plot is homoscedastic. Linear model is OK. *(b)* Heteroscedastic. Try a power transformation. *(c)* Discernible trend to residuals. Try a power transformation, or use multiple regression. *(d)* Outlier. Examine the offending data point to see if it is an error. If not, compute the least-squares line both with and without the outlier to see if it makes a noticeable difference.

Checking Independence and Normality

If the plot of residuals versus fitted values looks good, it may be advisable to perform additional diagnostics to further check the fit of the linear model. In particular, when the observations occur in a definite time order, it is desirable to plot the residuals against the order in which the observations were made. If there are trends in the plot, it indicates that the relationship between x and y may be varying with time. In these cases a variable representing time, or other variables related to time, should be included in the model as additional independent variables, and a multiple regression should be performed.

Sometimes a plot of residuals versus time shows that the residuals oscillate with time. This indicates that the value of each error is influenced by the errors in previous observations, so therefore the errors are not independent. When this feature is severe, linear regression should not be used, and the methods of time series analysis should be used instead. A good reference on time series analysis is Brockwell and Davis (2002).

To check that the errors are normally distributed, a normal probability plot of the residuals can be made. If the probability plot has roughly the appearance of a straight line, the residuals are approximately normally distributed. It can be a good idea to make a probability plot when variables are transformed, since one sign of a good transformation is that the residuals are approximately normally distributed. As previously mentioned, the assumption of normality is not so important when the number of data points is large.

Unfortunately, when the number of data points is small, it can be difficult to detect departures from normality.

Empirical Models and Physical Laws

How do we know whether the relationship between two variables is linear? In some cases, physical laws, such as Hooke's law, give us assurance that a linear model is correct. In other cases, such as the relationship between the log of the volume of fracture fluid pumped into a gas well and the log of its monthly production, there is no known physical law. In these cases, we use a linear model simply because it appears to fit the data well. A model that is chosen because it appears to fit the data, in the absence of physical theory, is called an **empirical model**. In real life, most data analysis is based on empirical models. It is less often that a known physical law applies. Of course, many physical laws started out as empirical models. If an empirical model is tested on many different occasions, under a wide variety of circumstances, and is found to hold without exception, it can gain the status of a physical law.

There is an important difference between the interpretation of results based on physical laws and the interpretation of results based on empirical models. A physical law may be regarded as *true*, whereas the best we can hope for from an empirical model is that it is *useful*. For example, in the Hooke's law data, we can be sure that the relationship between the load on the spring and its length is truly linear. We are sure that when we place another weight on the spring, the length of the spring can be accurately predicted from the linear model. For the gas well data, on the other hand, while the linear relationship describes the data well, we cannot be sure that it captures the true relationship between fracture fluid volume and production.

Here is a simple example that illustrates the point. Figure 8.6 presents 20 triangles of varying shapes. Assume that we do not know the formula for the area of a triangle.

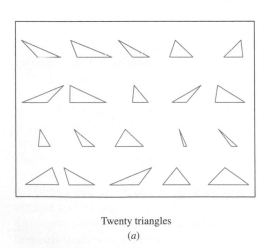

Twenty triangles

(a)

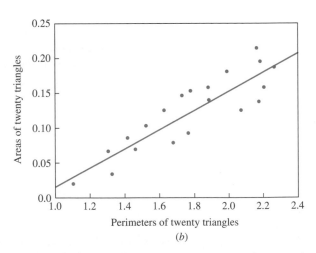

(b)

FIGURE 8.6 *(a)* Twenty triangles. *(b)* Area versus perimeter for 20 triangles. The correlation between perimeter and area is 0.88.

We notice, however, that triangles with larger perimeters seem to have larger areas, so we fit a linear model:

$$\text{Area} = \beta_0 + \beta_1 \,(\text{Perimeter}) + \varepsilon$$

The scatterplot of area versus perimeter, with the least-squares line superimposed, is shown to the right in Figure 8.6. The equation of the least-squares line is

$$\text{Area} = -1.232 + 1.373 \,(\text{Perimeter})$$

The units in this equation are arbitrary. The correlation between area and perimeter is $r = 0.88$, which is strongly positive. The linear model appears to fit well. We could use this model to predict, for example, that a triangle with perimeter equal to 5 will have an area of 5.633.

Now while this linear model may be useful, it is not true. The linear model correctly shows that there is a strong tendency for triangles with larger perimeters to have larger areas. In the absence of a better method, it may be of some use in estimating the areas of triangles. But it does not help to reveal the true mechanism behind the determination of area. The true mechanism, of course, is given by the law

$$\text{Area} = 0.5 \times \text{base} \times \text{height}$$

The results predicted by an empirical model may not hold up under replication. For example, a collection of triangles could be designed in such a way that the ones with the larger perimeters had smaller areas. In another collection, the area might appear to be proportional to the square of the perimeter, or to its logarithm. We cannot determine by statistical analysis of the triangle data how well the empirical model will apply to a triangle not yet observed. Deciding whether it is appropriate to apply the results of an empirical model to future observations is a matter of scientific judgment rather than statistics.

Summary

- Physical laws are applicable to all future observations.
- An empirical model is valid only for the data to which it is fit. It may or may not be useful in predicting outcomes for subsequent observations.
- Determining whether to apply an empirical model to a future observation requires scientific judgment rather than statistical analysis.

Exercises for Section 8.2

1. The following MINITAB output is for the least-squares fit of the model $\ln y = \beta_0 + \beta_1 \ln x + \varepsilon$, where y represents the monthly production of a gas well and x represents the volume of fracture fluid pumped in. (A scatterplot of these data is presented in Figure 8.4.)

```
Regression Analysis: LN PROD versus LN FLUID

The regression equation is
LN PROD = - 0.444 + 0.798 LN FLUID

Predictor         Coef        StDev            T        P
Constant       -0.4442       0.5853        -0.76    0.449
LN FLUID       0.79833      0.08010         9.97    0.000

S = 0.7459      R-Sq = 28.2%      R-Sq(adj) = 27.9%

Analysis of Variance

Source            DF          SS           MS        F        P
Regression         1      55.268       55.268    99.34    0.000
Residual Error   253     140.756        0.556
Total            254     196.024

Predicted Values for New Observations

New Obs     Fit        SE Fit           95.0% CI              95.0% PI
1        5.4457        0.0473    ( 5.3526,  5.5389)  ( 3.9738,  6.9176)

Values of Predictors for New Observations

New Obs    LN FLUID
1            7.3778
```

a. What is the equation of the least-squares line for predicting $\ln y$ from $\ln x$?
b. Predict the production of a well into which 2500 gal/ft of fluid have been pumped.
c. Predict the production of a well into which 1600 gal/ft of fluid have been pumped.
d. Find a 95% prediction interval for the production of a well into which 1600 gal/ft of fluid have been pumped. (*Note*: ln 1600 = 7.3778.)

2. The processing of raw coal involves "washing," in which coal ash (nonorganic, incombustible material) is removed. The article "Quantifying Sampling Precision for Coal Ash Using Gy's Discrete Model of the Fundamental Error" (*Journal of Coal Quality*, 1989:33–39) provides data relating the percentage of ash to the volume of a coal particle. The average percentage of ash for six volumes of coal particles was measured. The data are as follows:

Volume (cm³)	0.01	0.06	0.58	2.24	15.55	276.02
Percent ash	3.32	4.05	5.69	7.06	8.17	9.36

a. Compute the least-squares line for predicting percent ash (y) from volume (x). Plot the residuals versus the fitted values. Does the linear model seem appropriate? Explain.
b. Compute the least-squares line for predicting percent ash from ln volume. Plot the residuals versus the fitted values. Does the linear model seem appropriate? Explain.

c. Compute the least-squares line for predicting percent ash from $\sqrt{\text{volume}}$. Plot the residuals versus the fitted values. Does the linear model seem appropriate? Explain.

d. Using the most appropriate model, predict the percent ash for particles with a volume of 50 m³.

e. Using the most appropriate model, construct a 95% confidence interval for the mean percent ash for particles with a volume of 50 m³.

3. The article "Application of Genetic Algorithms to Optimum Design of Elasto-damping Elements of a Half-Car Model Under Random Road Excitations" (M. Mirzael and R. Hassannejad, *Proceedings of the Institution of Mechanical Engineers*, 2007:515–526) presents values of a coefficient (*y*), a unitless quantity that measures the road impact on an automobile suspension, and the time (*t*) for the car to travel a distance equal to the length between the front and rear axles. The results are as follows:

t	y	t	y	t	y
0.0	1.000	0.5	0.995	1.0	0.987
0.1	0.999	0.6	0.993	1.1	0.985
0.2	0.998	0.7	0.992	1.2	0.983
0.3	0.997	0.8	0.990	1.3	0.981
0.4	0.996	0.9	0.988	1.4	0.979

a. Compute the least-squares line for predicting road impact (*y*) from time (*t*). Plot the residuals versus the fitted values.

b. Compute the least-squares line for predicting ln *y* from *t*. Plot the residuals versus the fitted values.

c. Compute the least-squares line for predicting *y* from $t^{1.5}$. Plot the residuals versus the fitted values.

d. Which of the three models (a) through (c) fits best? Explain.

e. Using the best model, estimate the road impact for a time of 0.75.

4. Two radon detectors were placed in different locations in the basement of a home. Each provided an hourly measurement of the radon concentration, in units of pCi/L. The data are presented in the following table.

R_1	R_2	R_1	R_2	R_1	R_2	R_1	R_2
1.2	1.2	3.4	2.0	4.0	2.6	5.5	3.6
1.3	1.5	3.5	2.0	4.0	2.7	5.8	3.6
1.3	1.6	3.6	2.1	4.3	2.7	5.9	3.9
1.3	1.7	3.6	2.1	4.3	2.8	6.0	4.0
1.5	1.7	3.7	2.1	4.4	2.9	6.0	4.2
1.5	1.7	3.8	2.2	4.4	3.0	6.1	4.4
1.6	1.8	3.8	2.2	4.7	3.1	6.2	4.4
2.0	1.8	3.8	2.3	4.7	3.2	6.5	4.4
2.0	1.9	3.9	2.3	4.8	3.2	6.6	4.4
2.4	1.9	3.9	2.4	4.8	3.5	6.9	4.7
2.9	1.9	3.9	2.4	4.9	3.5	7.0	4.8
3.0	2.0	3.9	2.4	5.4	3.5		

a. Compute the least-squares line for predicting the radon concentration at location 2 from the concentration at location 1.

b. Plot the residuals versus the fitted values. Does the linear model seem appropriate?

c. Divide the data into two groups: points where $R_1 < 4$ in one group, points where $R_1 \geq 4$ in the other. Compute the least-squares line and the residual plot for each group. Does the line describe either group well? Which one?

d. Explain why it might be a good idea to fit a linear model to part of these data, and a nonlinear model to the other.

5. To determine the effect of temperature on the yield of a certain chemical process, the process is run 24 times at various temperatures. The temperature (in °C) and the yield (expressed as a percentage of a theoretical maximum) for each run are given in the following table. The results are presented in the order in which they were run, from earliest to latest.

Order	Temp	Yield	Order	Temp	Yield	Order	Temp	Yield
1	30	49.2	9	25	59.3	17	34	65.9
2	32	55.3	10	38	64.5	18	43	75.2
3	35	53.4	11	39	68.2	19	34	69.5
4	39	59.9	12	30	53.0	20	41	80.8
5	31	51.4	13	30	58.3	21	36	78.6
6	27	52.1	14	39	64.3	22	37	77.2
7	33	60.2	15	40	71.6	23	42	80.3
8	34	60.5	16	44	73.0	24	28	69.5

a. Compute the least-squares line for predicting yield (y) from temperature (x).

b. Plot the residuals versus the fitted values. Does the linear model seem appropriate? Explain.

c. Plot the residuals versus the order in which the observations were made. Is there a trend in the residuals over time? Does the linear model seem appropriate? Explain.

6. The article "Characteristics and Trends of River Discharge into Hudson, James, and Ungava Bays, 1964–2000" (S. Déry, M. Stieglitz, et al., *Journal of Climate*, 2005:2540–2557) presents measurements of discharge rate x (in km^3/yr) and peak flow y (in m^3/s) for 42 rivers that drain into the Hudson, James, and Ungava Bays. The data are shown in the following table:

Discharge	Peak Flow	Discharge	Peak Flow	Discharge	Peak Flow
94.24	4110.3	17.96	3420.2	3.98	551.8
66.57	4961.7	17.84	2655.3	3.74	288.9
59.79	10275.5	16.06	3470.3	3.25	295.2
48.52	6616.9	14.69	1561.6	3.15	500.1
40.00	7459.5	11.63	869.8	2.76	611.0
32.30	2784.4	11.19	936.8	2.64	1311.5
31.20	3266.7	11.08	1315.7	2.59	413.8
30.69	4368.7	10.92	1727.1	2.25	263.2
26.65	1328.5	9.94	768.1	2.23	490.7
22.75	4437.6	7.86	483.3	0.99	204.2
21.20	1983.0	6.92	334.5	0.84	491.7
20.57	1320.1	6.17	1049.9	0.64	74.2
19.77	1735.7	4.88	485.1	0.52	240.6
18.62	1944.1	4.49	289.6	0.30	56.6

a. Compute the least-squares line for predicting y from x. Make a plot of residuals versus fitted values.

b. Compute the least-squares line for predicting y from $\ln x$. Make a plot of residuals versus fitted values.

c. Compute the least-squares line for predicting $\ln y$ from $\ln x$. Make a plot of residuals versus fitted values.

d. Which of the three models (a) through (c) fits best? Explain.

e. Using the best model, predict the peak flow when the discharge is 50.0 km³/yr.

7. The article "Some Parameters of the Population Biology of Spotted Flounder (*Ciutharus linguatula* Linnaeus, 1758) in Edremit Bay (North Aegean Sea)" (D. Türker, B. Bayhan, et al., *Turkish Journal of Veterinary and Animal Science*, 2005:1013–1018) models the relationship between weight W and length L of spotted flounder as $W = aL^b$ where a and b are constants to be estimated from data. Transform this equation to produce a linear model.

8. The article "Mechanistic-Empirical Design of Bituminous Roads: An Indian Perspective" (A. Das and B. Pandey, *Journal of Transportation Engineering*, 1999:463–471) presents an equation of the form $y = a(1/x_1)^b(1/x_2)^c$ for predicting the number of repetitions for laboratory fatigue failure (y) in terms of the tensile strain at the bottom of the bituminous beam (x_1) and the resilient modulus (x_2). Transform this equation into a linear model, and express the linear model coefficients in terms of a, b, and c.

9. An engineer wants to determine the spring constant for a particular spring. She hangs various weights on one end of the spring and measures the length of the spring each time. A scatterplot of length (y) versus load (x) is depicted in the following figure.

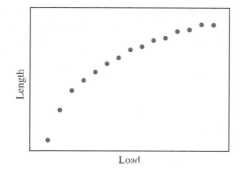

a. Is the model $y = \beta_0 + \beta_1 x$ an empirical model or a physical law?

b. Should she transform the variables to try to make the relationship more linear, or would it be better to redo the experiment? Explain.

8.3 Multiple Regression

The methods of simple linear regression apply when we wish to fit a linear model relating the value of a dependent variable y to the value of a single independent variable x. There are many situations, however, in which a single independent variable is not enough. For example, the degree of wear on a lubricated bearing in a machine may depend both on the load on the bearing and on the physical properties of the lubricant. An equation that expressed wear as a function of load alone or of lubricant properties alone would fail as a predictor. In situations like this, there are several independent variables, $x_1, x_2, \ldots, x_p$, that are related to a dependent variable y. If the relationship between the dependent and independent variables is linear, the technique of **multiple regression** can be used.

We describe the multiple regression model. Assume that we have a sample of n items, and that on each item we have measured a dependent variable y and p independent variables $x_1, \ldots, x_p$. The ith sample item thus gives rise to the ordered set $(y_i, x_{1i}, \ldots, x_{pi})$. We can then fit the **multiple regression model**

$$y_i = \beta_0 + \beta_1 x_{1i} + \cdots + \beta_p x_{pi} + \varepsilon_i \tag{8.17}$$

Several special cases of the multiple regression model (8.17) are often used in practice. One is the **polynomial regression model**, in which the independent variables are all powers of a single variable. The polynomial regression model of degree p is

$$y_i = \beta_0 + \beta_1 x_i + \beta_2 x_i^2 + \cdots + \beta_p x_i^p + \varepsilon_i \tag{8.18}$$

Multiple regression models can also be made up of powers of several variables. For example, a polynomial regression model of degree 2, also called a **quadratic model**, in two variables x_1 and x_2 is given by

$$y_i = \beta_0 + \beta_1 x_{1i} + \beta_2 x_{2i} + \beta_3 x_{1i} x_{2i} + \beta_4 x_{1i}^2 + \beta_5 x_{2i}^2 + \varepsilon_i \tag{8.19}$$

A variable that is the product of two other variables is called an **interaction**. In model (8.19), the variable $x_{1i} x_{2i}$ is the **interaction** between x_1 and x_2.

Models (8.18) and (8.19) are considered to be linear models, even though they contain nonlinear terms in the independent variables. The reason they are still linear models is that *they are linear in the coefficients β_i.*

Estimating the Coefficients

In any multiple regression model, the estimates $\hat{\beta}_0, \hat{\beta}_1, \ldots, \hat{\beta}_p$ are computed by least-squares, just as in simple linear regression. The equation

$$\hat{y} = \hat{\beta}_0 + \hat{\beta}_1 x_1 + \cdots + \hat{\beta}_p x_p \tag{8.20}$$

is called the **least-squares equation** or **fitted regression equation**. Now define $\hat{y}_i$ to be the y coordinate of the least-squares equation corresponding to the x values $(x_{1i}, \ldots, x_{pi})$. The residuals are the quantities $e_i = y_i - \hat{y}_i$, which are the differences between the observed y values and the y values given by the equation. We want to compute $\hat{\beta}_0, \hat{\beta}_1, \ldots, \hat{\beta}_p$ so as to minimize the sum of the squared residuals $\sum_{i=1}^{n} e_i^2$. To do this, we express e_i in terms of $\hat{\beta}_0, \hat{\beta}_1, \ldots, \hat{\beta}_p$:

$$e_i = y_i - \hat{\beta}_0 - \hat{\beta}_1 x_{1i} - \cdots - \hat{\beta}_p x_{pi} \tag{8.21}$$

Thus we wish to minimize the sum

$$\sum_{i=1}^{n} (y_i - \hat{\beta}_0 - \hat{\beta}_1 x_{1i} - \cdots - \hat{\beta}_p x_{pi})^2 \tag{8.22}$$

To do this, we can take partial derivatives of (8.22) with respect to $\hat{\beta}_0, \hat{\beta}_1, \ldots, \hat{\beta}_p$, set them equal to 0, and solve the resulting $p + 1$ equations in $p + 1$ unknowns. The expressions obtained for $\hat{\beta}_0, \hat{\beta}_1, \ldots, \hat{\beta}_p$ are complicated. Fortunately, they have been coded into many software packages, so that you can calculate them on the computer. For each estimated coefficient $\hat{\beta}_i$, there is an estimated standard deviation $s_{\hat{\beta}_i}$. Expressions for these quantities are complicated as well, so nowadays people rely on computers to calculate them.

Sums of Squares

Much of the analysis in multiple regression is based on three fundamental quantities. They are the **regression sum of squares** (SSR), the **error sum of squares** (SSE), and the **total sum of squares** (SST). We defined these quantities in Section 2.3, in our discussion of simple linear regression. The definitions hold for multiple regression as well. We repeat them here.

Definition

Sums of Squares

In the multiple regression model

$$y_i = \beta_0 + \beta_1 x_{1i} + \cdots + \beta_p x_{pi} + \varepsilon_i$$

the following sums of squares are defined:

- Regression sum of squares: $\text{SSR} = \sum_{i=1}^{n} (\hat{y}_i - \overline{y})^2$
- Error sum of squares: $\text{SSE} = \sum_{i=1}^{n} (y_i - \hat{y}_i)^2$
- Total sum of squares: $\text{SST} = \sum_{i=1}^{n} (y_i - \overline{y})^2$

It can be shown that

$$\text{SST} = \text{SSR} + \text{SSE} \tag{8.23}$$

Equation (8.23) is called the **analysis of variance identity**.

We will now see how these sums of squares are used to derive the statistics used in multiple regression. As we did for simple linear regression, we will restrict our discussion to the simplest case, in which four assumptions about the errors ε_i are satisfied. We repeat these assumptions here.

Assumptions for Errors in Linear Models

In the simplest situation, the following assumptions are satisfied:

1. The errors $\varepsilon_1, \ldots, \varepsilon_n$ are random and independent. In particular, the magnitude of any error ε_i does not influence the value of the next error ε_{i+1}.
2. The errors $\varepsilon_1, \ldots, \varepsilon_n$ all have mean 0.
3. The errors $\varepsilon_1, \ldots, \varepsilon_n$ all have the same variance, which we denote by σ^2.
4. The errors $\varepsilon_1, \ldots, \varepsilon_n$ are normally distributed.

Just as in simple linear regression, these assumptions imply that the observations y_i are independent random variables. To be specific, each y_i has a normal distribution with mean $\beta_0 + \beta_1 x_{1i} + \cdots + \beta_p x_{pi}$ and variance σ^2. Each coefficient β_i represents the change in the mean of y associated with an increase of one unit in the value of x_i, when the other x variables are held constant.

Summary

In the multiple regression model $y_i = \beta_0 + \beta_1 x_{1i} + \cdots + \beta_p x_{pi} + \varepsilon_i$, under assumptions 1 through 4, the observations $y_1, \ldots, y_n$ are independent random variables that follow the normal distribution. The mean and variance of y_i are given by

$$\mu_{y_i} = \beta_0 + \beta_1 x_{1i} + \cdots + \beta_p x_{pi}$$

$$\sigma_{y_i}^2 = \sigma^2$$

Each coefficient β_i represents the change in the mean of y associated with an increase of one unit in the value of x_i, when the other x variables are held constant.

The Statistics s^2, R^2, and F

The three statistics most often used in multiple regression are the estimated error variance s^2, the coefficient of determination R^2, and the F statistic. Each of these has an analog in simple linear regression. We discuss them in turn.

In simple linear regression, the estimated error variance is $\sum_{i=1}^{n}(y_i - \hat{y}_i)^2/(n-2)$. We divide by $n - 2$ rather than n because the residuals ($e_i = y_i - \hat{y}_i$) tend to be a little smaller than the errors ε_i. The reason that the residuals are a little smaller is that the two coefficients ($\hat{\beta}_0$ and $\hat{\beta}_1$) have been chosen to minimize $\sum_{i=1}^{n}(y_i - \hat{y}_i)^2$. Now in the case of multiple regression, we are estimating $p + 1$ coefficients rather than just two. Thus the residuals tend to be smaller still, so we must divide $\sum_{i=1}^{n}(y_i - \hat{y}_i)^2$ by a still smaller denominator. It turns out that the appropriate denominator is equal to the number of observations (n) minus the number of parameters in the model ($p + 1$). Therefore the estimated error variance is given by

$$s^2 = \frac{\sum_{i=1}^{n}(y_i - \hat{y}_i)^2}{n - p - 1} = \frac{\text{SSE}}{n - p - 1} \tag{8.24}$$

The estimated variance $s_{\hat{\beta}_i}^2$ of each least-squares coefficient $\hat{\beta}_i$ is computed by multiplying s^2 by a rather complicated function of the variables x_{ij}. In practice, the values of $s_{\hat{\beta}_i}^2$ are calculated on a computer. When assumptions 1 through 4 are satisfied, the quantity

$$\frac{\hat{\beta}_i - \beta_i}{s_{\hat{\beta}_i}}$$

has a Student's t distribution with $n - p - 1$ degrees of freedom. The number of degrees of freedom is equal to the denominator used to compute the estimated error variance s^2 (Equation 8.24). This statistic is used to compute confidence intervals and to perform hypothesis tests on the values β_i, just as in simple linear regression.

In simple linear regression, the coefficient of determination, r^2, measures the goodness of fit of the linear model. The goodness-of-fit statistic in multiple regression is a quantity denoted R^2, which is also called the coefficient of determination, or the proportion of variance explained by regression. The value of R^2 is calculated in the same way as is r^2 in simple linear regression (Equation 2.14, in Section 2.3). That is,

$$R^2 = \frac{\sum_{i=1}^n (y_i - \overline{y})^2 - \sum_{i=1}^n (y_i - \hat{y}_i)^2}{\sum_{i=1}^n (y_i - \overline{y})^2} = \frac{\text{SST} - \text{SSE}}{\text{SST}} = \frac{\text{SSR}}{\text{SST}} \qquad (8.25)$$

In simple linear regression, a test of the null hypothesis $\beta_1 = 0$ is almost always made. If this hypothesis is not rejected, then the linear model may not be useful. The analogous null hypothesis in multiple regression is $H_0: \beta_1 = \beta_2 = \cdots = \beta_p = 0$. This is a very strong hypothesis. It says that none of the independent variables has any linear relationship with the dependent variable. In practice, the data usually provide sufficient evidence to reject this hypothesis. The test statistic for this hypothesis is

$$F = \frac{\left[\sum_{i=1}^n (y_i - \overline{y})^2 - \sum_{i=1}^n (y_i - \hat{y}_i)^2\right]/p}{\left[\sum_{i=1}^n (y_i - \hat{y}_i)^2\right]/(n - p - 1)} = \frac{[\text{SST} - \text{SSE}]/p}{\text{SSE}/(n - p - 1)} = \frac{\text{SSR}/p}{\text{SSE}/(n - p - 1)}$$

$$(8.26)$$

This is an F statistic; its null distribution is $F_{p,n-p-1}$. Note that the denominator of the F statistic is s^2 (Equation 8.24). The subscripts p and $n - p - 1$ are the **degrees of freedom** for the F statistic.

Slightly different versions of the F statistic can be used to test weaker null hypotheses. In particular, given a model with independent variables $x_1, \ldots, x_p$, we sometimes want to test the null hypothesis that some of them (say $x_{k+1}, \ldots, x_p$) are not linearly related to the dependent variable. To do this, a version of the F statistic can be constructed that will test the null hypothesis $H_0 : \beta_{k+1} = \cdots = \beta_p = 0$. We will discuss this further in Section 8.4.

An Example

Let us now look at an example of multiple regression. We first describe the data. A mobile ad hoc computer network consists of several computers (nodes) that move within a network area. Often messages are sent from one node to another. When the receiving node is out of range, the message must be sent to a nearby node, which then forwards it from node to node along a routing path toward its destination. The routing path is determined by a routine known as a routing protocol. The effectiveness of a routing protocol is measured by the percentage of messages that are successfully delivered, which is called the goodput. Goodput is affected by various aspects of the movement of the nodes, in particular the average node speed and the length of time that the nodes pause at each destination. Table 8.2 (page 350) presents average node speed, average pause time, and goodput for 25 simulated mobile ad hoc networks. These data were generated for a study described in the article "Metrics to Enable Adaptive Protocols for Mobile Ad Hoc Networks" (J. Boleng, W. Navidi, and T. Camp, *Proceedings of the 2002 International Conference on Wireless Networks*, 2002:293–298).

To study the effectiveness of a routing protocol under a variety of conditions, a multiple regression model was designed to predict goodput from node speed and pause time. Specifically, the output on page 350 (from MINITAB) presents the results of fitting the model

$$\text{Goodput} = \beta_0 + \beta_1 \text{ Speed} + \beta_2 \text{ Pause} + \beta_3 \text{ Speed} \cdot \text{Pause}$$
$$+ \beta_4 \text{ Speed}^2 + \beta_5 \text{ Pause}^2 + \varepsilon$$

TABLE 8.2 Average node speed, pause time, and goodput for computer networks

Speed (m/s)	Pause Time (s)	Goodput (%)	Speed (m/s)	Pause Time (s)	Goodput (%)
5	10	95.111	20	40	87.800
5	20	94.577	20	50	89.941
5	30	94.734	30	10	62.963
5	40	94.317	30	20	76.126
5	50	94.644	30	30	84.855
10	10	90.800	30	40	87.694
10	20	90.183	30	50	90.556
10	30	91.341	40	10	55.298
10	40	91.321	40	20	78.262
10	50	92.104	40	30	84.624
20	10	72.422	40	40	87.078
20	20	82.089	40	50	90.101
20	30	84.937			

```
The regression equation is
Goodput = 96.0 − 1.82 Speed + 0.565 Pause + 0.0247 Speed*Pause
          + 0.0140 Speed^2 − 0.0118 Pause^2


Predictor        Coef       StDev          T        P
Constant       96.024       3.946      24.34    0.000
Speed         −1.8245      0.2376      −7.68    0.000
Pause          0.5652      0.2256       2.51    0.022
Speed*Pa     0.024731    0.003249       7.61    0.000
Speed^2      0.014020    0.004745       2.95    0.008
Pause^2     −0.011793    0.003516      −3.35    0.003


S = 2.942       R-Sq = 93.2%     R-Sq(adj) = 91.4%

Analysis of Variance

Source           DF          SS         MS        F        P
Regression        5     2240.49     448.10    51.77    0.000
Residual Error   19      164.46       8.66
Total            24     2404.95

Predicted Values for New Observations

New
Obs     Fit    SE Fit        95% CI               95% PI
  1  74.272     1.175   (71.812, 76.732)   (67.641, 80.903)

Values of Predictors for New Observations

New
Obs   Speed  Pause   Speed*Pause   Speed^2   Pause^2
  1    25.0   15.0           375       625       225
```

Much of the output is analogous to that of simple linear regression. The fitted regression equation is presented near the top of the output. Below that, the coefficient estimates $\hat{\beta}_i$ and their estimated standard deviations $s_{\hat{\beta}_i}$ are shown. Next to each standard deviation is the Student's t statistic for testing the null hypothesis that the true value of the coefficient is equal to 0. This statistic is equal to the quotient of the coefficient estimate and its standard deviation. Since there are $n = 25$ observations and $p = 5$ independent variables, the number of degrees of freedom for the Student's t statistic is $25 - 5 - 1 = 19$. The P-values for the tests are given in the next column. All the P-values are small, so it would be reasonable to conclude that each of the independent variables in the model is useful in predicting the goodput.

The quantity "S" is s, the estimated error standard deviation, and "R-sq" is the coefficient of determination R^2. The adjusted R^2, "R-sq(adj)," is primarily used in model selection. We will discuss this statistic in Section 8.4.

The analysis of variance table is analogous to the one found in simple linear regression. We'll go through it column by column. In the degrees of freedom column "DF," the degrees of freedom for regression is equal to the number of independent variables (5). Note that Speed^2, Pause^2, and $\text{Speed} \cdot \text{Pause}$ each count as separate independent variables, even though they can be computed from Speed and Pause. In the next row down, labeled "Residual Error," the number of degrees of freedom is 19, which represents the number of observations (25) minus the number of parameters estimated (6: the intercept, and coefficients for the five independent variables). Finally, the "Total" degrees of freedom is one less than the sample size of 25. Note that the total degrees of freedom is the sum of the degrees of freedom for regression and the degrees of freedom for error. Going down the column "SS," we find the regression sum of squares SSR, the error sum of squares SSE, and the total sum of squares SST. Note that $\text{SST} = \text{SSR} + \text{SSE}$. The column "MS" presents the **mean squares**, which are the sums of squares divided by their respective degrees of freedom. Note that the mean square for error is equal to s^2, the estimate for the error variance: ($s^2 = S^2 = 2.942^2 = 8.66$). The column labeled "F" presents the mean square for regression divided by the mean square for error ($448.10/8.66 = 51.77$, allowing for roundoff error). This is the F statistic shown in Equation (8.26), and it is used to test the null hypothesis that none of the independent variables are linearly related to the dependent variable. The P-value for this test is approximately 0.

The output under the heading "Predicted Values for New Observations" presents confidence intervals on the mean response and predicted intervals for values of the dependent variables specified by the user. The values of the dependent variables that have been specified are listed under the heading "Values of Predictors for New Observations." The values of the independent variables in this output are Speed $= 25$ and Pause $= 15$. The quantity 74.242, labeled "Fit," is the value of $\hat{y}$ obtained by substituting these values into the fitted regression equation. The quantity labeled "SE Fit" is the estimated standard deviation of $\hat{y}$, which is used to compute the 95% confidence interval, labeled "95% CI." The quantity labeled "95% PI" is the 95% prediction interval for a future observation of the dependent variable when the independent variables are set to the given values. Like the confidence interval, this interval is centered at $\hat{y}$, but it is wider, just as in simple linear regression.

Example 8.14

Use the multiple regression model to predict the goodput for a network with speed 12 m/s and pause time 25 s.

Solution
From the MINITAB output, the fitted model is

$$\text{Goodput} = 96.0 - 1.82\,\text{Speed} + 0.565\,\text{Pause} + 0.0247\,\text{Speed} \cdot \text{Pause}$$
$$+ 0.0140\,\text{Speed}^2 - 0.0118\,\text{Pause}^2$$

Substituting 12 for Speed and 25 for Pause, we find that the predicted goodput is 90.336.

Example 8.15

For the goodput data, find the residual for the point Speed = 20, Pause = 30.

Solution
The observed value of goodput (Table 8.2) is $y = 84.937$. The predicted value $\hat{y}$ is found by substituting Speed = 20 and Pause = 30 into the fitted model presented in the solution to Example 8.14. This yields a predicted value for goodput of $\hat{y} = 86.350$. The residual is given by $y - \hat{y} = 84.937 - 86.350 = -1.413$.

It is straightforward to compute confidence intervals and to test hypotheses regarding the least-squares coefficients, by using the computer output. Examples 8.16 through 8.18 provide illustrations.

Example 8.16

Find a 95% confidence interval for the coefficient of Speed in the multiple regression model.

Solution
From the output, the estimated coefficient is -1.8245, with a standard deviation of 0.2376. To find a confidence interval, we use the Student's t distribution with 19 degrees of freedom. The degrees of freedom for the t statistic is equal to the degrees of freedom for error. The t value for a 95% confidence interval is $t_{19,\,.025} = 2.093$. The 95% confidence interval is

$$-1.8245 \pm (2.093)(0.2376) = -1.8245 \pm 0.4973 = (-2.3218,\ -1.3272)$$

Example 8.17

Test the null hypothesis that the coefficient of Pause is less than or equal to 0.3.

Solution
The estimated coefficient of Pause is $\hat{\beta}_2 = 0.5652$, with standard deviation $s_{\hat{\beta}_2} = 0.2256$. The null hypothesis is $\beta_2 \leq 0.3$. Under H_0, we take $\beta_2 = 0.3$, so the quantity

$$t = \frac{\hat{\beta}_2 - 0.3}{0.2256}$$

has a Student's t distribution with 19 degrees of freedom. Note that the degrees of freedom for the t statistic is equal to the degrees of freedom for error. The value of the t statistic is $(0.5652 - 0.3)/0.2256 = 1.1755$. The P-value is between 0.10 and 0.25. It is plausible that $\beta_2 \leq 0.3$.

Find a 95% confidence interval for the mean response μ_{y_i}, and a 95% prediction interval for a future observation when Speed = 25 and Pause = 15.

Solution
From the output, under the heading "Predicted Values for New Observations," the 95% confidence interval is $(71.812, 76.732)$ and the 95% prediction interval is $(67.641, 80.903)$.

Checking Assumptions in Multiple Regression

In multiple regression, as in simple linear regression, it is important to test the validity of the assumptions for errors in linear models (presented at the beginning of this section). The diagnostics for these assumptions used in the case of simple linear regression can be used in multiple regression as well. These are plots of residuals versus fitted values, normal probability plots of residuals, and plots of residuals versus the order in which the observations were made. It is also a good idea to make plots of the residuals versus each of the independent variables. If the residual plots indicate a violation of assumptions, transformations of the variables may be tried to cure the problem, as in simple linear regression.

Figure 8.7 presents a plot of the residuals versus the fitted values for the goodput data. Figure 8.8 (page 354) and Figure 8.9 (page 354) present plots of the residuals versus speed and pause, respectively. The plot of residuals versus fitted values gives some

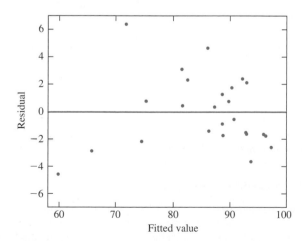

FIGURE 8.7 Plot of residuals versus fitted values for the Goodput data.

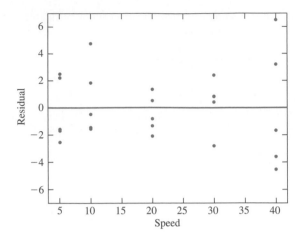

FIGURE 8.8 Plot of residuals versus Speed for the Goodput data.

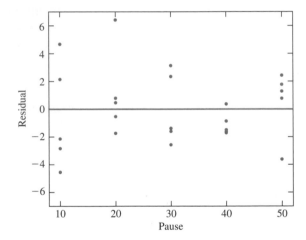

FIGURE 8.9 Plot of residuals versus Pause for the Goodput data.

impression of curvature, which is caused primarily by a few points at either end. The plots of residuals versus independent variables do not indicate any serious violations of the model assumptions. In practice, one might accept this model as fitting well enough, or one might use model selection techniques (discussed in Section 8.4) to explore alternative models.

Exercises for Section 8.3

1. In an experiment to determine the factors affecting tensile strength in steel plates, the tensile strength (in kg/mm^2), the manganese content (in parts per thousand), and the thickness (in mm) were measured for a sample of 20 plates. The following MINITAB output presents the results of fitting the model Tensile strength $= \beta_0 + \beta_1$ Manganese $+ \beta_2$ Thickness.

```
The regression equation is
Strength = 26.641 + 3.3201 Manganese - 0.4249 Thickness

Predictor        Coef        StDev           T         P
Constant         26.641      2.72340         9.78      0.000
Manganese        3.3201      0.33198         10.00     0.000
Thickness        -0.4249     0.12606         -3.37     0.004

S = 0.8228       R-Sq = 86.2%       R-Sq(adj) = 84.6%

Analysis of Variance

Source           DF          SS              MS          F         P
Regression       2           72.01           36.005      53.19     0.000
Residual Error   17          11.508          0.6769
Total            19          83.517
```

a. Predict the strength for a specimen that is 10 mm thick and contains 8.2 ppt manganese.

b. If two specimens have the same thickness, and one contains 10 ppt more manganese, by how much would you predict their strengths to differ?

c. If two specimens have the same proportion of manganese, and one is 5 mm thicker than the other, by how much would you predict their strengths to differ?

2. Refer to Exercise 1.

   a. Find a 95% confidence interval for the coefficient of Manganese.
   b. Find a 99% confidence interval for the coefficient of Thickness.
   c. Can you conclude that $\beta_1 > 3$? Perform the appropriate hypothesis test.
   d. Can you conclude that $\beta_2 < -0.1$? Perform the appropriate hypothesis test.

3. The data used to fit the model in Exercise 1 are presented in the following table, along with the residuals and the fitted values. Plot the residuals versus the fitted values. Does the plot indicate that the linear model is reasonable? Explain.

Strength	Manganese	Thickness	Residual	Fitted Value
47.7	7.4	8.0	−0.111	47.811
50.9	8.8	10.0	−0.709	51.609
51.7	8.8	10.0	0.091	51.609
51.9	8.8	10.0	0.291	51.609
50.0	8.1	7.1	−0.517	50.517
50.5	8.1	9.0	0.790	49.710
50.0	8.1	7.1	−0.517	50.517
49.7	8.1	9.0	−0.010	49.710
50.6	8.1	9.0	0.890	49.710
47.7	7.2	7.2	0.214	47.486
47.1	7.3	7.8	−0.464	47.564
45.0	7.3	11.8	−0.864	45.864

(Continued)

Strength	Manganese	Thickness	Residual	Fitted Value
47.6	7.3	8.0	0.121	47.479
45.7	7.3	11.8	−0.164	45.864
47.0	7.3	8.7	−0.181	47.181
45.7	7.3	11.7	−0.206	45.906
48.8	7.3	8.7	1.619	47.181
45.8	7.3	7.8	−1.764	47.564
48.5	7.3	9.0	1.446	47.054
48.6	7.6	7.8	0.040	48.560

4. The article "Application of Analysis of Variance to Wet Clutch Engagement" (M. Mansouri, M. Khonsari, et al., *Proceedings of the Institution of Mechanical Engineers*, 2002:117–125) presents the following fitted model for predicting clutch engagement time in seconds (y) from engagement starting speed in m/s (x_1), maximum drive torque in N $\cdot$ m (x_2), system inertia in kg $\cdot$ m^2 (x_3), and applied force rate in kN/s (x_4):

$$y = -0.83 + 0.017x_1 + 0.0895x_2 + 42.771x_3 + 0.027x_4 - 0.0043x_2x_4$$

   The sum of squares for regression was SSR $= 1.08613$ and the sum of squares for error was SSE $= 0.036310$. There were 44 degrees of freedom for error.

   a. Predict the clutch engagement time when the starting speed is 20 m/s, the maximum drive torque is 17 N $\cdot$ m, the system inertia is 0.006 kg $\cdot$ m^2, and the applied force rate is 10 kN/s.
   b. Is it possible to predict the change in engagement time associated with an increase of 2 m/s in starting speed? If so, find the predicted change. If not, explain why not.
   c. Is it possible to predict the change in engagement time associated with an increase of 2 N $\cdot$ m in maximum drive torque? If so, find the predicted change. If not, explain why not.
   d. Compute the coefficient of determination R^2.
   e. Compute the F statistic for testing the null hypothesis that all the coefficients are equal to 0. Can this hypothesis be rejected?

5. In the article "Application of Statistical Design in the Leaching Study of Low-Grade Manganese Ore Using Aqueous Sulfur Dioxide" (P. Naik, L. Sukla, and S. Das, *Separation Science and Technology*, 2002:1375–1389), a fitted model for predicting the extraction of manganese in % (y) from particle size in mm (x_1), the amount of sulfur dioxide in multiples of the stoichiometric quantity needed for the dissolution of manganese (x_2), and the duration of leaching in minutes (x_3) is given as

$$y = 56.145 - 9.046x_1 - 33.421x_2 + 0.243x_3 - 0.5963x_1x_2 - 0.0394x_1x_3 + 0.6022x_2x_3$$
$$+ 0.6901x_1^2 + 11.7244x_2^2 - 0.0097x_3^2$$

There were a total of $n = 27$ observations, with SSE $= 209.55$ and SST $= 6777.5$.

   a. Predict the extraction percent when the particle size is 3 mm, the amount of sulfur dioxide is 1.5, and the duration of leaching is 20 minutes.
   b. Is it possible to predict the change in extraction percent when the duration of leaching increases by one minute? If so, find the predicted change. If not, explain why not.
   c. Compute the coefficient of determination R^2.
   d. Compute the F statistic for testing the null hypothesis that all the coefficients are equal to 0. Can this hypothesis be rejected?

6. The article "Earthmoving Productivity Estimation Using Linear Regression Techniques" (S. Smith, *Journal of Construction Engineering and Management*, 1999:133–141) presents the following linear model to predict earthmoving productivity (in m^3 moved per hour):

$$\text{Productivity} = -297.877 + 84.787x_1 + 36.806x_2 + 151.680x_3 - 0.081x_4 - 110.517x_5$$
$$- 0.267x_6 - 0.016x_1x_4 + 0.107x_4x_5 + 0.0009448x_4x_6 - 0.244x_5x_6$$

where x_1 = number of trucks,
x_2 = number of buckets per load,
x_3 = bucket volume, in m^3
x_4 = haul length, in m
x_5 = match factor (ratio of hauling capacity to loading capacity),
x_6 = truck travel time, in s

a. If the bucket volume increases by 1 m^3, while other independent variables are unchanged, can you determine the change in the predicted productivity? If so, determine it. If not, state what other information you would need to determine it.

b. If the haul length increases by 1 m, can you determine the change in the predicted productivity? If so, determine it. If not, state what other information you would need to determine it.

7. In a study of the lung function of children, the volume of air exhaled under force in one second is called FEV_1. (FEV_1 stands for forced expiratory volume in one second.) Measurements were made on a group of children each year for two years. A linear model was fit to predict this year's FEV_1 as a function of last year's FEV_1 (in liters), the child's gender (0 = Male, 1 = Female), the child's height (in m), and the ambient atmospheric pressure (in mm). The following MINITAB output presents the results of fitting the model

$$FEV_1 = \beta_0 + \beta_1 \text{ Last } FEV_1 + \beta_2 \text{ Gender} + \beta_3 \text{ Height} + \beta_4 \text{ Pressure} + \varepsilon$$

```
The regression equation is
FEV1 = −0.219 + 0.779 Last FEV − 0.108 Gender + 1.354 Height − 0.00134 Pressure

Predictor         Coef        StDev            T          P
Constant       −0.21947       0.4503        −0.49      0.627
Last FEV         0.779        0.04909       15.87      0.000
Gender        −0.10827        0.0352        −3.08      0.002
Height          1.3536        0.2880         4.70      0.000
Pressure     −0.0013431       0.0004722     −2.84      0.005

S = 0.22039        R-Sq = 93.5%       R-Sq(adj) = 93.3%

Analysis of Variance

Source           DF           SS           MS          F          P
Regression        4        111.31        27.826     572.89      0.000
Residual Error  160          7.7716       0.048572
Total           164        119.08
```

a. Predict the FEV_1 for a boy who is 1.4 m tall, if the measurement was taken at a pressure of 730 mm and last year's measurement was 2.113 L.

b. If two girls differ in height by 5 cm, by how much would you expect their FEV_1 measurements to differ, other things being equal?

c. The constant term β_0 is estimated to be negative. But FEV_1 must always be positive. Is something wrong? Explain.

8. Refer to Exercise 7.

a. Find a 95% confidence interval for the coefficient of Last FEV.

b. Find a 98% confidence interval for the coefficient of Height.

c. Can you conclude that $\beta_2 < -0.08$? Perform the appropriate hypothesis test.

d. Can you conclude that $\beta_3 > 0.5$? Perform the appropriate hypothesis test.

9. The article "Drying of Pulps in Sprouted Bed: Effect of Composition on Dryer Performance" (M. Medeiros, S. Rocha, et al., *Drying Technology*, 2002:865–881) presents measurements of pH, viscosity (in kg/m·s), density (in g/cm^3), and BRIX (in %). The following MINITAB output presents the results of fitting the model

$$pH = \beta_0 + \beta_1 \text{ Viscosity} + \beta_2 \text{ Density} + \beta_3 \text{ BRIX} + \varepsilon$$

```
The regression equation is
pH = -1.79 + 0.000266 Viscosity + 9.82 Density - 0.300 BRIX

Predictor          Coef      SE Coef        T        P
Constant        -1.7914       6.2339    -0.29    0.778
Viscosity     0.00026626   0.00011517     2.31    0.034
Density          9.8184       5.7173     1.72    0.105
BRIX            -0.29982     0.099039    -3.03    0.008

S = 0.379578    R-Sq = 50.0%    R-Sq(adj) = 40.6%

Predicted Values for New Observations

New
Obs      Fit   SE Fit        95% CI              95% PI
  1   3.0875   0.1351   (2.8010, 3.3740)   (2.2333, 3.9416)
  2   3.7351   0.1483   (3.4207, 4.0496)   (2.8712, 4.5990)
  3   2.8576   0.2510   (2.3255, 3.3896)   (1.8929, 3.8222)

Values of Predictors for New Observations

New
Obs  Viscosity  Density  BRIX
  1       1000     1.05  19.0
  2       1200     1.08  18.0
  3       2000     1.03  20.0
```

a. Predict the pH for a pulp with a viscosity of 1500 kg/m · s, a density of 1.04 g/cm^3, and a BRIX of 17.5%.

b. If two pulps differ in density by 0.01 g/cm^3, by how much would you expect them to differ in pH, other things being equal?

c. The constant term β_0 is estimated to be negative. But pulp pH must always be positive. Is something wrong? Explain.

d. Find a 95% confidence interval for the mean pH of pulps with viscosity 1200 kg/m · s, density 1.08 g/cm^3, and BRIX 18.0%.

e. Find a 95% prediction interval for the pH of a pulp with viscosity 1000 kg/m · s, density 1.05 g/cm^3, and BRIX 19.0%.

f. Pulp A has viscosity 2000, density 1.03, and BRIX 20.0. Pulp B has viscosity 1000, density 1.05, and BRIX 19.0. Which pulp will have its pH predicted with greater precision? Explain.

10. A scientist has measured quantities y, x_1, and x_2. She believes that y is related to x_1 and x_2 through the equation $y = \alpha e^{\beta_1 x_1 + \beta_2 x_2} \delta$, where δ is a random error that is always positive. Find a transformation of the data that will enable her to use a linear model to estimate β_1 and β_2.

11. The following MINITAB output is for a multiple regression. Something went wrong with the printer, so some of the numbers are missing. Fill them in.

Predictor	Coef	StDev	T	P
Constant	-0.58762	0.2873	(a)	0.086
X1	1.5102	(b)	4.30	0.005
X2	(c)	0.3944	-0.62	0.560
X3	1.8233	0.3867	(d)	0.003

S = 0.869   R-Sq = 90.2%   R-Sq(adj) = 85.3%

Analysis of Variance

Source	DF	SS	MS	F	P
Regression	3	41.76	(e)	(f)	0.000
Residual Error	6	(g)	0.76		
Total	(h)	46.30			

12. The following MINITAB output is for a multiple regression. Some of the numbers got smudged and are illegible. Fill in the missing numbers.

Predictor	Coef	StDev	T	P
Constant	(a)	1.4553	5.91	0.000
X1	1.2127	(b)	1.71	0.118
X2	7.8369	3.2109	(c)	0.035
X3	(d)	0.8943	-3.56	0.005

S = 0.82936   R-Sq = 78.0%   R-Sq(adj) =71.4

Source	DF	SS	MS	F	P
Regression	(e)	(f)	8.1292	11.818	0.001
Residual Error	10	6.8784	(g)		
Total	13	(h)			

13. The article "Evaluating Vent Manifold Inerting Requirements: Flash Point Modeling for Organic Acid-Water Mixtures" (R. Garland and M. Malcolm, *Process Safety Progress*, 2002:254–260) presents a model to predict the flash

point (in °F) of a mixture of water, acetic acid, propionic acid, and butyric acid from the concentrations (in weight %) of the three acids. The results are as follows. The variable "Butyric Acid $*$ Acetic Acid" is the interaction between butyric acid concentration and acetic acid concentration.

Predictor	Coef	StDev	T	P
Constant	267.53	11.306	23.66	0.000
Acetic Acid	−1.5926	0.1295	−12.30	0.000
Propionic Acid	−1.3897	0.1260	−11.03	0.000
Butyric Acid	−1.0934	0.1164	−9.39	0.000
Butyric Acid * Acetic Acid	−0.002658	0.001145	−2.32	0.034

a. Predict the flash point for a mixture that is 30% acetic acid, 35% propionic acid, and 30% butyric acid. (*Note*: In the model, 30% is represented by 30, not by 0.30.)

b. Someone asks by how much the predicted flash point will change if the concentration of acetic acid is increased by 10% while the other concentrations are kept constant. Is it possible to answer this question? If so, answer it. If not, explain why not.

c. Someone asks by how much the predicted flash point will change if the concentration of propionic acid is increased by 10% while the other concentrations are kept constant. Is it possible to answer this question? If so, answer it. If not, explain why not.

14. In the article "Low-Temperature Heat Capacity and Thermodynamic Properties of 1,1,1-trifluoro-2, 2-dichloroethane" (R. Varushchenko and A. Druzhinina, *Fluid Phase Equilibria*, 2002:109–119), the relationship between vapor pressure (p) and heat capacity (t) is given as $p = t^{\beta_3} \cdot e^{\beta_0 + \beta_1 t + \beta_2/t} \delta$, where δ is a random error that is always positive. Express this relationship as a linear model by using an appropriate transformation.

15. The following data were collected in an experiment to study the relationship between extrusion pressure (in KPa) and wear (in mg).

x	150	175	200	225	250	275
y	10.4	12.4	14.9	15.0	13.9	11.9

The least-squares quadratic model is $y = -32.445714 + 0.43154286x - 0.000982857x^2$.

a. Using this equation, compute the residuals.

b. Compute the error sum of squares SSE and the total sum of squares SST.

c. Compute the error variance estimate s^2.

d. Compute the coefficient of determination R^2.

e. Compute the value of the F statistic for the hypothesis $H_0: \beta_1 = \beta_2 = 0$. How many degrees of freedom does this statistic have?

f. Can the hypothesis $H_0: \beta_1 = \beta_2 = 0$ be rejected at the 5% level? Explain.

16. The following data were collected in an experiment to study the relationship between the speed of a cutting tool in m/s (x) and the lifetime of the tool in hours (y).

x	1	1.5	2	2.5	3
y	99	96	88	76	66

The least-squares quadratic model is $y = 101.4000 + 3.371429x - 5.142857x^2$.

a. Using this equation, compute the residuals.

b. Compute the error sum of squares SSE and the total sum of squares SST.
c. Compute the error variance estimate s^2.
d. Compute the coefficient of determination R^2.
e. Compute the value of the F statistic for the hypothesis $H_0 : \beta_1 = \beta_2 = 0$. How many degrees of freedom does this statistic have?
f. Can the hypothesis $H_0 : \beta_1 = \beta_2 = 0$ be rejected at the 5% level? Explain.

17. The November 24, 2001, issue of *The Economist* published economic data for 15 industrialized nations. Included were the percent changes in gross domestic product (GDP), industrial production (IP), consumer prices (CP), and producer prices (PP) from fall 2000 to fall 2001, and the unemployment rate in fall 2001 (UNEMP). An economist wants to construct a model to predict GDP from the other variables. A fit of the model

$$GDP = \beta_0 + \beta_1 IP + \beta_2 UNEMP + \beta_3 CP + \beta_4 PP + \varepsilon$$

yields the following output:

```
The regression equation is
GDP = 1.19 + 0.17 IP + 0.18 UNEMP + 0.18 CP - 0.18 PP

Predictor        Coef         StDev          T          P
Constant       1.18957      0.42180        2.82       0.018
IP             0.17326      0.041962       4.13       0.002
UNEMP          0.17918      0.045895       3.90       0.003
CP             0.17591      0.11365        1.55       0.153
PP            -0.18393      0.068808      -2.67       0.023
```

a. Predict the percent change in GDP for a country with IP = 0.5, UNEMP = 5.7, CP = 3.0, and PP = 4.1.
b. If two countries differ in unemployment rate by 1%, by how much would you predict their percent changes in GDP to differ, other things being equal?
c. CP and PP are both measures of the inflation rate. Which one is more useful in predicting GDP? Explain.
d. The producer price index for Sweden in September 2000 was 4.0, and for Austria it was 6.0. Other things being equal, for which country would you expect the percent change in GDP to be larger? Explain.

18. The article "Multiple Linear Regression for Lake Ice and Lake Temperature Characteristics" (S. Gao and H. Stefan, *Journal of Cold Regions Engineering*, 1999:59–77) presents data on maximum ice thickness in mm (y), average number of days per year of ice cover (x_1), average number of days the bottom temperature is lower than 8°C (x_2), and the average snow depth in mm (x_3) for 13 lakes in Minnesota. The data are presented in the following table.

y	x_1	x_2	x_3	y	x_1	x_2	x_3
730	152	198	91	730	157	204	90
760	173	201	81	650	136	172	47
850	166	202	69	850	142	218	59
840	161	202	72	740	151	207	88
720	152	198	91	720	145	209	60
730	153	205	91	710	147	190	63
840	166	204	70				

a. Fit the model $y = \beta_0 + \beta_1 x_1 + \beta_2 x_2 + \beta_3 x_3 + \varepsilon$. For each coefficient, find the P-value for testing the null hypothesis that the coefficient is equal to 0.

b. If two lakes differ by 2 in the average number of days per year of ice cover, with other variables being equal, by how much would you expect their maximum ice thicknesses to differ?

c. Do lakes with greater average snow depth tend to have greater or lesser maximum ice thickness? Explain.

19. In an experiment to estimate the acceleration of an object down an inclined plane, the object is released and its distance in meters (y) from the top of the plane is measured every 0.1 second from time $t = 0.1$ to $t = 1.0$. The data are presented in the following table.

t	y
0.1	0.03
0.2	0.1
0.3	0.27
0.4	0.47
0.5	0.73
0.6	1.07
0.7	1.46
0.8	1.89
0.9	2.39
1.0	2.95

The data follow the quadratic model $y = \beta_0 + \beta_1 t + \beta_2 t^2 + \varepsilon$, where β_0 represents the initial position of the object, β_1 represents the initial velocity of the object, and $\beta_2 = a/2$, where a is the acceleration of the object, assumed to be constant. In a perfect experiment, both the position and velocity of the object would be zero at time 0. However, due to experimental error, it is possible that the position and velocity at $t = 0$ are nonzero.

a. Fit the quadratic model $y = \beta_0 + \beta_1 t + \beta_2 t^2 + \varepsilon$.

b. Find a 95% confidence interval for β_2.

c. Find a 95% confidence interval for the acceleration a.

d. Compute the P-value for each coefficient.

e. Can you conclude that the initial position was not zero? Explain.

f. Can you conclude that the initial velocity was not zero? Explain.

8.4 Model Selection

There are many situations in which a large number of independent variables have been measured, and we need to decide which of them to include in a model. This is the problem of **model selection**, and it is a challenging one. In practice, model selection often proceeds by ad hoc methods, guided by whatever physical intuition may be available. We will not attempt a complete discussion of this extensive and difficult topic. Instead, we will be content to state some basic principles and to present some examples. An advanced reference such as Miller (2002) can be consulted for information on specific methods.

Good model selection rests on a basic principle known as Occam's razor. This principle is stated as follows:

> **Occam's Razor**
> The best scientific model is the simplest model that explains the observed facts.

In terms of linear models, Occam's razor implies the **principle of parsimony**:

> **The Principle of Parsimony**
> A model should contain the smallest number of variables necessary to fit the data.

There are some exceptions to the principle of parsimony:

> 1. A linear model should always contain an intercept, unless physical theory dictates otherwise.
> 2. If a power x^n of a variable is included in a model, all lower powers x, $x^2, \ldots, x^{n-1}$ should be included as well, unless physical theory dictates otherwise.
> 3. If a product $x_i x_j$ of two variables is included in a model, then the variables x_i and x_j should be included separately as well, unless physical theory dictates otherwise.

Models that contain only the variables that are needed to fit the data are called **parsimonious** models. Much of the practical work of multiple regression involves the development of parsimonious models.

We illustrate the principle of parsimony with the following example. The data in Table 8.3 were taken from the article "Capacities and Performance Characteristics of Jaw Crushers" (S. Sastri, *Minerals and Metallurgical Processing*, 1994:80–86). Feed rates and amounts of power drawn were measured for several industrial jaw crushers.

TABLE 8.3 Feed rates and power for industrial jaw crushers

Feed Rate (100 tons/h)	Power (kW)	Feed Rate (100 tons/h)	Power (kW)	Feed Rate (100 tons/h)	Power (kW)	Feed Rate (100 tons/h)	Power (kW)
0.10	11	0.20	15	0.91	45	1.36	58
1.55	60	2.91	84	0.59	12	2.36	45
3.00	40	0.36	30	0.27	24	2.95	75
3.64	150	0.14	16	0.55	49	1.09	44
0.38	69	0.91	30	0.68	45	0.91	58
1.59	77	4.27	150	4.27	150	2.91	149
4.73	83	4.36	144	3.64	100		

The following output (from MINITAB) presents the results for fitting the model

$$\text{Power} = \beta_0 + \beta_1 \text{FeedRate} + \varepsilon \qquad (8.27)$$

```
The regression equation is
Power = 21.0 + 24.6 FeedRate

Predictor        Coef        StDev           T          P
Constant       21.028        8.038        2.62      0.015
FeedRate       24.595        3.338        7.37      0.000

S = 26.20       R-Sq = 68.5%      R-Sq(adj) = 67.2%
```

From the output, we see that the fitted model is

$$\text{Power} = 21.028 + 24.595 \text{FeedRate} \qquad (8.28)$$

and that the coefficient for FeedRate is significantly different from 0 ($t = 7.37$, $P \approx 0$). We wonder whether a quadratic model might fit better than this linear one. So we fit

$$\text{Power} = \beta_0 + \beta_1 \text{FeedRate} + \beta_2 \text{FeedRate}^2 + \varepsilon \qquad (8.29)$$

The results are presented in the following output (from MINITAB). Note that the values for the intercept and for the coefficient of FeedRate are different than they were in the linear model. This is typical. Adding a new variable to a model can substantially change the coefficients of the variables already in the model.

```
The regression equation is
Power = 19.3 + 27.5 FeedRate − 0.64 FeedRate^2

Predictor        Coef        StDev           T          P
Constant        19.34       11.56        1.67      0.107
FeedRate        27.47       14.31        1.92      0.067
FeedRate^2    −0.6387        3.090       −0.21      0.838

S = 26.72       R-Sq = 68.5%      R-Sq(adj) = 65.9%
```

The most important point to notice is that the P-value for the coefficient of FeedRate^2 is large (0.838). Recall that this P-value is for the test of the null hypothesis that the coefficient is equal to 0. Thus the data provide no evidence that the coefficient of FeedRate^2 is different from 0. Note also that including FeedRate^2 in the model increases the value of the goodness-of-fit statistic R^2 only slightly, in fact so slightly that the first three digits are unchanged. It follows that there is no evidence that the quadratic model fits the data better than the linear model, so by the principle of parsimony, we should prefer the linear model.

Figure 8.10 provides a graphical illustration of the principle of parsimony. The scatterplot of power versus feed rate is presented, and both the least-squares line (8.28)

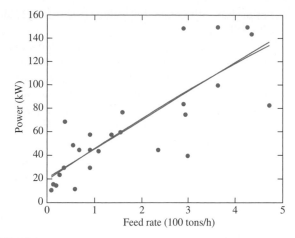

FIGURE 8.10 Scatterplot of power versus feed rate for 27 industrial jaw crushers. The least-squares line and best-fitting quadratic model are both superimposed. The two curves are practically identical, which reflects the fact that the coefficient of FeedRate^2 in the quadratic model does not differ significantly from 0.

and the fitted quadratic model (8.29) are superimposed. Even though the coefficients of the models are different, we can see that the two curves are almost identical. There is no reason to include the quadratic term in the model. It makes the model more complicated, without improving the fit.

Determining Whether Variables Can Be Dropped from a Model

It often happens that one has formed a model that contains a large number of independent variables, and one wishes to determine whether a given subset of them may be dropped from the model without significantly reducing the accuracy of the model. To be specific, assume that we know that the model

$$y_i = \beta_0 + \beta_1 x_{1i} + \cdots + \beta_k x_{ki} + \beta_{k+1} x_{k+1 i} + \cdots + \beta_p x_{pi} + \varepsilon_i \qquad (8.30)$$

is correct, in that it represents the true relationship between the x variables and y. We will call this model the "full" model.

We wish to test the null hypothesis

$$H_0 : \beta_{k+1} = \cdots = \beta_p = 0$$

If H_0 is true, the model will remain correct if we drop the variables $x_{k+1}, \ldots, x_p$, so we can replace the full model with the following reduced model:

$$y_i = \beta_0 + \beta_1 x_{1i} + \cdots + \beta_k x_{ki} + \varepsilon_i \qquad (8.31)$$

To develop a test statistic for H_0, we begin by computing the error sum of squares for both the full and the reduced models. We'll call them SSE_{full} and $\text{SSE}_{\text{reduced}}$. The number of degrees of freedom for SSE_{full} is $n - p - 1$, and the number of degrees of freedom for $SSE_{reduced}$ is $n - k - 1$.

Now since the full model is correct, we know that the quantity $SSE_{full}/(n - p - 1)$ is an estimate of the error variance σ^2; in fact it is just s^2. If H_0 is true, then the reduced model is also correct, so the quantity $SSE_{reduced}/(n - k - 1)$ is also an estimate of the error variance. Intuitively, SSE_{full} is close to $(n - p - 1)\sigma^2$, and if H_0 is true, $SSE_{reduced}$ is close to $(n - k - 1)\sigma^2$. It follows that if H_0 is true, the difference $(SSE_{reduced} - SSE_{full})$ is close to $(p - k)\sigma^2$, so the quotient $(SSE_{reduced} - SSE_{full})/(p - k)$ is close to σ^2. The test statistic is

$$f = \frac{(SSE_{reduced} - SSE_{full})/(p - k)}{SSE_{full}/(n - p - 1)} \tag{8.32}$$

Now if H_0 is true, both numerator and denominator of f are estimates of σ^2, so f is likely to be near 1. If H_0 is false, the quantity $SSE_{reduced}$ tends to be larger, so the value of f tends to be larger. The statistic f is an F statistic; its null distribution is $F_{p-k, n-p-1}$.

The method we have just described is very useful in practice for developing parsimonious models by removing unnecessary variables. However, the conditions under which it is formally valid are seldom met in practice. First, it is rarely the case that the full model is correct; there will be nonrandom quantities that affect the value of the dependent variable y that are not accounted for by the independent variables. Second, for the method to be formally valid, the subset of variables to be dropped must be determined independently of the data. This is usually not the case. More often, a large model is fit, some of the variables are seen to have fairly large P-values, and the F test is used to decide whether to drop them from the model. As we have said, this is a useful technique in practice, but, like most methods of model selection, it should be seen as an informal tool rather than a rigorous theory-based procedure.

We illustrate the method with an example. In mobile ad hoc computer networks, messages must be forwarded from computer to computer until they reach their destinations. The data overhead is the number of bytes of information that must be transmitted along with the messages to get them to the right places. A successful protocol will generally have a low data overhead. The overhead is affected by several features of the network, including the speed with which the computers are moving, the length of time they pause at each destination, and the link change rate. The link change rate for a given computer is the rate at which other computers in the network enter and leave the transmission range of the given computer. Table 8.4 presents average speed, pause time, link change rate (LCR), and data overhead for 25 simulated computer networks. These data were generated for a study published in the article "Metrics to Enable Adaptive Protocols for Mobile Ad Hoc Networks" (J. Boleng, W. Navidi, and T. Camp, *Proceedings of the 2002 International Conference on Wireless Networks*, 2002:293–298).

To study network performance under a variety of conditions, we begin by fitting a full quadratic model to these data, namely,

$$Overhead = \beta_0 + \beta_1\,Speed + \beta_2\,Pause + \beta_3\,LCR + \beta_4\,Speed \cdot Pause + \beta_5\,Speed \cdot LCR$$

$$+\beta_6\,Pause \cdot LCR + \beta_7\,Speed^2 + \beta_8\,Pause^2 + \beta_9\,LCR^2 + \varepsilon$$

TABLE 8.4 Data overhead, speed, pause time, and link change rate for a mobile computer network

Speed (m/s)	Pause Time (s)	LCR (100/s)	Data Overhead (kB)	Speed (m/s)	Pause Time (s)	LCR (100/s)	Data Overhead (kB)
5	10	9.426	428.90	20	40	12.117	501.48
5	20	8.318	443.68	20	50	10.284	519.20
5	30	7.366	452.38	30	10	33.009	445.45
5	40	6.744	461.24	30	20	22.125	489.02
5	50	6.059	475.07	30	30	16.695	506.23
10	10	16.456	446.06	30	40	13.257	516.27
10	20	13.281	465.89	30	50	11.107	508.18
10	30	11.155	477.07	40	10	37.823	444.41
10	40	9.506	488.73	40	20	24.140	490.58
10	50	8.310	498.77	40	30	17.700	511.35
20	10	26.314	452.24	40	40	14.064	523.12
20	20	19.013	475.97	40	50	11.691	523.36
20	30	14.725	499.67				

The results from fitting this model are as follows.

```
The regression equation is
Overhead = 436 + 0.56 Speed - 2.16 Pause - 2.25 LCR - 0.0481 Speed*Pause
         - 0.146 Speed*LCR + 0.364 Pause*LCR + 0.0511 Speed^2 + 0.0236 Pause^2
         + 0.0758 LCR^2

Predictor        Coef    SE Coef      T      P
Constant       436.03      25.78  16.92  0.000
Speed           0.560      2.349   0.24  0.815
Pause          -2.156      1.290  -1.67  0.115
LCR            -2.253      3.255  -0.69  0.499
Speed*Pause   -0.04813    0.03141 -1.53  0.146
Speed*LCR     -0.14594    0.08420 -1.73  0.104
Pause*LCR      0.36387    0.09421  3.86  0.002
Speed^2        0.05113    0.02551  2.00  0.063
Pause^2        0.02359    0.01290  1.83  0.087
LCR^2          0.07580    0.09176  0.83  0.422


S = 4.19902    R-Sq = 98.7%    R-Sq(adj) = 97.9%

Analysis of Variance

Source          DF        SS      MS      F      P
Regression       9   19859.9  2206.7  125.15  0.000
Residual Error  15     264.5    17.6
Total           24   20124.3
```

We will use the F test to determine whether the reduced model obtained by dropping Speed^2, Pause^2, and LCR^2 is a reasonable one. First, from the output for the full model, note that $\text{SSE}_{\text{full}} = 264.5$, and it has 15 degrees of freedom. The number of independent variables in the full model is $p = 9$.

We now drop Speed^2, Pause^2, and LCR^2, and fit the reduced model

$$\text{Overhead} = \beta_0 + \beta_1 \, \text{Speed} + \beta_2 \, \text{Pause} + \beta_3 \, \text{LCR} + \beta_4 \, \text{Speed} \cdot \text{Pause}$$
$$+ \beta_5 \, \text{Speed} \cdot \text{LCR} + \beta_6 \, \text{Pause} \cdot \text{LCR} + \varepsilon$$

The results from fitting this model are as follows.

```
The regression equation is
Overhead = 405 + 2.01 Speed + 0.246 Pause - 0.242 LCR - 0.0255 Speed*Pause
          - 0.0644 Speed*LCR + 0.185 Pause*LCR

Predictor         Coef   SE Coef        T      P
Constant        405.48     9.221    43.97  0.000
Speed            2.008    0.9646     2.08  0.052
Pause          0.24569    0.3228     0.76  0.456
LCR           -0.24165    0.5306    -0.46  0.654
Speed*Pa     -0.025539   0.01274    -2.01  0.060
Speed*LC     -0.064433   0.02166    -2.98  0.008
Pause*LC       0.18515   0.03425     5.41  0.000

S = 4.4778         R-Sq = 98.2%      R-Sq(adj) = 97.6%

Source           DF        SS       MS       F      P
Regression        6   19763.4   3293.9  164.28  0.000
Residual Error   18     360.9     20.1
Total            24   20124.3    838.5
```

From the output for this reduced model, we note that $\text{SSE}_{\text{reduced}} = 360.9$. The number of variables in this reduced model is $k = 6$.

Now we can compute the F statistic. Using Equation (8.32), we compute

$$f = \frac{(360.9 - 264.5)/(9 - 6)}{264.5/15} = 1.822$$

The null distribution is $F_{3,15}$. From the F table (Table A.6, in Appendix A), we find that $P > 0.10$. Since the P-value is large, the reduced model is plausible.

Best Subsets Regression

As we have mentioned, methods of model selection are often rather informal and ad hoc. There are a few tools, however, that can make the process somewhat more systematic. One of them is **best subsets regression**. The concept is simple. Assume that there are p independent variables, $x_1, \ldots, x_p$, that are available to be put into the model. Let's assume that we wish to find a good model that contains exactly four independent variables. We can simply fit every possible model containing four of the variables, and rank them in order of their goodness-of-fit, as measured by the coefficient of determination R^2. The subset of four variables that yields the largest value of R^2 is the "best" subset of size 4. One can repeat the process for subsets of other sizes, finding the best subsets of size 1, 2, ..., p. These best subsets can then be examined to see which provide a good fit, while being parsimonious.

The best subsets procedure is computationally intensive. When there are a lot of potential independent variables, there are a lot of models to fit. However, for most data sets, computers today are powerful enough to handle 30 or more independent variables, which is enough to cover many situations in practice. The following MINITAB output is for the best subsets procedure, applied to the data in Table 8.4. There are a total of nine independent variables being considered: Speed, Pause, LCR, Speed · Pause, Speed · LCR, Pause · LCR, Speed^2, Pause^2, and LCR^2.

Best Subsets Regression

Response is Overhead

Vars	R-Sq	R-Sq(adj)	Mallows C-p	S	Speed	Pause	LCR	Speed· Pause	Speed· LCR	Pause· LCR	Speed^2	Pause^2	LCR^2
1	85.5	84.9	144.5	11.264							X		
1	73.7	72.6	279.2	15.171		X							
2	97.7	97.5	6.9	4.5519							X		X
2	97.3	97.1	11.5	4.9448						X	X		
3	98.0	97.7	5.6	4.3590				X			X		X
3	97.8	97.5	7.8	4.5651						X	X		X
4	98.2	97.9	5.3	4.2354	X	X					X		X
4	98.2	97.8	5.8	4.2806						X	X	X	X
5	98.4	97.9	5.6	4.1599					X	X	X	X	X
5	98.3	97.9	6.4	4.2395	X					X	X	X	X
6	98.6	98.1	4.9	3.9434	X				X	X	X	X	X
6	98.4	97.9	7.4	4.2444				X	X	X	X	X	X
7	98.6	98.1	6.6	4.0265	X			X	X	X	X	X	X
7	98.6	98.0	6.8	4.0496	X	X			X	X	X	X	X
8	98.7	98.0	8.1	4.0734		X	X	X	X	X	X	X	X
8	98.6	98.0	8.5	4.1301	X	X		X	X	X	X	X	X
9	98.7	97.9	10.0	4.1990	X	X	X	X	X	X	X	X	X

In this output, both the best and the second-best subset are presented, for sizes 1 through 9. We emphasize that the term *best* means only that the model has the largest value of R^2, and does not guarantee that it is best in any practical sense. We'll explain the output column by column. The first column, labeled "Vars," presents the number of variables in the model. Thus the first row of the table describes the best model that can be made with one independent variable, and the second row describes the second-best such model. The third and fourth rows describe the best and second-best models that can be made with two variables, and so on. The second column presents the coefficient of determination, R^2, for each model. Note that the value of R^2 for the best subset increases as the number of variables increases. It is a mathematical fact that the best subset of $k+1$ variables will always have at least as large an R^2 as the best subset of k variables. We will skip over the next two columns for the moment. The column labeled "s" presents the estimate of the error standard deviation. It is the square root of the estimate s^2 (Equation 8.24 in Section 8.3). Finally, the columns on the right represent the independent variables that are candidates for inclusion into the model. The name of each variable is written vertically above its column. An "X" in the column means that the variable is included in the model. Thus, the best model containing three variables is the one with the variables LCR, Pause $\cdot$ LCR, and LCR^2.

Looking at the best subsets regression output, it is important to note how little difference there is in the fit between the best and second-best models of each size (except for size 1). It is also important to realize that the value of R^2 is a random quantity; it depends on the data. If the process were repeated and new data obtained, the values of R^2 for the various models would be somewhat different, and different models would be "best." For this reason, one should not use this procedure, or any other, to choose a single model. Instead, one should realize that there will be many models that fit the data about equally well.

Nevertheless, methods have been developed to choose a single model, presumably the "best" of the "best." We describe two of them here, with a caution not to take them too seriously. We begin by noting that if we simply choose the model with the highest value of R^2, we will always pick the one that contains all the variables, since the value of R^2 necessarily increases as the number of variables in the model increases. The methods for selecting a single model involve statistics that adjust the value of R^2, so as to eliminate this feature.

The first is the **adjusted R^2**. Let n denote the number of observations, and let k denote the number of independent variables in the model. The adjusted R^2 is defined as follows:

$$\text{Adjusted } R^2 = R^2 - \left(\frac{k}{n-k-1}\right)(1-R^2) \tag{8.33}$$

The adjusted R^2 is always smaller than R^2, since a positive quantity is subtracted from R^2. As the number of variables k increases, R^2 will increase, but the amount subtracted from it will increase as well. The value of k for which the value of adjusted R^2 is a maximum can be used to determine the number of variables in the model, and the best subset of that size can be chosen as the model. In the preceding output, we can see that the adjusted R^2 reaches its maximum (98.1%) at the six-variable model containing the variables Pause, Speed · Pause, Speed · LCR, Pause · LCR, Speed^2, and Pause^2. (There is a seven-variable model whose adjusted R^2 is also 98.1% to three significant digits, but is in fact slightly smaller than that of the six-variable model.)

Another commonly used statistic is **Mallows' C_p**. To compute this quantity, let n be the number of observations, let p be the total number of independent variables under consideration, and let k be the number of independent variables in a subset. As before, let SSE_{full} denote the error sum of squares for the full model containing all p variables, and let $SSE_{reduced}$ denote the error sum of squares for the model containing only the subset of k variables. Mallows' C_p is defined as

$$C_p = \frac{(n - p - 1)SSE_{reduced}}{SSE_{full}} - (n - 2k - 2) \qquad (8.34)$$

A number of criteria have been proposed for model selection using the C_p statistic; the most common one is to choose the model with the smallest value of C_p. In the preceding output, the model with the minimum C_p is the six-variable model whose C_p is 4.9. This model was also chosen by the adjusted R^2 criterion.

For these data, the C_p criterion and the adjusted R^2 criterion both chose the same model. In many cases, these criteria will choose different models. In any event, these criteria should be interpreted as suggesting models that may be appropriate, and it should be kept in mind that in most cases several models will fit the data about equally well.

Stepwise Regression

Stepwise regression is perhaps the most widely used model selection technique. Its main advantage over best subsets regression is that it is less computationally intensive, so it can be used in situations where there are a very large number of candidate independent variables and too many possible subsets for every one of them to be examined. The version of stepwise regression that we will describe is based on the P-values of the t statistics for the independent variables. An equivalent version is based on the F statistic (which is the square of the t statistic). Before running the algorithm, the user chooses two threshold P-values, α_{in} and α_{out}, with $\alpha_{in} \leq \alpha_{out}$. Stepwise regression begins with a step called a **forward selection** step, in which the independent variable with the smallest P-value is selected, provided that it satisfies $P < \alpha_{in}$. This variable is entered into the model, creating a model with a single independent variable. Call this variable x_1. In the next step, also a forward selection step, the remaining variables are examined one at a time as candidates for the second variable in the model. The one with the smallest P-value is added to the model, again provided that $P < \alpha_{in}$.

Now it is possible that adding the second variable to the model has increased the P-value of the first variable. In the next step, called a **backward elimination** step, the first variable is dropped from the model if its P-value has grown to exceed the value α_{out}. The algorithm then continues by alternating forward selection steps with backward elimination steps: at each forward selection step adding the variable with the smallest P-value if $P < \alpha_{in}$, and at each backward elimination step dropping the variable with the largest P-value if $P > \alpha_{out}$. The algorithm terminates when no variables meet the criteria for being added to or dropped from the model.

The following output is from the MINITAB stepwise regression procedure, applied to the data in Table 8.4. The threshold P-values are $\alpha_{in} = \alpha_{out} = 0.15$. There are a total of nine independent variables being considered: Speed, Pause, LCR, Speed · Pause, Speed · LCR, Pause · LCR, Speed^2, Pause^2, and LCR^2.

```
Alpha-to-Enter = 0.15    Alpha-to-Remove = 0.15

Response is Overhead on 9 predictors, with N = 25


Step                 1         2         3         4
Constant         407.7     416.0     410.0     394.9

Pause*LCR       0.1929    0.1943    0.1902    0.1606
T-Value          11.65     29.02     27.79      7.71
P-Value          0.000     0.000     0.000     0.000

LCR^2                     -0.0299   -0.0508   -0.0751
T-Value                   -10.90     -4.10     -3.72
P-Value                    0.000     0.001     0.001

LCR                                    0.90      2.38
T-Value                                1.73      2.15
P-Value                                0.098     0.044

Pause                                            0.37
T-Value                                          1.50
P-Value                                          0.150

S                 11.3      4.55      4.36      4.24
R-Sq             85.50     97.73     98.02     98.22
R-Sq(adj)        84.87     97.53     97.73     97.86
Mallows C-p      144.5       6.9       5.6       5.3
```

In step 1, the variable Pause · LCR had the smallest P-value (0.000) among the seven, so it was the first variable in the model. In step 2, LCR^2 had the smallest P-value (0.000) among the remaining variables, so it was added next. The P-value for Pause · LCR remained at 0.000 after the addition of Pause to the model; since it did not rise to a value greater than $\alpha_{out} = 0.15$, it is not dropped from the model. In steps 3, and 4, the variables LCR and Pause are added in turn. At no point does the P-value of a variable in the model exceed the threshold $\alpha_{out} = 0.15$, so no variables are dropped. After five steps, none of the variables remaining have P-values less than $\alpha_{in} = 0.15$, so the algorithm terminates. The final model contains the variables Pause · LCR, LCR^2, LCR, and Pause.

Model Selection Procedures Sometimes Find Models When They Shouldn't

When constructing a model to predict the value of a dependent variable, it might seem reasonable to try to start with as many candidate independent variables as possible, so that a model selection procedure has a very large number of models to choose from. Unfortunately, this is not a good idea, as we will now demonstrate.

A correlation coefficient can be computed between any two variables. Sometimes, two variables that have no real relationship will be strongly correlated, just by chance. For example, the statistician George Udny Yule noticed that the annual birthrate in Great Britain was almost perfectly correlated ($r = -0.98$) with the annual production of pig iron in the United States for the years 1875–1920. Yet no one would suggest trying to predict one of these variables from the other. This illustrates a difficulty shared by all model selection procedures. The more candidate independent variables that are provided, the more likely it becomes that some of them will exhibit meaningless correlations with the dependent variable, just by chance.

We illustrate this phenomenon with a simulation. We generated a simple random sample $y_1, \ldots, y_{30}$ of size 30 from a $N(0, 1)$ distribution. We will denote this sample by y. Then we generated 20 more independent samples of size 30 from a $N(0, 1)$ distribution; we will denote these samples by $x_1, \ldots, x_{20}$. To make the notation clear, the sample x_i contains 30 values $x_{i1}, \ldots, x_{i30}$. We then applied both stepwise regression and best subsets regression to these simulated data. None of the x_i are related to y; they were all generated independently. Therefore the ideal output from a model selection procedure would be to produce a model with no dependent variables at all. The actual behavior was quite different. The following two MINITAB outputs are for the stepwise regression and best subsets procedures. The stepwise regression method recommends a model containing six variables, with an adjusted R^2 of 41.89%. The best subsets procedure produces the best-fitting model for each number of variables from 1 to 20. Using the adjusted R^2 criterion, the best subsets procedure recommends a 12-variable model, with an adjusted R^2 of 51.0%. Using the minimum Mallows' C_p criterion, the "best" model is a five-variable model.

Anyone taking this output at face value would believe that some of the independent variables might be useful in predicting the dependent variable. But none of them are. All the apparent relationships are due entirely to chance.

```
Stepwise Regression: Y versus X1, X2, ...

   Alpha-to-Enter: 0.15  Alpha-to-Remove: 0.15

Response is Y on 20 predictors, with N = 30

Step                 1        2        3        4        5        6
Constant       0.14173  0.11689  0.12016  0.13756  0.09070  0.03589

X15              -0.38    -0.38    -0.28    -0.32    -0.28    -0.30
T-Value          -2.08    -2.19    -1.60    -1.87    -1.69    -1.89
```

	1	2	3	4	5	6
P-Value	0.047	0.037	0.122	0.073	0.105	0.071
X6		0.39	0.55	0.57	0.57	0.52
T-Value		2.04	2.76	2.99	3.15	2.87
P-Value		0.051	0.010	0.006	0.004	0.009
X16			−0.43	−0.43	−0.55	−0.73
T-Value			−1.98	−2.06	−2.60	−3.07
P-Value			0.058	0.050	0.016	0.005
X12				0.33	0.42	0.49
T-Value				1.79	2.29	2.66
P-Value				0.086	0.031	0.014
X3					−0.42	−0.52
T-Value					−1.83	−2.23
P-Value					0.080	0.035
X17						0.35
T-Value						1.53
P-Value						0.140
S	1.15	1.09	1.04	0.998	0.954	0.928
R-Sq	13.33	24.92	34.75	42.15	49.23	53.91
R-Sq(adj)	10.24	19.36	27.22	32.90	38.66	41.89
Mallows C-p	5.5	3.3	1.7	1.0	0.4	0.7

Best Subsets Regression: Y versus X1, X2, ...

Response is Y

```
                                          X X X X X X X X X X X
                          Mallows         X X X X X X X X X X 1 1 1 1 1 1 1 1 1 1 2
Vars  R-Sq  R-Sq(adj)     C-p      S       1 2 3 4 5 6 7 8 9 0 1 2 3 4 5 6 7 8 9 0
  1   13.3    10.2        5.5   1.1539                             X
  2   28.3    23.0        2.0   1.0685               X                 X
  3   34.8    27.2        1.7   1.0390               X               X X
  4   43.2    34.1        0.6   0.98851        X     X           X     X
  5   49.2    38.7        0.4   0.95391        X     X           X   X X
  6   53.9    41.9        0.7   0.92844          X   X           X   X X X
  7   57.7    44.3        1.3   0.90899  X X       X           X   X X       X
  8   61.2    46.4        2.1   0.89168        X   X         X X   X X X     X
  9   65.0    49.3        2.7   0.86747        X     X X     X X   X X X     X
 10   67.6    50.5        3.8   0.85680        X     X X     X X   X X X X X X
 11   69.2    50.4        5.2   0.85803  X X X     X X     X X     X X X     X
 12   71.3    51.0        6.4   0.85267  X X X     X X       X X   X X X X   X
 13   72.4    49.9        8.0   0.86165  X X X       X X   X X X   X X X X   X
```

14	73.0	47.8	9.8	0.87965	X X X		X X X	X X X	X X X X	X
15	74.2	46.5	11.4	0.89122	X X X		X X X	X X X	X X X X X X	
16	74.5	43.1	13.3	0.91886	X X X	X X X X	X X X	X X X X X X	
17	74.8	39.2	15.1	0.94985	X X X	X X X X	X X X X X X X X X	
18	75.1	34.2	17.1	0.98777	X X X	X X X X	X X X X X X X X X X X X	
19	75.1	27.9	19.0	1.0344	X X X	X X X X X X X X X X X X X X X X	
20	75.2	20.1	21.0	1.0886	X X X X X X X X X X X X X X X X X X X X	

How can one determine which variables, if any, in a selected model are really related to the dependent variable, and which were selected only by chance? Statistical methods are not much help here. The most reliable method is to repeat the experiment, collecting more data on the dependent variable and on the independent variables that were selected for the model. Then the independent variables suggested by the selection procedure can be fit to the dependent variable using the new data. If some of these variables fit well in the new data, the evidence of a real relationship becomes more convincing.

We summarize our discussion of model selection by emphasizing four points.

Summary

When selecting a regression model, keep the following in mind:

- When there is little or no theory to rely on, many different models will fit the data about equally well.

- The methods for choosing a model involve statistics (R^2, the F statistic, C_p), whose values depend on the data. Therefore if the experiment is repeated, these statistics will come out differently, and different models may appear to be "best."

- Some or all of the independent variables in a selected model may not really be related to the dependent variable. Whenever possible, experiments should be repeated to test these apparent relationships.

- Model selection is an art, not a science.

Exercises for Section 8.4

1. True or false:
   a. For any set of data, there is always one best model.
   b. When there is no physical theory to specify a model, there is usually no best model, but many that are about equally good.
   c. Model selection methods such as best subsets and stepwise regression, when properly used, are scientifically designed to find the best available model.
   d. Model selection methods such as best subsets and stepwise regression, when properly used, can suggest models that fit the data well.

2. The article "Experimental Design Approach for the Optimization of the Separation of Enantiomers in Preparative Liquid Chromatography" (S. Lai and Z. Lin, *Separation Science and Technology*, 2002: 847–875) describes an

experiment involving a chemical process designed to separate enantiomers. A model was fit to estimate the cycle time (y) in terms of the flow rate (x_1), sample concentration (x_2), and mobile-phase composition (x_3). The results of a least-squares fit are presented in the following table. (The article did not provide the value of the t statistic for the constant term.)

Predictor	Coefficient	T	P
Constant	1.603		
x_1	−0.619	−22.289	0.000
x_2	0.086	3.084	0.018
x_3	0.306	11.011	0.000
x_1^2	0.272	8.542	0.000
x_2^2	0.057	1.802	0.115
x_3^2	0.105	3.300	0.013
$x_1 x_2$	−0.022	−0.630	0.549
$x_1 x_3$	−0.036	−1.004	0.349
$x_2 x_3$	0.036	1.018	0.343

Of the following, which is the best next step in the analysis?

i. Nothing needs to be done. This model is fine.
ii. Drop x_1^2, x_2^2, and x_3^2 from the model, and then perform an F test.
iii. Drop $x_1 x_2$, $x_1 x_3$, and $x_2 x_3$ from the model, and then perform an F test.
iv. Drop x_1 and x_1^2 from the model, and then perform an F test.
v. Add cubic terms x_1^3, x_2^3, and x_3^3 to the model to try to improve the fit.

3. The article "Simultaneous Optimization of Mechanical Properties of Steel by Maximizing Exponential Desirability Functions" (K. J. Kim and D. K. J. Lin, *Journal of the Royal Statistical Society Series C, Applied Statistics*, 2000: 311–325) presents measurements on 72 steel plates. The following MINITAB output presents the results of a study to determine the relationship between yield strength (in kg/mm^2), and the proportion of carbon, manganese, and silicon, each measured in percent. The model fit is

$$\text{Yield strength} = \beta_0 + \beta_1 \text{ Carbon} + \beta_2 \text{ Manganese} + \beta_3 \text{ Silicon} + \varepsilon$$

```
The regression equation is
Yield Strength = 24.677 - 19.402 Carbon + 14.720 Manganese + 70.720 Silicon

Predictor          Coef           StDev            T              P
Constant          24.677         5.8589           4.21           0.000
Carbon           -19.402        28.455           -0.68           0.498
Manganese         14.720         5.6237           2.62           0.011
Silicon           70.720        45.675            1.55           0.126
```

Of the following, which is the best next step in the analysis? Explain your reasoning.

i. Add interaction terms Carbon · Manganese and Manganese · Silicon to try to find more variables to put into the model.
ii. Add the interaction term Carbon · Silicon to try to find another variable to put into the model.
iii. Nothing needs to be done. This model is fine.

iv. Drop Carbon and Silicon, and then perform an F test.

v. Drop Manganese, and then perform an F test.

4. The following MINITAB output is for a best subsets regression involving five dependent variables $X_1, \ldots, X_5$. The two models of each size with the highest values of R^2 are listed.

```
Best Subsets Regression: Y versus X1, X2, X3, X4, X5

Response is Y

                       Mallows          X X X X X
Vars  R-Sq  R-Sq(adj)    C-p        S   1 2 3 4 5
  1   77.3    77.1      133.6    1.4051  X
  1   10.2     9.3      811.7    2.7940        X
  2   89.3    89.0       14.6    0.97126  X    X
  2   77.8    77.3      130.5    1.3966  X X
  3   90.5    90.2        3.6    0.91630  X    X X
  3   89.4    89.1       14.6    0.96763  X X X
  4   90.7    90.3        4.3    0.91446  X X X X
  4   90.6    90.2        5.3    0.91942  X    X X X
  5   90.7    90.2        6.0    0.91805  X X X X X
```

a. Which variables are in the model selected by the minimum C_p criterion?

b. Which variables are in the model selected by the adjusted R^2 criterion?

c. Are there any other good models?

5. The following is supposed to be the result of a best subsets regression involving five independent variables $X_1, \ldots, X_5$. The two models of each size with the highest values of R^2 are listed. Something is wrong. What is it?

```
Best Subsets Regression

Response is Y

              Adj.               X X X X X
Vars  R-Sq   R-Sq   C-p     S    1 2 3 4 5

  1   69.1   68.0  101.4  336.79         X
  1   60.8   59.4  135.4  379.11   X
  2   80.6   79.2   55.9  271.60   X   X
  2   79.5   77.9   60.7  279.59   X X
  3   93.8   92.8   13.4  184.27   X X   X
  3   93.7   92.7   18.8  197.88   X X X
  4   91.4   90.4    5.5  159.59   X X   X X
  4   90.1   88.9    5.6  159.81   X X X   X
  5   94.2   93.0    6.0  157.88   X X X X X
```

6. In a study to determine the effect of vehicle weight in tons (x_1) and engine displacement in in^3 (x_2) on fuel economy in miles per gallon (y), these quantities were measured for ten automobiles. The full quadratic model $y = \beta_0 + \beta_1 x_1 + \beta_2 x_2 + \beta_3 x_1^2 + \beta_4 x_2^2 + \beta_5 x_1 x_2 + \varepsilon$ was fit to the data, and the sum of squares for error was SSE $= 62.068$. Then the reduced model $y = \beta_0 + \beta_1 x_1 + \beta_2 x_2 + \varepsilon$ was fit, and the sum of squares for error was SSE $= 66.984$. Is it reasonable to use the reduced model, rather than the full quadratic model, to predict fuel economy? Explain.

7. (Continues Exercise 7 in Section 8.3.) To try to improve the prediction of FEV_1, additional independent variables are included in the model. These new variables are Weight (in kg), the product (interaction) of Height and Weight, and the ambient temperature (in °C). The following MINITAB output presents results of fitting the model

$$FEV_1 = \beta_0 + \beta_1 \text{ Last FEV}_1 + \beta_2 \text{ Gender} + \beta_3 \text{ Height} + \beta_4 \text{ Weight} + \beta_5 \text{ Height} \cdot \text{Weight}$$
$$+ \beta_6 \text{ Temperature} + \beta_7 \text{ Pressure} + \varepsilon$$

```
The regression equation is
FEV1 = −0.257 + 0.778 Last FEV − 0.105 Gender + 1.213 Height − 0.00624 Weight
+ 0.00386 Height*Weight − 0.00740 Temp − 0.00148 Pressure

Predictor              Coef            StDev            T            P
Constant            −0.2565           0.7602         −0.34        0.736
Last FEV            0.77818           0.05270         14.77        0.000
Gender             −0.10479           0.03647         −2.87        0.005
Height              1.2128            0.4270          2.84        0.005
Weight             −0.0062446         0.01351         −0.46        0.645
Height*Weight       0.0038642         0.008414         0.46        0.647
Temp               −0.007404          0.009313        −0.79        0.428
Pressure           −0.0014773         0.0005170       −2.86        0.005

S = 0.22189        R-Sq = 93.5%       R-Sq(adj) = 93.2%

Analysis of Variance

Source             DF             SS             MS            F            P
Regression          7          111.35          15.907       323.06       0.000
Residual Error    157          7.7302          0.049237
Total             164          119.08
```

a. The following MINITAB output, reproduced from Exercise 7 in Section 8.3, is for a reduced model in which Weight, Height · Weight, and Temp have been dropped. Compute the F statistic for testing the plausibility of the reduced model.

```
The regression equation is
FEV1 = −0.219 + 0.779 Last FEV − 0.108 Gender + 1.354 Height − 0.00134 Pressure

Predictor           Coef            StDev            T            P
Constant         −0.21947          0.4503         −0.49        0.627
Last FEV          0.779            0.04909         15.87        0.000
Gender           −0.10827          0.0352          −3.08        0.002
Height            1.3536           0.2880          4.70        0.000
Pressure         −0.0013431        0.0004722       −2.84        0.005
```

```
S = 0.22039        R-Sq = 93.5%      R-Sq(adj) = 93.3%

Analysis of Variance

Source            DF          SS          MS          F          P
Regression         4       111.31      27.826      572.89      0.000
Residual Error   160       7.7716      0.048572
Total            164       119.08
```

b. How many degrees of freedom does the F statistic have?

c. Find the P-value for the F statistic. Is the reduced model plausible?

d. Someone claims that since each of the variables being dropped had large P-values, the reduced model must be plausible, and it was not necessary to perform an F test. Is this correct? Explain why or why not.

e. The total sum of squares is the same in both models, even though the independent variables are different. Is there a mistake? Explain.

8. The article "Optimization of Enterocin P Production by Batch Fermentation of *Enterococcus faecium* P13 at Constant pH" (C. Herran, J. Martinez, et al., *Applied Microbiology and Biotechnology*, 2001:378–383) described a study involving the growth rate of the bacterium *Enterococcus faecium* in media with varying pH. The log of the maximum growth rate for various values of pH are presented in the following table:

ln Growth rate	−2.12	−1.51	−0.89	−0.33	−0.05	−0.11	0.39	−0.25
pH	4.7	5.0	5.3	5.7	6.0	6.2	7.0	8.5

a. Fit the linear model: ln Growth rate $= \beta_0 + \beta_1 \, pH + \varepsilon$. For each coefficient, find the P-value for the null hypothesis that the coefficient is equal to 0. In addition, compute the analysis of variance (ANOVA) table.

b. Fit the quadratic model: ln Growth rate $= \beta_0 + \beta_1 \, pH + \beta_2 \, pH^2 + \varepsilon$. For each coefficient, find the P-value for the null hypothesis that the coefficient is equal to 0. In addition, compute the ANOVA table.

c. Fit the cubic model: ln Growth rate $= \beta_0 + \beta_1 \, pH + \beta_2 \, pH^2 + \beta_3 \, pH^3 + \varepsilon$. For each coefficient, find the P-value for the null hypothesis that the coefficient is equal to 0. In addition, compute the ANOVA table.

d. Which of these models do you prefer, and why?

9. The article "Vehicle-Arrival Characteristics at Urban Uncontrolled Intersections" (V. Rengaraju and V. Rao, *Journal of Transportation Engineering*, 1995:317–323) presents data on traffic characteristics at 10 intersections in Madras, India. The following table provides data on road width in m (x_1), traffic volume in vehicles per lane per hour (x_2), and median speed in km/h (x_3).

y	x_1	x_2	y	x_1	x_2
35.0	76	370	26.5	75	842
37.5	88	475	27.5	92	723
26.5	76	507	28.0	90	923
33.0	80	654	23.5	86	1039
22.5	65	917	24.5	80	1120

a. Fit the model $y = \beta_0 + \beta_1 x_1 + \beta_2 x_2 + \varepsilon$. Find the P-values for testing that the coefficients are equal to 0.

b. Fit the model $y = \beta_0 + \beta_1 x_1 + \varepsilon$. Find the P-values for testing that the coefficients are equal to 0.

c. Fit the model $y = \beta_0 + \beta_1 x_2 + \varepsilon$. Find the P-values for testing that the coefficients are equal to 0.

d. Which of the models (a) through (c) do you think is best? Why?

10. The following table presents measurements of mean noise levels in dBA (y), roadway width in m (x_1), and mean speed in km/h (x_2), for 10 locations in Bangkok, Thailand, as reported in the article "Modeling of Urban Area Stop-and-Go Traffic Noise" (P. Pamanikabud and C. Tharasawatipipat, *Journal of Transportation Engineering*, 1999:152–159).

y	x_1	x_2	y	x_1	x_2
78.1	6.0	30.61	78.1	12.0	28.26
78.1	10.0	36.55	78.6	6.5	30.28
79.6	12.0	36.22	78.5	6.5	30.25
81.0	6.0	38.73	78.4	9.0	29.03
78.7	6.5	29.07	79.6	6.5	33.17

Construct a good linear model to predict mean noise levels using roadway width, mean speed, or both, as predictors. Provide the standard deviations of the coefficient estimates and the P-values for testing that they are different from 0. Explain how you chose your model.

11. The article "Modeling Resilient Modulus and Temperature Correction for Saudi Roads" (H. Wahhab, I. Asi, and R. Ramadhan, *Journal of Materials in Civil Engineering*, 2001:298–305) describes a study designed to predict the resilient modulus of pavement from physical properties. The following table presents data for the resilient modulus at 40°C in 10^6 kPa (y), the surface area of the aggregate in m^2/kg (x_1), and the softening point of the asphalt in °C (x_2).

y	x_1	x_2	y	x_1	x_2	y	x_1	x_2
1.48	5.77	60.5	3.06	6.89	65.3	1.88	5.93	63.2
1.70	7.45	74.2	2.44	8.64	66.2	1.90	8.17	62.1
2.03	8.14	67.6	1.29	6.58	64.1	1.76	9.84	68.9
2.86	8.73	70.0	3.53	9.10	68.6	2.82	7.17	72.2
2.43	7.12	64.6	1.04	8.06	58.8	1.00	7.78	54.1

The full quadratic model is $y = \beta_0 + \beta_1 x_1 + \beta_2 x_2 + \beta_3 x_1 x_2 + \beta_4 x_1^2 + \beta_5 x_2^2 + \varepsilon$. Which submodel of this full model do you believe is most appropriate? Justify your answer by fitting two or more models and comparing the results.

12. The article "Models for Assessing Hoisting Times of Tower Cranes" (A. Leung and C. Tam, *Journal of Construction Engineering and Management*, 1999:385–391) presents a model constructed by a stepwise regression procedure to predict the time needed for a tower crane hoisting operation. Twenty variables were considered, and the stepwise procedure chose a nine-variable model. The adjusted R^2 for the selected model was 0.73. True or false:

a. The value 0.73 is a reliable measure of the goodness of fit of the selected model.

b. The value 0.73 may exaggerate the goodness of fit of the model.

c. A stepwise regression procedure selects only variables that are of some use in predicting the value of the dependent variable.

d. It is possible for a variable that is of no use in predicting the value of a dependent variable to be part of a model selected by a stepwise regression procedure.

Supplementary Exercises for Chapter 8

1. The article "Effect of Temperature on the Marine Immersion Corrosion of Carbon Steels" (R. Melchers, *Corrosion*, 2002:768–781) presents measurements of corrosion loss (in mm) for copper-bearing steel specimens immersed in seawater in 14 different locations. For each location, the average corrosion loss (in mm) was recorded, along with the mean water temperature (in °C). The results, after one year of immersion, are presented in the following table.

Corrosion	Mean Temp.	Corrosion	Mean Temp.
0.2655	23.5	0.2200	26.5
0.1680	18.5	0.0845	15.0
0.1130	23.5	0.1860	18.0
0.1060	21.0	0.1075	9.0
0.2390	17.5	0.1295	11.0
0.1410	20.0	0.0900	11.0
0.3505	26.0	0.2515	13.5

a. Compute the least-squares line for predicting corrosion loss (y) from mean temperature (x).
b. Find a 95% confidence interval for the slope.
c. Find a 95% confidence interval for the mean corrosion loss at a mean temperature of 20°C.
d. Find a 90% prediction interval for the corrosion loss of a specimen immersed at a mean temperature of 20°C.

2. The article "Measurements of the Thermal Conductivity and Thermal Diffusivity of Polymer Melts with the Short-Hot-Wire Method" (X. Zhang, W. Hendro, et al., *International Journal of Thermophysics*, 2002:1077–1090) presents measurements of the thermal conductivity and diffusivity of several polymers at several temperatures. The results for the thermal diffusivity of polycarbonate (in $10^{-7} \ m^2 s^{-1}$) are given in the following table.

Diff.	Temp. (°C)	Diff.	Temp. (°C)
1.43	28	1.05	159
1.53	38	1.13	169
1.43	61	1.03	181
1.34	83	1.06	204
1.36	107	1.26	215
1.34	119	0.86	225
1.29	130	1.01	237
1.36	146	0.98	248

a. Compute the least-squares line for predicting diffusivity (y) from temperature (x).
b. Find a 95% confidence interval for the slope.
c. Find a 95% confidence interval for the diffusivity of polycarbonate at a temperature of 100°C.
d. Find a 95% prediction interval for the diffusivity of polycarbonate at a temperature of 100°C.
e. Which is likely to be the more useful, the confidence interval or the prediction interval? Explain.

3. The article "Copper Oxide Mounted on Activated Carbon as Catalyst for Wet Air Oxidation of Aqueous Phenol. 1. Kinetic and Mechanistic Approaches" (P. Alvarez, D. McLurgh, and P. Plucinski, *Industrial Engineering and Chemistry Research*, 2002: 2147–2152) reports the results of experiments to describe the mechanism of the catalytic wet air oxidation of aqueous phenol. In one set of experiments, the initial oxidation rate (in kilograms of phenol per

kilogram of catalyst per hour) and the oxygen concentration (in mol/m^3) were measured. The results (read from a graph) are presented in the following table.

Rate (y)	0.44	0.49	0.60	0.64	0.72
O$_2$ concentration (x)	3.84	4.76	6.08	7.06	8.28

a. It is known that x and y are related by an equation of the form $y = kx^r$, where r is the oxygen reaction order. Make appropriate transformations to express this as a linear equation.
b. Estimate the values of k and r by computing the least-squares line.
c. Based on these data, is it plausible that the oxygen reaction order is equal to 0.5? Explain.

4. A materials scientist is experimenting with a new material with which to make beverage cans. She fills cans with liquid at room temperature, and then refrigerates them to see how fast they cool. According to Newton's law of cooling, if t is the time refrigerated and y is the temperature drop at time t, then y is related to t by an equation of the form

$$\ln y = \beta_0 + \beta_1 t$$

where β_0 is a constant that depends on the initial temperature of the can and the ambient temperature of the refrigerator, and β_1 is a constant that depends on the physical properties of the can. The scientist measures the temperature at regular intervals, and then fits this model to the data. The results are shown in the following figure. A scatterplot, with the least-squares line superimposed, is on the left, and the residual plot is on the right.

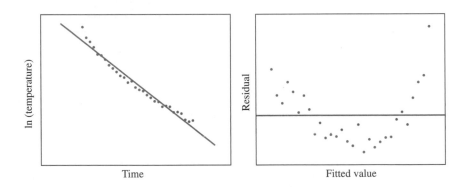

What should the scientist do next?
  i. Try to find a transformation that makes the relationship more linear.
 ii. Use the model as is, because Newton's law of cooling is a physical law.
iii. Use the model as is, because it fits well enough.
 iv. Carefully examine the experimental setup to see what might have gone wrong.

5. Eruptions of the Old Faithful geyser in Yellowstone National Park typically last from 1.5 to 5 minutes. Between eruptions are dormant periods, which typically last from 50 to 100 minutes. A dormant period can also be thought of as the waiting time between eruptions. The durations in minutes for 60 consecutive dormant periods are given in the following table. It is desired to predict the length of a dormant period from the length of the dormant period immediately preceding it. To express this in symbols, denote the sequence of dormant periods $T_1, \ldots, T_{60}$. It is desired to predict T_{i+1} from T_i.

i	T_i	i	T_i	i	T_i	i	T_i	i	T_i	i	T_i
1	80	11	56	21	82	31	88	41	72	51	67
2	84	12	80	22	51	32	51	42	75	52	81
3	50	13	69	23	76	33	80	43	75	53	76
4	93	14	57	24	82	34	49	44	66	54	83
5	55	15	90	25	84	35	82	45	84	55	76
6	76	16	42	26	53	36	75	46	70	56	55
7	58	17	91	27	86	37	73	47	79	57	73
8	74	18	51	28	51	38	67	48	60	58	56
9	75	19	79	29	85	39	68	49	86	59	83
10	80	20	53	30	45	40	86	50	71	60	57

a. Construct a scatterplot of the points (T_i, T_{i+1}), for $i = 1, \ldots, 59$.

b. Compute the least-squares line for predicting T_{i+1} from T_i. (*Hint:* The values of the independent variable (x) are $T_1, \ldots, T_{59}$, and the values of the dependent variable (y) are $T_2, \ldots, T_{60}$.)

c. Find a 95% confidence interval for the slope β_1.

d. If the waiting time before the last eruption was 70 minutes, what is the predicted waiting time before the next eruption?

e. Find a 98% confidence interval for the mean waiting time before the next eruption when the time before the last eruption was 70 minutes.

f. Find a 99% prediction interval for the waiting time before the next eruption, if the time before the last eruption was 70 minutes.

6. Refer to Exercise 5.

a. Plot the residuals versus the fitted values. Does the plot indicate any serious violations of the standard assumptions?

b. Plot the residuals versus the order of the data. Does the plot indicate any serious violations of the standard assumptions?

7. A chemist is calibrating a spectrophotometer that will be used to measure the concentration of carbon monoxide (CO) in atmospheric samples. To check the calibration, samples of known concentration are measured. The true concentrations (x) and the measured concentrations (y) are given in the following table. Because of random error, repeated measurements on the same sample will vary. The machine is considered to be in calibration if its mean response is equal to the true concentration.

True Concentration (ppm)	Measured Concentration (ppm)
0	1
10	11
20	21
30	28
40	37
50	48
60	56
70	68
80	75
90	86
100	96

To check the calibration, the linear model $y = \beta_0 + \beta_1 x + \varepsilon$ is fit. Ideally, the value of β_0 should be 0 and the value of β_1 should be 1.

a. Compute the least-squares estimates $\hat{\beta}_0$ and $\hat{\beta}_1$.
b. Can you reject the null hypothesis $H_0: \beta_0 = 0$?
c. Can you reject the null hypothesis $H_0: \beta_1 = 1$?
d. Do the data provide sufficient evidence to conclude that the machine is out of calibration?
e. Compute a 95% confidence interval for the mean measurement when the true concentration is 20 ppm.
f. Compute a 95% confidence interval for the mean measurement when the true concentration is 80 ppm.
g. Someone claims that the machine is in calibration for concentrations near 20 ppm. Do these data provide sufficient evidence for you to conclude that this claim is false? Explain.

8. The article "Approach to Confidence Interval Estimation for Curve Numbers" (R. McCuen, *Journal of Hydrologic Engineering*, 2002:43–48) discusses the relationship between rainfall depth and runoff depth at several locations. At one particular location, rainfall depth and runoff depth were recorded for 13 rainstorms. Following is MINITAB output for a fit of the least-squares line to predict runoff depth from rainfall depth (both measured in inches).

```
The regression equation is
Runoff = -0.23 + 0.73 Rainfall

Predictor          Coef      SE Coef            T        P
Constant       -0.23429      0.23996        -0.98    0.350
Rainfall        0.72868      0.06353        11.47    0.000

S = 0.40229          R-Sq = 92.3%         R-Sq(adj) = 91.6%

Analysis of Variance

Source            DF           SS           MS           F        P
Regression         1       21.290       21.290      131.55    0.000
Residual Error    11        1.780      0.16184
Total             12       23.070
```

a. Predict the runoff for a storm with 2.5 in. of rainfall.
b. Someone claims that if two storms differ in their rainfall by 1 in., then their runoffs will differ, on the average, by 1 in. as well. Is this a plausible claim? Explain.
c. It is a fact that if the rainfall is 0, the runoff is 0. Is the least-squares line consistent with this fact? Explain.

9. Refer to Exercise 8. Someone wants to compute a 95% confidence interval for the mean runoff when the rainfall is 3 in. Can this be computed from the information in the MINITAB output shown in Exercise 8? Or is more information needed? Choose the best answer.

   i. Yes, it can be computed from the MINITAB output.
   ii. No, we also need to know the rainfall values that were used to compute the least-squares line.
   iii. No, we also need to know the runoff values that were used to compute the least-squares line.
   iv. No, we also need to know both the rainfall and the runoff values that were used to compute the least-squares line.

10. During the production of boiler plate, test pieces are subjected to a load, and their elongations are measured. In one particular experiment, five tests will be made, at loads (in MPa) of 11, 37, 54, 70, and 93. The least-squares line will

be computed to predict elongation from load. Confidence intervals for the mean elongation will be computed for several different loads. Which of the following intervals will be the widest? Which will be the narrowest?

 i. The 95% confidence interval for the mean elongation under a load of 53 MPa.

 ii. The 95% confidence interval for the mean elongation under a load of 72 MPa.

 iii. The 95% confidence interval for the mean elongation under a load of 35 MPa.

11. The article "Low-Temperature Heat Capacity and Thermodynamic Properties of 1,1,1-trifluoro-2,2-dichloroethane" (R. Varushchenko and A. Druzhinina, *Fluid Phase Equilibria*, 2002:109–119) describes an experiment in which samples of Freon R-123 were melted in a calorimeter. Various quantities of energy were supplied to the calorimeter for melting. The equilibrium melting temperatures (t) and fractions melted (f) were measured. The least-squares line was fit to the model $t = \beta_0 + \beta_1(1/f) + \varepsilon$, where $1/f$ is the reciprocal fraction. The results of the fit are as follows.

```
The regression equation is
Temperature = 145.74 - 0.052 Reciprocal Frac

Predictor          Coef     SE Coef            T       P
Constant        145.736     0.00848      17190.1   0.000
Recip Frac     -0.05180     0.00226      -22.906   0.000

S = 0.019516        R-Sq = 97.6%       R-Sq(adj) = 97.4%

Analysis of Variance

Source             DF          SS          MS        F       P
Regression          1       0.200       0.200   524.70   0.000
Residual Error     13     0.00495    0.000381
Total              14       0.205
```

a. Estimate the temperature at which half of the sample will melt (i.e., $f = 1/2$).

b. Can you determine the correlation coefficient between equilibrium temperature and reciprocal of the fraction melted from this output? If so, determine it. If not, explain what additional information is needed.

c. The triple-point temperature is the lowest temperature at which the whole sample will melt (i.e., $f = 1$). Estimate the triple-point temperature.

12. The article "Two Different Approaches for RDC Modelling When Simulating a Solvent Deasphalting Plant" (J. Aparicio, M. Heronimo, et al., *Computers and Chemical Engineering*, 2002:1369–1377) reports flow rate (in dm^3/h) and specific gravity measurements for a sample of paraffinic hydrocarbons. The natural logs of the flow rates (y) and the specific gravity measurements (x) are presented in the following table.

y	x	y	x
−1.204	0.8139	1.311	0.8264
−0.580	0.8171	1.959	0.8294
0.049	0.8202	2.614	0.8323
0.673	0.8233	3.270	0.8352

a. Fit the linear model $y = \beta_0 + \beta_1 x + \varepsilon$. For each coefficient, test the hypothesis that the coefficient is equal to 0.

b. Fit the quadratic model $y = \beta_0 + \beta_1 x + \beta_2 x^2 + \varepsilon$. For each coefficient, test the hypothesis that the coefficient is equal to 0.

c. Fit the cubic model $y = \beta_0 + \beta_1 x + \beta_2 x^2 + \beta_3 x^3 + \varepsilon$. For each coefficient, test the hypothesis that the coefficient is equal to 0.

d. Which of the models in parts (a) through (c) is the most appropriate? Explain.

e. Using the most appropriate model, estimate the flow rate when the specific gravity is 0.83.

13. The article "Advances in Oxygen Equivalence Equations for Predicting the Properties of Titanium Welds" (D. Harwig, W. Ittiwattana, and II. Castner, *The Welding Journal*, 2001:126s–136s) reports an experiment to predict various properties of titanium welds. Among other properties, the elongation (in %) was measured, along with the oxygen content and nitrogen content (both in %). The following MINITAB output presents results of fitting the model

$$\text{Elongation} = \beta_0 + \beta_1 \text{ Oxygen} + \beta_2 \text{ Nitrogen} + \beta_3 \text{ Oxygen} \cdot \text{Nitrogen}$$

```
The regression equation is
Elongation = 46.80 - 130.11 Oxygen - 807.1 Nitrogen + 3580.5 Oxy*Nit

Predictor        Coef        StDev           T          P
Constant       46.802        3.702       12.64      0.000
Oxygen        -130.11       20.467       -6.36      0.000
Nitrogen      -807.10      158.03       -5.107      0.000
Oxy*Nit        3580.5      958.05        3.737      0.001

S = 2.809        R-Sq = 74.5%      R-Sq(adj) = 72.3%

Analysis of Variance

Source            DF           SS          MS          F          P
Regression         3       805.43      268.48      34.03      0.000
Residual Error    35       276.11        7.89
Total             38      1081.54
```

a. Predict the elongation for a weld with an oxygen content of 0.15% and a nitrogen content of 0.01%.

b. If two welds both have a nitrogen content of 0.006%, and their oxygen content differs by 0.05%, what would you predict their difference in elongation to be?

c. Two welds have identical oxygen contents, and nitrogen contents that differ by 0.005%. Is this enough information to predict their difference in elongation? If so, predict the elongation. If not, explain what additional information is needed.

14. Refer to Exercise 13.

a. Find a 95% confidence interval for the coefficient of Oxygen.

b. Find a 99% confidence interval for the coefficient of Nitrogen.

c. Find a 98% confidence interval for the coefficient of the interaction term Oxygen · Nitrogen.

d. Can you conclude that $\beta_1 < -75$? Find the P-value.

e. Can you conclude that $\beta_2 > -1000$? Find the P-value.

15. The following MINITAB output is for a multiple regression. Some of the numbers got smudged, becoming illegible. Fill in the missing numbers.

Predictor	Coef	StDev	T	P
Constant	(a)	0.3501	0.59	0.568
X1	1.8515	(b)	2.31	0.040
X2	2.7241	0.7124	(c)	0.002

S = (d)   R-Sq = 83.4%   R-Sq(adj) = 80.6%

Analysis of Variance

Source	DF	SS	MS	F	P
Regression	(e)	(f)	(g)	(h)	0.000
Residual Error	12	17.28	1.44		
Total	(i)	104.09			

16. An engineer tries three different methods for selecting a linear model. First she uses an informal method based on the F statistic, as described in Section 8.4. Then she runs the best subsets routine, and finds the model with the best adjusted R^2 and the one with the best Mallows C_p. It turns out that all three methods select the same model. The engineer says that since all three methods agree, this model must be the best one. One of her colleagues says that other models might be equally good. Who is right? Explain.

17. In a simulation of 30 mobile computer networks, the average speed, pause time, and number of neighbors were measured. A "neighbor" is a computer within the transmission range of another. The data are presented in the following table.

Neighbors	Speed	Pause	Neighbors	Speed	Pause	Neighbors	Speed	Pause
10.17	5	0	9.36	5	10	8.92	5	20
8.46	5	30	8.30	5	40	8.00	5	50
10.20	10	0	8.86	10	10	8.28	10	20
7.93	10	30	7.73	10	40	7.56	10	50
10.17	20	0	8.24	20	10	7.78	20	20
7.44	20	30	7.30	20	40	7.21	20	50
10.19	30	0	7.91	30	10	7.45	30	20
7.30	30	30	7.14	30	40	7.08	30	50
10.18	40	0	7.72	40	10	7.32	40	20
7.19	40	30	7.05	40	40	6.99	40	50

a. Fit the model with Neighbors as the dependent variable, and independent variables Speed, Pause, Speed · Pause, $Speed^2$, and $Pause^2$.

b. Construct a reduced model by dropping any variables whose P-values are large, and test the plausibility of the model with an F test.

c. Plot the residuals versus the fitted values for the reduced model. Are there any indications that the model is inappropriate? If so, what are they?

d. Someone suggests that a model containing Pause and $Pause^2$ as the only dependent variables is adequate. Do you agree? Why or why not?

e. Using a best subsets software package, find the two models with the highest R^2 value for each model size from one to five variables. Compute C_p and adjusted R^2 for each model.

f. Which model is selected by minimum C_p? By adjusted R^2? Are they the same?

18. The voltage output (y) of a battery was measured over a range of temperatures (x) from 0°C to 50°C. The following figure is a scatterplot of voltage versus temperature, with three fitted curves superimposed. The curves are the linear model $y = \beta_0 + \beta_1 x + \varepsilon$, the quadratic model $y = \beta_0 + \beta_1 x = \beta_2 x^2 + \varepsilon$, and the cubic model $y = \beta_0 + \beta_1 x + \beta_2 x^2 + \beta_3 x^3 + \varepsilon$. Based on the plot, which of the models should be used to describe the data? Explain.

   i. The linear model.

   ii. The quadratic model.

   iii. The cubic model.

   iv. All three appear to be about equally good.

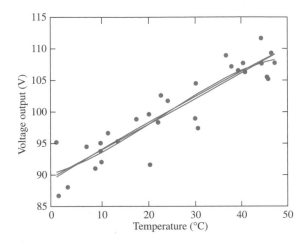

19. The data presented in the following table give the tensile strength in psi (y) of paper as a function of the percentage of hardwood content (x).

Hardwood Content	Tensile Strength	Hardwood Content	Tensile Strength
1.0	26.8	7.0	52.1
1.5	29.5	8.0	56.1
2.0	36.6	9.0	63.1
3.0	37.8	10.0	62.0
4.0	38.2	11.0	62.5
4.5	41.5	12.0	58.0
5.0	44.8	13.0	52.9
5.5	44.7	14.0	38.2
6.0	48.5	15.0	32.9
6.5	50.1	16.0	21.9

  a. Fit polynomial models of degrees 1, 2, and so on, to predict tensile strength from hardwood content. For each one, use the F test to compare it with the model of degree one less. Stop when the P-value of the F test is greater than 0.05. What is the degree of the polynomial model chosen by this method?

  b. Using the model from part (a), estimate the hardwood concentration that produces the highest tensile strength.

20. The article "Electrical Impedance Variation with Water Saturation in Rock" (Q. Su, Q. Feng, and Z. Shang, *Geophysics*, 2000:68–75) reports measurements of permeabilities (in $10^{-3}\,\mu m^2$), porosities (in %), and surface area per unit volume of pore space (in 10^4 cm^{-1}) for several rock samples. The results are presented in the following table, denoting ln Permeability by y, porosity by x_1, and surface area per unit volume by x_2.

y	x_1	x_2	y	x_1	x_2
−0.27	19.83	9.55	0.58	10.52	20.03
2.58	17.93	10.97	−0.56	18.92	13.10
3.18	21.27	31.02	−0.49	18.55	12.78
1.70	18.67	28.12	−0.01	13.72	40.28
−1.17	7.98	52.35	−1.71	9.12	53.67
−0.27	10.16	32.82	−0.12	14.39	26.75
−0.53	17.86	57.66	−0.92	11.38	75.62
−0.29	13.48	21.10	2.18	16.59	9.95
4.94	17.49	9.15	4.46	16.77	7.88
1.94	14.18	11.72	2.11	18.55	88.10
3.74	23.88	5.43	−0.04	18.02	10.95

a. Fit the model $y = \beta_0 + \beta_1 x_1 + \beta_2 x_2 + \beta_3 x_1 x_2 + \varepsilon$. Compute the analysis of variance table.
b. Fit the model $y = \beta_0 + \beta_1 x_1 + \beta_2 x_2 + \varepsilon$. Compute the analysis of variance table.
c. Fit the model $y = \beta_0 + \beta_1 x_1 + \varepsilon$. Compute the analysis of variance table.
d. Compute the F statistics for comparing the models in parts (b) and (c) with the model in part (a). Which model do you prefer? Why?

21. The article "Groundwater Electromagnetic Imaging in Complex Geological and Topographical Regions: A Case Study of a Tectonic Boundary in the French Alps" (S. Houtot, P. Tarits, et al., *Geophysics*, 2002:1048–1060) presents measurements of concentrations of several chemicals (in mmol/L) and electrical conductivity (in 10^{-2} S/m) for several water samples in various locations near Gittaz Lake in the French Alps. The results for magnesium and calcium are presented in the following table. Two outliers have been deleted.

Conductivity	Magnesium	Calcium	Conductivity	Magnesium	Calcium
2.77	0.037	1.342	1.10	0.027	0.487
3.03	0.041	1.500	1.11	0.039	0.497
3.09	0.215	1.332	2.57	0.168	1.093
3.29	0.166	1.609	3.27	0.172	1.480
3.37	0.100	1.627	2.28	0.044	1.093
0.88	0.031	0.382	3.32	0.069	1.754
0.77	0.012	0.364	3.93	0.188	1.974
0.97	0.017	0.467	4.26	0.211	2.103

a. To predict conductivity (y) from the concentrations of magnesium (x_1) and calcium (x_2), fit the full quadratic model $y = \beta_0 + \beta_1 x_1 + \beta_2 x_2 + \beta_3 x_1^2 + \beta_4 x_2^2 + \beta_5 x_1 x_2 + \varepsilon$. Compute the analysis of variance table.
b. Use the F test to investigate some submodels of the full quadratic model. State which model you prefer and why.

c. Use a best subsets routine to find the submodels with the maximum adjusted R^2 and the minimum Mallows C_p. Are they the same model? Comment on the appropriateness of this (these) model(s).

22. The article "Low-Temperature Heat Capacity and Thermodynamic Properties of 1,1,1-trifluoro-2, 2-dichloroethane" (R. Varushchenko and A. Druzhinina, *Fluid Phase Equilibria*, 2002:109–119) presents measurements of the molar heat capacity (y) of 1,1,1-trifluoro-2,2-dichloroethane (in $J \cdot K^{-1} \cdot mol^{-1}$) at several temperatures (x) in units of 10 K. The results for every tenth measurement are presented in the following table.

y	x	y	x
5.7037	1.044	60.732	6.765
16.707	1.687	65.042	7.798
29.717	2.531	71.283	9.241
41.005	3.604	75.822	10.214
48.822	4.669	80.029	11.266
55.334	5.722		

a. Fit the simple linear model $y = \beta_0 + \beta_1 x + \varepsilon$. Make a residual plot, and comment on the appropriateness of the model.

b. Fit the simple linear model $y = \beta_0 + \beta_1 \ln x + \varepsilon$. Make a residual plot, and comment on the appropriateness of the model.

c. Compute the coefficients and their standard deviations for polynomials of degrees 2, 3, 4, and 5. Make residual plots for each.

d. The article cited at the beginning of this exercise recommends the quartic model $y = \beta_0 + \beta_1 x + \beta_2 x^2 + \beta_3 x^3 + \beta_4 x^4 + \varepsilon$. Does this seem reasonable? Why or why not?

23. The article "Lead Dissolution from Lead Smelter Slags Using Magnesium Chloride Solutions" (A. Xenidis, T. Lillis, and I. Hallikia, *The Aus IMM Proceedings*, 1999:37–44) discusses an investigation of leaching rates of lead in solutions of magnesium chloride. The data in the following table (read from a graph) present the percentage of lead that has been extracted at various times (in minutes).

Time (t)	4	8	16	30	60	120
Percent extracted (y)	1.2	1.6	2.3	2.8	3.6	4.4

a. The article suggests fitting a quadratic model $y = \beta_0 + \beta_1 t + \beta_2 t^2 + \varepsilon$ to these data. Fit this model, and compute the standard deviations of the coefficients.

b. The reaction rate at time t is given by the derivative $dy/dt = \beta_1 + 2\beta_2 t$. Estimate the time at which the reaction rate will be equal to 0.05.

c. The reaction rate at $t = 0$ is equal to β_1. Find a 95% confidence interval for the reaction rate at $t = 0$.

d. Can you conclude that the reaction rate is decreasing with time? Explain.

24. The article "Seismic Hazard in Greece Based on Different Strong Ground Motion Parameters" (S. Koutrakis, G. Karakaisis, et al., *Journal of Earthquake Engineering*, 2002:75–109) presents a study of seismic events in Greece during the period 1978–1997. Of interest is the duration of "strong ground motion," which is the length of time that the acceleration of the ground exceeds a specified value. For each event, measurements of the duration of strong ground motion were made at one or more locations. Table SE24 presents, for each of 121 such measurements, the data for

the duration of time y (in seconds) that the ground acceleration exceeded twice the acceleration due to gravity, the magnitude m of the earthquake, the distance d (in km) of the measurement from the epicenter, and two indicators of the soil type s_1 and s_2, defined as follows: $s_1 = 1$ if the soil consists of soft alluvial deposits, $s_1 = 0$ otherwise, and $s_2 = 1$ if the soil consists of tertiary or older rock, $s_2 = 0$ otherwise. Cases where both $s_1 = 0$ and $s_2 = 0$ correspond to intermediate soil conditions. The article presents repeated measurements at some locations, which we have not included here.

TABLE SE24 Data for Exercise 24

y	m	d	s_1	s_2	y	m	d	s_1	s_2	y	m	d	s_1	s_2
8.82	6.4	30	1	0	4.31	5.3	6	0	0	5.74	5.6	15	0	0
4.08	5.2	7	0	0	28.27	6.6	31	1	0	5.13	6.9	128	1	0
15.90	6.9	105	1	0	17.94	6.9	33	0	0	3.20	5.1	13	0	0
6.04	5.8	15	0	0	3.60	5.4	6	0	0	7.29	5.2	19	1	0
0.15	4.9	16	1	0	7.98	5.3	12	1	0	0.02	6.2	68	1	0
5.06	6.2	75	1	0	16.23	6.2	13	0	0	7.03	5.4	10	0	0
0.01	6.6	119	0	1	3.67	6.6	85	1	0	2.17	5.1	45	0	1
4.13	5.1	10	1	0	6.44	5.2	21	0	0	4.27	5.2	18	1	0
0.02	5.3	22	0	1	10.45	5.3	11	0	1	2.25	4.8	14	0	1
2.14	4.5	12	0	1	8.32	5.5	22	1	0	3.10	5.5	15	0	0
4.41	5.2	17	0	0	5.43	5.2	49	0	1	6.18	5.2	13	0	0
17.19	5.9	9	0	0	4.78	5.5	1	0	0	4.56	5.5	1	0	0
5.14	5.5	10	1	0	2.82	5.5	20	0	1	0.94	5.0	6	0	1
0.05	4.9	14	1	0	3.51	5.7	22	0	0	2.85	4.6	21	1	0
20.00	5.8	16	1	0	13.92	5.8	34	1	0	4.21	4.7	20	1	0
12.04	6.1	31	0	0	3.96	6.1	44	0	0	1.93	5.7	39	1	0
0.87	5.0	65	1	0	6.91	5.4	16	0	0	1.56	5.0	44	1	0
0.62	4.8	11	1	0	5.63	5.3	6	1	0	5.03	5.1	2	1	0
8.10	5.4	12	1	0	0.10	5.2	21	1	0	0.51	4.9	14	1	0
1.30	5.8	34	1	0	5.10	4.8	16	1	0	13.14	5.6	5	1	0
11.92	5.6	5	0	0	16.52	5.5	15	1	0	8.16	5.5	12	1	0
3.93	5.7	65	1	0	19.84	5.7	50	1	0	10.04	5.1	28	1	0
2.00	5.4	27	0	1	1.65	5.4	27	1	0	0.79	5.4	35	0	0
0.43	5.4	31	0	1	1.75	5.4	30	0	1	0.02	5.4	32	1	0
14.22	6.5	20	0	1	6.37	6.5	90	1	0	0.10	6.5	61	0	1
0.06	6.5	72	0	1	2.78	4.9	8	0	0	5.43	5.2	9	0	0
1.48	5.2	27	0	0	2.14	5.2	22	0	0	0.81	4.6	9	0	0
3.27	5.1	12	0	0	0.92	5.2	29	0	0	0.73	5.2	22	0	0
6.36	5.2	14	0	0	3.18	4.8	15	0	0	11.18	5.0	8	0	0
0.18	5.0	19	0	0	1.20	5.0	19	0	0	2.54	4.5	6	0	0
0.31	4.5	12	0	0	4.37	4.7	5	0	0	1.55	4.7	13	0	1
1.90	4.7	12	0	0	1.02	5.0	14	0	0	0.01	4.5	17	0	0
0.29	4.7	5	1	0	0.71	4.8	4	1	0	0.21	4.8	5	0	1
6.26	6.3	9	1	0	4.27	6.3	9	0	1	0.04	4.5	3	1	0
3.44	5.4	4	1	0	3.25	5.4	4	0	1	0.01	4.5	1	1	0
2.32	5.4	5	1	0	0.90	4.7	4	1	0	1.19	4.7	3	1	0
1.49	5.0	4	1	0	0.37	5.0	4	0	1	2.66	5.4	1	1	0
2.85	5.4	1	0	1	21.07	6.4	78	0	1	7.47	6.4	104	0	0
0.01	6.4	86	0	1	0.04	6.4	105	0	1	30.45	6.6	51	1	0
9.34	6.6	116	0	1	15.30	6.6	82	0	1	12.78	6.6	65	1	0
10.47	6.6	117	0	0										

Use the data in Table SE24 to construct a linear model to predict duration y from some or all of the variables m, d, s_1, and s_2. Be sure to consider transformations of the variables, as well as powers of and interactions between the independent variables. Describe the steps taken to construct your model. Plot the residuals versus the fitted values to verify that your model satisfies the necessary assumptions. In addition, note that the data is presented in chronological order, reading down the columns. Make a plot to determine whether time should be included as an independent variable.

25. The article "Estimating Resource Requirements at Conceptual Design Stage Using Neural Networks" (A. Elazouni, I. Nosair, et al., *Journal of Computing in Civil Engineering*, 1997:217–223) suggests that certain resource requirements in the construction of concrete silos can be predicted from a model. These include the quantity of concrete in m^3 (y), the number of crew-days of labor (z), or the number of concrete mixer hours (w) needed for a particular job. Table SE25A defines 23 potential independent variables that can be used to predict y, z, or w. Values of the dependent and independent variables, collected on 28 construction jobs, are presented in Table SE25B and Table SE25C (page 394). Unless otherwise stated, lengths are in meters, areas in m^2, and volumes in m^3.

a. Using best subsets regression, find the model that is best for predicting y according to the adjusted R^2 criterion.

b. Using best subsets regression, find the model that is best for predicting y according to the minimum Mallows C_p criterion.

c. Find a model for predicting y using stepwise regression. Explain the criterion you are using to determine which variables to add to or drop from the model.

d. Using best subsets regression, find the model that is best for predicting z according to the adjusted R^2 criterion.

e. Using best subsets regression, find the model that is best for predicting z according to the minimum Mallows C_p criterion.

f. Find a model for predicting z using stepwise regression. Explain the criterion you are using to determine which variables to add to or drop from the model.

g. Using best subsets regression, find the model that is best for predicting w according to the adjusted R^2 criterion.

h. Using best subsets regression, find the model that is best for predicting w according to the minimum Mallows C_p criterion.

i. Find a model for predicting w using stepwise regression. Explain the criterion you are using to determine which variables to add to or drop from the model.

TABLE SE25A Descriptions of Variables for Exercise 25

x_1	Number of bins	x_{13}	Breadth-to-thickness ratio
x_2	Maximum required concrete per hour	x_{14}	Perimeter of complex
x_3	Height	x_{15}	Mixer capacity
x_4	Sliding rate of the slipform (m/day)	x_{16}	Density of stored material
x_5	Number of construction stages	x_{17}	Waste percent in reinforcing steel
x_6	Perimeter of slipform	x_{18}	Waste percent in concrete
x_7	Volume of silo complex	x_{19}	Number of workers in concrete crew
x_8	Surface area of silo walls	x_{20}	Wall thickness (cm)
x_9	Volume of one bin	x_{21}	Number of reinforcing steel crews
x_{10}	Wall-to-floor areas	x_{22}	Number of workers in forms crew
x_{11}	Number of lifting jacks	x_{23}	Length-to-breadth ratio
x_{12}	Length-to-thickness ratio		

TABLE SE25B Data for Exercise 25

y	z	w	x_1	x_2	x_3	x_4	x_5	x_6	x_7	x_8	x_9	x_{10}	x_{11}
1,850	9,520	476	33	4.5	19.8	4.0	4	223	11,072	14,751	335	26.1	72
932	4,272	268	24	3.5	22.3	4.0	2	206	2,615	8,875	109	27.9	64
556	3,296	206	18	2.7	20.3	5.0	2	130	2,500	5,321	139	28.4	48
217	1,088	68	9	3.2	11.0	4.5	1	152	1,270	1,675	141	11.6	40
199	2,587	199	2	1.0	23.8	5.0	1	79	1,370	7,260	685	17.1	21
56	1,560	120	2	0.5	16.6	5.0	1	43	275	1,980	137	22.0	15
64	1,534	118	2	0.5	18.4	5.0	1	43	330	825	165	23.6	12
397	2,660	133	14	3.0	16.0	4.0	1	240	5,200	18,525	371	12.8	74
1,926	11,020	551	42	3.5	16.0	4.0	4	280	15,500	3,821	369	12.8	88
724	3,090	103	15	7.8	15.0	3.5	1	374	4,500	5,600	300	12.2	114
711	2,860	143	25	5.0	16.0	3.5	1	315	2,100	6,851	87	24.8	60
1,818	9,900	396	28	4.8	22.0	4.0	3	230	13,500	13,860	482	17.6	44
619	2,626	202	12	3.0	18.0	5.0	1	163	1,400	2,935	115	26.4	36
375	2,060	103	12	5.8	15.0	3.5	1	316	4,200	4,743	350	11.8	93
214	1,600	80	12	3.5	15.0	4.5	1	193	1,300	2,988	105	20.6	40
300	1,820	140	6	2.1	14.0	5.0	1	118	800	1,657	133	17.0	24
771	3,328	256	30	3.0	14.0	5.0	3	165	2,800	2,318	92	19.9	43
189	1,456	91	12	4.0	17.0	4.5	1	214	2,400	3,644	200	13.6	53
494	4,160	320	27	3.3	20.0	4.5	3	178	6,750	3,568	250	14.0	44
389	1,520	95	6	4.1	19.0	4.0	1	158	2,506	3,011	401	11.8	38
441	1,760	110	6	4.0	22.0	5.0	1	154	2,568	3,396	428	14.1	35
768	3,040	152	12	5.0	24.0	4.0	1	275	5,376	6,619	448	14.5	65
797	3,180	159	9	5.0	25.0	4.0	1	216	4,514	5,400	501	14.8	52
261	1,131	87	3	3.0	17.5	4.0	1	116	1,568	2,030	522	10.5	24
524	1,904	119	6	4.4	18.8	4.0	1	190	3,291	3,572	548	9.8	42
1,262	5,070	169	15	7.0	24.6	3.5	1	385	8,970	9,490	598	12.9	92
839	7,080	354	9	5.2	25.5	4.0	1	249	5,845	6,364	649	13.9	60
1,003	3,500	175	9	5.7	27.7	4.0	1	246	6,095	6,248	677	15.1	60

TABLE SE25C Data for Exercise 25

X_{12}	X_{13}	X_{14}	X_{15}	X_{16}	X_{17}	X_{18}	X_{19}	X_{20}	X_{21}	X_{22}	X_{23}
19.6	17.6	745	0.50	800	6.00	5.50	10	24	7	20	1.12
16.0	16.0	398	0.25	600	7.00	5.00	10	20	6	20	1.00
15.3	13.5	262	0.25	850	7.00	4.50	8	20	5	18	1.13
17.0	13.8	152	0.25	800	5.00	4.00	8	25	6	16	1.23
28.1	27.5	79	0.15	800	7.50	3.50	5	20	4	14	1.02
20.3	20.0	43	0.15	600	5.00	4.00	5	15	1	12	1.02
24.0	18.3	43	0.15	600	5.05	4.25	5	15	2	12	1.31
27.5	23.0	240	0.25	600	6.00	4.00	8	20	7	22	1.20
27.5	23.0	1121	0.25	800	8.00	4.00	10	20	9	24	1.20
21.2	18.4	374	0.75	800	5.00	3.50	10	25	12	24	1.15
10.6	10.0	315	0.50	800	6.00	4.00	10	25	11	20	1.06
20.0	20.0	630	0.50	800	7.00	5.00	10	25	9	18	1.00
13.7	13.9	163	0.25	600	6.00	4.50	8	18	11	18	1.20
20.4	20.4	316	0.50	800	6.50	3.50	10	25	6	14	1.00
13.6	10.2	193	0.50	800	5.00	3.50	10	25	4	14	1.33
13.6	12.8	118	0.25	800	5.00	3.75	8	25	6	14	1.06
13.6	9.6	424	0.25	800	5.00	3.75	8	25	6	14	1.42
18.5	16.0	214	0.50	600	6.00	4.00	8	20	4	14	1.15
19.5	16.0	472	0.25	600	6.50	4.50	10	20	3	14	1.20
21.0	12.8	158	0.50	800	5.50	3.50	6	25	8	14	1.30
20.8	16.0	154	0.50	800	7.00	4.00	8	36	8	14	1.35
23.4	17.3	275	0.50	600	7.50	5.50	8	22	11	16	1.40
16.8	15.4	216	0.50	800	8.00	5.50	8	28	12	16	1.10
26.8	17.8	116	0.25	850	6.50	3.00	6	25	5	14	1.50
23.6	16.1	190	0.50	850	6.50	4.50	5	28	9	16	1.45
23.6	16.6	385	0.75	800	8.00	6.50	15	25	16	20	1.43
25.6	16.0	249	0.50	600	8.00	5.50	12	25	13	16	1.60
22.3	14.3	246	0.50	800	8.50	6.00	8	28	16	16	1.55

26. The article referred to in Exercise 25 presents values for the dependent and independent variables for 10 additional construction jobs. These values are presented in Tables SE26A and SE26B.

a. Using the equation constructed in part (a) of Exercise 25, predict the concrete quantity (y) for each of these 10 jobs.

b. Denoting the predicted values by $\hat{y}_1, \ldots, \hat{y}_{10}$ and the observed values by $y_1, \ldots, y_{10}$, compute the quantities $y_i - \hat{y}_i$. These are the *prediction errors*.

c. Now compute the fitted values $\hat{y}_1, \ldots, \hat{y}_{28}$ from the data in Exercise 25. Using the observed values $y_1, \ldots, y_{28}$ from those data, compute the residuals $y_i - \hat{y}_i$.

d. On the whole, which are larger, the residuals or the prediction errors? Why will this be true in general?

TABLE SE26A Data for Exercise 26

y	z	w	x_1	x_2	x_3	x_4	x_5	x_6	x_7	x_8	x_9	x_{10}	x_{11}
1,713	3,400	170	6	4.2	27.0	4.0	1	179	4,200	4,980	700.0	15.1	42
344	1,616	101	3	3.4	20.0	5.0	1	133	2,255	2,672	751.5	16.7	30
474	2,240	140	3	3.4	28.0	5.0	1	116	2,396	3,259	798.8	17.0	24
1,336	5,700	190	15	7.0	26.0	3.5	1	344	12,284	9,864	818.9	16.0	86
1,916	9,125	365	18	5.6	26.5	3.5	2	307	15,435	8,140	852.5	12.4	68
1,280	11,980	599	9	2.1	28.3	4.0	1	283	8,064	8,156	896.0	14.0	68
1,683	6,390	213	12	7.9	29.0	3.5	1	361	11,364	10,486	947.0	13.4	87
901	2,656	166	6	5.4	29.5	4.5	1	193	5,592	5,696	932.0	14.8	39
460	2,943	150	3	3.0	30.0	5.0	1	118	2,943	3,540	981.0	17.2	26
826	3,340	167	6	4.9	29.8	4.5	1	211	6,000	6,293	1,000.0	15.1	50

TABLE SE26B Data for Exercise 26

x_{12}	x_{13}	x_{14}	x_{15}	x_{16}	x_{17}	x_{18}	x_{19}	x_{20}	x_{21}	x_{22}	x_{23}
22.5	14.8	179	0.50	850	8.0	5.0	6	28	11	16	1.52
32.0	18.8	133	0.25	800	7.5	3.0	10	25	7	14	1.70
24.6	15.0	116	0.25	800	9.0	4.0	10	28	9	14	1.65
20.2	21.1	344	0.75	850	8.5	6.5	12	28	19	18	1.72
30.0	13.2	540	0.50	600	6.5	7.0	15	25	12	18	1.75
25.3	14.3	283	0.25	800	7.5	6.5	14	30	20	16	1.80
22.7	14.0	361	0.75	800	9.0	7.0	10	30	25	18	1.42
20.5	16.0	193	0.50	850	9.5	5.5	10	30	15	16	1.20
26.0	20.1	118	0.25	600	10.0	4.0	10	25	8	14	1.30
32.0	20.0	211	0.50	600	9.5	5.0	10	25	13	16	1.90

Chapter 9

Factorial Experiments

Introduction

Experiments are essential to the development and improvement of engineering and scientific methods. Many experiments involve varying the values of one or more quantities to determine their effect on a response. For example, agriculture scientists, trying to maximize the yield of a crop, may use any of several types of fertilizer and any of several watering schedules. They will plant the crop on several plots, treat the plots with various combinations of fertilizer and watering schedule, and measure the yield on each plot. Experiments like these are often carried out in industrial settings as well. Examples include studying the effect of catalyst concentration on the yield of a chemical reaction, the effects of cutting speed and blade angle on the life of a tool, and the effects of sand type, cement type, and curing time on the strength of concrete blocks.

The quantities that are varied in these experiments are called **factors**, and the experiments are called **factorial experiments**. For factorial experiments to be useful, they must be designed properly, and the data they produce must be analyzed appropriately. In Sections 9.1 and 9.2, we will discuss experiments in which only one factor is varied. In Sections 9.3 and 9.4, we will discuss two-factor experiments, and in Section 9.5, we will introduce experiments with more than two factors.

9.1 One-Factor Experiments

We begin with an example. The article "An Investigation of the $CaCO_3$-CaF_2-K_2SiO_3-SiO_2-Fe Flux System Using the Submerged Arc Welding Process on HSLA-100 and AISI-1081 Steels" (G. Fredrickson, M.S. thesis, Colorado School of Mines, 1992) describes an experiment in which welding fluxes with differing chemical compositions were prepared. Several welds using each flux were made on AISI-1018 steel base metal.

TABLE 9.1 Brinell hardness of welds using four different fluxes

Flux	Sample Values					Sample Mean	Sample Standard Deviation
A	250	264	256	260	239	253.8	9.7570
B	263	254	267	265	267	263.2	5.4037
C	257	279	269	273	277	271.0	8.7178
D	253	258	262	264	273	262.0	7.4498

The results of hardness measurements, on the Brinell scale, of five welds using each of four fluxes are presented in Table 9.1.

Figure 9.1 presents dotplots for the hardnesses using the four fluxes. Each sample mean is marked with an "X." It is clear that the sample means differ. In particular, the welds made using flux C have the largest sample mean and those using flux A have the smallest. Of course, there is random variation in the sample means, and the question is whether the sample means differ from each other by a greater amount than could be accounted for by random variation alone. Another way to phrase the question is this: Can we conclude that there are differences in the population means among the four flux types?

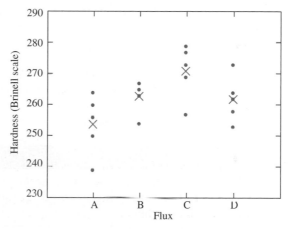

FIGURE 9.1 Dotplots for each sample in Table 9.1. Each sample mean is marked with an "X." The sample means differ somewhat, but the sample values overlap considerably.

This is an example of a factorial experiment. In general, a factorial experiment involves several variables. One variable is the **response variable**, which is sometimes called the **outcome variable** or the **dependent variable**. The other variables are called **factors**. The question addressed by a factorial experiment is whether varying the levels of the factors produces a difference in the mean of the response variable. In the experiment described in Table 9.1, the hardness is the response, and there is one factor: flux type. Since there is only one factor, this is a **one-factor experiment**. There are four different values for the flux-type factor in this experiment. These different values are called the **levels** of the factor and can also be called **treatments**. Finally, the objects on which measurements are made are called **experimental units**. The units assigned to a given treatment are called **replicates**. In the preceding experiment, the welds are the experimental units, and there are five replicates for each treatment.

In this welding experiment, the four particular flux compositions were chosen deliberately by the experimenter, rather than at random from a larger population of fluxes. Such an experiment is said to follow a **fixed effects model**. In some experiments, treatments are chosen at random from a population of possible treatments. In this case the experiment is said to follow a **random effects model**. The methods of analysis for these two models are essentially the same, although the conclusions to be drawn from them differ. We will focus on fixed effects models. Later in this section, we will discuss some of the differences between fixed and random effects models.

Completely Randomized Experiments

In this welding experiment, a total of 20 welds were produced, five with each of the four fluxes. Each weld was produced on a different steel base plate. Therefore, to run the experiment, the experimenter had to choose, from a total of 20 base plates, a group of 5 to be welded with flux A, another group of 5 to be welded with flux B, and so on. Now the base plates vary somewhat in their physical properties, so some base plates may tend to produce harder welds. It is important to avoid any systematic tendency to assign the better base plates to a particular flux. Therefore, the best way to assign the base plates to the fluxes is at random. In this way, the experimental design will not favor any one treatment over another. For example, the experimenter could number the plates from 1 to 20, and then generate a random ordering of the integers from 1 to 20. The plates whose numbers correspond to the first five numbers on the list are assigned to flux A, and so on. This is an example of a **completely randomized experiment**.

Definition

A factorial experiment in which experimental units are assigned to treatments at random, with all possible assignments being equally likely, is called a **completely randomized experiment**.

In many situations, the results of an experiment can be affected by the order in which the observations are taken. For example, the performance of a machine used to make measurements may change over time, due, for example, to calibration drift, or to warm-up effects. In cases like this, the ideal procedure is to take the observations in random order. This requires switching from treatment to treatment as observations are taken, rather than running all the observations that correspond to a given treatment consecutively. In some cases changing treatments involves considerable time or expense, so it is not feasible to switch back and forth. In these cases, the treatments should be run in a random order, with all the observations corresponding to the first randomly chosen treatment being run first, and so on.

In a completely randomized experiment, it is appropriate to think of each treatment as representing a population, and the responses observed for the units assigned to that treatment as a simple random sample from that population. The data from the experiment thus consist of several random samples, each from a different population. The population means are called **treatment means**. The questions of interest concern the treatment means—whether they are all equal, and if not, which ones are different, how big the differences are, and so on.

One-Way Analysis of Variance

To make a formal determination as to whether the treatment means differ, a hypothesis test is needed. We begin by introducing the notation. We have I samples, each from a different treatment. The treatment means are denoted

$$\mu_1, \ldots, \mu_I$$

It is not necessary that the sample sizes be equal, although it is desirable, as we will discuss later in this section. The sample sizes are denoted

$$J_1, \ldots, J_I$$

The total number in all the samples combined is denoted by N.

$$N = J_1 + J_2 + \cdots + J_I$$

The hypotheses we wish to test are

$$H_0 : \mu_1 = \cdots = \mu_I \quad \text{versus} \quad H_1 : \text{two or more of the } \mu_i \text{ are different}$$

If there were only two samples, we could use the two-sample t test (Section 7.3) to test the null hypothesis. Since there are more than two samples, we use a method known as **one-way analysis of variance** (ANOVA). To define the test statistic for one-way ANOVA, we first develop the notation for the sample observations. Since there are several samples, we use a double subscript to denote the observations. Specifically, we let X_{ij} denote the jth observation in the ith sample. The sample mean of the ith sample is denoted $\overline{X}_{i.}$.

$$\overline{X}_{i.} = \frac{\sum_{j=1}^{J_i} X_{ij}}{J_i} \tag{9.1}$$

The **sample grand mean**, denoted $\overline{X}_{..}$, is the average of all the sampled items taken together:

$$\overline{X}_{..} = \frac{\sum_{i=1}^{I} \sum_{j=1}^{J_i} X_{ij}}{N} \tag{9.2}$$

With a little algebra, it can be shown that the sample grand mean is also a weighted average of the sample means:

$$\overline{X}_{..} = \frac{\sum_{i=1}^{I} J_i \overline{X}_{i.}}{N} \tag{9.3}$$

xample

9.1

For the data in Table 9.1, find I, $J_1, \ldots, J_I$, N, X_{23}, $\overline{X}_3$, $\overline{X}_{..}$.

Solution

There are four samples, so $I = 4$. Each sample contains five observations, so $J_1 = J_2 = J_3 = J_4 = 5$. The total number of observations is $N = 20$. The quantity X_{23} is the third observation in the second sample, which is 267. The quantity $\overline{X}_3$ is the sample mean of the third sample. This value is $\overline{X}_3 = 271.0$. Finally, we use Equation (9.3) to compute the sample grand mean $\overline{X}_{..}$.

$$\overline{X}_{..} = \frac{(5)(253.8) + (5)(263.2) + (5)(271.0) + (5)(262.0)}{20}$$

$$= 262.5$$

Figure 9.2 presents the idea behind one-way ANOVA. The figure illustrates several hypothetical samples from different treatments, along with their sample means and the sample grand mean. The sample means are spread out around the sample grand mean. One-way ANOVA provides a way to measure this spread. If the sample means are highly spread out, then it is likely that the treatment means are different, and we will reject H_0.

The variation of the sample means around the sample grand mean is measured by a quantity called the **treatment sum of squares** (SSTr for short), which is given by

$$\text{SSTr} = \sum_{i=1}^{I} J_i (\overline{X}_{i.} - \overline{X}_{..})^2 \tag{9.4}$$

Each term in SSTr involves the distance from the sample means to the sample grand mean. Note that each squared distance is multiplied by the sample size corresponding to its sample mean, so that the means for the larger samples count more. SSTr provides an indication of how different the treatment means are from each other. If SSTr is large, then the sample means are spread out widely, and it is reasonable to conclude that the

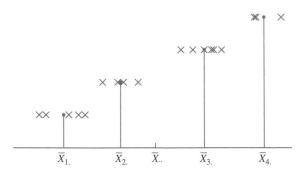

FIGURE 9.2 The variation of the sample means around the sample grand mean can be due both to random variation and to differences among the treatment means. The variation within a given sample around its own sample mean is due only to random variation.

treatment means differ and to reject H_0. If SSTr is small, then the sample means are all close to the sample grand mean and therefore to each other, so it is plausible that the treatment means are equal.

An equivalent formula for SSTr, which is a bit easier to compute by hand, is

$$\text{SSTr} = \sum_{i=1}^{I} J_i \overline{X}_{i.}^{\,2} - N\overline{X}_{..}^{\,2} \tag{9.5}$$

In order to determine whether SSTr is large enough to reject H_0, we compare it to another sum of squares, called the **error sum of squares** (SSE for short). SSE measures the variation in the individual sample points around their respective sample means. This variation is measured by summing the squares of the distances from each point to its own sample mean. SSE is given by

$$\text{SSE} = \sum_{i=1}^{I} \sum_{j=1}^{J_i} (X_{ij} - \overline{X}_{i.})^2 \tag{9.6}$$

The quantities $X_{ij} - \overline{X}_{i.}$ are called the **residuals**, so SSE is the sum of the squared residuals. SSE, unlike SSTr, depends only on the distances of the sample points from their own means and is not affected by the location of treatment means relative to one another. SSE therefore measures only the underlying random variation in the process being studied. It is analogous to the error sum of squares in regression.

An equivalent formula for SSE, which is a bit easier to compute by hand, is

$$\text{SSE} = \sum_{i=1}^{I} \sum_{j=1}^{J_i} X_{ij}^2 - \sum_{i=1}^{I} J_i \overline{X}_{i.}^{\,2} \tag{9.7}$$

Another equivalent formula for SSE is based on the sample variances. Let s_i^2 denote the sample variance of the ith sample. Then

$$s_i^2 = \frac{\sum_{j=1}^{J_i} (X_{ij} - \overline{X}_{i.})^2}{J_i - 1} \tag{9.8}$$

It follows from Equation (9.8) that $\sum_{j=1}^{J_i} (X_{ij} - \overline{X}_{i.})^2 = (J_i - 1)s_i^2$. Substituting into Equation (9.6) yields

$$\text{SSE} = \sum_{i=1}^{I} (J_i - 1)s_i^2 \tag{9.9}$$

Example

9.2

For the data in Table 9.1, compute SSTr and SSE.

Solution

The sample means are presented in Table 9.1. They are

$$\overline{X}_{1.} = 253.8 \qquad \overline{X}_{2.} = 263.2 \qquad \overline{X}_{3.} = 271.0 \qquad \overline{X}_{4.} = 262.0$$

The sample grand mean was computed in Example 9.1 to be $\overline{X}_{..} = 262.5$. We now use Equation (9.4) to calculate SSTr.

$$\text{SSTr} = 5(253.8-262.5)^2+5(263.2-262.5)^2+5(271.0-262.5)^2+5(262.0-262.5)^2$$
$$= 743.4$$

To compute SSE we will use Equation (9.9), since the sample standard deviations s_i have already been presented in Table 9.1.

$$\text{SSE} = (5 - 1)(9.7570)^2 + (5 - 1)(5.4037)^2 + (5 - 1)(8.7178)^2 + (5 - 1)(7.4498)^2$$
$$= 1023.6$$

We can use SSTr and SSE to construct a test statistic, provided the following two assumptions are met.

Assumptions for One-Way ANOVA

The standard one-way ANOVA hypothesis tests are valid under the following conditions:

1. The treatment populations must be normal.

2. The treatment populations must all have the same variance, which we will denote by σ^2.

Before presenting the test statistic, we will explain how it works. If the two assumptions for one-way ANOVA are approximately met, we can compute the means of SSE and SSTr. The mean of SSTr depends on whether H_0 is true, because SSTr tends to be smaller when H_0 is true and larger when H_0 is false. The mean of SSTr satisfies the condition

$$\mu_{\text{SSTr}} = (I - 1)\sigma^2 \quad \text{when } H_0 \text{ is true} \tag{9.10}$$

$$\mu_{\text{SSTr}} > (I - 1)\sigma^2 \quad \text{when } H_0 \text{ is false} \tag{9.11}$$

The likely size of SSE, and thus its mean, does not depend on whether H_0 is true. The mean of SSE is given by

$$\mu_{\text{SSE}} = (N - I)\sigma^2 \quad \text{whether or not } H_0 \text{ is true} \tag{9.12}$$

The quantities $I - 1$ and $N - I$ are the **degrees of freedom** for SSTr and SSE, respectively. When a sum of squares is divided by its degrees of freedom, the quantity obtained is called a **mean square**. The **treatment mean square** is denoted MSTr, and the **error mean square** is denoted MSE. They are defined by

$$\text{MSTr} = \frac{\text{SSTr}}{I - 1} \qquad \text{MSE} = \frac{\text{SSE}}{N - I} \tag{9.13}$$

It follows from Equations (9.10) through (9.13) that

$$\mu_{\text{MSTr}} = \sigma^2 \quad \text{when } H_0 \text{ is true} \tag{9.14}$$

$$\mu_{\text{MSTr}} > \sigma^2 \quad \text{when } H_0 \text{ is false} \tag{9.15}$$

$$\mu_{\text{MSE}} = \sigma^2 \quad \text{whether or not } H_0 \text{ is true} \tag{9.16}$$

Equations (9.14) and (9.16) show that when H_0 is true, MSTr and MSE have the same mean. Therefore, when H_0 is true, we would expect their quotient to be near 1. This quotient is in fact the test statistic. The test statistic for testing $H_0 : \mu_1 = \cdots = \mu_I$ is

$$F = \frac{\text{MSTr}}{\text{MSE}} \tag{9.17}$$

When H_0 is true, the numerator and denominator of F are on average the same size, so F tends to be near 1. In fact, when H_0 is true, this test statistic has an F distribution with $I - 1$ and $N - I$ degrees of freedom, denoted $F_{I-1,N-I}$. When H_0 is false, MSTr tends to be larger, but MSE does not, so F tends to be greater than 1.

The F test for One-Way ANOVA

To test $H_0 : \mu_1 = \cdots = \mu_I$ versus H_1 : two or more of the μ_i are different:

1. Compute $\text{SSTr} = \sum_{i=1}^{I} J_i(\overline{X}_{i.} - \overline{X}_{..})^2 = \sum_{i=1}^{I} J_i \overline{X}_{i.}^2 - N\overline{X}_{..}^2$.

2. Compute $\text{SSE} = \sum_{i=1}^{I}\sum_{j=1}^{J_i}(X_{ij}-\overline{X}_{i.})^2 = \sum_{i=1}^{I}\sum_{j=1}^{J_i} X_{ij}^2 - \sum_{i=1}^{I} J_i \overline{X}_{i.}^2 = \sum_{i=1}^{I}(J_i - 1)s_i^2$.

3. Compute $\text{MSTr} = \dfrac{\text{SSTr}}{I - 1}$ and $\text{MSE} = \dfrac{\text{SSE}}{N - I}$.

4. Compute the test statistic: $F = \dfrac{\text{MSTr}}{\text{MSE}}$.

5. Find the P-value by consulting the F table (Table A.6 in Appendix A) with $I - 1$ and $N - I$ degrees of freedom.

We now apply the method of analysis of variance to the example with which we introduced this section.

Example

9.3

For the data in Table 9.1, compute MSTr, MSE, and F. Find the P-value for testing the null hypothesis that all the means are equal. What do you conclude?

Solution

From Example 9.2, $\text{SSTr} = 743.4$ and $\text{SSE} = 1023.6$. We have $I = 4$ samples and $N = 20$ observations in all the samples taken together. Using Equation (9.13),

$$\text{MSTr} = \frac{743.4}{4 - 1} = 247.8 \qquad \text{MSE} = \frac{1023.6}{20 - 4} = 63.975$$

The value of the test statistic F is therefore

$$F = \frac{247.8}{63.975} = 3.8734$$

To find the P-value, we consult the F table (Table A.6). The degrees of freedom are $4 - 1 = 3$ for the numerator and $20 - 4 = 16$ for the denominator. Under H_0, F has an $F_{3,16}$ distribution. Looking at the F table under 3 and 16 degrees of freedom, we find that the upper 5% point is 3.24 and the upper 1% point is 5.29. Therefore the P-value is between 0.01 and 0.05 (see Figure 9.3; a computer software package gives a value of 0.029 accurate to two significant digits). It is reasonable to conclude that the population means are not all equal and thus that flux composition does affect hardness.

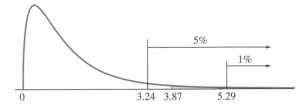

FIGURE 9.3 The observed value of the test statistic is 3.87. The upper 5% point of the $F_{3,16}$ distribution is 3.24. The upper 1% point of the $F_{3,16}$ distribution is 5.29. Therefore the P-value is between 0.01 and 0.05. A computer software package gives a value of 0.029.

The ANOVA Table

The results of an analysis of variance are usually summarized in an analysis of variance (ANOVA) table. This table is much like the analysis of variance table produced in multiple regression. The following output (from MINITAB) shows the analysis of variance for the weld data presented in Table 9.1.

```
One-way ANOVA: A, B, C, D

Source  DF      SS       MS      F      P
Factor   3   743.40   247.800   3.87  0.029
Error   16  1023.60    63.975
Total   19  1767.00

S = 7.998   R-Sq = 42.07%   R-Sq(adj) = 31.21%


                                  Individual 95% CIs For Mean Based on
                                  Pooled StDev
Level  N    Mean   StDev   ----+---------+---------+---------+-----
A      5   253.80   9.76   (-------*------)
B      5   263.20   5.40             (------*-------)
C      5   271.00   8.72                     (------*-------)
D      5   262.00   7.45          (-------*-------)
                               ----+---------+---------+---------+-----
                                 250       260       270       280

Pooled StDev = 8.00
```

In the ANOVA table, the column labeled "DF" presents the number of degrees of freedom for both the treatment ("Factor") and error ("Error") sum of squares. The column labeled "SS" presents SSTr (in the row labeled "Factor") and SSE (in the row labeled "Error"). The row labeled "Total" contains the **total sum of squares**, which is the sum of SSTr and SSE. The column labeled "MS" presents the mean squares MSTr and MSE. The column labeled "F" presents the F statistic for testing the null hypothesis that all the population means are equal. Finally, the column labeled "P" presents the P-value for the F test. Below the ANOVA table, the value "S" is the pooled estimate of the error standard deviation σ, computed by taking the square root of MSE. The quantity "R-sq" is R^2, the coefficient of determination, which is equal to the quotient SSTr/SST. This is analogous to the multiple regression situation (see Equation 8.25 in Section 8.3). The value "R-Sq(adj)" is the adjusted R^2, equal to $R^2 - [(I - 1)/(N - I)](1 - R^2)$, again analogous to multiple regression. The quantities R^2 and adjusted R^2 are not used as much in analysis of variance as they are in multiple regression. Finally, sample means and standard deviations are presented for each treatment group, along with a graphic that illustrates a 95% confidence interval for each treatment mean.

$8\ 27.842$

Example
9.4

In the article "Review of Development and Application of CRSTER and MPTER Models" (R. Wilson, *Atmospheric Environment*, 1993:41–57), several measurements of the maximum hourly concentrations (in $\mu g/m^3$) of SO_2 are presented for each of four power plants. The results are as follows (two outliers have been deleted):

$\text{one}\ 4$
$SSr =$

$606.75 \quad I = 4$
$DF = 18$

Plant 1:	438	619	732	638	
Plant 2:	857	1014	1153	883	1053
Plant 3:	925	786	1179	786	
Plant 4:	893	891	917	695	675	595

The following output (from MINITAB) presents results for a one-way ANOVA. Can you conclude that the maximum hourly concentrations differ among the plants?

```
One-way ANOVA: Plant 1, Plant 2, Plant 3, Plant 4

Source   DF       SS       MS      F      P
Plant     3   378610   126203   6.21  0.006
Error    15   304838    20323
Total    18   683449

S = 142.6    R-Sq = 55.40%    R-Sq(adj) = 46.48%


                  Individual 95% CIs For Mean Based on
                  Pooled StDev
```

```
Level  N   Mean   StDev  -------+---------+---------+---------+--
  1    4  606.8   122.9  (------*-------)
  2    5  992.0   122.7                          (------*-----)
  3    4  919.0   185.3                    (-------*-------)
  4    6  777.7   138.8              (-----*-----)
                              -------+---------+---------+---------+--
                                   600       800      1000      1200

Pooled StDev = 142.6
```

Solution

In the ANOVA table, the P-value for the null hypothesis that all treatment means are equal is 0.006. Therefore we conclude that not all the treatment means are equal.

Checking the Assumptions

As previously mentioned, the methods of analysis of variance require the assumptions that the observations on each treatment are a sample from a normal population and that the normal populations all have the same variance. A good way to check the normality assumption is with a normal probability plot. If the sample sizes are large enough, one can construct a separate probability plot for each sample. This is rarely the case in practice. When the sample sizes are not large enough for individual probability plots to be informative, the residuals $X_{ij} - \overline{X}_{i.}$ can all be plotted together in a single plot. When the assumptions of normality and constant variance are satisfied, these residuals will be normally distributed with mean zero and should plot approximately on a straight line. Figure 9.4 presents a normal probability plot of the residuals for the weld data of Table 9.1. There is no evidence of a serious violation of the assumption of normality.

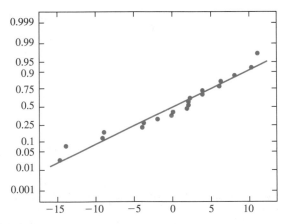

FIGURE 9.4 Probability plot for the residuals from the weld data. There is no evidence of a serious violation of the assumption of normality.

The assumption of equal variances can be difficult to check, because with only a few observations in each sample, the sample standard deviations can differ greatly (by a factor of 2 or more) even when the assumption holds. For the weld data, the sample standard deviations range from 5.4037 to 9.7570. It is reasonable to proceed as though the variances were equal.

The spreads of the observations within the various samples can be checked visually by making a residual plot. This is done by plotting the residuals $X_{ij} - \overline{X}_{i.}$ versus the fitted values, which are the sample means $\overline{X}_{i.}$. If the spreads differ considerably among the samples, the assumption of equal variances is suspect. If one or more of the samples contain outliers, the assumption of normality is suspect as well. Figure 9.5 presents a residual plot for the weld data. There are no serious outliers, and the spreads do not differ greatly among samples.

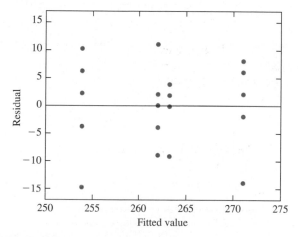

FIGURE 9.5 Residual plot of the values $X_{ij} - \overline{X}_{i.}$ versus $\overline{X}_{i.}$ for the weld data. The spreads do not differ greatly from sample to sample, and there are no serious outliers.

Balanced versus Unbalanced Designs

When equal numbers of units are assigned to each treatment, the design is said to be **balanced**. Although one-way analysis of variance can be used with both balanced and unbalanced designs, balanced designs offer a big advantage. A balanced design is much less sensitive to violations of the assumption of equality of variance than an unbalanced one. Since moderate departures from this assumption can be difficult to detect, it is best to use a balanced design whenever possible, so that undetected violations of the assumption will not seriously compromise the validity of the results. When a balanced design is impossible to achieve, a slightly unbalanced design is preferable to a severely unbalanced one.

Summary

■ With a balanced design, the effect of unequal variances is generally not great.

■ With an unbalanced design, the effect of unequal variances can be substantial.

■ The more unbalanced the design, the greater the effect of unequal variances.

The Analysis of Variance Identity

In both linear regression and analysis of variance, a quantity called the total sum of squares is obtained by subtracting the sample grand mean from each observation, squaring these deviations, and then summing them. An analysis of variance identity is an equation that expresses the total sum of squares as a sum of other sums of squares. We have presented an analysis of variance identity for multiple regression (Equation 8.23 in Section 8.3).

The total sum of squares for one-way ANOVA is given by

$$\text{SST} = \sum_{i=1}^{I} \sum_{j=1}^{J_i} (X_{ij} - \overline{X}_{..})^2 \tag{9.18}$$

An equivalent formula is given by

$$\text{SST} = \sum_{i=1}^{I} \sum_{j=1}^{J_i} X_{ij}^2 - N\overline{X}_{..}^2 \tag{9.19}$$

Examining Equations (9.5), (9.7), and (9.19) shows that the total sum of squares is equal to the treatment sum of squares plus the error sum of squares. This is the analysis of variance identity for one-way analysis of variance.

The Analysis of Variance Identity

$$\text{SST} = \text{SSTr} + \text{SSE} \tag{9.20}$$

An Alternate Parameterization

Our presentation of one-way analysis of variance, as a method to compare several treatment means by using random samples drawn from each treatment population, is one natural way to view the subject. There is another way to express these same ideas, in somewhat different notation, that is sometimes useful.

For each observation X_{ij}, define $\varepsilon_{ij} = X_{ij} - \mu_i$, the difference between the observation and its mean. By analogy with linear regression, the quantities ε_{ij} are called **errors**. It is clearly true that

$$X_{ij} = \mu_i + \varepsilon_{ij} \tag{9.21}$$

Now since X_{ij} is normally distributed with mean μ_i and variance σ^2, it follows that ε_{ij} is normally distributed with mean 0 and variance σ^2.

In a single-factor experiment, we are interested in determining whether the treatment means are all equal. Given treatment means $\mu_1, \ldots, \mu_I$, the quantity

$$\mu = \frac{1}{I} \sum_{i=1}^{I} \mu_i \tag{9.22}$$

is the average of all the treatment means. The quantity μ is called the **population grand mean**. The ith **treatment effect**, denoted α_i, is the difference between the ith treatment mean and the population grand mean:

$$\alpha_i = \mu_i - \mu \tag{9.23}$$

It follows from the definition of α_i that $\sum_{i=1}^{I} \alpha_i = 0$.

We can now decompose the treatment means as follows:

$$\mu_i = \mu + \alpha_i \tag{9.24}$$

Combining Equations (9.21) and (9.24) yields the **one-way analysis of variance model**:

$$X_{ij} = \mu + \alpha_i + \varepsilon_{ij} \tag{9.25}$$

The null hypothesis $H_0 : \mu_1 = \cdots = \mu_I$ is equivalent to $H_0 : \alpha_1 = \cdots = \alpha_I = 0$.

In one-way ANOVA, it is possible to work with the treatment means μ_i, as we have done, rather than with the treatment effects α_i. In multifactor experiments, however, the treatment means by themselves are not sufficient and must be decomposed in a manner analogous to the one described here. We will discuss this further in Section 9.3.

Random Effects Models

In many factorial experiments, the treatments are chosen deliberately by the experimenter. These experiments are said to follow a **fixed effects model**. In some cases, the treatments are chosen at random from a population of possible treatments. In these cases the experiments are said to follow a **random effects model**. In a fixed effects model, the interest is on the specific treatments chosen for the experiment. In a random effects model, the interest is in the whole population of possible treatments, and there is no particular interest in the ones that happened to be chosen for the experiment.

The article describing the weld experiment states that the treatments were chosen deliberately and do not represent a random sample from a larger population of flux compositions. This experiment therefore follows a fixed effects model. The four power plants in Example 9.4 are a sample of convenience; they are plants at which measurements were readily available. In some cases it is appropriate to treat a sample of convenience as if it were a simple random sample (see the discussion in Section 1.1). If these conditions hold, then the power plant experiment may be considered to follow a random effects model; otherwise it must be treated as a fixed effects model.

There is an important difference in interpretation between the results of a fixed effects model and those of a random effects model. In a fixed effects model, the only conclusions that can be drawn are conclusions about the treatments actually used in the experiment. In a random effects model, however, since the treatments are a simple random sample from a population of treatments, conclusions can be drawn concerning the whole population, including treatments not actually used in the experiment.

This difference in interpretations results in a difference in the null hypotheses to be tested. In the fixed effects model, the null hypothesis of interest is $H_0: \mu_1 = \cdots = \mu_I$. In the random effects model, the null hypothesis of interest is

$$H_0: \text{the treatment means are equal for every treatment in the population}$$

In the random effects model, the alternate hypothesis is that the population of treatment means is normally distributed, and therefore not the same for all treatments.

Interestingly enough, although the null hypothesis for the random effects model differs from that of the fixed effects model, the hypothesis test is exactly the same. The F test previously described is used for the random effects model as well as for the fixed effects model.

In Example 9.4, assume that it is reasonable to treat the four power plants as a random sample from a large population of power plants, and furthermore, assume that the SO_2 concentrations in the population of plants are normally distributed. Can we conclude that there are differences in SO_2 concentrations among the power plants in the population?

Solution

This is a random effects model, so we can use the F test to test the null hypothesis that all the treatment means in the population are the same. The results of the F test are shown in Example 9.4. The P-value is 0.006. We therefore reject the null hypothesis and conclude that there are differences in mean SO_2 concentrations among the power plants in the population.

Exercises for Section 9.1

1. A study is made of the effect of curing temperature on the compressive strength of a certain type of concrete. Five concrete specimens are cured at each of four temperatures, and the compressive strength of each specimen is measured (in MPa). The results are as follows:

Temperature (°C)	Strengths				
0	31.2	29.6	30.8	30.0	31.4
10	30.0	27.7	31.1	31.3	30.6
20	35.9	36.8	35.0	34.6	36.5
30	38.3	37.0	37.5	36.1	38.4

   a. Construct an ANOVA table. You may give a range for the P-value.
   b. Can you conclude that the mean strength differs with temperature?

2. The yield strength of CP titanium welds was measured for welds cooled at rates of 10 °C/s, 15 °C/s, and 28 °C/s. The results are presented in the following table. (Based on the article "Advances in Oxygen Equivalence Equations for Predicting the Properties of Titanium Welds," D. Harwig, W. Ittiwattana, and H. Castner, *The Welding Journal*, 2001:126s–136s.)

Cooling Rate	Yield Strengths
10	71.00 75.00 79.67 81.00 75.50 72.50 73.50 78.50 78.50
15	63.00 68.00 73.00 76.00 79.67 81.00
28	68.65 73.70 78.40 84.40 91.20 87.15 77.20 80.70 84.85 88.40

a. Construct an ANOVA table. You may give a range for the P-value.

b. Can you conclude that the yield strength of CP titanium welds varies with the cooling rate?

3. The removal of ammoniacal nitrogen is an important aspect of treatment of leachate at landfill sites. The rate of removal (in % per day) is recorded for several days for each of several treatment methods. The results are presented in the following table. (Based on the article "Removal of Ammoniacal Nitrogen from Landfill Leachate by Irrigation onto Vegetated Treatment Planes," S. Tyrrel, P. Leeds-Harrison, and K. Harrison, *Water Research*, 2002:291–299.)

Treatment	Rate of Removal
A	5.21 4.65
B	5.59 2.69 7.57 5.16
C	6.24 5.94 6.41
D	6.85 9.18 4.94
E	4.04 3.29 4.52 3.75

a. Construct an ANOVA table. You may give a range for the P-value.

b. Can you conclude that the treatment methods differ in their rates of removal?

4. In the article "Calibration of an FTIR Spectrometer" (P. Pankratz, *Statistical Case Studies for Industrial and Process Improvement*, SIAM-ASA, 1997:19–38), a spectrometer was used to make five measurements of the carbon content (in ppb) of a certain silicon wafer on four consecutive days. The results are as follows:

Day 1:	358	390	380	372	366
Day 2:	373	376	413	367	368
Day 3:	336	360	370	368	352
Day 4:	368	359	351	349	343

a. Construct an ANOVA table. You may give a range for the P-value.

b. Can you conclude that the calibration of the spectrometer differs among the four days?

5. The article "Quality of the Fire Clay Coal Bed, Southeastern Kentucky" (J. Hower, W. Andrews, et al., *Journal of Coal Quality*, 1994:13–26) contains measurements on samples of coal from several sites in Kentucky. The data on percent TiO_2 ash are as follows (one outlier has been deleted):

Buckeye Hollow:	0.96	0.86	0.94	0.91	0.70	1.28	1.19	1.04	1.42	0.82
	0.89	1.45	1.66	1.68	2.10	2.19				
Bear Branch:	0.91	1.42	2.54	2.23	2.20	1.44	1.70	1.53	1.84	
Defeated Creek:	1.30	1.39	2.58	1.49	1.49	2.07	1.87	1.39	1.02	0.91
	0.82	0.67	1.34	1.51						
Turkey Creek:	1.20	1.60	1.32	1.24	1.08	2.33	1.81	1.76	1.25	0.81

(continued Turkey Creek row values: 0.95 1.92)

a. Construct an ANOVA table. You may give a range for the P-value.

b. Can you conclude that there are differences in TiO_3 content among these sites?

6. Archaeologists can determine the diets of ancient civilizations by measuring the ratio of carbon-13 to carbon-12 in bones found at burial sites. Large amounts of carbon-13 suggest a diet rich in grasses such as maize, while small amounts suggest a diet based on herbaceous plants. The article "Climate and Diet in Fremont Prehistory: Economic Variability and Abandonment of Maize Agriculture in the Great Salt Lake Basin" (J. Coltrain and S. Leavitt, *American Antiquity*, 2002:453–485) reports ratios, as a difference from a standard in units of parts per thousand, for bones from individuals in several age groups. The data are presented in the following table.

Age Group (years)	Ratio
0–11	17.2 18.4 17.9 16.6 19.0 18.3 13.6 13.5 18.5 19.1 19.1 13.4
12–24	14.8 17.6 18.3 17.2 10.0 11.3 10.2 17.0 18.9 19.2
25–45	18.4 13.0 14.8 18.4 12.8 17.6 18.8 17.9 18.5 17.5 18.3 15.2 10.8 19.8 17.3 19.2 15.4 13.2
46+	15.5 18.2 12.7 15.1 18.2 18.0 14.4 10.2 16.7

   a. Construct an ANOVA table. You may give a range for the P-value.
   b. Can you conclude that the concentration ratios differ among the age groups?

7. The article "Secretion of Parathyroid Hormone Oscillates Depending on the Change in Serum Ionized Calcium During Hemodialysis and May Affect Bone Metabolism" (T. Kitahara, K. Ueki, et al., *Nephron Clinical Practice*, 2005:c9–c17) presents measurements of basal ionized calcium (Ca) levels for four groups of patients with differing levels of basal intact parathyroid hormone (PTH). The following results are consistent with means and standard deviations presented in the article.

Group	Ca levels (mM)
I	1.23 1.02 1.33 1.36 1.11 1.51 1.51 1.30 1.36 1.34 1.27 1.43 1.21 1.69 1.28 1.33 1.49
II	1.18 1.16 1.07 1.21 1.01 1.26 1.37 1.09 1.28 1.33 0.98 0.99 1.24 1.12 1.26 1.27 1.26 1.33 1.26 1.32
III	1.04 1.32 1.29 0.95 1.38 1.08 1.65 1.14 1.44 1.37 1.11 0.82 1.31 1.09 1.46
IV	1.35 1.67 1.38 1.05 1.32 0.95 1.21 1.21 1.22 1.13 1.51 0.72 1.33 1.46

   a. Construct an ANOVA table. You may give a range for the P-value.
   b. Can you conclude that there are differences among the mean Ca levels?

8. The article "Impact of Free Calcium Oxide Content of Fly Ash on Dust and Sulfur Dioxide Emissions in a Lignite-Fired Power Plant" (D. Sotiropoulos, A. Georgakopoulos, and N. Kolovos, *Journal of Air and Waste Management*, 2005:1042–1049) presents measurements of dust emissions, in mg/m^3, for four power plants. Thirty measurements were taken for each plant. The sample means and standard deviations are presented in the following table:

	Mean	Standard Deviation	Sample Size
Plant 1	211.50	24.85	30
Plant 2	214.00	35.26	30
Plant 3	211.75	33.53	30
Plant 4	236.08	23.09	30

a. Construct an ANOVA table. You may give a range for the *P*-value.
b. Can you conclude that there are differences among the mean emission levels?

9. A certain chemical reaction was run three times at each of three temperatures. The yields, expressed as a percent of a theoretical maximum, were as follows:

Temperature (°C)	Yields		
70	81.1	82.6	77.4
80	93.3	88.9	86.0
90	87.8	89.2	88.5

a. Construct an ANOVA table. You may give a range for the *P*-value.
b. Can you conclude that the mean yield differs with temperature?

10. An experiment to compare the lifetimes of four different brands of spark plug was carried out. Five plugs of each brand were used, and the number of miles until failure was recorded for each. Following is part of the MINITAB output for a one-way ANOVA.

```
One-way Analysis of Variance

Analysis of Variance
Source    DF          SS      MS        F      P
Brand      3     176.482    (a)      (b)    (c)
Error    (d)        (e)    (f)
Total     19     235.958
```

Fill in the missing numbers in the table. You may give a range for the *P*-value.

11. Refer to Exercise 10. Is it plausible that the brands of spark plug all have the same mean lifetime?

12. Four different types of solar energy collectors were tested. Each was tested at five randomly chosen times, and the power (in watts) was measured. The results were as follows.

Collector	Power				
A	1.9	1.6	2.0	1.8	1.6
B	1.7	1.9	1.8	1.7	1.7
C	1.2	0.9	1.2	0.9	1.4
D	1.5	1.0	1.4	1.3	1.4

a. Construct an ANOVA table. You may give a range for the *P*-value.
b. Can you conclude that the mean power differs for different collectors?

13. An experiment was performed to determine whether the annealing temperature of ductile iron affects its tensile strength. Five specimens were annealed at each of four temperatures. The tensile strength (in ksi) was measured for each. The results are presented in the following table.

Temperature (°C)	Sample Values				
750	19.72	20.88	19.63	18.68	17.89
800	16.01	20.04	18.10	20.28	20.53
850	16.66	17.38	14.49	18.21	15.58
900	16.93	14.49	16.15	15.53	13.25

   a. Construct an ANOVA table. You may give a range for the P-value.
   b. Can you conclude that there are differences among the mean strengths?

14. The article "The Lubrication of Metal-on-Metal Total Hip Joints: A Slide Down the Stribeck Curve" (S. Smith, D. Dowson, and A. Goldsmith, *Proceedings of the Institution of Mechanical Engineers*, 2001:483–493) presents results from wear tests done on metal artificial hip joints. Joints with several different diameters were tested. The data presented in the following table on head roughness are consistent with the means and standard deviations reported in the article.

Diameter (mm)	Head Roughness (nm)				
16	0.83	2.25	0.20	2.78	3.93
28	2.72	2.48	3.80		
36	5.99	5.32	4.59		

   a. Construct an ANOVA table. You may give a range for the P-value.
   b. Can you conclude that mean roughness varies with diameter? Explain.

15. The article "Mechanical Grading of Oak Timbers" (D. Kretschmann and D. Green, *Journal of Materials in Civil Engineering*, 1999:91–97) presents measurements of the modulus of rupture, in MPa, for green mixed oak 7 by 9 timbers from West Virginia and Pennsylvania. The sample means, standard deviations, and sample sizes for four different grades of lumber are presented in the following table.

Grade	Mean	Standard Deviation	Sample Size
Select	45.1	8.52	32
No. 1	42.0	5.50	11
No. 2	33.2	6.71	15
Below grade	38.1	8.04	42

   a. Construct an ANOVA table. You may give a range for the P-value.
   b. Can you conclude that the mean modulus of rupture differs for different grades of lumber?

16. The article "Withdrawal Strength of Threaded Nails" (D. Rammer, S. Winistorfer, and D. Bender, *Journal of Structural Engineering*, 2001:442–449) describes an experiment comparing the withdrawal strengths for several types of nails. The data presented in the following table are consistent with means and standard deviations reported in the article for

three types of nails: annularly threaded, helically threaded, and smooth shank. All nails had diameters within 0.1 mm of each other, and all were driven into the same type of lumber.

Nail Type	Withdrawal Strength (N/mm)
Annularly threaded	36.57 29.67 43.38 26.94 12.03 21.66 41.79 31.50 35.84 40.81
Helically threaded	14.66 24.22 23.83 21.80 27.22 38.25 28.15 36.35 23.89 28.44
Smooth shank	12.61 25.71 17.69 24.69 26.48 19.35 28.60 42.17 25.11 19.98

a. Construct an ANOVA table. You may give a range for the P-value.
b. Can you conclude that the mean withdrawal strength is different for different nail types?

17. The article "Solid-Phase Chemical Fractionation of Selected Trace Metals in Some Northern Kentucky Soils" (A. Karathanasis and J. Pils, *Soil and Sediment Contamination*, 2005:293–308) presents pH measurements of soil specimens taken from three different types of soils. The results in the following table are consistent with means and standard deviations reported in the article.

Soil Type	pH Measurements
Alluvium	6.53, 6.03, 6.75, 6.82, 6.24
Glacial till	6.07, 6.07, 5.36, 5.57, 5.48, 5.27, 5.80, 5.03, 6.65
Residuum	6.03, 6.16, 6.63, 6.13, 6.05, 5.68, 6.25, 5.43, 6.46, 6.91, 5.75, 6.53

a. Construct an ANOVA table. You may give a range for the P-value.
b. Can you conclude that there are differences among the mean pH levels?

9.2 Pairwise Comparisons in One-Factor Experiments

In a one-way ANOVA, an F test is used to test the null hypothesis that all the treatment means are equal. If this hypothesis is rejected, we can conclude that the treatment means are not all the same. But the test does not tell us which ones are different from the rest. Sometimes an experimenter may want to determine all the pairs of means that can be concluded to differ from each other. In this case a type of procedure called a **multiple comparisons method** must be used. We will discuss the Tukey-Kramer method, which is the most often used multiple comparisons method designed for ANOVA.

The Tukey-Kramer Method of Multiple Comparisons

The Tukey-Kramer method involves computing confidence intervals for the differences between every pair of treatment means $\mu_i - \mu_j$. With ordinary confidence intervals, a level $1 - \alpha$ confidence interval has probability $1 - \alpha$ of covering the true value, so that if several confidence intervals are computed, the probability that they all cover their true values is generally less than $1 - \alpha$. When Tukey-Kramer intervals are computed at level

$1 - \alpha$, however, the probability that all of the confidence intervals simultaneously cover their true values is equal to $1 - \alpha$. For this reason these confidence intervals are called **simultaneous confidence intervals**.

To construct the Tukey-Kramer confidence intervals, the point estimate for each pair $\mu_i - \mu_j$ is $\overline{X}_{i.} - \overline{X}_{j.}$. The critical values come from a distribution called the **Studentized range distribution**. This distribution has two values for degrees of freedom, which for the Tukey-Kramer method are I and $N - I$. (In comparison, the F test uses $I - 1$ and $N - I$ degrees of freedom.) The critical value for level $1 - \alpha$ confidence intervals is the $1 - \alpha$ quantile; this quantity is denoted $q_{I,N-I,\alpha}$. Table A.7 (in Appendix A) presents values of $q_{I,N-I,\alpha}$ for various values of I, N, and α. The margin of error for the confidence interval for $\mu_i - \mu_j$ is $q_{I,N-I,\alpha}\sqrt{(\text{MSE}/2)(1/J_i + 1/J_j)}$. This quantity is sometimes called the **honestly significant difference** (HSD).

The Tukey-Kramer method can also be used to construct **simultaneous hypothesis tests**. When simultaneous hypothesis tests are conducted for all null hypotheses of the form $H_0: \mu_i - \mu_j = 0$, then we can reject, at level α, every null hypothesis whose P-value is less than α.

The Tukey-Kramer Method for Simultaneous Confidence Intervals and Hypothesis Tests

The Tukey-Kramer level $100(1 - \alpha)\%$ simultaneous confidence intervals for all differences $\mu_i - \mu_j$ are

$$\overline{X}_{i.} - \overline{X}_{j.} \pm q_{I,N-I,\alpha}\sqrt{\frac{\text{MSE}}{2}\left(\frac{1}{J_i} + \frac{1}{J_j}\right)} \qquad (9.26)$$

We are $100(1 - \alpha)\%$ confident that the Tukey-Kramer confidence intervals contain the true value of the difference $\mu_i - \mu_j$ for every i and j.

To test all null hypotheses $H_0: \mu_i - \mu_j = 0$ simultaneously, the test statistics are

$$\frac{\overline{X}_{i.} - \overline{X}_{j.}}{\sqrt{\dfrac{\text{MSE}}{2}\left(\dfrac{1}{J_i} + \dfrac{1}{J_j}\right)}}$$

The P-value for each test is found by consulting the Studentized range table (Table A.7) with I and $N - I$ degrees of freedom.

For every pair of levels i and j for which

$$|\overline{X}_{i.} - \overline{X}_{j.}| > q_{I,N-I,\alpha}\sqrt{\frac{\text{MSE}}{2}\left(\frac{1}{J_i} + \frac{1}{J_j}\right)}$$

the null hypothesis $H_0: \mu_i - \mu_j = 0$ is rejected at level α.

A note on terminology: When the design is balanced, with all sample sizes equal to J, the quantity $\sqrt{(\text{MSE}/2)(1/J_i + 1/J_j)}$ is equal to $\sqrt{\text{MSE}/J}$ for all pairs of levels. In this case, the method is often simply called Tukey's method.

Example
9.6

The following output (from MINITAB) presents the ANOVA table for the weld data in Table 9.1 (in Section 9.1). Which pairs of fluxes, if any, can be concluded, at the 5% level, to differ in their effect on hardness?

```
One-way ANOVA: A, B, C, D

Source   DF      SS        MS       F      P
Factor    3    743.40   247.800   3.87   0.029
Error    16   1023.60    63.975
Total    19   1767.00

S = 7.998    R-Sq = 42.07%    R-Sq(adj) = 31.21%
```

Solution

There are $I = 4$ levels, with $J = 5$ observations at each level, for a total of $N = 20$ observations in all. To test at level $\alpha = 0.05$, we consult the Studentized range table (Table A.7) to find $q_{4,16,.05} = 4.05$.

The value of MSE is 63.975. Therefore $q_{I,N-I,\alpha}\sqrt{MSE/J} = 4.05\sqrt{63.975/5} = 14.49$. The four sample means (from Table 9.1) are as follows:

Flux	A	B	C	D
Mean hardness	253.8	263.2	271.0	262.0

There is only one pair of sample means, 271.0 and 253.8, whose difference is greater than 14.49. We therefore conclude that welds produced with flux A have a different mean hardness than welds produced with flux C. None of the other differences are significant at the 5% level.

The following output (from MINITAB) presents the Tukey-Kramer 95% simultaneous confidence intervals for the weld data.

```
Tukey 95% Simultaneous Confidence Intervals
All Pairwise Comparisons

Individual confidence level = 98.87%


A subtracted from:

     Lower   Center   Upper   ------+---------+---------+---------+---
B   -5.087    9.400  23.887                   (--------*---------)
C    2.713   17.200  31.687                        (--------*---------)
D   -6.287    8.200  22.687               (--------*---------)
                             ------+---------+---------+---------+---
                                 -15        0        15       30
```

B subtracted from:

```
        Lower   Center   Upper    ------+---------+---------+---------+---
C      -6.687    7.800  22.287               (--------*---------)
D     -15.687   -1.200  13.287         (--------*---------)
                                 ------+---------+---------+---------+---
                                     -15         0        15        30
```

C subtracted from:

```
        Lower   Center   Upper    ------+---------+---------+---------+---
D     -23.487   -9.000   5.487    (---------*---------)
                                 ------+---------+---------+---------+---
                                     -15         0        15        30
```

The values labeled "Center" are the differences between pairs of treatment means. The quantities labeled "Lower" and "Upper" are the lower and upper bounds, respectively, of the confidence interval. We are 95% confident that every one of these confidence intervals contains the true difference in treatment means. Note that the "Individual confidence level" is 98.87%. This means that we are 98.87% confident that any one specific confidence interval contains its true value.

Example
9.7

In Example 9.4 (in Section 9.1), several measurements of the maximum hourly concentrations (in $\mu g/m^3$) of SO_2 were presented for each of four power plants, and it was concluded that the mean concentrations at the four plants were not all the same. The following output (from MINITAB) presents the Tukey-Kramer 95% simultaneous confidence intervals for mean concentrations at the four plants. Which pairs of plants, if any, can you conclude with 95% confidence to have differing means?

```
Tukey 95% Simultaneous Confidence Intervals
All Pairwise Comparisons

Individual confidence level = 98.87%


1 subtracted from:

        Lower   Center   Upper    -----+---------+---------+---------+----
2      109.4    385.3   661.1                    (--------*--------)
3       21.4    312.3   603.1                  (--------*---------)
4      -94.6    170.9   436.4             (--------*--------)
                                 -----+---------+---------+---------+----
                                    -300         0       300       600
```

2 subtracted from:

```
      Lower  Center  Upper   -----+---------+---------+---------+----
 3   -348.9  -73.0   202.9         (---------*--------)
 4   -463.4  -214.3   34.7   (--------*-------)
                             -----+---------+---------+---------+----
                              -300          0        300       600
```

3 subtracted from:

```
      Lower  Center  Upper   -----+---------+---------+---------+----
 4   -406.8  -141.3  124.1        (--------*--------)
                             -----+---------+---------+---------+----
                              -300          0        300       600
```

Solution

Among the simultaneous confidence intervals, there are two that do not contain 0. These are the intervals for $\mu_1 - \mu_2$ and for $\mu_1 - \mu_3$. Therefore we conclude that the mean concentrations differ between plants 1 and 2 and between plants 1 and 3.

Exercises for Section 9.2

1. The article "Organic Recycling for Soil Quality Conservation in a Sub-Tropical Plateau Region" (K. Chakrabarti, B. Sarkar, et al., *J. Agronomy and Crop Science*, 2000:137–142) reports an experiment in which soil specimens were treated with six different treatments, with two replicates per treatment, and the acid phosphate activity (in μmol *p*-nitrophenol released per gram of oven-dry soil per hour) was recorded. An ANOVA table for a one-way ANOVA is presented in the following box.

```
One-way ANOVA:  Treatments A, B, C, D, E, F

Source      DF      SS        MS       F       P
Treatment    5   1.18547   0.23709  46.64   0.000
Error        6   0.03050   0.00508
Total       11   1.21597
```

The treatment means were

Treatment	A	B	C	D	E	F
Mean	0.99	1.99	1.405	1.63	1.395	1.22

a. Can you conclude that there are differences in acid phosphate activity among the treatments?

b. Use the Tukey-Kramer method to determine which pairs of treatment means, if any, are different at the 5% level.

2. The article "Optimum Design of an A-pillar Trim with Rib Structures for Occupant Head Protection" (H. Kim and S. Kang, *Proceedings of the Institution of Mechanical Engineers*, 2001:1161–1169) discusses a study in which several types of A-pillars were compared to determine which provided the greatest protection to occupants of automobiles during a collision. Following is a one-way ANOVA table, where the treatments are three levels of longitudinal spacing

of the rib (the article also discussed two insignificant factors, which are omitted here). There were nine replicates at each level. The response is the head injury criterion (HIC), which is a unitless quantity that measures the impact energy absorption of the pillar.

```
One-way ANOVA: Spacing

Source    DF        SS        MS        F       P
Spacing    2    50946.6   25473.3   5.071   0.015
Error     24   120550.9    5023.0
Total     26   171497.4
```

The treatment means were

Treatment	A	B	C
Mean	930.87	873.14	979.41

    a. Can you conclude that the longitudinal spacing affects the absorption of impact energy?
    b. Use the Tukey-Kramer method to determine which pairs of treatment means, if any, are different at the 5% level.

3. Refer to Exercise 5 in Section 9.1. Use the Tukey-Kramer method to determine which pairs of treatments, if any, differ at the 5% level.

4. Refer to Exercise 9 in Section 9.1. Use the Tukey-Kramer method to determine which pairs of temperatures, if any, differ at the 5% level.

5. Refer to Exercise 14 in Section 9.1. Use the Tukey-Kramer method to determine which pairs of diameters, if any, differ at the 5% level.

6. Refer to Exercise 16 in Section 9.1. Use the Tukey-Kramer method to determine which pairs of nail types, if any, differ at the 5% level.

7. In an experiment to determine the effect of catalyst on the yield of a certain reaction, the mean yields for reactions run with each of four catalysts were $\overline{X}_{1.} = 89.88$, $\overline{X}_{2.} = 89.51$, $\overline{X}_{3.} = 86.98$, and $\overline{X}_{4.} = 85.79$. Assume that five runs were made with each catalyst.

    a. If MSE $= 3.85$, compute the value of the F statistic for testing the null hypothesis that all four catalysts have the same mean yield. Can this null hypothesis be rejected at the 5% level?
    b. Use the Tukey-Kramer method to determine which pairs of catalysts, if any, may be concluded to differ at the 5% level.

8. In an experiment to determine the effect of curing time on the compressive strength of a certain type of concrete, the mean strengths, in MPa, for specimens cured for each of four curing times were $\overline{X}_{1.} = 1316$, $\overline{X}_{2.} = 1326$, $\overline{X}_{3.} = 1375$, and $\overline{X}_{4.} = 1389$. Assume that four specimens were cured for each curing time.

    a. If MSE $= 875.2$, compute the value of the F statistic for testing the null hypothesis that all four curing times have the same mean strength. Can this null hypothesis be rejected at the 5% level?
    b. Use the Tukey-Kramer method to determine which pairs of curing times, if any, may be concluded to differ at the 5% level.

9. For some data sets, the F statistic will reject the null hypothesis of no difference in mean yields, but the Tukey-Kramer method will not find any pair of means that can be concluded to differ. For the four sample means given in Exercise 7, assuming a sample size of 5 for each treatment, find a value of MSE so that the F statistic rejects the null hypothesis

of no difference at the 5% level, while the Tukey-Kramer method does not find any pair of means to differ at the 5% level.

10. For some data sets, the F statistic will reject the null hypothesis of no difference in mean yields, but the Tukey-Kramer method will not find any pair of means that can be concluded to differ. For the four sample means given in Exercise 8, assuming a sample size of 4 for each treatment, find a value of MSE so that the F statistic rejects the null hypothesis of no difference at the 5% level, while the Tukey-Kramer method does not find any pair of means to differ at the 5% level.

9.3 Two-Factor Experiments

In one-factor experiments, discussed in Sections 9.1 and 9.2, the purpose is to determine whether varying the level of a single factor affects the response. Many experiments involve varying several factors, each of which may affect the response. In this section, we will discuss the case in which there are two factors. The experiments, naturally enough, are called **two-factor experiments**. We illustrate with an example.

A chemical engineer is studying the effects of various reagents and catalysts on the yield of a certain process. Yield is expressed as a percentage of a theoretical maximum. Four runs of the process were made for each combination of three reagents and four catalysts. The results are presented in Table 9.2. In this experiment there are two factors, the catalyst and the reagent. The catalyst is called the **row factor**, since its value varies from row to row in the table, while the reagent is called the **column factor**. These designations are arbitrary, in that the table could just as easily have been presented with the rows representing the reagents and the columns representing the catalysts.

TABLE 9.2 Yields for runs of a chemical process with various combinations of reagent and catalyst

Catalyst	Reagent		
	1	2	3
A	86.8 82.4 86.7 83.5	93.4 85.2 94.8 83.1	77.9 89.6 89.9 83.7
B	71.9 72.1 80.0 77.4	74.5 87.1 71.9 84.1	87.5 82.7 78.3 90.1
C	65.5 72.4 76.6 66.7	66.7 77.1 76.7 86.1	72.7 77.8 83.5 78.8
D	63.9 70.4 77.2 81.2	73.7 81.6 84.2 84.9	79.8 75.7 80.5 72.9

In general, there are I levels of the row factor and J levels of the column factor. (In Table 9.2, $I = 4$ and $J = 3$.) There are therefore IJ different combinations of the two factors. The terminology for these factor combinations is not standardized. We will refer to each combination of factors as a **treatment**, but some authors use the term **treatment combination**. Recall that the units assigned to a given treatment are called replicates. When the number of replicates is the same for each treatment, we will denote this number by K. Thus in Table 9.2, $K = 4$.

When observations are taken on every possible treatment, the design is called a **complete design** or a **full factorial design**. Incomplete designs, in which there are no data for one or more treatments, can be difficult to interpret, except for some special cases.

When possible, complete designs should be used. When the number of replicates is the same for each treatment, the design is said to be **balanced**. For one-factor experiments, we did not need to assume that the design was balanced. With two-factor experiments, unbalanced designs are more difficult to analyze than balanced designs. We will restrict our discussion to balanced designs. As with one-factor experiments, the factors may be fixed or random. The methods that we will describe apply to models where both effects are fixed. Later we will briefly describe models where one or both factors are random.

In a completely randomized design, each treatment represents a population, and the observations on that treatment are a simple random sample from that population. We will denote the sample values for the treatment corresponding to the ith level of the row factor and the jth level of the column factor by $X_{ij1}, \ldots, X_{ijK}$. We will denote the population mean outcome for this treatment by μ_{ij}. The values μ_{ij} are often called the **treatment means**. In general, the purpose of a two-factor experiment is to determine whether the treatment means are affected by varying either the row factor, the column factor, or both. The method of analysis appropriate for two-factor experiments is called **two-way analysis of variance**.

Parameterization for Two-Way Analysis of Variance

In a two-way analysis of variance, we wish to determine whether varying the level of the row or column factors changes the value of μ_{ij}. To do this, we must express μ_{ij} in terms of parameters that describe the row and column factors separately. We'll begin this task by describing some notation for the averages of the treatment means for the different levels of the row and column factors.

For any level i of the row factor, the average of all the treatment means μ_{ij} in the ith row is denoted $\overline{\mu}_{i.}$. We express $\overline{\mu}_{i.}$ in terms of the treatment means as follows:

$$\overline{\mu}_{i.} = \frac{1}{J} \sum_{j=1}^{J} \mu_{ij} \tag{9.27}$$

Similarly, for any level j of the column factor, the average of all the treatment means μ_{ij} in the jth column is denoted $\overline{\mu}_{.j}$. We express $\overline{\mu}_{.j}$ in terms of the treatment means as follows:

$$\overline{\mu}_{.j} = \frac{1}{I} \sum_{i=1}^{I} \mu_{ij} \tag{9.28}$$

Finally, we define the **population grand mean**, denoted by μ, which represents the average of all the treatment means μ_{ij}. The population grand mean can also be expressed as the average of the quantities $\overline{\mu}_{i.}$ or of the quantities $\overline{\mu}_{.j}$:

$$\mu = \frac{1}{I} \sum_{i=1}^{I} \overline{\mu}_{i.} = \frac{1}{J} \sum_{j=1}^{J} \overline{\mu}_{.j} = \frac{1}{IJ} \sum_{i=1}^{I} \sum_{j=1}^{J} \mu_{ij} \tag{9.29}$$

TABLE 9.3 Treatment means and their averages across rows and down columns

Row Level	Column Level 1	2	$\cdots$	J	Row Mean
1	μ_{11}	μ_{12}	$\cdots$	μ_{1J}	$\overline{\mu}_{1.}$
2	μ_{21}	μ_{22}	$\cdots$	μ_{2J}	$\overline{\mu}_{2.}$
$\vdots$	$\vdots$	$\vdots$	$\cdots$	$\vdots$	$\vdots$
I	μ_{I1}	μ_{I2}	$\cdots$	μ_{IJ}	$\overline{\mu}_{I.}$
Column Mean	$\overline{\mu}_{.1}$	$\overline{\mu}_{.2}$	$\cdots$	$\overline{\mu}_{.J}$	μ

Table 9.3 illustrates the relationships among μ_{ij}, $\overline{\mu}_{i.}$, $\overline{\mu}_{.j}$, and μ.

Using the quantities $\overline{\mu}_{i.}$, $\overline{\mu}_{.j}$, and μ, we can decompose the treatment mean μ_{ij} as follows:

$$\mu_{ij} = \mu + (\overline{\mu}_{i.} - \mu) + (\overline{\mu}_{.j} - \mu) + (\mu_{ij} - \overline{\mu}_{i.} - \overline{\mu}_{.j} + \mu) \qquad (9.30)$$

Equation (9.30) expresses the treatment mean μ_{ij} as a sum of four terms. In practice, simpler notation is used for the three rightmost terms in Equation (9.30):

$$\alpha_i = \overline{\mu}_{i.} - \mu \qquad (9.31)$$

$$\beta_j = \overline{\mu}_{.j} - \mu \qquad (9.32)$$

$$\gamma_{ij} = \mu_{ij} - \overline{\mu}_{i.} - \overline{\mu}_{.j} + \mu \qquad (9.33)$$

Each of quantities μ, α_i, β_j, and γ_{ij} has an important interpretation:

■ The quantity μ is the population grand mean, which is the average of all the treatment means.

■ The quantity $\alpha_i = \overline{\mu}_{i.} - \mu$ is called the ith **row effect**. It is the difference between the average treatment mean for the ith level of the row factor and the population grand mean. The value of α_i indicates the degree to which the ith level of the row factor tends to produce outcomes that are larger or smaller than the population grand mean.

■ The quantity $\beta_j = \overline{\mu}_{.j} - \mu$ is called the jth **column effect**. It is the difference between the average treatment mean for the jth level of the column factor and the population grand mean. The value of β_j indicates the degree to which the jth level of the column factor tends to produce outcomes that are larger or smaller than the population grand mean.

■ The quantity $\gamma_{ij} = \mu_{ij} - \overline{\mu}_{i.} - \overline{\mu}_{.j} + \mu$ is called the ij **interaction**. The effect of a level of a row (or column) factor may depend on which level of the column (or row) factor it is paired with. The interaction terms measure the degree to which this occurs. For example, assume that level 1 of the row factor tends to produce a large outcome when paired with column level 1, but a small outcome when paired with column level 2. In this case $\gamma_{1,1}$ would be positive, and $\gamma_{1,2}$ would be negative.

Both row effects and column effects are called **main effects** to distinguish them from the interactions. Note that there are I row effects, one for each level of the row factor, J column effects, one for each level of the column factor, and IJ interactions, one for each treatment. Furthermore, it follows from the definitions of quantities $\overline{\mu}_{i.}$,

$\overline{\mu}_{.j}$, and μ in Equations (9.27) through (9.29) that the row effects, column effects, and interactions satisfy the following constraints:

$$\sum_{i=1}^{I} \alpha_i = 0 \qquad \sum_{j=1}^{J} \beta_j = 0 \qquad \sum_{i=1}^{I} \gamma_{ij} = \sum_{j=1}^{J} \gamma_{ij} = 0 \qquad (9.34)$$

We now can express the treatment means μ_{ij} in terms of α_i, β_j, and γ_{ij}. From Equation (9.30) it follows that

$$\mu_{ij} = \mu + \alpha_i + \beta_j + \gamma_{ij} \qquad (9.35)$$

When the interactions γ_{ij} are all equal to 0, the **additive model** is said to apply. Under the additive model, Equation (9.35) becomes

$$\mu_{ij} = \mu + \alpha_i + \beta_j \qquad (9.36)$$

Under the additive model, the treatment mean μ_{ij} is equal to the population grand mean μ, plus an amount α_i that results from using row level i plus an amount β_j that results from using column level j. In other words, the combined effect of using row level i along with column level j is found by adding the individual main effects of the two levels. When some or all of the interactions are not equal to 0, the additive model does not hold, and the combined effect of a row level and a column level cannot be determined from their individual main effects.

Using Two-Way ANOVA to Test Hypotheses

A two-way analysis of variance is designed to address three main questions:

1. Does the additive model hold?
2. If so, is the mean outcome the same for all levels of the row factor?
3. If so, is the mean outcome the same for all levels of the column factor?

In general, we ask questions 2 and 3 only when we believe that the additive model may hold. We will discuss this further later in this section. The three questions are addressed by performing hypothesis tests. The null hypotheses for these tests are as follows:

1. To test whether the additive model holds, we test the null hypothesis that all the interactions are equal to 0:

$$H_0 : \gamma_{11} = \gamma_{12} = \cdots = \gamma_{IJ} = 0$$

If this null hypothesis is true, the additive model holds.

2. To test whether the mean outcome is the same for all levels of the row factor, we test the null hypothesis that all the row effects are equal to 0:

$$H_0 : \alpha_1 = \alpha_2 = \cdots = \alpha_I = 0$$

If this null hypothesis is true, then the mean outcome is the same for all levels of the row factor.

3. To test whether the mean outcome is the same for all levels of the column factor, we test the null hypothesis that all the column effects are equal to 0:

$$H_0 : \beta_1 = \beta_2 = \cdots = \beta_J = 0$$

If this null hypothesis is true, then the mean outcome is the same for all levels of the column factor.

The test statistics for these hypotheses are based on various averages of the data X_{ijk}, which we now define, using the data in Table 9.2 as an example. Table 9.4 presents the average yield for the four runs for each reagent and catalyst in Table 9.2.

TABLE 9.4 Average yields $\overline{X}_{ij.}$ for runs of a chemical process using different combinations of reagent and catalyst

Catalyst	Reagent 1	2	3	Row Mean $\overline{X}_{i..}$
A	84.85	89.13	85.28	86.42
B	75.35	79.40	84.65	79.80
C	70.30	76.65	78.20	75.05
D	73.18	81.10	77.23	77.17
Column Mean $\overline{X}_{.j.}$	75.92	81.57	81.34	Sample Grand Mean $\overline{X}_{...} = 79.61$

Each number in the body of Table 9.4 is the average of the four numbers in the corresponding cell of Table 9.2. These are called the **cell means**. They are denoted $\overline{X}_{ij.}$ and are defined by

$$\overline{X}_{ij.} = \frac{1}{K} \sum_{k=1}^{K} X_{ijk} \tag{9.37}$$

Averaging the cell means across the rows produces the **row means** $\overline{X}_{i..}$:

$$\overline{X}_{i..} = \frac{1}{J} \sum_{j=1}^{J} \overline{X}_{ij.} = \frac{1}{JK} \sum_{j=1}^{J} \sum_{k=1}^{K} X_{ijk} \tag{9.38}$$

Averaging the cell means down the columns produces the **column means** $\overline{X}_{.j.}$:

$$\overline{X}_{.j.} = \frac{1}{I} \sum_{i=1}^{I} \overline{X}_{ij.} = \frac{1}{IK} \sum_{i=1}^{I} \sum_{k=1}^{K} X_{ijk} \tag{9.39}$$

The sample grand mean $\overline{X}_{...}$ can be found by computing the average of the row means, the average of the column means, the average of the cell means, or the average of all the observations:

$$\overline{X}_{...} = \frac{1}{I} \sum_{i=1}^{I} \overline{X}_{i..} = \frac{1}{J} \sum_{j=1}^{J} \overline{X}_{.j.} = \frac{1}{IJ} \sum_{i=1}^{I} \sum_{j=1}^{J} \overline{X}_{ij.} = \frac{1}{IJK} \sum_{i=1}^{I} \sum_{j=1}^{J} \sum_{k=1}^{K} X_{ijk} \tag{9.40}$$

We now describe the standard tests for these null hypotheses. For the tests to be valid, the following conditions must hold:

Assumptions for Two-Way ANOVA

The standard two-way ANOVA hypothesis tests are valid under the following conditions:

1. The design must be complete.
2. The design must be balanced.
3. The number of replicates per treatment, K, must be at least 2.
4. Within any treatment, the observations $X_{ij1}, \ldots, X_{ijK}$ are a simple random sample from a normal population.
5. The population variance is the same for all treatments. We denote this variance by σ^2.

Just as in one-way ANOVA, the standard tests for these null hypotheses are based on sums of squares. Specifically, they are the row sum of squares (SSA), the column sum of squares (SSB), the interaction sum of squares (SSAB), and the error sum of squares (SSE). Also of interest is the total sum of squares (SST), which is equal to the sum of the others. Each of these sums of squares has a number of degrees of freedom associated with it. Table 9.5 presents the degrees of freedom for each sum of squares, along with a formula.

TABLE 9.5 ANOVA table for two-way ANOVA

Source	Degrees of Freedom	Sum of Squares
Rows (SSA)	$I - 1$	$JK \sum_{i=1}^{I} (\overline{X}_{i..} - \overline{X}_{...})^2$
Columns (SSB)	$J - 1$	$IK \sum_{j=1}^{J} (\overline{X}_{.j.} - \overline{X}_{...})^2$
Interactions (SSAB)	$(I - 1)(J - 1)$	$K \sum_{i=1}^{I} \sum_{j=1}^{J} (\overline{X}_{ij.} - \overline{X}_{i..} - \overline{X}_{.j.} + \overline{X}_{...})^2$
Error (SSE)	$IJ(K - 1)$	$\sum_{i=1}^{I} \sum_{j=1}^{J} \sum_{k=1}^{K} (X_{ijk} - \overline{X}_{ij.})^2$
Total (SST)	$IJK - 1$	$\sum_{i=1}^{I} \sum_{j=1}^{J} \sum_{k=1}^{K} (X_{ijk} - \overline{X}_{...})^2$

When SSA is large, we reject the null hypothesis that all the row effects are equal to 0. When SSB is large, we reject the null hypothesis that all the column effects are equal to 0. When SSAB is large, we reject the null hypothesis that all the interactions are equal to 0.

We can determine whether SSA, SSB, and SSAB are sufficiently large by comparing them to the error sum of squares, SSE. As in one-way ANOVA (Section 9.1), SSE depends only on the distances between the observations and their own cell means. SSE therefore measures only the random variation inherent in the process and is not affected by the values of the row effects, column effects, or interactions. To compare SSA, SSB, and SSAB with SSE, we first divide each sum of squares by its degrees of freedom, producing quantities known as **mean squares**. The mean squares, denoted MSA, MSB, MSAB, and MSE, are defined as follows:

$$\text{MSA} = \frac{\text{SSA}}{I-1} \qquad \text{MSB} = \frac{\text{SSB}}{J-1} \qquad \text{MSAB} = \frac{\text{SSAB}}{(I-1)(J-1)}$$

$$\text{MSE} = \frac{\text{SSE}}{IJ(K-1)} \tag{9.41}$$

The test statistics for the three null hypotheses are the quotients of MSA, MSB, and MSAB with MSE. The null distributions of these test statistics are F distributions. Specifically,

- Under $H_0: \alpha_1 = \cdots = \alpha_I = 0$, the statistic $\dfrac{\text{MSA}}{\text{MSE}}$ has an $F_{I-1,\,IJ(K-1)}$ distribution.

- Under $H_0: \beta_1 = \cdots = \beta_J = 0$, the statistic $\dfrac{\text{MSB}}{\text{MSE}}$ has an $F_{J-1,\,IJ(K-1)}$ distribution.

- Under $H_0: \gamma_{11} = \cdots = \gamma_{IJ} = 0$, the statistic $\dfrac{\text{MSAB}}{\text{MSE}}$ has an $F_{(I-1)(J-1),\,IJ(K-1)}$ distribution.

In practice, the sums of squares, mean squares, and test statistics are usually calculated with the use of a computer. The following output (from MINITAB) presents the ANOVA table for the data in Table 9.2.

```
Two-way ANOVA: Yield versus Catalyst, Reagent

Source        DF       SS        MS       F       P
Catalyst       3    877.56   292.521    9.36   0.000
Reagent        2    327.14   163.570    5.23   0.010
Interaction    6    156.98    26.164    0.84   0.550
Error         36   1125.33    31.259
Total         47   2487.02

S = 5.591    R-sq = 54.75%    R-Sq(adj) = 40.93%
```

The labels DF, SS, MS, F, and P refer to degrees of freedom, sum of squares, mean square, F statistic, and P-value, respectively. As in one-way ANOVA, the mean square for error (MSE) is an estimate of the error variance σ^2, the quantity labeled "S" is the square root of MSE and is an estimate of the error standard deviation σ. The quantities "R-sq" and "R-sq(adj)" are computed with formulas analogous to those in one-way ANOVA.

Use the preceding ANOVA table to determine whether the additive model is plausible for the yield data. If the additive model is plausible, can we conclude that either the catalyst or the reagent affects the yield?

Solution

We first check to see if the additive model is plausible. The P-value for the interactions is 0.55, which is not small. We therefore do not reject the null hypothesis that all the interactions are equal to 0, and we conclude that the additive model is plausible. Since the additive model is plausible, we now ask whether the row or column factors affect the outcome. We see from the table that the P-value for the row effects (Catalyst) is approximately 0, so we conclude that the catalyst does affect the yield. Similarly, the P-value for the column effects (Reagent) is small (0.010), so we conclude that the reagent affects the yield as well.

The article "Uncertainty in Measurements of Dermal Absorption of Pesticides" (W. Navidi and A. Bunge, *Risk Analysis*, 2002:1175–1182) describes an experiment in which a pesticide was applied to skin at various concentrations and for various lengths of time. The outcome is the amount of the pesticide that was absorbed into the skin. The following output (from MINITAB) presents the ANOVA table. Is the additive model plausible? If so, do either the concentration or the duration affect the amount absorbed?

```
Two-way ANOVA: Absorbed versus Concentration, Duration

Source        DF       SS       MS       F       P
Concent        2   49.991   24.996  107.99   0.000
Duration       2   19.157    9.579   41.38   0.000
Interaction    4    0.337    0.084    0.36   0.832
Error         27    6.250    0.231
Total         35   75.735
```

Solution

The P-value for the interaction is 0.832, so we conclude that the additive model is plausible. The P-values for both concentration and dose are very small. Therefore we can conclude that both concentration and duration affect the amount absorbed.

Checking the Assumptions

A residual plot can be used to check the assumption of equal variances, and a normal probability plot of the residuals can be used to check normality. The residual plot plots the residuals $X_{ijk} - \overline{X}_{ij\cdot}$ versus the fitted values, which are the sample means $\overline{X}_{ij\cdot}$. Figures 9.6 and 9.7 present both a normal probability plot and a residual plot for the yield data found in Table 9.2. The assumptions appear to be well satisfied.

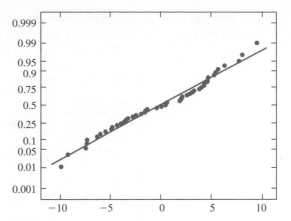

FIGURE 9.6 Normal probability plot for the residuals from the yield data. There is no evidence of a strong departure from normality.

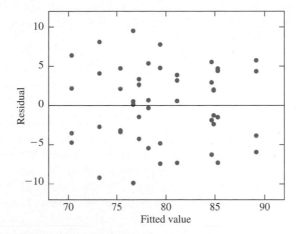

FIGURE 9.7 Residual plot for the yield data. There is no evidence against the assumption of equal variances.

Don't Interpret the Main Effects When the Additive Model Doesn't Hold

When the interactions are small enough so that the additive model is plausible, interpretation of the main effects is fairly straightforward, as shown in Examples 9.8 and 9.9. When the additive model does not hold, however, it is not always easy to interpret the main effects. Here is a hypothetical example to illustrate the point. Assume that a process is run under conditions obtained by varying two factors at two levels each.

Two runs are made at each of the four combinations of row and column levels. The yield of the process is measured each time, with the results presented in the following table.

Row Level	Column 1	Level 2
1	51, 49	43, 41
2	43, 41	51, 49

Clearly, if it is desired to maximize yield, the row and column factors matter—we want either row level 1 paired with column level 1 or row level 2 paired with column level 2.

Now look at the following ANOVA table.

Source	DF	SS	MS	F	P
Row	1	0.0000	0.0000	0.00	1.000
Column	1	0.0000	0.0000	0.00	1.000
Interaction	1	128.00	128.00	64.00	0.001
Error	4	8.0000	2.0000		
Total	7	136.00			

The main effects sum of squares for both the row and column main effects are equal to 0, and their P-values are equal to 1, which is as large as a P-value can be. If we follow the procedure used in Examples 9.8 and 9.9, we would conclude that neither the row factor nor the column factor affects the yield. But it is clear from the data that the row and column factors do affect the yield. What is happening is that the row and column factors do not matter *on the average*. Level 1 of the row factor is better if level 1 of the column factor is used, and level 2 of the row factor is better if level 2 of the column factor is used. When averaged over the two levels of the column factor, the levels of the row factor have the same mean yield. Similarly, the column levels have the same mean yield when averaged over the levels of the row factor. When the effects of the row levels depend on which column levels they are paired with, and vice versa, the main effects can be misleading.

It is the P-value for the interactions that tells us not to try to interpret the main effects. This P-value is quite small, so we reject the additive model. Then we know that some of the interactions are nonzero, so the effects of the row levels depend on the column levels, and vice versa. For this reason, when the additive model is rejected, we should not try to interpret the main effects. We need to look at the cell means themselves in order to determine how various combinations of row and column levels affect the outcome.

> ### Summary
>
> In a two-way analysis of variance:
>
> - If the additive model *is not* rejected, then hypothesis tests for the main effects can be used to determine whether the row or column factors affect the outcome.
>
> - If the additive model *is* rejected, then hypothesis tests for the main effects should not be used. Instead, the cell means must be examined to determine how various combinations of row and column levels affect the outcome.

Example 9.10

The thickness of the silicon dioxide layer on a semiconductor wafer is crucial to its performance. In the article "Virgin Versus Recycled Wafers for Furnace Qualification: Is the Expense Justified?" (V. Czitrom and J. Reece, *Statistical Case Studies for Process Improvement*, SIAM-ASA, 1997:87–103), oxide layer thicknesses were measured for three types of wafers: virgin wafers, wafers recycled in-house, and wafers recycled by an external supplier. In addition, several furnace locations were used to grow the oxide layer. A two-way ANOVA for three runs at one wafer site for the three types of wafers at three furnace locations was performed. The data are presented in the following table, followed by the results (from MINITAB).

Furnace Location	Wafer Type	Oxide Layer Thickness (Å)		
1	Virgin	90.1	90.7	89.4
1	In-house	90.4	88.8	90.6
1	External	92.6	90.0	93.3
2	Virgin	91.9	88.6	89.7
2	In-house	90.3	91.9	91.5
2	External	88.3	88.2	89.4
3	Virgin	88.1	90.2	86.6
3	In-house	91.0	90.4	90.2
3	External	91.5	89.8	89.8

```
Two-way ANOVA for Thickness versus Wafer, Location

Source        DF       SS       MS       F        P
Wafer          2   5.8756   2.9378   2.07   0.155
Location       2   4.1089   2.0544   1.45   0.262
Interaction    4   21.349   5.3372   3.76   0.022
Error         18   25.573   1.4207
Total         26   56.907
```

Since recycled wafers are cheaper, the company hopes that there is no difference in the oxide layer thickness among the three types of chips. If possible, determine whether the data are consistent with the hypothesis of no difference. If not possible, explain why not.

Solution

The P-value for the interactions is 0.022, which is small. Therefore the additive model is not plausible, so we cannot interpret the main effects. A good thing to do is to make a table of the cell means. Table 9.6 presents the sample mean for each treatment.

TABLE 9.6 Sample means for each treatment

Furnace Location	Wafer Type			Row Mean
	Virgin	In-House	External	
1	90.067	89.933	91.967	90.656
2	90.067	91.233	88.633	89.978
3	88.300	90.533	90.367	89.733
Column Mean	89.478	90.566	90.322	

From Table 9.6, it can be seen that the thicknesses do vary among wafer types, but no one wafer type consistently produces the thickest, or the thinnest, oxide layer. For example, at furnace location 1 the externally recycled wafers produce the thickest layer while the in-house recycled wafers produce the thinnest. At furnace location 2 the order is reversed: The in-house wafers produce the thickest layer while the external ones produce the thinnest. This is due to the interaction of furnace location and wafer type.

A Two-Way ANOVA Is Not the Same as Two One-Way ANOVAs

Example 9.10 presented a two-way ANOVA with three row levels and three column levels, for a total of nine treatments. If separate one-way ANOVAs were run on the row and column factors separately, there would be only six treatments. This means that in practice, running separate one-way ANOVAs on each factor may be less costly than running a two-way ANOVA. Unfortunately, this "one-at-a-time" design is sometimes used in practice for this reason. It is important to realize that running separate one-way analyses on the individual factors can give results that are misleading when interactions are present. To see this, look at Table 9.6. Assume that an engineer is trying to find the combination of furnace and location that will produce the thinnest oxide layer. He first runs the process once at each furnace location, using in-house recycled wafers, because those wafers are the ones currently being used in production. Furnace location 1 produces the thinnest layer for in-house wafers. Now the engineer runs the process once for each wafer type, all at location 1, which was the best for the in-house wafers. Of the three wafer types, in-house wafers produce the thinnest layer at location 1. So the conclusion drawn from the one-at-a-time analysis is that the thinnest layers are produced by the combination of in-house wafers at furnace location 1. A look at Table 9.6 shows that the conclusion is false. There are two combinations of furnace location and wafer type that produce thinner layers than this.

The one-at-a-time method assumes that the wafer that produces the thinnest layers at one location will produce the thinnest layers at all locations, and that the location that produces the thinnest layers for one wafer type will produce the thinnest layers for all types. This is equivalent to assuming that there are no interactions between the factors, which in the case of the wafers and locations is incorrect. In summary, the one-at-a-time method fails because it cannot detect interactions between the factors.

Summary

- When there are two factors, a two-factor design must be used.
- Examining one factor at a time cannot reveal interactions between the factors.

Interaction Plots

Interaction plots can help to visualize interactions. Figure 9.8 presents an interaction plot for the wafer data. The vertical axis represents the response, which is layer thickness. One factor is chosen to be represented on the horizontal axis. We chose furnace location; it would have been equally acceptable to have chosen wafer type. Now we proceed through the levels of the wafer-type factor. We'll start with external wafers. The three cell means for external wafers, as shown in Table 9.6, are 91.967, 88.633, and 90.367, corresponding to furnace locations 1, 2, and 3, respectively. These values are plotted above their respective furnace locations and are connected with line segments. This procedure is repeated for the other two wafer types to complete the plot.

For the wafer data, the means for external wafers follow a substantially different pattern than those for the other two wafer types. This is the source of the significant interaction and is the reason that the main effects of wafer and furnace type cannot be easily interpreted. In comparison, for perfectly additive data, for which the interaction

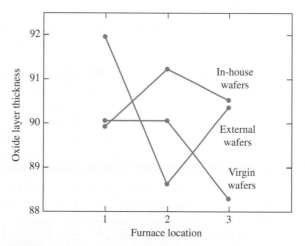

FIGURE 9.8 Interaction plot for the wafer data. The lines are far from parallel, indicating substantial interaction between the factors.

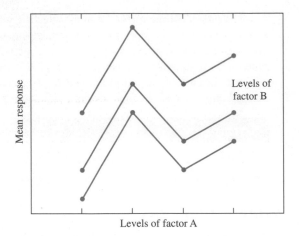

FIGURE 9.9 Interaction plot for hypothetical data with interaction estimates $\hat{\gamma}_{ij}$ equal to 0. The line segments are parallel.

estimates $\hat{\gamma}_{ij}$ are equal to 0, the line segments in the interaction plot are parallel. Figure 9.9 illustrates this hypothetical case.

Figure 9.10 presents an interaction plot for the yield data. The cell means were presented in Table 9.4. The lines are not parallel, but their slopes match better than those for the wafer data. This indicates that the interaction estimates are nonzero but smaller than those for the wafer data. In fact, the P-value for the test of the null hypothesis of no interaction was 0.550 (see page 427). The deviation from parallelism exhibited in Figure 9.10 is therefore small enough to be consistent with the hypothesis of no interaction.

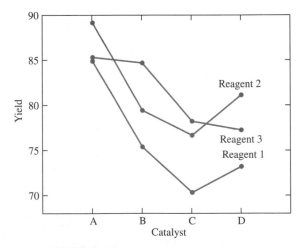

FIGURE 9.10 Interaction plot for yield data.

Two-Way ANOVA when $K = 1$

The F tests we have presented require the assumption that the sample size K for each treatment be at least 2. The reason for this is that when $K = 1$, the error sum of squares (SSE) will be equal to 0, since $X_{ijk} = \overline{X}_{ij.}$ for each i and j. In addition, the degrees of freedom for SSE, which is $IJ(K - 1)$, is equal to 0 when $K = 1$.

When $K = 1$, a two-way ANOVA cannot be performed unless it is certain that the additive model holds. In this case, since the interactions are assumed to be zero, the mean square for interaction (MSAB; see Equation 9.41) and its degrees of freedom can be used in place of MSE to test the main row and column effects.

Random Factors

Our discussion of two-factor experiments has focused on the case where both factors are fixed. Such an experiment is said to follow a **fixed effects model**. Experiments can also be designed in which one or both factors are random. If both factors are random, the experiment is said to follow a **random effects model**. If one factor is fixed and one is random, the experiment is said to follow a **mixed model**.

In the one-factor case, the analysis is the same for both fixed and random effects models, while the null hypothesis being tested differs. In the two-factor case, both the methods of analysis and the null hypotheses differ among fixed effects models, random effects models, and mixed models. Methods for models in which one or more effects are random can be found in more advanced texts, such as Hocking (2003).

Unbalanced Designs

We have assumed that the design is balanced—that is, that the number of replications is the same for each treatment. The methods described here do not apply to unbalanced designs. However, unbalanced designs that are complete may be analyzed with the methods of multiple regression. An advanced text such as Draper and Smith (1998) may be consulted for details.

Exercises for Section 9.3

1. To assess the effect of piston ring type and oil type on piston ring wear, three types of piston ring and four types of oil were studied. Three replications of an experiment, in which the number of milligrams of material lost from the ring in four hours of running was measured, were carried out for each of the 12 combinations of oil type and piston ring type. With oil type as the row effect and piston ring type as the column effect, the following sums of squares were observed: SSA = 1.0926, SSB = 0.9340, SSAB = 0.2485, SSE = 1.7034.

   a. How many degrees of freedom are there for the effect of oil type?
   b. How many degrees of freedom are there for the effect of piston ring type?
   c. How many degrees of freedom are there for interactions?
   d. How many degrees of freedom are there for error?
   e. Construct an ANOVA table. You may give ranges for the P-values.
   f. Is the additive model plausible? Provide the value of the test statistic and the P-value.
   g. Is it plausible that the main effects of oil type are all equal to 0? Provide the value of the test statistic and the P-value.

   h. Is it plausible that the main effects of piston ring type are all equal to 0? Provide the value of the test statistic and the P-value.

2. A study was done to assess the effect of water type and glycerol content on the latherability of soap. There were two levels of water type, de-ionized and tap, and three levels of glycerol, low, medium, and high. Five replications of the experiment were conducted, and the amount of lather, in mL, was measured. With water type as the row effect and glycerol level as the column effect, the following sums of squares were observed: SSA = 2584.3, SSB = 726.5, SSAB = 67.7, SSE = 3751.4.

   a. How many degrees of freedom are there for the water type effect?
   b. How many degrees of freedom are there for the glycerol effect?
   c. How many degrees of freedom are there for interactions?
   d. How many degrees of freedom are there for error?
   e. Construct an ANOVA table. You may give ranges for the P-values.
   f. Is the additive model plausible? Provide the value of the test statistic and the P-value.
   g. Is it plausible that the main effects of water type are all equal to 0? Provide the value of the test statistic and the P-value.
   h. Is it plausible that the main effects of glycerol level are all equal to 0? Provide the value of the test statistic and the P-value.

3. An experiment to determine the effect of mold temperature on tensile strength involved three different alloys and five different mold temperatures. Four specimens of each alloy were cast at each mold temperature. With mold temperature as the row factor and alloy as the column factor, the sums of squares were: SSA = 69,738, SSB = 8958, SSAB = 7275, and SST = 201,816.

   a. Construct an ANOVA table. You may give ranges for the P-values.
   b. Is the additive model plausible? Explain.
   c. Is it plausible that the main effects of mold temperature are all equal to 0? Provide the value of the test statistic and the P-value.
   d. Is it plausible that the main effects of alloy are all equal to 0? Provide the value of the test statistic and the P-value.

4. An experiment was performed to measure the effects of two factors on the ability of cleaning solution to remove oil from cloth. The factors were the fraction of lauric acid and the concentration of soap (in percent) in the solution. The levels of the lauric acid fraction were 5, 10, 20, and 30, and the levels of the soap concentration were 15, 20, and 25. Three replications of the experiment were performed. With lauric acid fraction as the row effect and concentration of soap as the column effect, the following sums of squares were observed: SSA = 127.8, SSB = 86.4, SSAB = 641.7, and SST = 1088.3.

   a. Construct an ANOVA table. You may give ranges for the P-values.
   b. Is the additive model plausible? Explain.
   c. Can the effect of lauric acid fraction be described by interpreting the main effects of lauric acid? If so, interpret the main effects, including the appropriate test statistic and P-value. If not, explain why not.
   d. Can the effect of soap concentration be described by interpreting the main effects of soap concentration? If so, interpret the main effects, including the appropriate test statistic and P-value. If not, explain why not.

5. The article "Change in Creep Behavior of Plexiform Bone with Phosphate Ion Treatment" (R. Regimbal, C. DePaula, and N. Guzelsu, *Bio-Medical Materials and Engineering*, 2003:11–25) describes an experiment to study the effects of saline and phosphate ion solutions on mechanical properties of plexiform bone. The following table presents the yield stress measurements for six specimens treated with either saline (NaCl) or phosphate ion (Na_2HPO_4) solution, at a temperature of either 25°C or 37°C. (The article presents means and standard deviations only; the values in the table are consistent with these.)

Solution	Temperature	Yield Stress (MPa)					
NaCl	25°C	138.40	130.89	94.646	96.653	116.90	88.215
NaCl	37°C	92.312	147.28	116.48	88.802	114.37	90.737
Na_2HPO_4	25°C	120.18	129.43	139.76	132.75	137.23	121.73
Na_2HPO_4	37°C	123.50	128.94	102.86	99.941	161.68	136.44

a. Construct an ANOVA table. You may give ranges for the P-values.
b. Is the additive model plausible? Provide the value of the test statistic and the P-value.
c. Can the effect of solution (NaCl versus Na_2HPO_4) on yield stress be described by interpreting the main effects of solution? If so, interpret the main effects, including the appropriate test statistic and P-value. If not, explain why not.
d. Can the effect of temperature on yield stress be described by interpreting the main effects of temperature? If so, interpret the main effects, including the appropriate test statistic and P-value. If not, explain why not.

6. The lifetime of a tool was investigated under three settings for feed rate and three settings for speed. Four tools were tested under each combination of settings. The results (in hours) were as follows.

Feed Rate	Speed	Lifetime			
Light	Slow	60.6	57.0	61.4	59.7
Light	Medium	57.8	59.4	62.8	58.2
Light	Fast	56.5	52.3	58.1	53.9
Medium	Slow	51.2	53.1	48.3	51.6
Medium	Medium	49.6	48.1	49.8	51.1
Medium	Fast	45.7	48.6	45.0	49.2
Heavy	Slow	44.8	46.7	41.9	51.3
Heavy	Medium	46.6	41.4	38.3	37.9
Heavy	Fast	37.2	32.8	39.9	35.9

a. Construct an ANOVA table. You may give ranges for the P-values.
b. Is the additive model plausible? Provide the value of a test statistic and the P-value.
c. Can the effect of feed rate on lifetime be described by interpreting the main effects of feed rate? If so, interpret the main effects, including the appropriate test statistic and P-value. If not, explain why not.
d. Can the effect of the speed on lifetime be described by interpreting the main effects of distance? If so, interpret the main effects, including the appropriate test statistic and P-value. If not, explain why not.

7. The effect of curing pressure on bond strength (in MPa) was tested for two different adhesives. There were three levels of curing pressure. Three replications were performed for each combination of curing pressure and adhesive. The results are presented in the following table.

Adhesive	Curing Pressure	Bond Strength		
A	Low	8.9	9.6	7.1
A	Medium	7.2	7.0	8.7
A	High	1.1	4.7	0.2
B	Low	4.3	5.2	2.9
B	Medium	2.3	4.6	5.0
B	High	5.9	2.2	4.6

a. Construct an ANOVA table. You may give ranges for the P-values.

b. Is the additive model plausible? Provide the value of the test statistic and the P-value.

c. Can the effect of adhesive on the bond strength be described by interpreting the main effects of adhesive? If so, interpret the main effects. If not, explain why not.

d. Can the effect of curing pressure on the bond strength be described by interpreting the main effects of curing pressure? If so, interpret the main effects. If not, explain why not.

8. The article "A 4-Year Sediment Trap Record of Alkenones from the Filamentous Upwelling Region off Cape Blanc, NW Africa and a Comparison with Distributions in Underlying Sediments" (P. Müller and G. Fischer, *Deep Sea Research*, 2001:1877–1903) reports on a study of sediment trap records to evaluate the transfer of surface water signals into the geological record. The data in the following table are measurements of the total mass flux (in mg/m² per day) for traps at two different locations and depths.

Location	Depth	Flux
A	Upper	109.8 86.5 150.5 69.8 63.2 107.8 72.4 74.4
A	Lower	163.7 139.4 176.9 170.6 123.5 142.9 130.3 111.6
B	Upper	93.2 123.6 143.9 163.2 82.6 63.0 196.7 160.1
B	Lower	137.0 88.3 53.4 104.0 78.0 39.3 117.9 143.0

a. Construct an ANOVA table. You may give ranges for the P-values.

b. Is the additive model plausible? Provide the value of the test statistic and the P-value.

c. Compute all the cell means. Use them to describe the way in which depth and location affect flux.

9. Artificial joints consist of a ceramic ball mounted on a taper. The article "Friction in Orthopaedic Zirconia Taper Assemblies" (W. Macdonald, A. Aspenberg, et al., *Proceedings of the Institution of Mechanical Engineers*, 2000: 685–692) presents data on the coefficient of friction for a push-on load of 2 kN for taper assemblies made from two zirconium alloys and employing three different neck lengths. Five measurements were made for each combination of material and neck length. The results presented in the following table are consistent with the cell means and standard deviations presented in the article.

Taper Material	Neck Length	Coefficient of Friction
CPTi-ZrO_2	Short	0.254 0.195 0.281 0.289 0.220
CPTi-ZrO_2	Medium	0.196 0.220 0.185 0.259 0.197
CPTi-ZrO_2	Long	0.329 0.481 0.320 0.296 0.178
TiAlloy-ZrO_2	Short	0.150 0.118 0.158 0.175 0.131
TiAlloy-ZrO_2	Medium	0.180 0.184 0.154 0.156 0.177
TiAlloy-ZrO_2	Long	0.178 0.198 0.201 0.199 0.210

a. Construct an ANOVA table. You may give ranges for the P-values.

b. Is the additive model plausible? Provide the value of the test statistic, its null distribution, and the P-value.

c. Can the effect of material on the coefficient of friction be described by interpreting the main effects of material? If so, interpret the main effects. If not, explain why not.

10. The article "Anodic Fenton Treatment of Treflan MTF" (D. Saltmiras and A. Lemley, *Journal of Environmental Science and Health*, 2001:261–274) describes a two-factor experiment designed to study the sorption of the herbicide trifluralin. The factors are the initial trifluralin concentration and the Fe^2:H_2O_2 delivery ratio. There were three

replications for each treatment. The results presented in the following table are consistent with the means and standard deviations reported in the article.

Initial Concentration (*M*)	Delivery Ratio	Sorption (%)		
15	1:0	10.90	8.47	12.43
15	1:1	3.33	2.40	2.67
15	1:5	0.79	0.76	0.84
15	1:10	0.54	0.69	0.57
40	1:0	6.84	7.68	6.79
40	1:1	1.72	1.55	1.82
40	1:5	0.68	0.83	0.89
40	1:10	0.58	1.13	1.28
100	1:0	6.61	6.66	7.43
100	1:1	1.25	1.46	1.49
100	1:5	1.17	1.27	1.16
100	1:10	0.93	0.67	0.80

a. Construct an ANOVA table. You may give ranges for the *P*-values.
b. Is the additive model plausible? Provide the value of the test statistic, its null distribution, and the *P*-value.

11. Refer to Exercise 10. The treatments with a delivery ratio of 1:0 were controls, or blanks. It was discovered after the experiment that the high apparent levels of sorption in these controls was largely due to volatility of the trifluralin. Eliminate the control treatments from the data.

a. Construct an ANOVA table. You may give ranges for the *P*-values.
b. Is the additive model plausible? Provide the value of the test statistic, its null distribution, and the *P*-value.
c. Construct an interaction plot. Explain how the plot illustrates the degree to which interactions are present.

12. The article "Use of Taguchi Methods and Multiple Regression Analysis for Optimal Process Development of High Energy Electron Beam Case Hardening of Cast Iron" (M. Jean and Y. Tzeng, *Surface Engineering*, 2003:150–156) describes a factorial experiment designed to determine factors in a high-energy electron beam process that affect hardness in metals. Results for two factors, each with three levels, are presented in the following table. Factor *A* is the travel speed in mm/s, and factor *B* is accelerating voltage in volts. The outcome is Vickers hardness. There were six replications for each treatment. In the article, a total of seven factors were studied; the two presented here are those that were found to be the most significant.

A	*B*	Hardness					
10	10	875	896	921	686	642	613
10	25	712	719	698	621	632	645
10	50	568	546	559	757	723	734
20	10	876	835	868	812	796	772
20	25	889	876	849	768	706	615
20	50	756	732	723	681	723	712
30	10	901	926	893	856	832	841
30	25	789	801	776	845	827	831
30	50	792	786	775	706	675	568

a. Construct an ANOVA table. You may give ranges for the P-values.

b. Is the additive model plausible? Provide the value of the test statistic and the P-value.

13. Each of three operators made two weighings of several silicon wafers. Results are presented in the following table for three of the wafers. All the wafers had weights very close to 54 g, so the weights are reported in units of μg above 54 g. (Based on the article "Revelation of a Microbalance Warmup Effect," J. Buckner, B. Chin, et al., *Statistical Case Studies for Industrial Process Improvement*, SIAM-ASA, 1997:39–45.)

Wafer	Operator 1	Operator 2	Operator 3
1	11 15	10 6	14 10
2	210 208	205 201	208 207
3	111 113	102 105	108 111

a. Construct an ANOVA table. You may give ranges for the P-values.

b. Can it be determined from the ANOVA table whether there are differences in the measured weights among the operators? If so, provide the value of the test statistic and the P-value. If not, explain why not.

14. Refer to Exercise 13. It turns out that the measurements of operator 2 were taken in the morning, shortly after the balance had been powered up. A new policy was instituted to leave the balance powered up continuously. The three operators then made two weighings of three different wafers. The results are presented in the following table.

Wafer	Operator 1	Operator 2	Operator 3
1	152 156	156 155	152 157
2	443 440	442 439	435 439
3	229 227	229 232	225 228

a. Construct an ANOVA table. You may give ranges for the P-values.

b. Compare the ANOVA table in part (a) with the ANOVA table in part (a) of Exercise 13. Would you recommend leaving the balance powered up continuously? Explain your reasoning.

15. The article "Cellulose Acetate Microspheres Prepared by O/W Emulsification and Solvent Evaporation Method" (K. Soppinmath, A. Kulkarni, et al., *Journal of Microencapsulation*, 2001:811–817) describes a study of the effects of the concentrations of polyvinyl alcohol (PVAL) and dichloromethane (DCM) on the encapsulation efficiency in a process that produces microspheres containing the drug ibuprofen. There were three concentrations of PVAL (measured in units of % w/v) and three of DCM (in mL). The results presented in the following table are consistent with the means and standard deviations presented in the article.

PVAL	DCM = 50	DCM = 40	DCM = 30
0.5	98.983 99.268 95.149	96.810 94.572 86.718	75.288 74.949 72.363
1.0	89.827 94.136 96.537	82.352 79.156 80.891	76.625 76.941 72.635
2.0	95.095 95.153 92.353	86.153 91.653 87.994	80.059 79.200 77.141

a. Construct an ANOVA table. You may give ranges for the P-values.

b. Discuss the relationships among PVAL concentration, DCM concentration, and encapsulation efficiency.

9.4 Randomized Complete Block Designs

In some experiments, there are factors that vary and have an effect on the response, but whose effects are not of interest to the experimenter. For example, in one commonly occurring situation, it is impossible to complete an experiment in a single day, so the observations have to be spread out over several days. If conditions that can affect the outcome vary from day to day, then the day becomes a factor in the experiment, even though there may be no interest in estimating its effect.

For a more specific example, imagine that three types of fertilizer are to be evaluated for their effect on yield of fruit in an orange grove, and that three replicates will be performed, for a total of nine observations. An area is divided into nine plots, in three rows of three plots each. Now assume there is a water gradient along the plot area, so that the rows receive differing amounts of water. The amount of water is now a factor in the experiment, even though there is no interest in estimating the effect of water amount on the yield of oranges.

If the water factor is ignored, a one-factor experiment could be carried out with fertilizer as the only factor. Each of the three fertilizers would be assigned to three of the plots. In a completely randomized experiment, the treatments would be assigned to the plots at random. Figure 9.11 presents two possible random arrangements. In the arrangement on the left, the plots with fertilizer A get more water than those with the other two fertilizers. In the plot on the right, the plots with fertilizer C get the most water. When the treatments for one factor are assigned completely at random, it is likely that they will not be distributed evenly over the levels of another factor.

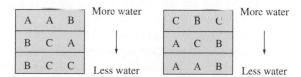

FIGURE 9.11 Two possible arrangements for three fertilizers, A, B, and C, assigned to nine plots completely at random. It is likely that the amounts of water will differ for the different fertilizers.

If the amount of water in fact has a negligible effect on the response, then the completely randomized one-factor design is appropriate. There is no reason to worry about a factor that does not affect the response. But now assume that the water level does have a substantial impact on the response. Then Figure 9.11 shows that in any one experiment, the estimated effects of the treatments are likely to be thrown off the mark, or biased, by the differing levels of water. Different arrangements of the treatments bias the estimates in different directions. If the experiment is repeated several times, the estimates are likely to vary greatly from repetition to repetition. For this reason, the completely randomized one-factor design produces estimated effects that have large standard deviations.

A better design for this experiment is a two-factor design, with water as the second factor. Since the effects of water are not of interest, water is called a **blocking factor**, rather than a treatment factor. In the two-factor experiment, there are nine treatment–block combinations, corresponding to the three fertilizer treatment levels and the three water block levels. With nine experimental units (the nine plots), it is necessary to assign one plot to each combination of fertilizer and water. Figure 9.12 presents two possible arrangements.

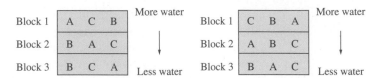

FIGURE 9.12 Two possible arrangements for three fertilizers, A, B, and C, with the restriction that each fertilizer must appear once at each water level (block). The distribution of water levels is always the same for each fertilizer.

In the two-factor design, each treatment appears equally often (once, in this example) in each block. As a result, the effect of the blocking factor does not contribute to random variation in the estimate of the main effects of the treatment factor. Because each treatment must appear equally often in each block, the only randomization in the assignment of treatments to experimental units is the order in which the treatments appear in each block. This is not a completely randomized design; it is a design in which the treatments are **randomized within blocks**. Since every possible combination of treatments and blocks is included in the experiment, the design is **complete**. For this reason the design is called a **randomized complete block design**.

Randomized complete block designs can be constructed with several treatment factors and several blocking factors. We will restrict our discussion to the case where there is one treatment factor and one blocking factor. The data from a randomized complete block design are analyzed with a two-way ANOVA, in the same way that data from any complete, balanced two-factor design would be. There is one important consideration, however. The only effects of interest are the main effects of the treatment factor. In order to interpret these main effects, **there must be no interaction between treatment and blocking factors**.

Example
9.11

Three fertilizers are studied for their effect on yield in an orange grove. Nine plots of land are used, divided into blocks of three plots each. A randomized complete block design is used, with each fertilizer applied once in each block. The results, in pounds of harvested fruit, are presented in the following table, followed by MINITAB output for the two-way ANOVA. Can we conclude that the mean yields differ among fertilizers? What assumption is made about interactions between fertilizer and plot? How is the sum of squares for error computed?

Fertilizer	Plot 1	Plot 2	Plot 3
A	430	542	287
B	367	463	253
C	320	421	207

```
Two-way ANOVA: Yield versus Block, Fertilizer

Source       DF      SS        MS       F       P
Fertilizer    2   16213.6   8106.778  49.75  0.001
Block         2   77046.9   38523.44  236.4  0.000
Error         4   651.778   162.9444
Total         8   93912.2
```

Solution

The *P*-value for the fertilizer factor is 0.001, so we conclude that fertilizer does have an effect on yield. The assumption is made that there is no interaction between the fertilizer and the blocking factor (plot), so that the main effects of fertilizer can be interpreted. Since there is only one observation for each treatment–block combination (i.e., $K = 1$), the sum of squares for error (SSE) reported in the output is really SSAB, the sum of squares for interaction, and the error mean square (MSE) is actually MSAB. (See the discussion on page 435.)

A closer look at the ANOVA table in Example 9.11 shows that in this experiment, blocking was necessary to detect the fertilizer effect. To see this, consider the experiment to be a one-factor experiment. The sum of squares for error (SSE) would then be the sum of SSE for the blocked design plus the sum of squares for blocks, or $651.778 + 77,046.9 = 77,698.7$. The degrees of freedom for error would be equal to the sum of the degrees of freedom for error in the blocked design plus the degrees of freedom for blocks, or $2 + 4 = 6$. The error mean square (MSE) would then be $77,698.7/6 \approx 12,950$ rather than 162.9444, and the F statistic for the fertilizer effect would be less than 1, which would result in a failure to detect an effect.

In general, using a blocked design reduces the degrees of freedom for error, which by itself tends to reduce the power to detect an effect. However, unless the blocking factor has very little effect on the response, this will usually be more than offset by a reduction in the sum of squares for error. Failing to include a blocking factor that affects the response can reduce the power greatly, while including a blocking factor that does not affect the response reduces the power only modestly in most cases. For this reason it is a good idea to use a blocked design whenever it is thought that the blocking factor may possibly be related to the response.

Summary

- A two-factor randomized complete block design is a complete balanced two-factor design in which the effects of one factor (the treatment factor) are of interest, while the effects of the other factor (the blocking factor) are not of interest. The blocking factor is included to reduce the random variation in the main effect estimates of the treatment factor.

- Since the object of a randomized complete block design is to estimate the main effects of the treatment factor, there must be no interaction between the treatment factor and the blocking factor.

- A two-way analysis of variance is used to estimate effects and to perform hypothesis tests on the main effects of the treatment factor.

- A randomized complete block design provides a great advantage over a completely randomized design when the blocking factor strongly affects the response and provides a relatively small disadvantage when the blocking factor has little or no effect. Therefore, when in doubt, use a blocked design.

*E*xample
9.12

The article "Experimental Design for Process Settings in Aircraft Manufacturing" (R. Sauter and R. Lenth, *Statistical Case Studies: A Collaboration Between Academe and Industry*, SIAM–ASA, 1998:151–157) describes an experiment in which the quality of holes drilled in metal aircraft parts was studied. One important indicator of hole quality is "excess diameter," which is the difference between the diameter of the drill bit and the diameter of the hole. Small excess diameters are better than large ones. Assume we are interested in the effect of the rotational speed of the drill on the excess diameter of the hole. Holes will be drilled in six test pieces (coupons), at three speeds: 6000, 10,000, and 15,000 rpm. The excess diameter can be affected not only by the speed of the drill, but also by the physical properties of the test coupon. Describe an appropriate design for this experiment.

Solution

A randomized complete block design is appropriate, with drill speed as the treatment factor, and test coupon as the blocking factor. Since six observations can be made in each block, each drill speed should be used twice in each block. The order of the speeds within each block should be chosen at random.

*E*xample
9.13

The design suggested in Example 9.12 has been adopted, and the experiment has been carried out. The results (from MINITAB) follow. Does the output indicate any violation of necessary assumptions? What do you conclude regarding the effect of drill speed on excess diameter?

```
Two-way ANOVA: Excess Diameter versus Block, Speed

Source        DF      SS         MS       F      P
Block          5   0.20156   0.0403117   1.08   0.404
Speed          2   0.07835   0.0391750   1.05   0.370
Interaction   10   0.16272   0.0162717   0.44   0.909
Error         18   0.67105   0.0372806
Total         35   1.11368

S = 0.1931    R-Sq = 39.74%    R-Sq(adj) = 0.00%
```

Solution

In a randomized complete block design, there must be no interaction between the treatment factor and the blocking factor, so that the main effect of the treatment factor may be interpreted. The P-value for interactions is 0.909, which is consistent with the hypothesis of no interactions. Therefore there is no indication in the output of any violation of assumptions. The P-value for the main effect of speed is 0.370, which is not small. Therefore we cannot conclude that excess hole diameter is affected by drill speed.

For more information on randomized block designs, a text on design of experiments, such as Montgomery (2009a), can be consulted.

Exercises for Section 9.4

1. The article "Methods for Evaluation of Easily-Reducible Iron and Manganese in Paddy Soils" (M. Borges, J. de Mello, et al., *Communication in Soil Science and Plant Analysis*, 2001:3009–3022) describes an experiment in which pH levels of four alluvial soils were measured. Various levels of liming were applied to each soil. The main interest centers on differences among the soils; there is not much interest in the effect of the liming. The results are presented in the following table.

Soil	Liming Level				
	1	2	3	4	5
A	5.8	5.9	6.1	6.7	7.1
B	5.2	5.7	6.0	6.4	6.8
C	5.5	6.0	6.2	6.7	7.0
D	6.0	6.6	6.7	6.7	7.5

a. Which is the blocking factor, and which is the treatment factor?
b. Construct an ANOVA table. You may give ranges for the P-values.
c. Can you conclude that the soils have differing pH levels?

2. A study was done to see which of four machines is fastest in performing a certain task. There are three operators; each performs the task twice on each machine. A randomized block design is employed. The MINITAB output follows.

```
Source           DF          SS        MS       F       P
Machine         (i)      257.678     (ii)    (iii)   0.021
Block           (iv)     592.428     (v)     (vi)    0.000
Interaction  (vii)       (viii)      (ix)    (x)     0.933
Error           (xi)     215.836   17.986
Total           (xii)   1096.646
```

a. Fill in the missing numbers (i) through (xii) in the output.
b. Does the output indicate that the assumptions for the randomized block design are satisfied? Explain.
c. Can you conclude that there are differences among the machines? Explain.

3. Four lighting methods were used in each of three rooms. For each method and each room, the illuminance (in lux) was measured in three separate occasions, resulting in three replications of the experiment. The only effect of interest is the lighting type; the room is a blocking factor. The following sums of squares were calculated: sum of squares for blocks = 11432, sum of squares for treatments = 9943, sum of squares for interactions = 6135, total sum of squares = 51376.

a. Construct an ANOVA table. You may give ranges for the P-values.
b. Are the assumptions for a randomized complete block design satisfied? Explain.
c. Does the ANOVA table provide evidence that lighting type affects illuminance? Explain.

4. Three different corrosion-resistant coatings are being considered for use on iron pipes. Ten pieces of pipe are marked off in six equal segments. For each pipe, two segments received coating A, two received coating B, and the remaining two received coating C. The pipes were placed in a corrosive environment for a period of time, then the depth of the deepest pit (in mm) caused by corrosion was measured for each segment on each pipe. The effect of interest is the coating; the pipe is a blocking factor, and there were two replications on each pipe. The following sums of squares were calculated: sum of squares for blocks = 11.2, sum of squares for treatments = 4.8, sum of squares for interactions = 12.4, total sum of squares = 44.7.

a. Construct an ANOVA table. You may give ranges for the P-values.
b. Are the assumptions for a randomized complete block design satisfied? Explain.
c. Does the ANOVA table provide evidence that mean pit depth differs among coatings? Explain.

5. The article "Genotype-Environment Interactions and Phenotypic Stability Analyses of Linseed in Ethiopia" (W. Adguna and M. Labuschagne, *Plant Breeding*, 2002:66–71) describes a study in which seed yields of 10 varieties of linseed were compared. Each variety was grown on six different plots. The yields, in kilograms per hectare, are presented in the following table.

Variety	Plot 1	2	3	4	5	6
A	2032	1377	1343	1366	1276	1209
B	1815	1126	1338	1188	1566	1454
C	1739	1311	1340	1250	1473	1617
D	1812	1313	1044	1245	1090	1280
E	1781	1271	1308	1220	1371	1361

(*Continued*)

			Plot			
Variety	1	2	3	4	5	6
F	1703	1089	1256	1385	1079	1318
G	1476	1333	1162	1363	1056	1096
H	1745	1308	1190	1269	1251	1325
I	1679	1216	1326	1271	1506	1368
J	1903	1382	1373	1609	1396	1366

a. Construct an ANOVA table. You may give ranges for the P-values.
b. Can you conclude that the varieties have differing mean yields?

6. The article "Application of Fluorescence Technique for Rapid Identification of IOM Fractions in Source Waters" (T. Marhaba and R. Lippincott, *Journal of Environmental Engineering*, 2000:1039–1044) presents measurements of concentrations of dissolved organic carbon (in mg/L) at six locations (A, B, C, D, E, F) along the Millstone River in central New Jersey. Measurements were taken at four times of the year: in January, April, July, and October. It is of interest to determine whether the concentrations vary among locations. The data are presented in the following table.

	January	April	July	October
A	3.9	3.7	3.7	4.1
B	4.0	3.5	3.4	5.7
C	4.2	3.4	3.0	4.8
D	4.1	3.3	2.9	4.6
E	4.1	3.4	3.0	3.4
F	4.2	3.5	2.8	4.7

a. Construct an ANOVA table. You may give ranges for the P-values.
b. Can you conclude that the concentration varies among locations?

7. You have been given the task of designing a study concerning the lifetimes of five different types of electric motor. The initial question to be addressed is whether there are differences in mean lifetime among the five types. There are 20 motors, four of each type, available for testing. A maximum of five motors can be tested each day. The ambient temperature differs from day to day, and this can affect motor lifetime.

a. Describe how you would choose the five motors to test each day. Would you use a completely randomized design? Would you use any randomization at all?
b. If X_{ij} represents the measured lifetime of a motor of type i tested on day j, express the test statistic for testing the null hypothesis of equal lifetimes in terms of the X_{ij}.

8. An engineering professor wants to determine which subject engineering students find most difficult among statistics, physics, and chemistry. She obtains the final exam grades for four students who took all three courses last semester and who were in the same sections of each class. The results are presented in the following table.

		Student		
Course	1	2	3	4
Statistics	82	94	78	70
Physics	75	70	81	83
Chemistry	93	82	80	70

a. The professor proposes a randomized complete block design, with the students as the blocks. Give a reason that this is likely not to be appropriate.

b. Describe the features of the data in the preceding table that suggest that the assumptions of the randomized complete block design are violated.

9.5 2^p Factorial Experiments

When an experimenter wants to study several factors simultaneously, the number of different treatments can become quite large. In these cases, preliminary experiments are often performed in which each factor has only two levels. One level is designated as the "high" level, and the other is designated as the "low" level. If there are p factors, there are then 2^p different treatments. Such experiments are called **2^p factorial experiments**. Often, the purpose of a 2^p experiment is to determine which factors have an important effect on the outcome. Once this is determined, more elaborate experiments can be performed, in which the factors previously found to be important are varied over several levels. We will begin by describing 2^3 factorial experiments.

Notation for 2^3 Factorial Experiments

In a 2^3 factorial experiment, there are three factors and $2^3 = 8$ treatments. The **main effect** of a factor is defined to be the difference between the mean response when the factor is at its high level and the mean response when the factor is at its low level. It is common to denote the main effects by A, B, and C. As with any factorial experiment, there can be interactions between the factors. With three factors, there are three two-way interactions, one for each pair of factors, and one three-way interaction. The two-way interactions are denoted by AB, AC, and BC, and the three-way interaction by ABC. The treatments are traditionally denoted with lowercase letters, with a letter indicating that a factor is at its high level. For example, ab denotes the treatment in which the first two factors are at their high level and the third factor is at its low level. The symbol "1" is used to denote the treatment in which all factors are at their low levels.

Estimating Effects in a 2^3 Factorial Experiment

Assume that there are K replicates for each treatment in a 2^3 factorial experiment. For each treatment, the cell mean is the average of the K observations for that treatment. The formulas for the effect estimates can be easily obtained from the 2^3 **sign table**, presented as Table 9.7.

The signs are placed in the table as follows. For the main effects A, B, C, the sign is $+$ for treatments in which the factor is at its high level, and $-$ for treatments where the factor is at its low level. So for the main effect A, the sign is $+$ for treatments a, ab, ac, and abc, and $-$ for the rest. For the interactions, the signs are computed by taking the product of the signs in the corresponding main effects columns. For example, the signs for the two-way interaction AB are the products of the signs in columns A and B, and the signs for the three-way interaction ABC are the products of the signs in columns A and B and C.

TABLE 9.7 Sign table for a 2^3 factorial experiment

Treatment	Cell Mean	A	B	C	AB	AC	BC	ABC
1	$\overline{X}_1$	−	−	−	+	+	+	−
a	$\overline{X}_a$	+	−	−	−	−	+	+
b	$\overline{X}_b$	−	+	−	−	+	−	+
ab	$\overline{X}_{ab}$	+	+	−	+	−	−	−
c	$\overline{X}_c$	−	−	+	+	−	−	+
ac	$\overline{X}_{ac}$	+	−	+	−	+	−	−
bc	$\overline{X}_{bc}$	−	+	+	−	−	+	−
abc	$\overline{X}_{abc}$	+	+	+	+	+	+	+

Estimating main effects and interactions is done with the use of the sign table. We illustrate how to estimate the main effect of factor A. Factor A is at its high level in the rows of the table where there is a "+" sign in column A. Each of the cell means $\overline{X}_a$, $\overline{X}_{ab}$, $\overline{X}_{ac}$, and $\overline{X}_{abc}$ is an average response for runs made with factor A at its high level. We estimate the mean response for factor A at its high level to be the average of these cell means.

$$\text{Estimated mean response for } A \text{ at high level} = \frac{1}{4}(\overline{X}_a + \overline{X}_{ab} + \overline{X}_{ac} + \overline{X}_{abc})$$

Similarly, each row with a "−" sign in column A represents a treatment with factor A set to its low level. We estimate the mean response for factor A at its low level to be the average of the cell means in these rows.

$$\text{Estimated mean response for } A \text{ at low level} = \frac{1}{4}(\overline{X}_1 + \overline{X}_b + \overline{X}_c + \overline{X}_{bc})$$

The estimate of the main effect of factor A is the difference in the estimated mean response between its high and low levels.

$$A \text{ effect estimate} = \frac{1}{4}(-\overline{X}_1 + \overline{X}_a - \overline{X}_b + \overline{X}_{ab} - \overline{X}_c + \overline{X}_{ac} - \overline{X}_{bc} + \overline{X}_{abc})$$

The quantity inside the parentheses is called the **contrast** for factor A. It is computed by adding and subtracting the cell means, using the signs in the appropriate column of the sign table. Note that the number of plus signs is the same as the number of minus signs, so the sum of the coefficients is equal to 0. The effect estimate is obtained by dividing the contrast by half the number of treatments, which is $2^3/2$, or 4. Estimates for other main effects and interactions are computed in an analogous manner. To illustrate, we present the effect estimates for the main effect C and for the two-way interaction AB:

$$C \text{ effect estimate} = \frac{1}{4}(-\overline{X}_1 - \overline{X}_a - \overline{X}_b - \overline{X}_{ab} + \overline{X}_c + \overline{X}_{ac} + \overline{X}_{bc} + \overline{X}_{abc})$$

$$AB \text{ interaction estimate} = \frac{1}{4}(\overline{X}_1 - \overline{X}_a - \overline{X}_b + \overline{X}_{ab} + \overline{X}_c - \overline{X}_{ac} - \overline{X}_{bc} + \overline{X}_{abc})$$

Summary

The **contrast** for any main effect or interaction is obtained by adding and subtracting the cell means, using the signs in the appropriate column of the sign table.

For a 2^3 factorial experiment,

$$\text{Effect estimate} = \frac{\text{contrast}}{4} \tag{9.42}$$

Example 9.14

A 2^3 factorial experiment was conducted to estimate the effects of three factors on the compressive strength of concrete cylinders. The factors were A: type of sand (fine or coarse), B: type of concrete (standard formulation or new formulation), and C: amount of water (low or high). Three replicates were obtained for each treatment. The strengths, presented in the following table, are measured in MPa. Estimate all effects and interactions.

Treatment	Yield			Cell Mean
1	16.157	18.777	21.351	18.7617
a	18.982	14.843	18.154	17.3263
b	18.760	20.661	20.409	19.9433
ab	23.459	29.009	23.051	25.1730
c	24.209	15.112	15.833	18.3847
ac	17.427	18.606	19.143	18.3920
bc	14.571	18.383	19.120	17.3580
abc	30.510	21.710	30.745	27.6550

Solution

We use the sign table (Table 9.7) to find the appropriate sums and differences of the cell means. We present the calculations for the main effect A, the two-way interaction BC, and the three-way interaction ABC:

$$
\begin{aligned}
A \text{ effect estimate} = \frac{1}{4}(&-18.7617 + 17.3263 - 19.9433 + 25.1730 \\
&- 18.3846 + 18.3920 - 17.3580 + 27.6550) = 3.5247
\end{aligned}
$$

$$
\begin{aligned}
BC \text{ interaction estimate} = \frac{1}{4}(&18.7617 + 17.3263 - 19.9433 - 25.1730 \\
&- 18.3846 - 18.3920 + 17.3580 + 27.6550) = -0.1980
\end{aligned}
$$

$$
\begin{aligned}
ABC \text{ interaction estimate} = \frac{1}{4}(&-18.7617 + 17.3263 + 19.9433 - 25.1730 \\
&+ 18.3846 - 18.3920 - 17.3580 + 27.6550) = 0.9062
\end{aligned}
$$

We present all the estimated effects in the following table, rounded off to the same precision as the data:

Term	Effect
A	3.525
B	4.316
C	0.146
AB	4.239
AC	1.628
BC	−0.198
ABC	0.906

For each effect, we can test the null hypothesis that the effect is equal to 0. When the null hypothesis is rejected, this provides evidence that the factors involved actually affect the outcome. To test these null hypotheses, an ANOVA table is constructed containing the appropriate sums of squares. For the tests we present to be valid, the number of replicates must be the same for each treatment and must be at least 2. In addition, the observations in each treatment must constitute a random sample from a normal population, and the populations must all have the same variance.

We compute the error sum of squares (SSE) by adding the sums of squared deviations from the sample means for all the treatments. To express this in an equation, let $s_1^2, \ldots, s_8^2$ denote the sample variances of the observations in each of the eight treatments, and let K be the number of replicates per treatment. Then

$$\text{SSE} = (K - 1) \sum_{i=1}^{8} s_i^2 \tag{9.43}$$

Each main effect and interaction has its own sum of squares as well. These are easy to compute. The sum of squares for any effect or interaction is computed by squaring its contrast, multiplying by the number of replicates K, and dividing by the total number of treatments, which is $2^3 = 8$.

$$\text{Sum of squares for an effect} = \frac{K(\text{contrast})^2}{8} \tag{9.44}$$

When using Equation (9.44), it is best to keep as many digits in the effect estimates as possible, in order to obtain maximum precision in the sum of squares. For presentation in a table, effect estimates and sums of squares may be rounded to the same precision as the data.

The sums of squares for the effects and interactions have one degree of freedom each. The error sum of squares has $8(K - 1)$ degrees of freedom. The method for computing mean squares and F statistics is the same as the one presented in Section 9.3

for a two-way ANOVA table. Each mean square is equal to its sum of squares divided by its degrees of freedom. The test statistic for testing the null hypothesis that an effect or interaction is equal to 0 is computed by dividing the mean square for the effect estimate by the mean square for error. When the null hypothesis is true, the test statistic has an $F_{1, 8(K-1)}$ distribution.

Example 9.15

Refer to Example 9.14. Construct an ANOVA table. For each effect and interaction, test the null hypothesis that it is equal to 0. Which factors, if any, seem most likely to have an effect on the outcome?

Solution

The ANOVA table follows. The sums of squares for the effects and interactions were computed by using Equation (9.44). The error sum of squares was computed by applying Equation (9.43) to the data in Example 9.14. Each F statistic is the quotient of the mean square with the mean square for error. Each F statistic has 1 and 16 degrees of freedom.

Source	Effect	DF	Sum of Squares	Mean Square	F	P
A	3.525	1	74.540	74.540	7.227	0.016
B	4.316	1	111.776	111.776	10.838	0.005
C	0.146	1	0.128	0.128	0.012	0.913
A B	4.239	1	107.798	107.798	10.452	0.005
A C	1.628	1	15.893	15.893	1.541	0.232
B C	−0.198	1	0.235	0.235	0.023	0.882
A B C	0.906	1	4.927	4.927	0.478	0.499
Error		16	165.020	10.314		
Total		23	480.316			

The main effects of factors A and B, as well as the AB interaction, have small P-values. This suggests that these effects are not equal to 0 and that factors A and B do affect the outcome. There is no evidence that the main effect of factor C, or any of its interactions, differ from 0. Further experiments might focus on factors A and B. Perhaps a two-way ANOVA would be conducted, with each of the factors A and B evaluated at several levels, to get more detailed information about their effects on the outcome.

Interpreting Computer Output

In practice, analyses of factorial designs are usually carried out on a computer. The following output (from MINITAB) presents the results of the analysis described in Examples 9.14 and 9.15.

```
Factorial Fit: Yield versus A, B, C

Estimated Effects and Coefficients for Yield (coded units)

Term        Effect       Coef  SE Coef       T      P
Constant             20.3742   0.6555   31.08  0.000
A           3.5247    1.7623   0.6555    2.69  0.016
B           4.3162    2.1581   0.6555    3.29  0.005
C           0.1463    0.0732   0.6555    0.11  0.913
A*B         4.2387    2.1193   0.6555    3.23  0.005
A*C         1.6275    0.8138   0.6555    1.24  0.232
B*C        -0.1980   -0.0990   0.6555   -0.15  0.882
A*B*C       0.9062    0.4531   0.6555    0.69  0.499


S = 3.2115    R-Sq = 65.64%   R-Sq(adj) = 50.61%


Analysis of Variance for Yield (coded units)

Source               DF   Seq SS   Adj SS   Adj MS      F      P
Main Effects          3  186.4439 186.4439  62.1480   6.03  0.006
2-Way Interactions    3  123.9255 123.9255  41.3085   4.01  0.026
3-Way Interactions    1    4.9268   4.9268   4.9268   0.48  0.499
Residual Error       16  165.0201 165.0201  10.3138
  Pure Error         16  165.0201 165.0201  10.3138
Total                23  480.3164
```

The table at the top of the output presents estimated effects and coefficients. The phrase "coded units" means that the values 1 and −1, rather than the actual values, are used to represent the high and low levels of each factor. The estimated effects are listed in the column labeled "Effect." In the next column are the estimated **coefficients**, each of which is equal to one-half the corresponding effect. While the effect represents the difference in the mean response between the high and low levels of a factor, the coefficient represents the difference between the mean response at the high level and the grand mean response, which is half as much. The coefficient labeled "Constant" is the mean of all the observations, that is, the sample grand mean. Every coefficient estimate has the same standard deviation, which is shown in the column labeled "SE Coef."

MINITAB uses the Student's t statistic, rather than the F statistic, to test the hypotheses that the effects are equal to zero. The column labeled "T" presents the value of the Student's t statistic, which is equal to the quotient of the coefficient estimate (Coef) and its standard deviation. Under the null hypothesis, the t statistic has a Student's t distribution with $8(K − 1)$ degrees of freedom. The P-values are presented in the column labeled "P." The t test performed by MINITAB is equivalent to the F test described in Example 9.15. The $t_{8(K-1)}$ statistic can be computed by taking the square root of the $F_{1,\,8(K-1)}$ statistic and applying the sign of the effect estimate. The P-values are identical.

We'll discuss the analysis of variance table next. The column labeled "DF" presents degrees of freedom. The columns labeled "Seq SS" (sequential sum of squares) and "Adj SS" (adjusted sum of squares) will be identical in all the examples we will consider and will contain sums of squares. The column labeled "Adj MS" contains mean squares, or sums of squares divided by their degrees of freedom. We will now explain the rows involving error. The row labeled "Pure Error" is concerned with the error sum of squares (SSE) (Equation 9.43). There are $8(K-1)=16$ degrees of freedom (DF) for pure error. The sum of squares for pure error, found in each of the next two columns is the error sum of squares (SSE). Under the column "Adj MS" is the mean square for error. The row above the pure error row is labeled "Residual Error." The sum of squares for residual error is equal to the sum of squares for pure error, plus the sums of squares for any main effects or interactions that are not included in the model. The degrees of freedom for the residual error sum of squares is equal to the degrees of freedom for pure error, plus the degrees of freedom (one each) for each main effect or interaction not included in the model. Since in this example all main effects and interactions are included in the model, the residual error sum of squares and its degrees of freedom are equal to the corresponding quantities for pure error. The row labeled "Total" contains the total sum of squares (SST). The total sum of squares and its degrees of freedom are equal to the sums of the corresponding quantities for all the effects, interactions, and residual error.

Going back to the top of the table, the first row is labeled "Main Effects." There are three degrees of freedom for main effects, because there are three main effects (A, B, and C), with one degree of freedom each. The sequential sum of squares is the sum of the sums of squares for each of the three main effects. The mean square (Adj MS) is the sum of squares divided by its degrees of freedom. The column labeled "F" presents the F statistic for testing the null hypothesis that all the main effects are equal to zero. The value of the F statistic (6.03) is equal to the quotient of the mean square for main effects (62.1480) and the mean square for (pure) error (10.3138). The degrees of freedom for the F statistic are 3 and 16, corresponding to the degrees of freedom for the two mean squares. The column labeled "P" presents the P-value for the F test. In this case the P-value is 0.006, which indicates that not all the main effects are zero.

The rows labeled "2-Way Interactions" and "3-Way Interactions" are analogous to the row for main effects. The P-value for two-way interactions is 0.026, which is reasonably strong evidence that at least some of the two-way interactions are not equal to zero. Since there is only one three-way interaction ($A*B*C$), the P-value in the row labeled "3-Way Interactions" is the same (0.499) as the P-value in the table at the top of the MINITAB output for $A*B*C$.

Recall that the hypothesis tests are performed under the assumption that all the observations have the same standard deviation σ. The quantity labeled "S" is the estimate of σ and is equal to the square root of the mean square for error (MSE). The quantities "R-sq" and "R-sq(adj)" are the coefficients of determination R^2 and the adjusted R^2, respectively, and are computed by methods analogous to those in one-way ANOVA.

Estimating Effects in a 2^p Factorial Experiment

A sign table can be used to obtain the formulas for computing effect estimates in any 2^p factorial experiment. The method is analogous to the 2^3 case. The treatments are listed in a column. The sign for any main effect is $+$ in the rows corresponding to treatments where the factor is at its high level, and $-$ in rows corresponding to treatments where the factor is at its low level. Signs for the interactions are found by multiplying the signs corresponding to the factors in the interaction. The estimate for any effect or interaction is found by adding and subtracting the cell means for the treatments, using the signs in the appropriate columns, to compute a contrast. The contrast is then divided by half the number of treatments, or 2^{p-1}, to obtain the effect estimate.

Summary

For a 2^p factorial experiment:

$$\text{Effect estimate} = \frac{\text{contrast}}{2^{p-1}} \qquad (9.45)$$

As an example, Table 9.8 (page 456) presents a sign table for a 2^5 factorial experiment. We list the signs for the main effects and for selected interactions.

Sums of squares are computed by a method analogous to that for a 2^3 experiment. To compute the error sum of squares (SSE), let $s_1, \ldots, s_{2^p}$ be the sample variances of the observations in each of the 2^p treatments. Then

$$\text{SSE} = (K - 1) \sum_{i=1}^{2^p} s_i^2$$

The degrees of freedom for error is $2^p(K - 1)$, where K is the number of replicates per treatment. The sum of squares for each effect and interaction is equal to the square of the contrast, multiplied by the number of replicates K and divided by the number of treatments 2^p. The sums of squares for the effects and interactions have one degree of freedom each.

$$\text{Sum of squares for an effect} = \frac{K(\text{contrast})^2}{2^p} \qquad (9.46)$$

F statistics for main effects and interactions are computed by dividing the sum of squares for the effect by the mean square for error. The null distribution of the F statistic is $F_{1,\,2^p(K-1)}$.

Factorial Experiments without Replication

When the number of factors p is large, it is often not feasible to perform more than one replicate for each treatment. In this case, it is not possible to compute SSE, so the hypothesis tests previously described cannot be performed. If it is reasonable to assume that some of the higher-order interactions are equal to 0, then the sums of squares for those interactions can be added together and treated like an error sum of squares. Then the main effects and lower-order interactions can be tested.

TABLE 9.8 Sign table for the main effects and selected interactions for a 2^5 factorial experiment

Treatment	A	B	C	D	E	AB	CDE	ABDE	ABCDE
1	−	−	−	−	−	+	−	+	−
a	+	−	−	−	−	−	−	−	+
b	−	+	−	−	−	−	−	−	+
ab	+	+	−	−	−	+	−	+	−
c	−	−	+	−	−	+	+	+	+
ac	+	−	+	−	−	−	+	−	−
bc	−	+	+	−	−	−	+	−	−
abc	+	+	+	−	−	+	+	+	+
d	−	−	−	+	−	+	+	−	+
ad	+	−	−	+	−	−	+	+	−
bd	−	+	−	+	−	−	+	+	−
abd	+	+	−	+	−	+	+	−	+
cd	−	−	+	+	−	+	−	−	−
acd	+	−	+	+	−	−	−	+	+
bcd	−	+	+	+	−	−	−	+	+
abcd	+	+	+	+	−	+	−	−	−
e	−	−	−	−	+	+	+	−	+
ae	+	−	−	−	+	−	+	+	−
be	−	+	−	−	+	−	+	+	−
abe	+	+	−	−	+	+	+	−	+
ce	−	−	+	−	+	+	−	−	−
ace	+	−	+	−	+	−	−	+	+
bce	−	+	+	−	+	−	−	+	+
abce	+	+	+	−	+	+	−	−	−
de	−	−	−	+	+	+	−	+	−
ade	+	−	−	+	+	−	−	−	+
bde	−	+	−	+	+	−	−	−	+
abde	+	+	−	+	+	+	−	+	−
cde	−	−	+	+	+	+	+	+	+
acde	+	−	+	+	+	−	+	−	−
bcde	−	+	+	+	+	−	+	−	−
abcde	+	+	+	+	+	+	+	+	+

A 2^5 factorial experiment was conducted to estimate the effects of five factors on the quality of lightbulbs manufactured by a certain process. The factors were A: plant (1 or 2), B: machine type (low or high speed), C: shift (day or evening), D: lead wire material (standard or new), and E: method of loading materials into the assembler (manual or automatic). One replicate was obtained for each treatment. Table 9.9 presents the results. Compute estimates of the main effects and interactions, and their sums of squares. Assume that the third-, fourth-, and fifth-order interactions are negligible, and add their sums of squares to use as a substitute for an error sum of squares. Use this substitute to test hypotheses concerning the main effects and second-order interactions.

TABLE 9.9

Treatment	Outcome	Treatment	Outcome	Treatment	Outcome	Treatment	Outcome
1	32.07	d	35.64	e	25.10	de	40.60
a	39.27	ad	35.91	ae	39.25	ade	37.57
b	34.81	bd	47.75	be	37.77	bde	47.22
ab	43.07	abd	51.47	abe	46.69	$abde$	56.87
c	31.55	cd	33.16	ce	32.55	cde	34.51
ac	36.51	acd	35.32	ace	32.56	$acde$	36.67
bc	28.80	bcd	48.26	bce	28.99	$bcde$	45.15
abc	43.05	$abcd$	53.28	$abce$	48.92	$abcde$	48.72

TABLE 9.10

Term	Effect	Sum of Squares	Term	Effect	Sum of Squares
A	6.33	320.05	ABD	−0.29	0.67
B	9.54	727.52	ABE	0.76	4.59
C	−2.07	34.16	ACD	0.11	0.088
D	6.70	358.72	ACE	−0.69	3.75
E	0.58	2.66	ADE	−0.45	1.60
AB	2.84	64.52	BCD	0.76	4.67
AC	0.18	0.27	BCE	−0.82	5.43
AD	−3.39	91.67	BDE	−2.17	37.63
AE	0.60	2.83	CDE	−1.25	12.48
BC	−0.49	1.95	$ABCD$	−2.83	63.96
BD	4.13	136.54	$ABCE$	0.39	1.22
BE	0.65	3.42	$ABDE$	0.22	0.37
CD	−0.18	0.26	$ACDE$	0.18	0.24
CE	−0.81	5.23	$BCDE$	−0.25	0.52
DE	0.24	0.46	$ABCDE$	−1.73	23.80
ABC	1.35	14.47			

Solution

We compute the effects, using the rules for adding and subtracting observations given by the sign table, and the sums of squares, using Equation (9.46). See Table 9.10.

Note that none of the three-, four-, or five-way interactions are among the larger effects. If some of them were, it would not be wise to combine their sums of squares. As it is, we add the sums of squares of the three-, four-, and five-way interactions. The results are presented in the following output (from MINITAB).

```
Factorial Fit: Response versus A, B, C, D, E

Estimated Effects and Coefficients for Response (coded units)

Term        Effect    Coef   SE Coef     T      P
Constant            39.658   0.5854   67.74   0.000
A            6.325   3.163   0.5854    5.40   0.000
B            9.536   4.768   0.5854    8.14   0.000
```

```
C          -2.066  -1.033   0.5854  -1.76   0.097
D           6.696   3.348   0.5854   5.72   0.000
E           0.576   0.288   0.5854   0.49   0.629
A*B         2.840   1.420   0.5854   2.43   0.027
A*C         0.183   0.091   0.5854   0.16   0.878
A*D        -3.385  -1.693   0.5854  -2.89   0.011
A*E         0.595   0.298   0.5854   0.51   0.618
B*C        -0.494  -0.247   0.5854  -0.42   0.679
B*D         4.131   2.066   0.5854   3.53   0.003
B*E         0.654   0.327   0.5854   0.56   0.584
C*D        -0.179  -0.089   0.5854  -0.15   0.881
C*E        -0.809  -0.404   0.5854  -0.69   0.500
D*E         0.239   0.119   0.5854   0.20   0.841


S = 3.31179    R-Sq = 90.89%    R-Sq(adj) = 82.34%


Analysis of Variance for Response (coded units)

Source               DF  Seq SS  Adj SS  Adj MS      F      P
Main Effects          5  1443.1  1443.1  288.62  26.31  0.000
2-Way Interactions   10   307.1   307.1   30.71   2.80  0.032
Residual Error       16   175.5   175.5   10.97
Total                31  1925.7
```

The estimates have not changed for the main effects or two-way interactions. The residual error sum of squares (175.5) in the analysis of variance table is found by adding the sums of squares for all the higher-order interactions that were dropped from the model. The number of degrees of freedom (16) is equal to the sum of the degrees of freedom (one each) for the 16 higher-order interactions. There is no sum of squares for pure error (SSE), because there is only one replicate per treatment. The residual error sum of squares is used as a substitute for SSE to compute all the quantities that require an error sum of squares.

We conclude from the output that factors A, B, and D are likely to affect the outcome. There seem to be interactions between some pairs of these factors as well. It might be appropriate to plan further experiments to focus on factors A, B, and D.

Using Probability Plots to Detect Large Effects

An informal method that has been suggested to help determine which effects are large is to plot the effect and interaction estimates on a normal probability plot. If in fact none of the factors affect the outcome, then the effect and interaction estimates form a simple random sample from a normal population and should lie approximately on a straight line.

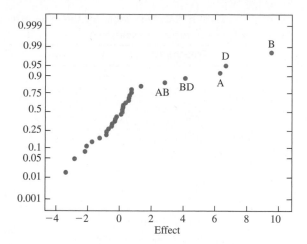

FIGURE 9.13 Normal probability plot of the effect estimates from the data in Example 9.16. The main effects of factors A, B, and D stand out as being larger than the rest.

In many cases, most of the estimates will fall approximately on a line, and a few will plot far from the line. The main effects and interactions whose estimates plot far from the line are the ones most likely to be important. Figure 9.13 presents a normal probability plot of the main effect and interaction estimates from the data in Example 9.16. It is clear from the plot that the main effects of factors A, B, and D, and the AB and BD interactions, stand out from the rest.

Fractional Factorial Experiments

When the number of factors is large, it may not be feasible to perform even one replicate for each treatment. In these cases, observations may be taken only for some fraction of the treatments. If these treatments are chosen correctly, it is still possible to obtain information about the factors.

When each factor has two levels, the fraction must always be a power of 2 (i.e., one-half, one-quarter, etc.). An experiment in which half the treatments are used is called a **half-replicate**; if one-quarter of the treatments are used, it is a **quarter-replicate**, and so on. A half-replicate of a 2^p experiment is often denoted 2^{p-1}, to indicate that while there are p factors, there are only 2^{p-1} treatments being considered. Similarly, a quarter-replicate is often denoted 2^{p-2}. We will focus on half-replicate experiments.

We present a method for choosing a half-replicate of a 2^5 experiment. Such an experiment will have 16 treatments chosen from the 32 in the 2^5 experiment. To choose the 16 treatments, start with a sign table for a 2^4 design that shows the signs for the main effects and the highest-order interaction. This is presented as Table 9.11 (page 460).

Table 9.11 has the right number of treatments (16), but only four factors. To transform it into a half-replicate for a 2^5 design, we must introduce a fifth factor, E. We do this by replacing the highest-order interaction by E. This establishes the signs for

TABLE 9.11 Sign table for the main effects and four-way interaction in a 2^4 factorial experiment

Treatment	A	B	C	D	ABCD
1	−	−	−	−	+
a	+	−	−	−	−
b	−	+	−	−	−
ab	+	+	−	−	+
c	−	−	+	−	−
ac	+	−	+	−	+
bc	−	+	+	−	+
abc	+	+	+	−	−
d	−	−	−	+	−
ad	+	−	−	+	+
bd	−	+	−	+	+
abd	+	+	−	+	−
cd	−	−	+	+	+
acd	+	−	+	+	−
bcd	−	+	+	+	−
abcd	+	+	+	+	+

the main effect of E. Then in each row where the sign for E is $+$, we add the letter e to the treatment, indicating that factor E is to be set to its high level for that treatment. Where the sign for E is $-$, factor E is set to its low level. The resulting design is called the **principal fraction** of the 2^5 design. Table 9.12 presents the signs for the main effects and selected interactions of this design.

TABLE 9.12 Sign table for the main effects and selected interactions for the principal fraction of a 2^5 factorial experiment

Treatment	A	B	C	D	E = ABCD	AB	CDE	ACDE
e	−	−	−	−	+	+	+	−
a	+	−	−	−	−	−	−	−
b	−	+	−	−	−	−	−	+
abe	+	+	−	−	+	+	+	+
c	−	−	+	−	−	+	+	−
ace	+	−	+	−	+	−	−	−
bce	−	+	+	−	+	−	−	+
abc	+	+	+	−	−	+	+	+
d	−	−	−	+	−	+	+	−
ade	+	−	−	+	+	−	−	−
bde	−	+	−	+	+	−	−	+
abd	+	+	−	+	−	+	+	+
cde	−	−	+	+	+	+	+	−
acd	+	−	+	+	−	−	−	−
bcd	−	+	+	+	−	−	−	+
abcde	+	+	+	+	+	+	+	+

There is a price to be paid for using only half of the treatments. To see this, note that in Table 9.12 the AB interaction has the same signs as the CDE interaction, and the $ACDE$ interaction has the same signs as the main effect for B. When two effects have the same signs, they are said to be **aliased**. In fact, the main effects and interactions in a half-fraction form pairs in which each member of the pair is aliased with the other. The alias pairs for this half-fraction of the 2^5 design are

$$\{A, BCDE\} \quad \{B, ACDE\} \quad \{C, ABDE\} \quad \{D, ABCE\} \quad \{E, ABCD\}$$

$$\{AB, CDE\} \quad \{AC, BDE\} \quad \{AD, BCE\} \quad \{AE, BCD\} \quad \{BC, ADE\}$$

$$\{BD, ACE\} \quad \{BE, ACD\} \quad \{CD, ABE\} \quad \{CE, ABD\} \quad \{DE, ABC\}$$

When two effects are aliased, their effect estimates are the same, because they involve the same signs. In fact, when the principal fraction of a design is used, the estimate of any effect actually represents the sum of that effect and its alias. Therefore for the principal fraction of a 2^5 design, each main effect estimate actually represents the sum of the main effect plus its aliased four-way interaction, and each two-way interaction estimate represents the sum of the two-way interaction and its aliased three-way interaction.

In many cases, it is reasonable to assume that the higher-order interactions are small. In the 2^5 half-replicate, if the four-way interactions are negligible, the main effect estimates will be accurate. If in addition the three-way interactions are negligible, the two-way interaction estimates will be accurate as well.

In a fractional design without replication, there is often no good way to compute an error sum of squares, and therefore no rigorous way to test the hypotheses that the effects are equal to 0. In many cases, the purpose of a fractional design is simply to identify a few factors that appear to have the greatest impact on the outcome. This information may then be used to design more elaborate experiments to investigate these factors. For this purpose, it may be enough simply to choose those factors whose effects or two-way interactions are unusually large, without performing hypothesis tests. This can be done by listing the estimates in decreasing order, and then looking to see if there are a few that are noticeably larger than the rest. Another method is to plot the effect and interaction estimates on a normal probability plot, as previously discussed.

Example
9.17

In an emulsion liquid membrane system, an emulsion (internal phase) is dispersed into an external liquid medium containing a contaminant. The contaminant is removed from the external liquid through mass transfer into the emulsion. Internal phase leakage occurs when portions of the extracted material spill into the external liquid. In the article "Leakage and Swell in Emulsion Liquid Membrane Systems: Batch Experiments" (R. Pfeiffer, W. Navidi, and A. Bunge, *Separation Science and Technology*, 2003: 519–539), the effects of five factors were studied to determine the effect on leakage in a certain system. The five factors were A: surfactant concentration, B: internal phase lithium hydroxide concentration, C: membrane phase, D: internal phase volume fraction, and E: extraction vessel stirring rate. A half-fraction of a 2^5 design was used. The data are presented in the following table (in the actual experiment, each point actually represented the average of two measurements). Leakage is measured in units of percent. Assume that the third-, fourth-, and fifth-order interactions are negligible. Estimate the main effects and two-way interactions. Which, if any, stand out as being noticeably larger than the rest?

Treatment	Leakage	Treatment	Leakage	Treatment	Leakage	Treatment	Leakage
e	0.61	c	0.35	d	2.03	cde	1.45
a	0.13	ace	0.075	ade	0.64	acd	0.31
b	2.23	bce	7.31	bde	11.72	bcd	1.33
abe	0.095	abc	0.080	abd	0.56	$abcde$	6.24

Solution
Using the sign table (Table 9.12), we compute estimates for the main effects and two-way interactions, shown in the following table.

Term	Effect	Term	Effect
A	−2.36	AE	−1.15
B	3.00	BC	0.20
C	−0.11	BD	0.86
D	1.68	BE	2.65
E	2.64	CD	−1.30
AB	−1.54	CE	0.61
AC	1.43	DE	1.32
AD	0.17		

Note that we do not bother to compute sums of squares for the estimates, because we have no SSE to compare them to. To determine informally which effects may be most worthy of further investigation, we rank the estimates in order of their absolute values: B: 3.00, BE: 2.65, E: 2.64, A: −2.36, D: 1.68, and so forth. It seems reasonable to decide that there is a fairly wide gap between the A and D effects, and therefore that factors A, B, and E are most likely to be important.

Exercises for Section 9.5

1. Construct a sign table for the principal fraction for a 2^4 design. Then indicate all the alias pairs.

2. Give an example of a factorial experiment in which failure to randomize can produce incorrect results.

3. A chemical reaction was run using two levels each of temperature (A), reagent concentration (B), and pH (C). For each factor, the high level is denoted 1, and the low level is denoted -1. The reaction was run twice for each combination of levels, and the yield (in percent) was recorded. The results were as follows.

A	B	C	Yields	Mean Yield
-1	-1	-1	74, 71	72.5
1	-1	-1	73, 74	73.5
-1	1	-1	78, 74	76.0
1	1	-1	86, 89	87.5
-1	-1	1	71, 68	69.5
1	-1	1	77, 77	77.0
-1	1	1	75, 85	80.0
1	1	1	92, 82	87.0

a. Compute estimates of the main effects and interactions, and the sum of squares and P-value for each.
b. Which main effects and interactions, if any, are important?
c. Other things being equal, will the mean yield be higher when the temperature is high or low? Explain.

4. The article "Efficient Pyruvate Production by a Multi-Vitamin Auxotroph of *Torulopsis glabrata:* Key Role and Optimization of Vitamin Levels" (Y. Li, J. Chen, et al., *Applied Microbiology and Biotechnology*, 2001:680–685) investigates the effects of the levels of several vitamins in a cell culture on the yield (in g/L) of pyruvate, a useful organic acid. The data in the following table are presented as two replicates of a 2^3 design. The factors are A: nicotinic acid, B: thiamine, and C: biotin. (Two statistically insignificant factors have been dropped. In the article, each factor was tested at four levels; we have collapsed these to two.)

A	B	C	Yields	Mean Yield
-1	-1	-1	0.55, 0.49	0.520
1	-1	-1	0.60, 0.42	0.510
-1	1	-1	0.37, 0.28	0.325
1	1	-1	0.30, 0.28	0.290
-1	-1	1	0.54, 0.54	0.540
1	-1	1	0.54, 0.47	0.505
-1	1	1	0.44, 0.33	0.385
1	1	1	0.36, 0.20	0.280

a. Compute estimates of the main effects and interactions, and the sum of squares and P-value for each.
b. Is the additive model appropriate?
c. What conclusions about the factors can be drawn from these results?

5. The article cited in Exercise 4 also investigated the effects of the factors on glucose consumption (in g/L). A single measurement is provided for each combination of factors (in the article, there was some replication). The results are presented in the following table.

A	B	C	Glucose Consumption
−1	−1	−1	68.0
1	−1	−1	77.5
−1	1	−1	98.0
1	1	−1	98.0
−1	−1	1	74.0
1	−1	1	77.0
−1	1	1	97.0
1	1	1	98.0

a. Compute estimates of the main effects and the interactions.
b. Is it possible to compute an error sum of squares? Explain.
c. Are any of the interactions among the larger effects? If so, which ones?
d. Assume that it is known from past experience that the additive model holds. Add the sums of squares for the interactions, and use that result in place of an error sum of squares to test the hypotheses that the main effects are equal to 0.

6. A metal casting process for the production of turbine blades was studied. Three factors were varied. They were A: the temperature of the metal, B: the temperature of the mold, and C: the pour speed. The outcome was the thickness of the blades, in mm. The results are presented in the following table.

A	B	C	Thickness
−1	−1	−1	4.61
1	−1	−1	4.51
−1	1	−1	4.60
1	1	−1	4.54
−1	−1	1	4.61
1	−1	1	4.61
−1	1	1	4.48
1	1	1	4.51

a. Compute estimates of the main effects and the interactions.
b. Is it possible to compute an error sum of squares? Explain.
c. Are any of the interactions among the larger effects? If so, which ones?
d. Someone claims that the additive model holds. Do the results tend to support this statement? Explain.

7. The measurement of the resistance of a tungsten-coated wafer used in integrated circuit manufacture may be affected by several factors, including A: which of two types of cassette was used to hold the wafer, B: whether the wafer is loaded on the top or on the bottom of the cassette, and C: whether the front or the back cassette station was used. Data from a 2^3 factorial experiment with one replicate are presented in the following table. (Based on the article

"Prometrix RS35e Gauge Study in Five Two-Level Factors and One Three-Level Factor," J. Buckner, B. Chin, and J. Henri, *Statistical Case Studies for Industrial Process Improvement*, SIAM-ASA, 1997:9–18.)

A	B	C	Resistance (mΩ)
−1	−1	−1	85.04
1	−1	−1	84.49
−1	1	−1	82.15
1	1	−1	86.37
−1	−1	1	82.60
1	−1	1	85.14
−1	1	1	82.87
1	1	1	86.44

a. Compute estimates of the main effects and the interactions.

b. Is it possible to compute an error sum of squares? Explain.

c. Plot the estimates on a normal probability plot. Does the plot show that some of the factors influence the resistance? If so, which ones?

8. In a 2^p design with one replicate per treatment, it sometimes happens that the observation for one of the treatments is missing, due to experimental error or to some other cause. When this happens, one approach is to replace the missing value with the value that makes the highest-order interaction equal to 0. Refer to Exercise 7. Assume the observation for the treatment where A, B, and C are all at their low level (−1) is missing.

a. What value for this observation makes the three-way interaction equal to 0?

b. Using this value, compute estimates for the main effects and the interactions.

9. Safety considerations are important in the design of automobiles. The article "An Optimum Design Methodology Development Using a Statistical Technique for Vehicle Occupant Safety" (J. Hong, M. Mun, and S. Song, *Proceedings of the Institution of Mechanical Engineers*, 2001:795–801) presents results from an occupant simulation study. The outcome variable is chest acceleration (in g) 3 ms after impact. Four factors were considered. They were A: the airbag vent characteristic, B: the airbag inflator trigger time, C: the airbag inflator mass flow rate, and D: the stress–strain relationship of knee foam. The results (part of a larger study) are presented in the following table. There is one replicate per treatment.

Treatment	Outcome	Treatment	Outcome	Treatment	Outcome	Treatment	Outcome
1	85.2	c	66.0	d	85.0	cd	62.6
a	79.2	ac	69.0	ad	82.0	acd	65.4
b	84.3	bc	68.5	bd	84.7	bcd	66.3
ab	89.0	abc	76.4	abd	82.2	abcd	69.0

a. Compute estimates of the main effects and the interactions.

b. If you were to design a follow-up study, which factor or factors would you focus on? Explain.

10. The article "Experimental Study of Workpiece-Level Variability in Blind via Electroplating" (G. Poon, J. Chan, and D. Williams, *Proceedings of the Institution of Mechanical Engineers*, 2001:521–530) describes a factorial experiment

that was carried out to determine which of several factors influence variability in the thickness of an acid copper electroplate deposit. The results of the experiment are presented here as a complete 2^5 design; two statistically insignificant factors have been omitted. There was one replicate per treatment. The factors are A: Sulphuric acid concentration (g/L), B: Copper sulphate concentration (g/L), C: Average current density (mA/cm^2), D: Electrode separation (cm), and E: Interhole distance (cm). The outcome is the variability in the thickness (in μm), measured by computing the standard deviation of thicknesses measured at several points on a test board, after dropping the largest and smallest measurement. The data are presented in the following table.

Treatment	Outcome	Treatment	Outcome	Treatment	Outcome	Treatment	Outcome
1	1.129	d	1.760	e	1.224	de	1.674
a	0.985	ad	1.684	ae	1.092	ade	1.215
b	1.347	bd	1.957	be	1.280	bde	1.275
ab	1.151	abd	1.656	abe	1.381	$abde$	1.446
c	2.197	cd	2.472	ce	1.859	cde	2.585
ac	1.838	acd	2.147	ace	1.865	$acde$	2.587
bc	1.744	bcd	2.142	bce	1.867	$bcde$	2.339
abc	2.101	$abcd$	2.423	$abce$	2.005	$abcde$	2.629

a. Compute estimates of the main effects and the interactions.
b. If you were to design a follow-up experiment, which factors would you focus on? Why?

11. The article "Factorial Design for Column Flotation of Phosphate Wastes" (N. Abdel-Khalek, *Particulate Science and Technology*, 2000:57–70) describes a 2^3 factorial design to investigate the effect of superficial air velocity (A), frothier concentration (B), and superficial wash water velocity (C) on the percent recovery of P_2O_5. There were two replicates. The data are presented in the following table.

A	B	C	Percent Recovery
-1	-1	-1	56.30, 54.85
1	-1	-1	70.10, 72.70
-1	1	-1	65.60, 63.60
1	1	-1	80.20, 78.80
-1	-1	1	50.30, 48.95
1	-1	1	65.30, 66.00
-1	1	1	60.53, 59.50
1	1	1	70.63, 69.86

a. Compute estimates of the main effects and interactions, along with their sums of squares and P-values.
b. Which factors seem to be most important? Do the important factors interact? Explain.

12. The article "An Application of Fractional Factorial Designs" (M. Kilgo, *Quality Engineering*, 1988:19–23) describes a 2^{5-1} design (half-replicate of a 2^5 design) involving the use of carbon dioxide (CO_2) at high pressure to extract oil from peanuts. The outcomes were the solubility of the peanut oil in the CO_2 (in mg oil/liter CO_2), and the yield of

peanut oil (in %). The five factors were A: CO_2 pressure, B: CO_2 temperature, C: peanut moisture, D: CO_2 flow rate, and E: peanut particle size. The results are presented in the following table.

Treatment	Solubility	Yield	Treatment	Solubility	Yield
e	29.2	63	d	22.4	23
a	23.0	21	ade	37.2	74
b	37.0	36	bde	31.3	80
abe	139.7	99	abd	48.6	33
c	23.3	24	cde	22.9	63
ace	38.3	66	acd	36.2	21
bce	42.6	71	bcd	33.6	44
abc	141.4	54	$abcde$	172.6	96

a. Assuming third- and higher-order interactions to be negligible, compute estimates of the main effects and inter-actions for the solubility outcome.

b. Plot the estimates on a normal probability plot. Does the plot show that some of the factors influence the solubility? If so, which ones?

c. Assuming third- and higher-order interactions to be negligible, compute estimates of the main effects and inter-actions for the yield outcome.

d. Plot the estimates on a normal probability plot. Does the plot show that some of the factors influence the yield? If so, which ones?

13. In a 2^{5-1} design (such as the one in Exercise 12) what does the estimate of the main effect of factor A actually represent?

i. The main effect of A.

ii. The sum of the main effect of A and the $BCDE$ interaction.

iii. The difference between the main effect of A and the $BCDE$ interaction.

iv. The interaction between A and $BCDE$.

Supplementary Exercises for Chapter 9

1. The article "Gypsum Effect on the Aggregate Size and Geometry of Three Sodic Soils Under Reclamation" (I. Lebron, D. Suarez, and T. Yoshida, *Journal of the Soil Science Society of America*, 2002:92–98) reports on an experiment in which gypsum was added in various amounts to soil samples before leaching. One of the outcomes of interest was the pH of the soil. Gypsum was added in four different amounts. Three soil samples received each amount added. The pH measurements of the samples are presented in the following table.

Gypsum (g/kg)	pH		
0.00	7.88	7.72	7.68
0.11	7.81	7.64	7.85
0.19	7.84	7.63	7.87
0.38	7.80	7.73	8.00

Can you conclude that the pH differs with the amount of gypsum added? Provide the value of the test statistic and the P-value.

2. The article referred to in Exercise 1 also considered the effect of gypsum on the electric conductivity (in dS m^{-1}) of soil. Two types of soil were each treated with three different amounts of gypsum, with two replicates for each soil–gypsum combination. The data are presented in the following table.

	Soil Type	
Gypsum (g/kg)	las Animas	Madera
0.00	1.52 1.05	1.01 0.92
0.27	1.49 0.91	1.12 0.92
0.46	0.99 0.92	0.88 0.92

   a. Is there convincing evidence of an interaction between the amount of gypsum and soil type?
   b. Can you conclude that the conductivity differs among the soil types?
   c. Can you conclude that the conductivity differs with the amount of gypsum added?

3. Penicillin is produced by the *Penicillium* fungus, which is grown in a broth whose sugar content must be carefully controlled. Several samples of broth were taken on each of three successive days, and the amount of dissolved sugars (in mg/mL) was measured on each sample. The results were as follows:

Day 1:	4.8 5.1 5.1 4.8 5.2 4.9 5.0 4.9 5.0 4.8 4.8 5.1 5.0
Day 2:	5.4 5.0 5.0 5.1 5.2 5.1 5.3 5.2 5.2 5.1 5.4 5.2 5.4
Day 3:	5.7 5.1 5.3 5.5 5.3 5.5 5.1 5.6 5.3 5.2 5.5 5.3 5.4

   Can you conclude that the mean sugar concentration differs among the three days?

4. The following MINITAB output is for a two-way ANOVA. Something went wrong with the printer, and some of the numbers weren't printed.

```
Two-way Analysis of Variance

Analysis of Variance
Source        DF          SS        MS        F       P
Row            3     145.375       (d)       (g)     (j)
Column         2      15.042       (e)       (h)     (k)
Interaction    6         (b)    4.2000       (i)     (l)
Error        (a)         (c)       (f)
Total         23     217.870
```

Fill in the missing numbers in the table. You may give ranges for the P-values.

5. An experiment was performed to determine whether different types of chocolate take different amounts of time to dissolve. Forty people were divided into five groups. Each group was assigned a certain type of chocolate. Each person dissolved one piece of chocolate, and the dissolve time (in seconds) was recorded. For comparison, each person in each group also dissolved one piece of butterscotch candy; these pieces were identical for all groups. The data, which include the group, the dissolve times for both chocolate and butterscotch, the difference between the dissolve times, and the ratio of the dissolve times are presented in the following table. Note that the design is slightly unbalanced; group 3 has nine people and group 5 has only seven.

Group	Chocolate	Butterscotch	Difference	Ratio
1	135	60	75	2.25
1	865	635	230	1.36
1	122	63	59	1.94
1	110	75	35	1.47
1	71	37	34	1.92
1	81	58	23	1.40
1	2405	1105	1300	2.18
1	242	135	107	1.79
2	42	38	4	1.11
2	30	30	0	1.00
2	104	110	−6	0.95
2	124	118	6	1.05
2	75	40	35	1.88
2	80	91	−11	0.88
2	255	121	134	2.11
2	71	71	0	1.00
3	51	53	−2	0.96
3	47	40	7	1.18
3	90	155	−65	0.58
3	65	90	−25	0.72
3	27	33	−6	0.82
3	105	68	37	1.54
3	90	72	18	1.25
3	54	52	2	1.04
3	93	77	16	1.21
4	48	30	18	1.60
4	85	55	30	1.55
4	82	50	32	1.64
4	46	22	24	2.09
4	64	46	18	1.39
4	125	45	80	2.78
4	69	30	39	2.30
4	73	44	29	1.66
5	105	45	60	2.33
5	99	58	41	1.71
5	45	23	22	1.96
5	137	64	73	2.14
5	170	105	65	1.62
5	153	93	60	1.65
5	49	28	21	1.75

a. To test whether there are differences in the mean dissolve times for the different types of chocolate, someone suggests performing a one-way ANOVA, using the dissolve times for the chocolate data. Do these data appear to satisfy the assumptions for a one-way ANOVA? Explain.

b. Someone else suggests using the differences (Chocolate − Butterscotch). Do these data appear to satisfy the assumptions for a one-way ANOVA? Explain.

c. Perform a one-way analysis of variance using the ratios. Can you conclude that the mean ratio of dissolve times differs for different types of chocolate?

6. The article "Stability of Silico-Ferrite of Calcium and Aluminum (SFCA) in Air-Solid Solution Limits Between 1240°C and 1390°C and Phase Relationships Within the Fe_2O_3-CaO-Al_2O_3-SiO_2 (FCAS) System" (T. Patrick and M. Pownceby, *Metallurgical and Materials Transactions B*, 2002:79–90) investigates properties of silico-ferrites of calcium and aluminum (SFCA). The data in the following table present the ratio of the weights of Fe_2O_3 and CaO for SFCA specimens with several different weight percents of Al_2O_3 and C_4S_3.

Al_2O_3(%)	C_4S_3	Fe_2O_3/CaO			
1.0	Low (3%–6%)	7.25	6.92	6.60	6.31
1.0	Medium (7%–10%)	6.03	5.78	5.54	5.31
1.0	High (11%–14%)	5.10	4.90	4.71	4.53
5.0	Low (3%–6%)	6.92	6.59	6.29	6.01
5.0	Medium (7%–10%)	5.74	5.26	5.04	4.84
5.0	High (11%–14%)	4.84	4.65	4.47	4.29
10.0	Low (3%–6%)	6.50	6.18	5.89	5.63
10.0	Medium (7%–10%)	5.37	5.14	4.92	4.71
10.0	High (11%–14%)	4.52	4.33	4.16	3.99

a. Construct an ANOVA table. You may give ranges for the P-values.
b. Do the data indicate that there are any interactions between the weight percent of Al_2O_3 and the weight percent of C_4S_3? Explain.
c. Do the data convincingly demonstrate that the Fe_2O_3/CaO ratio depends on the weight percent of Al_2O_3? Explain.
d. Do the data convincingly demonstrate that the Fe_2O_3/CaO ratio depends on the weight percent of C_4S_3? Explain.

7. A component can be manufactured according to either of two designs and with either a more expensive or a less expensive material. Several components are manufactured with each combination of design and material, and the lifetimes of each are measured (in hours). A two-way analysis of variance was performed to estimate the effects of design and material on component lifetime. The cell means and main effect estimates are presented in the following table.

Cell Means			Main Effects	
	Design 1	**Design 2**		
			More expensive	14
More expensive	118	120	Less expensive	−14
Less expensive	60	122	Design 1	−16
			Design 2	16

ANOVA table

Source	DF	SS	MS	F	P
Material	1	2352.0	2352.0	10.45	0.012
Design	1	3072.0	3072.0	13.65	0.006
Interaction	1	2700.0	2700.0	12.00	0.009
Error	8	1800.0	225.00		
Total	11	9924.0			

The process engineer recommends that design 2 should be used along with the more expensive material. He argues that the main effects of both design 2 and the more expensive material are positive, so using this combination will result in the longest component life. Do you agree with the recommendation? Why or why not?

8. The article "Case Study Based Instruction of DOE and SPC" (J. Brady and T. Allen, *The American Statistician*, 2002:312–315) presents the result of a 2^{4-1} factorial experiment to investigate the effects of four factors on the yield of a process that manufactures printed circuit boards. The factors were A: transistor power output (upper or lower specification limit), B: transistor mounting approach (screwed or soldered), C: transistor heat sink type (current or alternative configuration), and D: screw position on the frequency adjustor (one-half or two turns). The results are presented in the following table. The yield is a percent of a theoretical maximum.

A	B	C	D	Yield
−1	−1	−1	−1	79.8
1	−1	−1	1	69.0
−1	1	−1	1	72.3
1	1	−1	−1	71.2
−1	−1	1	1	91.3
1	−1	1	−1	95.4
−1	1	1	−1	92.7
1	1	1	1	91.5

a. Estimate the main effects of each of the four factors.
b. Assuming all interactions to be negligible, pool the sums of squares for interaction to use in place of an error sum of squares.
c. Which of the four factors, if any, can you conclude to affect the yield? What is the P-value of the relevant test?

9. The article "Combined Analysis of Real-Time Kinematic GPS Equipment and Its Users for Height Determination" (W. Featherstone and M. Stewart, *Journal of Surveying Engineering*, 2001:31–51) presents a study of the accuracy of global positioning system (GPS) equipment in measuring heights. Three types of equipment were studied, and each was used to make measurements at four different base stations (in the article a fifth station was included, for which the results differed considerably from the other four). There were 60 measurements made with each piece of equipment at each base. The means and standard deviations of the measurement errors (in mm) are presented in the following table for each combination of equipment type and base station.

	Instrument A		Instrument B		Instrument C	
	Mean	Standard Deviation	Mean	Standard Deviation	Mean	Standard Deviation
Base 0	3	15	−24	18	−6	18
Base 1	14	26	−13	13	−2	16
Base 2	1	26	−22	39	4	29
Base 3	8	34	−17	26	15	18

a. Construct an ANOVA table. You may give ranges for the P-values.
b. The question of interest is whether the mean error differs among instruments. It is not of interest to determine whether the error differs among base stations. For this reason, a surveyor suggests treating this as a randomized complete block design, with the base stations as the blocks. Is this appropriate? Explain.

10. Vermont maple sugar producers sponsored a testing program to determine the benefit of a potential new fertilizer regimen. A random sample of 27 maple trees in Vermont were chosen and treated with one of three levels of fertilizer suggested by the chemical producer. In this experimental setup, nine trees (three in each of three climatic zones) were treated with each fertilizer level, and the amount of sap produced (in mL) by the trees in the subsequent season was measured. The results are presented in the following table.

	Southern Zone			Central Zone			Northern Zone		
Low fertilizer	76.2	80.4	74.2	79.4	87.9	86.9	84.5	85.2	80.1
Medium fertilizer	87.0	95.1	93.0	98.2	94.7	96.2	88.4	90.4	92.2
High fertilizer	84.2	87.5	83.1	90.3	89.9	93.2	81.4	84.7	82.2

a. Estimate the main effects of fertilizer levels and climatic zone, and their interactions.
b. Construct an ANOVA table. You may give ranges for the P-values.
c. Test the hypothesis that there is no interaction between fertilizer levels and climatic zone.
d. Test the hypothesis that there is no difference in sap production for the three fertilizer levels.

11. A civil engineer is interested in several designs for a drainage canal used to divert floodwaters from around a city. The drainage times of a reservoir attached to each of five different channel designs obtained from a series of experiments using similar initial flow conditions are given in the following table.

Channel Type	Drainage time (min)			
1	41.4	43.4	50.0	41.2
2	37.7	49.3	52.1	37.3
3	32.6	33.7	34.8	22.5
4	27.3	29.9	32.3	24.8
5	44.9	47.2	48.5	37.1

a. Can you conclude that there is a difference in the mean drainage times for the different channel designs?
b. Which pairs of designs, if any, can you conclude to differ in their mean drainage times?

12. A process that manufactures vinyl for automobile seat covers was studied. Three factors were varied: the proportion of a certain plasticizer (A), the rate of extrusion (B), and the temperature of drying (C). The outcome of interest was the thickness of the vinyl (in mils). A 2^3 factorial design with four replicates was employed. The results are presented in the following table. (Based on the article "Split-Plot Designs and Estimation Methods for Mixture Experiments with Process Variables," S. Kowalski, J. Cornell, and G. Vining, *Technometrics*, 2002:72–79.)

A	B	C	Thickness			
−1	−1	−1	7	5	6	7
1	−1	−1	6	5	5	5
−1	1	−1	8	8	4	6
1	1	−1	9	5	6	9
−1	−1	1	7	6	5	5
1	−1	1	7	7	11	10
−1	1	1	6	4	5	8
1	1	1	8	11	11	9

a. Estimate all main effects and interactions.
b. Construct an ANOVA table. You may give ranges for the P-values.
c. Is the additive model appropriate? Explain.
d. What conclusions about the factors can be drawn from these results?

13. In the article "Occurrence and Distribution of Ammonium in Iowa Groundwater" (K. Schilling, *Water Environment Research*, 2002:177–186), ammonium concentrations (in mg/L) were measured at a large number of wells in the state of Iowa. These included five types of bedrock wells. The number of wells of each type, along with the mean and standard deviation of the concentrations in those wells, is presented in the following table.

Well Type	Sample Size	Mean	Standard Deviation
Cretaceous	53	0.75	0.90
Mississippian	57	0.90	0.92
Devonian	66	0.68	1.03
Silurian	67	0.50	0.97
Cambrian-Ordovician	51	0.82	0.89

Can you conclude that the mean concentration differs among the five types of wells?

14. The article "Enthalpies and Entropies of Transfer of Electrolytes and Ions from Water to Mixed Aqueous Organic Solvents" (G. Hefter, Y. Marcus, and W. Waghorne, *Chemical Reviews*, 2002:2773–2836) presents measurements of entropy and enthalpy changes for many salts under a variety of conditions. The following table presents the results for enthalpy of transfer (in kJ/mol) from water to water + methanol of NaCl (table salt) for several concentrations of methanol. Four independent measurements were made at each concentration.

Concentration (%)	Enthalpy			
5	1.62	1.60	1.62	1.66
10	2.69	2.66	2.72	2.73
20	3.56	3.45	3.65	3.52
30	3.35	3.18	3.40	3.06

a. Is it plausible that the enthalpy is the same at all concentrations? Explain.
b. Which pairs of concentrations, if any, can you conclude to have differing enthalpies?

15. The article "Factorial Experiments in the Optimization of Alkaline Wastewater Pretreatment" (M. Prisciandaro, A. Del Borghi, and F. Veglio, *Industrial Engineering and Chemistry Research*, 2002:5034–5041) presents the results of several experiments to investigate methods of treating alkaline wastewater. One experiment was an unreplicated 2^4 design. The four factors were A: concentration of sulfuric acid, B: temperature, C: time, and D: concentration of calcium chloride. The outcome variable is the amount of precipitate in kg/m^3. The results are presented in the following table.

A	B	C	D	Outcome		A	B	C	D	Outcome
−1	−1	−1	−1	6.4		−1	−1	−1	1	11.9
1	−1	−1	−1	12.9		1	−1	−1	1	13.1
−1	1	−1	−1	8.6		−1	1	−1	1	12.1
1	1	−1	−1	12.9		1	1	−1	1	16.0
−1	−1	1	−1	7.4		−1	−1	1	1	12.4
1	−1	1	−1	12.0		1	−1	1	1	16.5
−1	1	1	−1	10.7		−1	1	1	1	15.3
1	1	1	−1	15.0		1	1	1	1	18.3

a. Estimate all main effects and interactions.
b. Which effects seem to be larger than the others?
c. Assume that all third- and higher-order interactions are equal to 0, and add their sums of squares. Use the result in place of an error sum of squares to compute F statistics and P-values for the main effects. Which factors can you conclude to have an effect on the outcome?
d. The article described some replicates of the experiment, in which the error mean square was found to be 1.04, with four degrees of freedom. Using this value, compute F statistics and P-values for all main effects and interactions.
e. Do the results of part (d) help to justify the assumption that the third- and higher-order interactions are equal to 0? Explain.
f. Using the results of part (d), which factors can you conclude to have an effect on the outcome?

16. The Williamsburg Bridge is a suspension bridge that spans the East River, connecting the boroughs of Brooklyn and Manhattan in New York City. An assessment of the strengths of its cables is reported in the article "Estimating Strength of the Williamsburg Bridge Cables" (R. Perry, *The American Statistician*, 2002:211–217). Each suspension cable consists of 7696 wires. From one of the cables, wires were sampled from 128 points. These points came from four locations along the length of the cable (I, II, III, IV). At each location there were eight equally spaced points around the circumference of the cable (A, B, C, D, E, F, G, H). At each of the eight points, wires were sampled from four depths: (1) the external surface of the cable, (2) two inches deep, (3) four inches deep, and (4) seven inches deep (the cable is 9.625 inches in radius). Under assumptions made in the article, it is appropriate to consider this as a two-factor experiment with circumferential position and depth as the factors, and with location providing four replicates for each combination of these factors. The minimum breaking strength (in lbf) is presented in the following table for each of the 128 points.

		Location			
Circumference	Depth	I	II	III	IV
A	1	6250	5910	5980	5800
A	2	6650	6690	6780	5540
A	3	5390	6080	6550	5690
A	4	6510	6580	6700	5980
B	1	6200	6240	6180	6740
B	2	6430	6590	6500	6110
B	3	5710	6230	6450	6310
B	4	6510	6600	6250	5660
C	1	5570	5700	6390	6170
C	2	6260	6290	5630	6990
C	3	6050	6120	6290	5800

(*Continued*)

Circumference	Depth	Location			
		I	II	III	IV
C	4	6390	6540	6590	6620
D	1	6140	6210	5710	5090
D	2	5090	6000	6020	6480
D	3	5280	5650	5410	5730
D	4	6300	6320	6650	6050
E	1	4890	4830	5000	6490
E	2	5360	5640	5920	6390
E	3	5600	5500	6250	6510
E	4	6640	6810	5760	5200
F	1	5920	5300	5670	6200
F	2	5880	5840	7270	5230
F	3	6570	6130	5800	6200
F	4	6120	6430	6100	6370
G	1	6070	6980	6570	6980
G	2	6180	6340	6830	6260
G	3	6360	6420	6370	6550
G	4	6340	6380	6480	7020
H	1	5950	5950	6450	5870
H	2	6180	6560	5730	6550
H	3	6560	6560	6450	6790
H	4	6700	6690	6670	6600

a. Construct an ANOVA table. You may give ranges for the *P*-values.
b. Can you conclude that there are interactions between circumferential position and depth? Explain.
c. Can you conclude that the strength varies with circumferential position? Explain.
d. Can you conclude that the strength varies with depth? Explain.

17. The article "Factorial Experiments in the Optimization of Alkaline Wastewater Pretreatment" (M. Prisciandaro, A. Del Borghi, and F. Veglio, *Industrial Engineering and Chemistry Research*, 2002:5034–5041) presents the results of an experiment to investigate the effects of the concentrations of sulfuric acid (H_2SO_4) and calcium chloride ($CaCl_2$) on the amount of black mud precipitate in the treatment of alkaline wastewater. There were three levels of each concentration, and two replicates of the experiment were made at each combination of levels. The results are presented in the following table (all measurements are in units of kg/m^3).

H_2SO_4	$CaCl_2$	Precipitate	
110	15	100.2	98.2
110	30	175.8	176.2
110	45	216.5	206.0
123	15	110.5	105.5
123	30	184.7	189.0
123	45	234.0	222.0
136	15	106.5	107.0
136	30	181.7	189.0
136	45	211.8	201.3

a. Construct an ANOVA table. You may give ranges for the *P*-values.
b. Is the additive model plausible? Explain.
c. Can you conclude that H_2SO_4 concentration affects the amount of precipitate? Explain.
d. Can you conclude that $CaCl_2$ concentration affects the amount of precipitate? Explain.

18. Fluid inclusions are microscopic volumes of fluid that are trapped in rock during rock formation. The article "Fluid Inclusion Study of Metamorphic Gold-Quartz Veins in Northwestern Nevada, U.S.A.: Characteristics of Tectonically Induced Fluid" (S. Cheong, *Geosciences Journal*, 2002:103–115) describes the geochemical properties of fluid inclusions in several different veins in northwest Nevada. The following table presents data on the maximum salinity (% NaCl by weight) of inclusions in several rock samples from several areas.

Area	Salinity					
Humboldt Range	9.2	10.0	11.2	8.8		
Santa Rosa Range	5.2	6.1	8.3			
Ten Mile	7.9	6.7	9.5	7.3	10.4	7.0
Antelope Range	6.7	8.4	9.9			
Pine Forest Range	10.5	16.7	17.5	15.3	20.0	

Can you conclude that the salinity differs among the areas?

10

Statistical Quality Control

Introduction

As the marketplace for industrial goods has become more global, manufacturers have realized that the quality and reliability of their products must be as high as possible for them to be competitive. It is now generally recognized that the most cost-effective way to maintain high quality is through constant monitoring of the production process. This monitoring is often done by sampling units of production and measuring a quality characteristic. Because the units are sampled from some larger population, these methods are inherently statistical in nature.

One of the early pioneers in the area of statistical quality control was Dr. Walter A. Shewart of the Bell Telephone Laboratories. In 1924, he developed the modern control chart, which remains one of the most widely used tools for quality control to this day. After World War II, W. Edwards Deming was instrumental in stimulating interest in quality control; first in Japan, and then in the United States and other countries. The Japanese scientist Genichi Taguchi played a major role as well, developing methods of experimental design with a view toward improving quality. In this chapter, we will focus on the Shewart control charts and on cumulative sum (CUSUM) charts, since these are among the most powerful of the commonly used tools for statistical quality control.

10.1 Basic Ideas

The basic principle of control charts is that in any process there will always be variation in the output. Some of this variation will be due to causes that are inherent in the process and are difficult or impossible to specify. These causes are called **common causes** or **chance causes**. When common causes are the only causes of variation, the process is said to be **in a state of statistical control** or, more simply, **in control**.

Sometimes special factors are present that produce additional variability. Malfunctioning machines, operator error, fluctuations in ambient conditions, and variations in the properties of raw materials are among the most common of these factors. These are called **special causes** or **assignable causes**. Special causes generally produce a higher level of variability than do common causes; this variability is considered to be unacceptable. When a process is operating in the presence of one or more special causes, it is said to be **out of statistical control**.

Control charts enable the quality engineer to decide whether a process appears to be in control, or whether one or more special causes are present. If the process is found to be out of control, the nature of the special cause must be determined and corrected, so as to return the process to a state of statistical control. There are several types of control charts; which ones are used depend on whether the quality characteristic being measured is a **continuous variable**, a **binary variable**, or a **count variable**. For example, when monitoring a process that manufactures aluminum beverage cans, the height of each can in a sample might be measured. Height is a continuous variable. In some situations, it might be sufficient simply to determine whether the height falls within some specification limits. In this case the quality measurement takes on one of only two values: conforming (within the limits) or nonconforming (not within the limits). This measurement is a binary variable, since it has two possible values. Finally, we might be interested in counting the number of flaws on the surface of the can. This is a count variable.

Control charts used for continuous variables are called **variables control charts**. Examples include the $\overline{X}$ chart, the R chart, and the S chart. Control charts used for binary or count variables are called **attribute control charts**. The p chart is most commonly used for binary variables, while the c chart is commonly used for count variables.

Collecting Data—Rational Subgroups

Data to be used in the construction of a control chart are collected in a number of samples, taken over a period of time. These samples are called **rational subgroups**. There are many different strategies for choosing rational subgroups. The basic principle to be followed is that all the variability within the units in a rational subgroup should be due to common causes, and none should be due to special causes. In general, a good way to choose rational subgroups is to decide which special causes are most important to detect, and then choose the rational subgroups to provide the best chance to detect them. The two most commonly used methods are

■   Sample at regular time intervals, with all the items in each sample manufactured near the time the sampling is done.

■   Sample at regular time intervals, with the items in each sample drawn from all the units produced since the last sample was taken.

For variables data, the number of units in each sample is typically small, often between three and eight. The number of samples should be at least 20. In general, many small samples taken frequently are better than a few large samples taken less frequently. For binary and for count data, samples must in general be larger.

Control versus Capability

It is important to understand the difference between process *control* and process *capability*. A process is in control if there are no special causes operating. The distinguishing feature of a process that is in control is that the values of the quality characteristic vary without any trend or pattern, since the common causes do not change over time. However, it is quite possible for a process to be in control, yet producing output that does not meet a given specification. For example, assume that a process produces steel rods whose lengths vary randomly between 19.9 and 20.1 cm, with no apparent pattern to the fluctuation. This process is in a state of control. However, if the design specification calls for a length between 21 and 21.2 cm, very little of the output would meet the specification. The ability of a process to produce output that meets a given specification is called the **capability** of the process. We will discuss the measurement of process capability in Section 10.5.

Process Control Must Be Done Continually

There are three basic phases to the use of control charts. First, data are collected. Second, these data are plotted to determine whether the process is in control. Third, once the process is brought into control, its capability may be assessed. Of course, a process that is in control and capable at a given time may go out of control at a later time, as special causes reoccur. For this reason, processes must be continually monitored.

Similarities Between Control Charts and Hypothesis Tests

Control charts function in many ways like hypothesis tests. The null hypothesis is that the process is in control. The control chart presents data that provide evidence about the truth of this hypothesis. If the evidence against the null hypothesis is sufficiently strong, the process is declared out of control. Understanding how to use control charts involves knowing what data to collect and knowing how to organize those data to measure the strength of the evidence against the hypothesis that the process is in control.

Exercises for Section 10.1

1. Indicate whether each of the following quality characteristics is a continuous, binary, or count variable.
   a. The number of flaws in a plate glass window.
   b. The length of time taken to perform a final inspection of a finished product.
   c. Whether the breaking strength of a bolt meets a specification.
   d. The diameter of a rivet head.

2. True or false:
   a. Control charts are used to determine whether special causes are operating.
   b. If no special causes are operating, then most of the output produced will meet specifications.
   c. Variability due to common causes does not increase or decrease much over short periods of time.
   d. Variability within the items sampled in a rational subgroup is due to special causes.
   e. If a process is in a state of statistical control, there will be almost no variation in the output.

3. Fill in the blank. The choices are: *is in control*; *has high capability*.
   a. If the variability in a process is approximately constant over time, the process _____ .
   b. If most units produced conform to specifications, the process _____ .

4. Fill in the blank: Once a process has been brought into a state of statistical control, _____

  i. it must still be monitored continually.

  ii. monitoring can be stopped for a while, since it is unlikely that the process will go out of control again right away.

  iii. the process need not be monitored again, unless it is redesigned.

5. True or false:

  a. When a process is in a state of statistical control, then most of the output will meet specifications.

  b. When a process is out of control, an unacceptably large proportion of the output will not meet specifications.

  c. When a process is in a state of statistical control, all the variation in the process is due to causes that are inherent in the process itself.

  d. When a process is out of control, some of the variation in the process is due to causes that are outside of the process.

6. Fill in the blank: When sampling units for rational subgroups, _____

  i. it is more important to choose large samples than to sample frequently, since large samples provide more precise information about the process.

  ii. it is more important to sample frequently than to choose large samples, so that special causes can be detected more quickly.

10.2 Control Charts for Variables

When a quality measurement is made on a continuous scale, the data are called **variables data**. For these data an R chart or S chart is first used to control the variability in the process, and then an $\overline{X}$ chart is used to control the process mean. The methods described in this section assume that the measurements follow an approximately normal distribution.

We illustrate with an example. The quality engineer in charge of a salt packaging process is concerned about the moisture content in packages of salt. To determine whether the process is in statistical control, it is first necessary to define the rational subgroups, and then to collect some data. Assume that for the salt packaging process, the primary concern is that variation in the ambient humidity in the plant may be causing variation in the mean moisture content in the packages over time. Recall that rational subgroups should be chosen so that the variation within each sample is due only to common causes, not to special causes. Therefore a good choice for the rational subgroups in this case is to draw samples of several packages each at regular time intervals. The packages in each sample will be produced as close to each other in time as possible. In this way, the ambient humidity will be nearly the same for each package in the sample, so the within-group variation will not be affected by this special cause. Assume that five packages of salt are sampled every 15 minutes for eight hours, and that the moisture content in each package is measured as a percentage of total weight. The data are presented in Table 10.1.

Since moisture is measured on a continuous scale, these are variables data. Each row of Table 10.1 presents the five moisture measurements in a given sample, along with their sample mean $\overline{X}$, their sample standard deviation s, and their sample range R (the difference between the largest and smallest value). The last row of the table contains the mean of the sample means $(\overline{\overline{X}})$, the mean of the sample ranges $(\overline{R})$, and the mean of the sample standard deviations $(\overline{s})$.

We assume that each of the 32 samples in Table 10.1 is a sample from a normal population with mean μ and standard deviation σ. The quantity μ is called the **process**

TABLE 10.1 Moisture content for salt packages, as a percentage of total weight

Sample	Sample Values					Mean ($\overline{X}$)	Range (R)	SD (s)
1	2.53	2.66	1.88	2.21	2.26	2.308	0.780	0.303
2	2.69	2.38	2.34	2.47	2.61	2.498	0.350	0.149
3	2.67	2.23	2.10	2.43	2.54	2.394	0.570	0.230
4	2.10	2.26	2.51	2.58	2.28	2.346	0.480	0.196
5	2.64	2.42	2.56	2.51	2.36	2.498	0.280	0.111
6	2.64	1.63	2.95	2.12	2.67	2.402	1.320	0.525
7	2.58	2.69	3.01	3.01	2.23	2.704	0.780	0.327
8	2.31	2.39	2.60	2.40	2.46	2.432	0.290	0.108
9	3.03	2.68	2.27	2.54	2.63	2.630	0.760	0.274
10	2.86	3.22	2.72	3.09	2.48	2.874	0.740	0.294
11	2.71	2.80	3.09	2.60	3.39	2.918	0.790	0.320
12	2.95	3.54	2.59	3.31	2.87	3.052	0.950	0.375
13	3.14	2.84	3.77	2.80	3.22	3.154	0.970	0.390
14	2.85	3.29	3.25	3.35	3.59	3.266	0.740	0.267
15	2.82	3.71	3.36	2.95	3.37	3.242	0.890	0.358
16	3.17	3.07	3.14	3.63	3.70	3.342	0.630	0.298
17	2.81	3.21	2.95	3.04	2.85	2.972	0.400	0.160
18	2.99	2.65	2.79	2.80	2.95	2.836	0.340	0.137
19	3.11	2.74	2.59	3.01	3.03	2.896	0.520	0.221
20	2.83	2.74	3.03	2.68	2.49	2.754	0.540	0.198
21	2.76	2.85	2.59	2.23	2.87	2.660	0.640	0.265
22	2.54	2.63	2.32	2.48	2.93	2.580	0.610	0.226
23	2.27	2.54	2.82	2.11	2.69	2.486	0.710	0.293
24	2.40	2.62	2.84	2.50	2.51	2.574	0.440	0.168
25	2.41	2.72	2.29	2.35	2.63	2.480	0.430	0.186
26	2.40	2.33	2.40	2.02	2.43	2.316	0.410	0.169
27	2.56	2.47	2.11	2.43	2.85	2.484	0.740	0.266
28	2.21	2.61	2.59	2.24	2.34	2.398	0.400	0.191
29	2.56	2.26	1.95	2.26	2.40	2.286	0.610	0.225
30	2.42	2.37	2.13	2.09	2.41	2.284	0.330	0.161
31	2.62	2.11	2.47	2.27	2.49	2.392	0.510	0.201
32	2.21	2.15	2.18	2.59	2.61	2.348	0.460	0.231

$$\overline{\overline{X}} = 2.6502 \qquad \overline{R} = 0.6066 \qquad \overline{s} = 0.2445$$

mean, and σ is called the **process standard deviation**. The idea behind control charts is that each value of $\overline{X}$ approximates the process mean during the time its sample was taken, while the values of R and s can be used to approximate the process standard deviation. If the process is in control, then the process mean and standard deviation are the same for each sample. If the process is out of control, the process mean μ or the process standard deviation σ, or both, will differ from sample to sample. Therefore the values of $\overline{X}$, R, and s will vary less when the process is in control than when the process is out of control. If the process is in control, the values of $\overline{X}$, R, and s will almost always be contained within certain limits, called **control limits**. If the process is out of control, the values of $\overline{X}$, R, and s will be more likely to exceed these limits. A control chart plots the values of $\overline{X}$, R, or s, along with the control limits, so that it can be easily seen whether the variation is large enough to conclude that the process is out of control.

Now let's see how to determine whether the salt packaging process is in a state of statistical control with respect to moisture content. Since the variation within each sample is supposed to be due only to common causes, this variation should not be too different from one sample to another. Therefore the first thing to do is to check to make sure that the amount of variation within each sample, measured by either the sample range or the sample standard deviation, does not vary too much from sample to sample. For this purpose the R chart can be used to assess variation in the sample range, or the S chart can be used to assess variation in the sample standard deviation. We will discuss the R chart first, since it is the more traditional. We will discuss the S chart at the end of this section.

Figure 10.1 presents the R chart for the moisture data. The horizontal axis represents the samples, numbered from 1 to 32. The sample ranges are plotted on the vertical axis. Most important are the three horizontal lines. The line in the center is plotted at the value $\overline{R}$ and is called the **center line**. The upper and lower lines indicate the 3σ upper and lower **control limits** (UCL and LCL, respectively). The control limits are drawn so that when the process is in control, almost all the points will lie within the limits. A point plotting outside the control limits is evidence that the process is out of control.

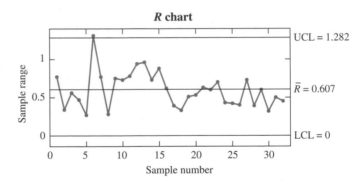

FIGURE 10.1 R chart for the moisture data.

To understand where the control limits are plotted, assume that the 32 sample ranges come from a population with mean μ_R and standard deviation σ_R. The values of μ_R and σ_R will not be known exactly, but it is known that for most populations, it is unusual to observe a value that differs from the mean by more than three standard deviations. For this reason, it is conventional to plot the control limits at points that approximate the values $\mu_R \pm 3\sigma_R$. It can be shown by advanced methods that the quantities $\mu_R \pm 3\sigma_R$ can be estimated with multiples of $\overline{R}$; these multiples are denoted D_3 and D_4. The quantity $\mu_R - 3\sigma_R$ is estimated with $D_3\overline{R}$, and the quantity $\mu_R + 3\sigma_R$ is estimated with $D_4\overline{R}$. The quantities D_3 and D_4 are constants whose values depend on the sample size n. A brief table of values of D_3 and D_4 follows. A more extensive tabulation is provided in Table A.8 (in Appendix A). Note that for sample sizes of 6 or less, the value of D_3 is 0. For these small sample sizes, the quantity $\mu_R - 3\sigma_R$ is negative. In these cases the lower control limit is set to 0, because it is impossible for the range to be negative.

n	2	3	4	5	6	7	8
D_3	0	0	0	0	0	0.076	0.136
D_4	3.267	2.575	2.282	2.114	2.004	1.924	1.864

Example 10.1

Compute the 3σ R chart upper and lower control limits for the moisture data in Table 10.1.

Solution
The value of $\overline{R}$ is 0.6066 (Table 10.1). The sample size is $n = 5$. From the table, $D_3 = 0$ and $D_4 = 2.114$. Therefore the upper control limit is $(2.114)(0.6066) = 1.282$, and the lower control limit is $(0)(0.6066) = 0$.

Summary

In an R chart, the center line and the 3σ upper and lower control limits are given by

$$3\sigma \text{ upper limit} = D_4\overline{R}$$
$$\text{Center line} = \overline{R}$$
$$3\sigma \text{ lower limit} = D_3\overline{R}$$

The values D_3 and D_4 depend on the sample size. Values are tabulated in Table A.8.

Once the control limits have been calculated and the points plotted, the R chart can be used to assess whether the process is in control with respect to variation. Figure 10.1 shows that the range for sample number 6 is above the upper control limit, providing evidence that a special cause was operating and that the process variation is not in control. The appropriate action is to determine the nature of the special cause, and then delete the out-of-control sample and recompute the control limits. Assume it is discovered that a technician neglected to close a vent, causing greater than usual variation in moisture content during the time period when the sample was chosen. Retraining the technician will correct that special cause. We delete sample 6 from the data and recompute the R chart. The results are shown in Figure 10.2 (page 484). The process variation is now in control.

Now that the process variation has been brought into control, we can assess whether the process mean is in control by plotting the $\overline{X}$ chart. The $\overline{X}$ chart is presented in Figure 10.3 (page 484). The sample means are plotted on the vertical axis. Note that sample 6 has not been used in this chart since it had to be deleted in order to bring the

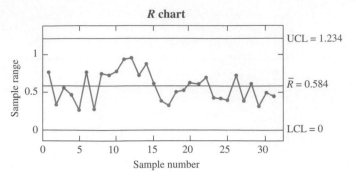

FIGURE 10.2 R chart for the moisture data, after deleting the out-of-control sample.

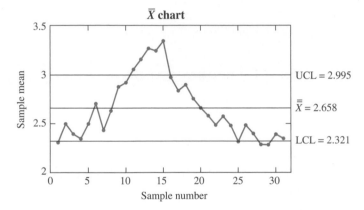

FIGURE 10.3 $\overline{X}$ chart for the moisture data. Sample 6 has been deleted to bring the process variation under control. However, the $\overline{X}$ chart shows that the process mean is out of control.

process variation under control. Like all control charts, the $\overline{X}$ chart has a center line and upper and lower control limits.

To compute the center line and the control limits, we can assume that the process standard deviation is the same for all samples, since the R chart has been used to bring the process variation into control. If the process mean μ is in control as well, then it, too, is the same for all samples. In that case the 32 sample means are drawn from a normal population with mean $\mu_{\overline{X}} = \mu$ and standard deviation $\sigma_{\overline{X}} = \sigma/\sqrt{n}$, where n is the sample size, equal to 5 in this case. Ideally, we would like to plot the center line at μ and the 3σ control limits at $\mu \pm 3\sigma_{\overline{X}}$. However, the values of μ and $\sigma_{\overline{X}}$ are usually unknown and have to be estimated from the data. We estimate μ with $\overline{\overline{X}}$, the average of the sample means. The center line is therefore plotted at $\overline{\overline{X}}$. The quantity $\sigma_{\overline{X}}$ can be estimated by using either the average range $\overline{R}$ or by using the sample standard deviations. We will use $\overline{R}$ here and discuss the methods based on the standard deviation at the end of the section, in conjunction with the discussion of S charts. It can be shown by advanced methods that the quantity $3\sigma_{\overline{X}}$ can be estimated with $A_2\overline{R}$, where A_2 is a constant whose value depends on the sample size. A short table of values of A_2 follows. A more extensive tabulation is provided in Table A.8.

n	2	3	4	5	6	7	8
A_2	1.880	1.023	0.729	0.577	0.483	0.419	0.373

Summary

In an $\overline{X}$ chart, when $\overline{R}$ is used to estimate $\sigma_{\overline{X}}$, the center line and the 3σ upper and lower control limits are given by

$$3\sigma \text{ upper limit} = \overline{\overline{X}} + A_2\overline{R}$$
$$\text{Center line} = \overline{\overline{X}}$$
$$3\sigma \text{ lower limit} = \overline{\overline{X}} - A_2\overline{R}$$

The value A_2 depends on the sample size. Values are tabulated in Table A.8.

Compute the 3σ $\overline{X}$ chart upper and lower control limits for the moisture data in Table 10.1.

Solution

With sample 6 deleted, the value of $\overline{\overline{X}}$ is 2.658, and the value of $\overline{R}$ is 0.5836. The sample size is $n = 5$. From the table, $A_2 = 0.577$. Therefore the upper control limit is $2.658 + (0.577)(0.5836) = 2.995$, and the lower control limit is $2.658 - (0.577)(0.5836) = 2.321$.

The $\overline{X}$ chart clearly shows that the process mean is not in control, as several points plot outside the control limits. The production manager installs a hygrometer to monitor the ambient humidity and determines that the fluctuations in moisture content are caused by fluctuations in ambient humidity. A dehumidifier is installed to stabilize the ambient humidity. After this special cause is remedied, more data are collected, and a new R chart and $\overline{X}$ chart are constructed. Figure 10.4 (page 486) presents the results. The process is now in a state of statistical control. Of course, the process must be continually monitored, since new special causes are bound to crop up from time to time and will need to be detected and corrected.

Note that while control charts can detect the presence of a special cause, they cannot determine its nature, nor how to correct it. It is necessary for the process engineer to have a good understanding of the process, so that special causes detected by control charts can be diagnosed and corrected.

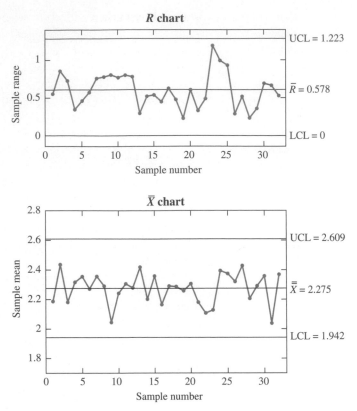

FIGURE 10.4 *R* chart and $\overline{X}$ chart after special cause is remedied. The process is now in a state of statistical control.

Summary

The steps in using the *R* chart and $\overline{X}$ chart are

1. Choose rational subgroups.
2. Compute the *R* chart.
3. Determine the special causes for any out-of-control points.
4. Recompute the *R* chart, omitting samples that resulted in out-of-control points.
5. Once the *R* chart indicates a state of control, compute the $\overline{X}$ chart, omitting samples that resulted in out-of-control points on the *R* chart.
6. If the $\overline{X}$ chart indicates that the process is not in control, identify and correct any special causes.
7. Continue to monitor $\overline{X}$ and *R*.

Control Chart Performance

There is a close connection between control charts and hypothesis tests. The null hypothesis is that the process is in a state of control. A point plotting outside the 3σ control limits presents evidence against the null hypothesis. As with any hypothesis test, it is possible to make an error. For example, a point will occasionally plot outside the 3σ limits even when the process is in control. This is called a **false alarm**. It can also happen that a process that is not in control may not exhibit any points outside the control limits, especially if it is not observed for a long enough time. This is called a **failure to detect**.

It is desirable for these errors to occur as infrequently as possible. We describe the frequency with which these errors occur with a quantity called the **average run length** (ARL). The ARL is the number of samples that must be observed, on average, before a point plots outside the control limits. We would like the ARL to be large when the process is in control, and small when the process is out of control. We can compute the ARL for an $\overline{X}$ chart if we assume that process mean μ and the process standard deviation σ are known. Then the center line is located at the process mean μ and the control limits are at $\mu \pm 3\sigma_{\overline{X}}$. We must also assume, as is always the case with the $\overline{X}$ chart, that the quantity being measured is approximately normally distributed. Examples 10.3 through 10.6 show how to compute the ARL.

Example
10.3

For an $\overline{X}$ chart with control limits at $\mu \pm 3\sigma_{\overline{X}}$, compute the ARL for a process that is in control.

Solution

Let $\overline{X}$ be the mean of a sample. Then $\overline{X} \sim N(\mu, \sigma_{\overline{X}}^2)$. The probability that a point plots outside the control limits is equal to $P(\overline{X} < \mu - 3\sigma_{\overline{X}}) + P(\overline{X} > \mu + 3\sigma_{\overline{X}})$. This probability is equal to $0.00135 + 0.00135 = 0.0027$ (see Figure 10.5). Therefore, on the average, 27 out of every 10,000 points will plot outside the control limits. This is equivalent to 1 every $10,000/27 = 370.4$ points. The average run length is therefore equal to 370.4.

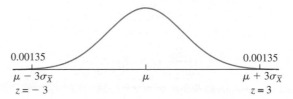

0.00135   0.00135

$\mu - 3\sigma_{\overline{X}}$ μ $\mu + 3\sigma_{\overline{X}}$
$z = -3$ $z = 3$

FIGURE 10.5 The probability that a point plots outside the 3σ control limits, when the process is in control, is 0.0027 (0.00135 + 0.00135).

The result of Example 10.3 can be interpreted as follows: If a process is in control, we expect to observe about 370 samples, on the average, before finding one that plots outside the control limits, causing a false alarm. Note also that the ARL in Example 10.3

was 10,000/27, which is equal to 1/0.0027, where 0.0027 is the probability that any given sample plots outside the control limits. This is true in general.

Summary

The average run length (ARL) is the number of samples that will be observed, on the average, before a point plots outside the control limits. If p is the probability that any given point plots outside the control limits, then

$$\text{ARL} = \frac{1}{p} \tag{10.1}$$

If a process is out of control, the ARL will be less than 370.4. Example 10.4 shows how to compute the ARL for a situation where the process shifts to an out-of-control condition.

*E*xample
10.4

A process has mean $\mu = 3$ and standard deviation $\sigma = 1$. Samples of size $n = 4$ are taken. If a special cause shifts the process mean to a value of 3.5, find the ARL.

Solution

We first compute the probability p that a given point plots outside the control limits. Then $\text{ARL} = 1/p$. The control limits are plotted on the basis of a process that is in control. Therefore they are at $\mu \pm 3\sigma_{\overline{X}}$, where $\mu = 3$ and $\sigma_{\overline{X}} = \sigma/\sqrt{n} = 1/\sqrt{4} = 0.5$. The lower control limit is thus at 1.5, and the upper control limit is at 4.5. If $\overline{X}$ is the mean of a sample taken after the process mean has shifted, then $\overline{X} \sim N(3.5, \ 0.5^2)$. The probability that $\overline{X}$ plots outside the control limits is equal to $P(\overline{X} < 1.5) + P(\overline{X} > 4.5)$. This probability is 0.0228 (see Figure 10.6). The ARL is therefore equal to $1/0.0228 = 43.9$. We will have to observe about 44 samples, on the average, before detecting this shift.

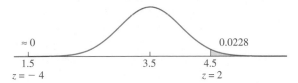

FIGURE 10.6 The process mean has shifted from $\mu = 3$ to $\mu = 3.5$. The upper control limit of 4.5 is now only $2\sigma_{\overline{X}}$ above the mean, indicated by the fact that $z = 2$. The lower limit is now $4\sigma_{\overline{X}}$ below the mean. The probability that a point plots outside the control limits is 0.0228 $(0 + 0.0228)$.

*E*xample
10.5

Refer to Example 10.4. An upward shift to what value can be detected with an ARL of 20?

Solution

Let m be the new mean to which the process has shifted. Since we have specified an upward shift, $m > 3$. In Example 10.4 we computed the control limits to be 1.5 and 4.5. If $\overline{X}$ is the mean of a sample taken after the process mean has shifted, then $\overline{X} \sim N(m, 0.5^2)$. The probability that $\overline{X}$ plots outside the control limits is equal to $P(\overline{X} < 1.5) + P(\overline{X} > 4.5)$ (see Figure 10.7). This probability is equal to $1/\text{ARL} = 1/20 = 0.05$. Since $m > 3$, m is closer to 4.5 than to 1.5. We will begin by assuming that the area to the left of 1.5 is negligible and that the area to the right of 4.5 is equal to 0.05. The z-score of 4.5 is then 1.645, so $(4.5 - m)/0.5 = 1.645$. Solving for m, we have $m = 3.68$. We finish by checking our assumption that the area to the left of 1.5 is negligible. With $m = 3.68$, the z-score for 1.5 is $(1.5 - 3.68)/0.5 = -4.36$. The area to the left of 1.5 is indeed negligible.

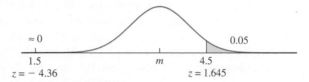

FIGURE 10.7 Solution to Example 10.5.

Example

10.6

Refer to Example 10.4. If the sample size remains at $n = 4$, what must the value of the process standard deviation σ be to produce an ARL of 10 when the process mean shifts to 3.5?

Solution

Let σ denote the new process standard deviation. The new control limits are $3 \pm 3\sigma/\sqrt{n}$, or $3 \pm 3\sigma/2$. If the process mean shifts to 3.5, then $\overline{X} \sim N(3.5, \sigma^2/4)$. The probability that $\overline{X}$ plots outside the control limits is equal to $P(\overline{X} < 3 - 3\sigma/2) + P(\overline{X} > 3 + 3\sigma/2)$. This probability is equal to $1/\text{ARL} = 1/10 = 0.10$ (see Figure 10.8). The process mean, 3.5, is closer to $3 + 3\sigma/2$ than to $3 - 3\sigma/2$. We will assume that the area to the left of $3 - 3\sigma/2$ is negligible and that the area to the right of $3 + 3\sigma/2$ is equal to 0.10. The z-score for $3 + 3\sigma/2$ is then 1.28, so

$$\frac{(3 + 3\sigma/2) - 3.5}{\sigma/2} = 1.28$$

Solving for σ, we obtain $\sigma = 0.58$. We finish by checking that the area to the left of $3 - 3\sigma/2$ is negligible. Substituting $\sigma = 0.58$, we obtain $3 - 3\sigma/2 = 2.13$. The z-score is $(2.13 - 3.5)/(0.58/2) = -4.72$. The area to the left of $3 - 3\sigma/2$ is indeed negligible.

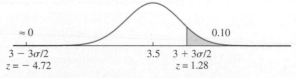

FIGURE 10.8 Solution to Example 10.6.

Examples 10.4 through 10.6 show that $\overline{X}$ charts do not usually detect small shifts quickly. In other words, the ARL is high when shifts in the process mean are small. In principle, one could reduce the ARL by moving the control limits closer to the centerline. This would reduce the size of the shift needed to detect an out-of-control condition, so that changes in the process mean would be detected more quickly. However, there is a trade-off. The false alarm rate would increase as well, because shifts outside the control limits would be more likely to occur by chance. The situation is much like that in fixed-level hypothesis testing. The null hypothesis is that the process is in control. The control chart performs a hypothesis test on each sample. When a point plots outside the control limits, the null hypothesis is rejected. With the control limits at $\pm 3\sigma_{\overline{X}}$, a type I error (rejection of a true null hypothesis) will occur about once in every 370 samples. The price to pay for this low false alarm rate is lack of power to reject the null hypothesis when it is false. Moving the control limits closer together is not the answer. Although it will increase the power, it will also increase the false alarm rate.

Two of the ways in which practitioners have attempted to improve their ability to detect small shifts quickly are by using the **Western Electric rules** to interpret the control chart and by using CUSUM charts. The Western Electric rules are described next. CUSUM charts are described in Section 10.4.

The Western Electric Rules

Figure 10.9 presents an $\overline{X}$ chart. While none of the points fall outside the 3σ control limits, the process is clearly not in a state of control, since there is a nonrandom pattern to the sample means. In recognition of the fact that a process can fail to be in control even when no points plot outside the control limits, engineers at the Western Electric company in 1956 suggested a list of conditions, any one of which could be used as evidence that

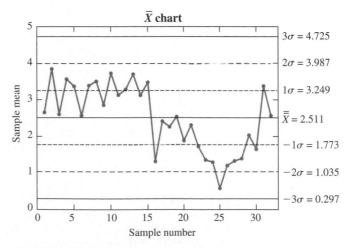

FIGURE 10.9 This $\overline{X}$ chart exhibits nonrandom patterns, indicating a lack of statistical control, even though no points plot outside the 3σ control limits. The 1σ and 2σ control limits are shown on this plot, so that the Western Electric rules can be applied.

a process is out of control. The idea behind these conditions is that if a trend or pattern in the control chart persists for long enough, it can indicate the absence of control, even if no point plots outside the 3σ control limits.

To apply the Western Electric rules, it is necessary to compute the 1σ and 2σ control limits. The 1σ control limits are given by $\overline{\overline{X}} \pm A_2\overline{R}/3$, and the 2σ control limits are given by $\overline{\overline{X}} \pm 2A_2\overline{R}/3$.

The Western Electric Rules

Any one of the following conditions is evidence that a process is out of control:

1. Any point plotting outside the 3σ control limits.
2. Two out of three consecutive points plotting above the upper 2σ limit, or two out of three consecutive points plotting below the lower 2σ limit.
3. Four out of five consecutive points plotting above the upper 1σ limit, or four out of five consecutive points plotting below the lower 1σ limit.
4. Eight consecutive points plotting on the same side of the center line.

In Figure 10.9, the Western Electric rules indicate that the process is out of control at sample number 8, at which time four out of five consecutive points have plotted above the upper 1σ control limit. For more information on using the Western Electric rules to interpret control charts, see Montgomery (2009b).

The S chart

The S chart is an alternative to the R chart. Both the S chart and the R chart are used to control the variability in a process. While the R chart assesses variability with the sample range, the S chart uses the sample standard deviation. Figure 10.10 presents the S chart for the moisture data in Table 10.1.

Note that the S chart for the moisture data is similar in appearance to the R chart (Figure 10.10), for the same data. Like the R chart, the S chart indicates that the variation was out of control in sample 6.

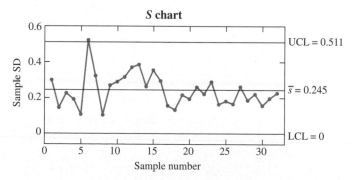

FIGURE 10.10 S chart for the moisture data. Compare with Figure 10.1.

To understand where the control limits are plotted, assume that the 32 sample standard deviations come from a population with mean μ_s and standard deviation σ_s. Ideally we would like to plot the center line at μ_s and the control limits at $\mu_s \pm 3\sigma_s$. These quantities are typically unknown. We approximate μ_s with $\bar{s}$, the average of the sample standard deviations. Thus the center line is plotted at $\bar{s}$. It can be shown by advanced methods that the quantities $\mu_s \pm 3\sigma_s$ can be estimated with multiples of $\bar{s}$; these multiples are denoted B_3 and B_4. The quantity $\mu_s - 3\sigma_s$ is estimated with $B_3\bar{s}$, while the quantity $\mu_s + 3\sigma_s$ is estimated with $B_4\bar{s}$. The quantities B_3 and B_4 are constants whose values depend on the sample size n. A brief table of values of B_3 and B_4 follows. A more extensive tabulation is provided in Table A.8 (Appendix A).

Note that for samples of size 5 or less, the value of B_3 is 0. For samples this small, the value of $\mu_s - 3\sigma_s$ is negative. In these cases the lower control limit is set to 0, because it is impossible for a standard deviation to be negative.

n	2	3	4	5	6	7	8
B_3	0	0	0	0	0.030	0.118	0.185
B_4	3.267	2.568	2.266	2.089	1.970	1.882	1.815

10.7

Compute the center line and the 3σ S chart upper and lower control limits for the moisture data in Table 10.1.

Solution

The value of $\bar{s}$ is 0.2445 (Table 10.1). The sample size is $n = 5$. From the table immediately preceding, $B_3 = 0$ and $B_4 = 2.089$. Therefore the upper control limit is $(2.089)(0.2445) = 0.5108$, and the lower control limit is $(0)(0.2445) = 0$.

Summary

In an S chart, the center line and the 3σ upper and lower control limits are given by

$$3\sigma \text{ upper limit} = B_4\bar{s}$$
$$\text{Center line} = \bar{s}$$
$$3\sigma \text{ lower limit} = B_3\bar{s}$$

The values B_3 and B_4 depend on the sample size. Values are tabulated in Table A.8.

The S chart in Figure 10.10 shows that the process variation is out of control in sample 6. We delete this sample and recompute the S chart. Figure 10.11 presents the results. The variation is now in control. Note that this S chart is similar in appearance to the R chart in Figure 10.2.

Once the variation is in control, we compute the $\overline{X}$ chart to assess the process mean. Recall that for the $\overline{X}$ chart, the center line is at $\overline{\overline{X}}$, and the upper and lower control

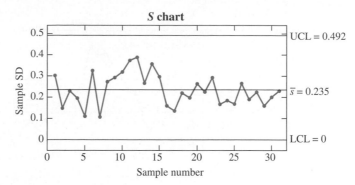

FIGURE 10.11 S chart for the moisture data, after deleting the out-of-control sample. Compare with Figure 10.2.

limits would ideally be located a distance $3\sigma_{\overline{X}}$ above and below the center line. Since we used the S chart to assess the process variation, we will estimate the quantity $3\sigma_{\overline{X}}$ with a multiple of $\overline{s}$. Specifically, we estimate $3\sigma_{\overline{X}}$ with $A_3\overline{s}$, where A_3 is a constant whose value depends on the sample size n. A brief table of values of A_3 follows. A more extensive tabulation is provided in Table A.8.

n	2	3	4	5	6	7	8
A_3	2.659	1.954	1.628	1.427	1.287	1.182	1.099

Summary

In an $\overline{X}$ chart, when $\overline{s}$ is used to estimate $\sigma_{\overline{X}}$, the center line and the 3σ upper and lower control limits are given by

$$3\sigma \text{ upper limit} = \overline{\overline{X}} + A_3\overline{s}$$
$$\text{Center line} = \overline{\overline{X}}$$
$$3\sigma \text{ lower limit} = \overline{\overline{X}} - A_3\overline{s}$$

The value A_3 depends on the sample size. Values are tabulated in Table A.8.

If Western Electric rules are to be used, 1σ and 2σ control limits must be computed. The 1σ limits are $\overline{\overline{X}} \pm A_3\overline{s}/3$; the 2σ limits are $\overline{\overline{X}} \pm 2A_3\overline{s}/3$.

*E*xample

10.8

Compute the 3σ $\overline{X}$ chart upper and lower control limits for the moisture data in Table 10.1.

Solution

With sample 6 deleted, the value of $\overline{\overline{X}}$ is 2.658, and the value of $\overline{s}$ is 0.2354. The sample size is $n = 5$. From the table, $A_3 = 1.427$. Therefore the upper control limit is $2.658 + (1.427)(0.2354) = 2.994$, and the lower control limit is $2.658 - (1.427)(0.2354) = 2.322$.

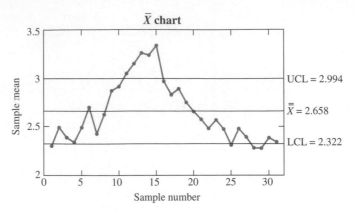

FIGURE 10.12 $\overline{X}$ chart for the moisture data. The control limits are based on the sample standard deviations rather than the sample ranges. Compare with Figure 10.3.

The $\overline{X}$ chart for the moisture data with sample 6 deleted is shown in Figure 10.12. The control limits are very similar to those calculated from the sample ranges, shown in Figure 10.3. Figure 10.12 indicates that the process is out of control. After taking corrective action, a new S chart and $\overline{X}$ chart are constructed. Figure 10.13 presents the results. The process is now in a state of statistical control.

In summary, the S chart is an alternative to the R chart, to be used in combination with the $\overline{X}$ chart. For the moisture data, it turned out that the two charts gave very similar results. This is true in many cases, but it will sometimes happen that the results differ.

Which Is Better, the *S* Chart or the *R* Chart?

Both the R chart and S chart have the same purpose: to estimate the process standard deviation and to determine whether it is in control. It seems more natural to estimate the process standard deviation with the sample standard deviation s than with the range R. In fact, when the population is normal, s is a more precise estimate of the process standard deviation than is R, because it has a smaller standard deviation. To see this intuitively, note that the computation of s involves all the measurements in each sample, while the computation of R involves only two measurements (the largest and the smallest). It turns out that the improvement in precision obtained with s as opposed to R increases as the sample size increases. It follows that the S chart is a better choice, especially for larger sample sizes (greater than 5 or so). The R chart is still widely used, largely through tradition. At one time, the R chart had the advantage that the sample range required less arithmetic to compute than did the sample standard deviation. Now that most calculations are done electronically, this advantage no longer holds. So the S chart is in general the better choice.

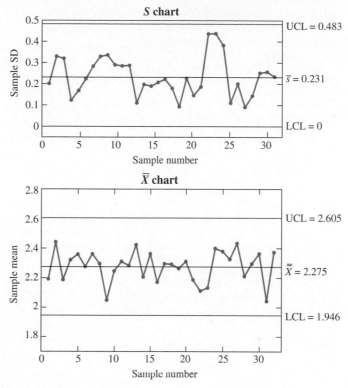

FIGURE 10.13 S chart and $\overline{X}$ chart after special cause is remedied. The process is now in a state of statistical control. Compare with Figure 10.4.

Samples of Size 1

Sometimes it is necessary to define rational subgroups in such a way that each sample can contain only one value. For example, if the production rate is very slow, it may not be convenient to wait to accumulate samples larger than $n = 1$. It is impossible to compute a sample range or a sample standard deviation for a sample of size 1, so R charts and S charts cannot be used. Several other methods are available. One method is the CUSUM chart, which is discussed in Section 10.4.

Exercises for Section 10.2

1. The quality control plan for a certain production process involves taking samples of size 4. The results from the last 30 samples can be summarized as follows:

$$\sum_{i=1}^{30} \overline{X}_i = 712.5 \quad \sum_{i=1}^{30} R_i = 143.7 \quad \sum_{i=1}^{30} s_i = 62.5$$

   a. Compute the 3σ control limits for the R chart.

   b. Compute the 3σ control limits for the S chart.

   c. Using the sample ranges, compute the 3σ control limits for the $\overline{X}$ chart.

   d. Using the sample standard deviations, compute the 3σ control limits for the $\overline{X}$ chart.

2. The following $\overline{X}$ chart depicts the last 50 samples taken from the output of a process. Using the

Western Electric rules, is the process detected to be out of control at any time? If so, specify at which sample the process is first detected to be out of control and which rule is violated.

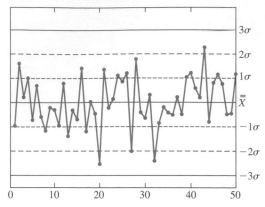

3. The thickness, in mm, of metal washers is measured on samples of size 5. The following table presents the means, ranges, and standard deviations for 20 consecutive samples.

Sample	$\overline{X}$	R	s
1	2.49	0.12	0.07
2	2.45	0.17	0.06
3	2.51	0.13	0.06
4	2.53	0.25	0.09
5	2.49	0.11	0.06
6	2.44	0.11	0.06
7	2.44	0.12	0.05
8	2.42	0.18	0.06
9	2.42	0.08	0.05
10	2.47	0.06	0.02
11	2.54	0.19	0.07
12	2.45	0.09	0.04
13	2.54	0.21	0.07
14	2.55	0.10	0.05
15	2.50	0.25	0.08
16	2.53	0.11	0.04
17	2.58	0.16	0.07
18	2.59	0.09	0.03
19	2.60	0.12	0.05
20	2.56	0.14	0.06

The means are $\overline{\overline{X}} = 2.505$, $\overline{R} = 0.1395$, and $\overline{s} = 0.057$.

a. Calculate the 3σ control limits for the R chart. Is the variance under control? If not, delete the

samples that are out of control and recompute $\overline{\overline{X}}$ and $\overline{R}$.

b. Based on the sample range R, calculate the 3σ control limits for the $\overline{X}$ chart. Based on the 3σ limits, is the process mean in control? If not, when is it first detected to be out of control?

c. Based on the Western Electric rules, is the process mean in control? If not, when is it first detected to be out of control?

4. Repeat Exercise 3, using the S chart in place of the R chart.

5. A process has mean 12 and standard deviation 3. The process is monitored by taking samples of size 5 at regular intervals. The process is declared to be out of control if a point plots outside the 3σ control limits on an $\overline{X}$ chart.

a. If the process mean shifts to 14, what is the average number of samples that will be drawn before the shift is detected on an $\overline{X}$ chart?

b. An upward shift to what value will be detected with an ARL of 4?

c. If the sample size remains at 5, to what value must the standard deviation be reduced to produce an ARL of 4 when the process mean shifts to 14?

d. If the standard deviation remains at 3, what sample size must be used to produce an ARL no greater than 4 when the process mean shifts to 14?

6. A process has mean 8 and standard deviation 2. The process is monitored by taking samples of size 4 at regular intervals. The process is declared to be out of control if a point plots outside the 3σ control limits on an $\overline{X}$ chart.

a. If the process mean shifts to 9, what is the average number of samples that will be drawn before the shift is detected on an $\overline{X}$ chart?

b. An upward shift to what value will be detected with an ARL of 8?

c. If the sample size remains at 4, to what value must the standard deviation be reduced to produce an ARL of 8 when the process mean shifts to 9?

d. If the standard deviation remains at 2, what sample size must be used to produce an ARL no greater than 8 when the process mean shifts to 9?

7. A process is monitored by taking samples at regular intervals and is declared to be out of control if a

point plots outside the 3σ control limits on an $\overline{X}$ chart. Assume the process is in control.

a. What is the probability that a false alarm will occur within the next 50 samples?

b. What is the probability that a false alarm will occur within the next 100 samples?

c. What is the probability that there will be no false alarm within the next 200 samples?

d. Fill in the blank: The probability is 0.5 that there will be a false alarm within the next _____ samples.

8. Samples of eight bolts are taken periodically, and their diameters (in mm) are measured. The following table presents the means, ranges, and standard deviations for 25 consecutive samples.

Sample	$\overline{X}$	R	s
1	9.99	0.28	0.09
2	10.02	0.43	0.13
3	10.10	0.16	0.05
4	9.90	0.26	0.09
5	9.92	0.22	0.07
6	10.05	0.40	0.15
7	9.97	0.08	0.03
8	9.93	0.48	0.15
9	10.01	0.25	0.09
10	9.87	0.29	0.10
11	9.90	0.39	0.14
12	9.97	0.27	0.08
13	10.02	0.20	0.07
14	9.99	0.37	0.13
15	9.99	0.20	0.06
16	10.04	0.26	0.09
17	10.07	0.23	0.07
18	10.04	0.35	0.12
19	9.95	0.25	0.09
20	9.98	0.15	0.06
21	9.98	0.30	0.10
22	10.02	0.14	0.06
23	9.94	0.24	0.07
24	10.04	0.13	0.04
25	10.04	0.24	0.07

The means are $\overline{\overline{X}} = 9.9892$, $\overline{R} = 0.2628$, and $\bar{s} = 0.0880$.

a. Calculate the 3σ control limits for the R chart. Is the variance under control? If not, delete the

samples that are out of control and recompute $\overline{\overline{X}}$ and $\overline{R}$.

b. Based on the sample range R, calculate the 3σ control limits for the $\overline{X}$ chart. Based on the 3σ limits, is the process mean in control? If not, when is it first detected to be out of control?

c. Based on the Western Electric rules, is the process mean in control? If not, when is it first detected to be out of control?

9. Repeat Exercise 8, using the S chart in place of the R chart.

10. A certain type of integrated circuit is connected to its frame by five wires. Thirty samples of five units each were taken, and the pull strength (in grams) of one wire on each unit was measured. The data are presented in Table E10 on page 498. The means are $\overline{\overline{X}} = 9.81$, $\overline{R} = 1.14$, and $\bar{s} = 0.4647$.

a. Compute the 3σ limits for the R chart. Is the variance out of control at any point? If so, delete the samples that are out of control and recompute $\overline{\overline{X}}$ and $\overline{R}$.

b. Compute the 3σ limits for the $\overline{X}$ chart. On the basis of the 3σ limits, is the process mean in control? If not, at what point is it first detected to be out of control?

c. On the basis of the Western Electric rules, is the process mean in control? If not, when is it first detected to be out of control?

11. Repeat Exercise 10, using the S chart in place of the R chart.

12. Copper wires are coated with a thin plastic coating. Samples of four wires are taken every hour, and the thickness of the coating (in mils) is measured. The data from the last 30 samples are presented in Table E12 on page 499. The means are $\overline{\overline{X}} = 150.075$, $\overline{R} = 6.97$, and $\bar{s} = 3.082$.

a. Compute the 3σ limits for the R chart. Is the variance out of control at any point? If so, delete the samples that are out of control and recompute $\overline{\overline{X}}$ and $\overline{R}$.

b. Compute the 3σ limits for the $\overline{X}$ chart. On the basis of the 3σ limits, is the process mean in control? If not, at what point is it first detected to be out of control?

c. On the basis of the Western Electric rules, is the process mean in control? If not, when is it first detected to be out of control?

13. Repeat Exercise 12, using the S chart in place of the R chart.

TABLE E10 Data for Exercise 10

Sample	Sample Values					$\overline{X}$	R	s
1	10.3	9.8	9.7	9.9	10.2	9.98	0.6	0.26
2	9.9	9.4	10.0	9.4	10.2	9.78	0.8	0.36
3	9.0	9.9	9.6	9.2	10.6	9.66	1.6	0.63
4	10.1	10.6	10.3	9.6	9.7	10.06	1.0	0.42
5	10.8	9.4	9.9	10.1	10.1	10.06	1.4	0.50
6	10.3	10.1	10.0	9.5	9.8	9.94	0.8	0.30
7	8.8	9.3	9.9	8.9	9.3	9.24	1.1	0.43
8	9.4	9.7	9.4	9.9	10.5	9.78	1.1	0.45
9	9.1	8.9	9.8	9.0	9.3	9.22	0.9	0.36
10	8.9	9.4	10.6	9.4	8.7	9.40	1.9	0.74
11	9.0	8.6	9.9	9.6	10.5	9.52	1.9	0.75
12	9.5	9.2	9.4	9.3	9.6	9.40	0.4	0.16
13	9.0	9.4	9.7	9.4	8.6	9.22	1.1	0.43
14	9.4	9.2	9.4	9.3	9.7	9.40	0.5	0.19
15	9.4	10.2	9.0	8.8	10.2	9.52	1.4	0.66
16	9.6	9.5	10.0	9.3	9.4	9.56	0.7	0.27
17	10.2	8.8	10.0	10.1	10.1	9.84	1.4	0.59
18	10.4	9.4	9.9	9.4	9.9	9.80	1.0	0.42
19	11.1	10.5	10.6	9.8	9.4	10.28	1.7	0.68
20	9.3	9.9	10.9	9.5	10.6	10.04	1.6	0.69
21	9.5	10.2	9.7	9.4	10.0	9.76	0.8	0.34
22	10.5	10.5	10.1	9.5	10.3	10.18	1.0	0.41
23	9.8	8.9	9.6	9.8	9.6	9.54	0.9	0.37
24	9.3	9.7	10.3	10.1	9.7	9.82	1.0	0.39
25	10.2	9.6	8.8	9.9	10.2	9.74	1.4	0.58
26	10.8	9.5	10.5	10.5	10.1	10.28	1.3	0.50
27	10.4	9.9	10.1	9.9	10.9	10.24	1.0	0.42
28	11.0	10.8	10.1	9.2	9.9	10.20	1.8	0.72
29	10.3	10.0	10.6	10.0	11.1	10.40	1.1	0.46
30	10.9	10.6	9.9	10.0	10.8	10.44	1.0	0.46

TABLE E12 Data for Exercise 12

Sample	Sample Values				$\overline{X}$	R	s
1	146.0	147.4	151.9	155.2	150.125	9.2	4.22
2	147.1	147.5	151.4	149.4	148.850	4.3	1.97
3	148.7	148.4	149.6	154.1	150.200	5.7	2.65
4	151.3	150.0	152.4	148.2	150.475	4.2	1.81
5	146.4	147.5	152.9	150.3	149.275	6.5	2.92
6	150.2	142.9	152.5	155.5	150.275	12.6	5.37
7	147.8	148.3	145.7	149.7	147.875	4.0	1.66
8	137.1	156.6	147.2	148.9	147.450	19.5	8.02
9	151.1	148.1	145.6	147.6	148.100	5.5	2.27
10	151.3	151.3	142.5	146.2	147.825	8.8	4.29
11	151.3	153.5	150.2	148.7	150.925	4.8	2.02
12	151.9	152.2	149.3	154.2	151.900	4.9	2.01
13	152.8	149.1	148.5	146.9	149.325	5.9	2.50
14	152.9	149.9	151.9	150.4	151.275	3.0	1.38
15	149.0	149.9	153.1	152.8	151.200	4.1	2.06
16	153.9	150.8	153.9	145.0	150.900	8.9	4.20
17	150.4	151.8	151.3	153.0	151.625	2.6	1.08
18	157.2	152.6	148.4	152.6	152.700	8.8	3.59
19	152.7	156.2	146.8	148.7	151.100	9.4	4.20
20	150.2	148.2	149.8	142.1	147.575	8.1	3.75
21	151.0	151.7	148.5	147.0	149.550	4.7	2.19
22	143.8	154.5	154.8	151.6	151.175	11.0	5.12
23	143.0	156.4	149.2	152.2	150.200	13.4	5.64
24	148.8	147.7	147.1	148.2	147.950	1.7	0.72
25	153.8	145.4	149.5	153.4	150.525	8.4	3.93
26	151.6	149.3	155.0	149.0	151.225	6.0	2.77
27	149.4	151.4	154.6	150.0	151.350	5.2	2.32
28	149.8	149.0	146.8	145.7	147.825	4.1	1.90
29	155.8	152.4	150.2	154.8	153.300	5.6	2.51
30	153.9	145.7	150.7	150.4	150.175	8.2	3.38

10.3 Control Charts for Attributes

The *p* Chart

The *p* chart is used when the quality characteristic being measured on each unit has only two possible values, usually "defective" and "not defective." In each sample, the proportion of defectives is calculated and plotted. We will now describe how the center line and control limits are calculated.

Let *p* be the probability that a given unit is defective. If the process is in control, this probability is constant over time. Let *k* be the number of samples. We will assume that all samples are the same size, and we will denote this size by *n*. Let X_i be the number of defective units in the *i*th sample, and let $\hat{p}_i = X_i/n$ be the proportion of defective items in the *i*th sample. Now $X_i \sim \text{Bin}(n, p)$, and if $np > 10$ the Central Limit Theorem specifies that $X \sim N(np, np(1-p))$, approximately. Since $\hat{p} = X/n$, it follows that $\hat{p}_i \sim N(p, p(1-p)/n)$. Since $\hat{p}_i$ has mean $\mu = p$ and standard deviation

$\sigma = \sqrt{p(1 - p)/n}$, it follows that the center line should be at p, and the 3σ control limits should be at $p \pm 3\sqrt{p(1 - p)/n}$. Usually p is not known and is estimated with $\overline{p} = \sum_{i=1}^{k} \hat{p}_i/k$, the average of the sample proportions $\hat{p}_i$.

Summary

In a p chart, where the number of items in each sample is n, the center line and the 3σ upper and lower control limits are given by

$$3\sigma \text{ upper limit} = \overline{p} + 3\sqrt{\frac{\overline{p}(1 - \overline{p})}{n}}$$

$$\text{Center line} = \overline{p}$$

$$3\sigma \text{ lower limit} = \overline{p} - 3\sqrt{\frac{\overline{p}(1 - \overline{p})}{n}}$$

These control limits will be valid if $n\,\overline{p} > 10$.

Example 10.9 illustrates these ideas.

Example

10.9

In the production of silicon wafers, 30 lots of size 500 are sampled, and the proportion of defective wafers is calculated for each sample. Table 10.2 presents the results. Compute the center line and 3σ control limits for the p chart. Plot the chart. Does the process appear to be in control?

Solution

The average of the 30 sample proportions is $\overline{p} = 0.050867$. The center line is therefore plotted at 0.050867. The control limits are plotted at $0.050867 \pm 3\sqrt{(0.050867)(0.949133)/500}$. The upper control limit is therefore 0.0803, and the lower control limit is 0.0214. Figure 10.14 presents the p chart. The process appears to be in control.

TABLE 10.2 Number and proportion defective, for Example 10.9

Sample	Number Defective	Proportion Defective ($\hat{p}$)	Sample	Number Defective	Proportion Defective ($\hat{p}$)
1	17	0.034	16	26	0.052
2	26	0.052	17	19	0.038
3	31	0.062	18	31	0.062
4	25	0.050	19	27	0.054
5	26	0.052	20	24	0.048
6	29	0.058	21	22	0.044
7	36	0.072	22	24	0.048
8	26	0.052	23	30	0.060
9	25	0.050	24	25	0.050
10	21	0.042	25	26	0.052
11	18	0.036	26	28	0.056
12	33	0.066	27	22	0.044
13	29	0.058	28	31	0.062
14	17	0.034	29	18	0.036
15	28	0.056	30	23	0.046

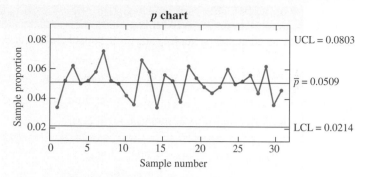

FIGURE 10.14 p chart for the data in Table 10.2.

The sample size needed to construct a p chart is usually much larger than that needed for an $\overline{X}$ chart. The reason is that the sample size must be large enough so that there will be several defective items in most of the samples. If defective items are not common, the sample size must be quite large.

Interpreting Out-of-Control Signals in Attribute Charts

When an attribute control chart is used to monitor the frequency of defective units, a point plotting above the upper control limit requires quite a different response than a point plotting below the lower control limit. Both conditions indicate that a special cause has changed the proportion of defective units. A point plotting above the upper control limit indicates that the proportion of defective units has increased, so action must be taken to identify and remove the special cause. A point plotting below the lower control limit, however, indicates that the special cause has *decreased* the proportion of defective units. The special cause still needs to be identified, but in this case, action should be taken to make it continue, so that the proportion of defective items can be decreased permanently.

The c Chart

The c chart is used when the quality measurement is a count of the number of defects, or flaws, in a given unit. A *unit* may be a single item, or it may be a group of items large enough so that the expected number of flaws is sufficiently large. Use of the c chart requires that the number of defects follows a Poisson distribution. Assume that k units are sampled, and let c_i denote the total number of defects in the ith unit. Let λ denote the mean total number of flaws per unit. Then $c_i \sim \text{Poisson}(\lambda)$. If the process is in control, the value of λ is constant over time. Now if λ is reasonably large, say $\lambda > 10$, the Central Limit Theorem specifies that $c_i \sim N(\lambda, \lambda)$, approximately. Note that the value of λ can in principle be made large enough by choosing a sufficiently large number of items per unit. The c chart is constructed by plotting the values c_i. Since c_i has mean λ and standard deviation equal to $\sqrt{\lambda}$, the center line should be plotted at λ and the 3σ control limits should be plotted at $\lambda \pm 3\sqrt{\lambda}$. Usually the value of λ is unknown and has to be estimated from the data. The appropriate estimate is $\overline{c} = \sum_{i=1}^{k} c_i / k$, the average number of defects per unit.

Summary

In a c chart, the center line and the 3σ upper and lower control limits are given by

$$3\sigma \text{ upper limit} = \bar{c} + 3\sqrt{\bar{c}}$$

$$\text{Center line} = \bar{c}$$

$$3\sigma \text{ lower limit} = \bar{c} - 3\sqrt{\bar{c}}$$

These control limits will be valid if $\bar{c} > 10$.

Example 10.10 illustrates these ideas.

Example 10.10

Rolls of sheet aluminum, used to manufacture cans, are examined for surface flaws. Table 10.3 presents the numbers of flaws in 40 samples of 100 m² each. Compute the center line and 3σ control limits for the c chart. Plot the chart. Does the process appear to be in control?

Solution

The average of the 40 counts is $\bar{c} = 12.275$. The center line is therefore plotted at 12.275. The 3σ control limits are plotted at $12.275 \pm 3\sqrt{12.275}$. The upper control limit is therefore 22.7857, and the lower control limit is 1.7643. Figure 10.15 presents the c chart. The process appears to be in control.

TABLE 10.3 Number of flaws, for Example 10.10

Sample	Number of Flaws (c)	Sample	Number of Flaws (c)	Sample	Number of Flaws (c)	Sample	Number of Flaws (c)
1	16	11	14	21	11	31	10
2	12	12	11	22	16	32	10
3	9	13	10	23	16	33	10
4	13	14	9	24	13	34	12
5	15	15	9	25	12	35	14
6	5	16	14	26	17	36	10
7	13	17	10	27	15	37	15
8	11	18	12	28	13	38	12
9	15	19	8	29	15	39	11
10	12	20	14	30	13	40	14

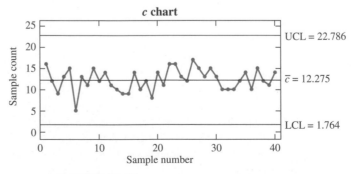

FIGURE 10.15 c chart for the data in Table 10.3.

Exercises for Section 10.3

1. A process is monitored for defective items by taking a sample of 300 items each day and calculating the proportion that are defective. Let p_i be the proportion of defective items in the ith sample. For the last 40 samples, the sum of the proportions is $\sum_{i=1}^{40} p_i = 1.42$. Calculate the center line and the 3σ upper and lower control limits for a p chart.

2. The target fill weight for a box of cereal is 350 g. Each day a sample of 300 boxes is taken, and the number that are underweight is counted. The number of underweight boxes for each of the last 25 days is as follows:

   23 14 17 19 20 19 21 27 26 23
   26 22 25 30 30 22 25 27 29 35
   39 43 41 39 29

   a. Compute the upper and lower 3σ limits for a p chart.
   b. Is the process in control? If not, when is it first detected to be out of control?

3. A process is monitored for defective items by periodically taking a sample of 200 items and counting the number that are defective. In the last 40 samples, there were a total of 748 defective items. Is this enough information to compute the 3σ control limits for a p chart? If so, compute the limits. If not, state what additional information would be required.

4. Refer to Exercise 3. In the last 40 samples, there were a total of 748 defective items. The largest number of defectives in any sample was 33, while the smallest number was 14. Is this enough information to determine whether the process was out of control at any time during the last 40 samples? If so, state whether or not the process was out of control. If not, state what additional information would be required to make the determination.

5. A newly designed quality control program for a certain process involves sampling 20 items each day and counting the number of defective items. The num-

bers of defectives in the first 10 samples are 0, 0, 1, 0, 1, 0, 0, 0, 1, 0. A member of the quality control team asks for advice, expressing concern that the numbers of defectives are too small to construct an accurate p chart. Which of the following is the best advice?

   i. Nothing needs to be changed. An accurate p chart can be constructed when the number of defective items is this small.
   ii. Since the proportion of items that are defective is so small, it isn't necessary to construct a p chart for this process.
   iii. Increase the value of p to increase the number of defectives per sample.
   iv. Increase the sample size to increase the number of defectives per sample.

6. A process that produces mirrors for automobiles is monitored by taking samples of 1500 mirrors and counting the total number of visual flaws on all the sample mirrors. Let c_i be the total number of flaws on the mirrors in the ith sample. For the last 70 samples, the quantity $\sum_{i=1}^{70} c_i = 876$ has been calculated. Compute the center line and the 3σ upper and lower control limits for a c chart.

7. Refer to Exercise 6. The number of flaws in the 34th sample was 27. Is it possible to determine whether the process was in control at this time? If so, state whether or not the process was in control. If not, state what additional information would be required to make the determination.

8. Each hour, a 10 m^2 section of fabric is inspected for flaws. The numbers of flaws observed for the last 20 hours are as follows:

   38 35 35 49 33 48 40 47 45 46
   41 53 36 41 51 63 35 58 55 57

   a. Compute the upper and lower 3σ limits for a c chart.
   b. Is the process in control? If not, when is it first detected to be out of control?

10.4 The CUSUM Chart

One purpose of an $\overline{X}$ chart is to detect a shift in the process mean. Unless a shift is fairly large, however, it may be some time before a point plots outside the 3σ control limits. Example 10.4 (in Section 10.2) showed that when a process mean shifts by an amount equal to $\sigma_{\overline{X}}$, the average run length (ARL) is approximately 44, which means that on the average 44 samples must be observed before the process is judged to be out of control. The Western Electric rules (Section 10.2) provide one method for reducing the ARL. CUSUM charts provide another.

One way that small shifts manifest themselves is with a run of points above or below the center line. The Western Electric rules are designed to respond to runs. Another way to detect smaller shifts is with **cumulative sums**. Imagine that a process mean shifts upward slightly. There will then be a tendency for points to plot above the center line. If we add the deviations from the center line as we go along and plot the cumulative sums, the points will drift upward and will exceed a control limit much sooner than they would in an $\overline{X}$ chart.

We now describe how to plot the points in a CUSUM chart. We assume that we have m samples of size n, with sample means $\overline{X}_1, \ldots, \overline{X}_m$. To begin, a target value μ must be specified for the process mean. Often μ is taken to be the value $\overline{\overline{X}}$. Then an estimate of $\sigma_{\overline{X}}$, the standard deviation of the sample means, is needed. This can be obtained either with sample ranges, using the estimate $\sigma_{\overline{X}} \approx A_2 \overline{R}/3$, or with sample standard deviations, using the estimate $\sigma_{\overline{X}} \approx A_3 \overline{s}/3$. If there is only one item per sample ($n = 1$), then an external estimate is needed. Even a rough guess can produce good results, so the CUSUM procedure can be useful when $n = 1$. Finally, two constants, usually called k and h, must be specified. Larger values for these constants result in longer average run lengths, and thus fewer false alarms, but also result in longer waiting times to discover that a process is out of control. The values $k = 0.5$ and $h = 4$ or 5 are often used, because they provide a reasonably long ARL when the process is in control but still have fairly good power to detect a shift of magnitude $1\sigma_{\overline{X}}$ or more in the process mean.

For each sample, the quantity $\overline{X}_i - \mu$ is the deviation from the target value. We define two cumulative sums, SH and SL. The sum SH is always either positive or zero and signals that the process mean has become greater than the target value. The sum SL is always either negative or zero and signals that the process mean has become less than the target value. Both these sums are computed recursively; in other words, the current value in the sequence is used to compute the next value. The initial values of SH and SL are

$$\mathrm{SH}_0 = 0 \qquad \mathrm{SL}_0 = 0 \tag{10.2}$$

For $i \geq 1$ the values are

$$\mathrm{SH}_i = \max[0, \ \overline{X}_i - \mu - k\sigma_{\overline{X}} + \mathrm{SH}_{i-1}] \tag{10.3}$$

$$\mathrm{SL}_i = \min[0, \ \overline{X}_i - \mu + k\sigma_{\overline{X}} + \mathrm{SL}_{i-1}] \tag{10.4}$$

If $\mathrm{SH}_i > h\sigma_{\overline{X}}$ for some i, it is concluded that the process mean has become greater than the target value. If $\mathrm{SL}_i < -h\sigma_{\overline{X}}$ for some i, it is concluded that the process mean has become less than the target value.

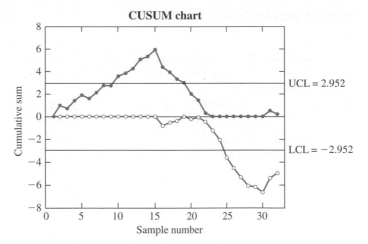

FIGURE 10.16 CUSUM chart for the data in Figure 10.9.

Figure 10.16 presents a CUSUM chart for the data in Figure 10.9 (in Section 10.2). The values $k = 0.5$ and $h = 4$ were used. The value 2.952 is the quantity $h\sigma_{\overline{X}} = 4(0.738)$. The CUSUM chart indicates an out-of-control condition on the tenth sample. For these data, the CUSUM chart performs about as well as the Western Electric rules, which determined that the process was out of control at the eighth sample (see Figure 10.9).

Summary

In a CUSUM chart, two cumulative sums, SH and SL, are plotted.
  The initial values are $SH_0 = SL_0 = 0$. For $i \geq 1$,

$$SH_i = \max[0, \overline{X}_i - \mu - k\sigma_{\overline{X}} + SH_{i-1}]$$
$$SL_i = \min[0, \overline{X}_i - \mu + k\sigma_{\overline{X}} + SL_{i-1}]$$

  The constants k and h must be specified. Good results are often obtained for the values $k = 0.5$ and $h = 4$ or 5.
  If for any i, $SH_i > h\sigma_{\overline{X}}$ or $SL_i < -h\sigma_{\overline{X}}$, the process is judged to be out of control.

There are several other methods for constructing CUSUM charts, which are equivalent, or nearly equivalent, to the method presented here. Some people define the deviations to be the z-scores $z_i = (\overline{X}_i - \mu)/\sigma_{\overline{X}}$, and then use z_i in place of $X_i - \mu$, and k in place of $k\sigma_{\overline{X}}$ in the formulas for SH and SL. With this definition, the control limits are plotted at $\pm h$ rather than $\pm h\sigma_{\overline{X}}$.

Other methods for graphing the CUSUM chart are available as well. The most common alternative is the "V-mask" approach. A text on statistical quality control, such as Montgomery (2009b), can be consulted for further information.

Exercises for Section 10.4

1. Refer to Exercise 3 in Section 10.2.

   a. Delete any samples necessary to bring the process variation under control. (You did this already if you did Exercise 3 in Section 10.2.)

   b. Use $\overline{R}$ to estimate $\sigma_{\overline{X}}$ ($\sigma_{\overline{X}}$ is the difference between $\overline{\overline{X}}$ and the 1σ control limit on an $\overline{X}$ chart).

   c. Construct a CUSUM chart, using $\overline{\overline{X}}$ for the target mean μ, and the estimate of $\sigma_{\overline{X}}$ found in part (b) for the standard deviation. Use the values $k = 0.5$ and $h = 4$.

   d. Is the process mean in control? If not, when is it first detected to be out of control?

   e. Construct an $\overline{X}$ chart, and use the Western Electric rules to determine whether the process mean is in control. (You did this already if you did Exercise 3 in Section 10.2.) Do the Western Electric rules give the same results as the CUSUM chart? If not, how are they different?

2. Refer to Exercise 8 in Section 10.2.

   a. Delete any samples necessary to bring the process variation under control. (You did this already if you did Exercise 8 in Section 10.2.)

   b. Use $\overline{R}$ to estimate $\sigma_{\overline{X}}$ ($\sigma_{\overline{X}}$ is the difference between $\overline{\overline{X}}$ and the 1σ control limit on an $\overline{X}$ chart).

   c. Construct a CUSUM chart, using $\overline{\overline{X}}$ for the target mean μ, and the estimate of $\sigma_{\overline{X}}$ found in part (b) for the standard deviation. Use the values $k = 0.5$ and $h = 4$.

   d. Is the process mean in control? If not, when is it first detected to be out of control?

   e. Construct an $\overline{X}$ chart, and use the Western Electric rules to determine whether the process mean is in control. (You did this already if you did Exercise 8 in Section 10.2.) Do the Western Electric rules give the same results as the CUSUM chart? If not, how are they different?

3. Refer to Exercise 10 in Section 10.2.

   a. Delete any samples necessary to bring the process variation under control. (You did this already if you did Exercise 10 in Section 10.2.)

   b. Use $\overline{R}$ to estimate $\sigma_{\overline{X}}$ ($\sigma_{\overline{X}}$ is the difference between $\overline{\overline{X}}$ and the 1σ control limit on an $\overline{X}$ chart).

   c. Construct a CUSUM chart, using $\overline{\overline{X}}$ for the target mean μ, and the estimate of $\sigma_{\overline{X}}$ found in part (b) for the standard deviation. Use the values $k = 0.5$ and $h = 4$.

   d. Is the process mean in control? If not, when is it first detected to be out of control?

   e. Construct an $\overline{X}$ chart, and use the Western Electric rules to determine whether the process mean is in control. (You did this already if you did Exercise 10 in Section 10.2.) Do the Western Electric rules give the same results as the CUSUM chart? If not, how are they different?

4. Refer to Exercise 12 in Section 10.2.

   a. Delete any samples necessary to bring the process variation under control. (You did this already if you did Exercise 12 in Section 10.2.)

   b. Use $\overline{R}$ to estimate $\sigma_{\overline{X}}$ ($\sigma_{\overline{X}}$ is the difference between $\overline{\overline{X}}$ and the 1σ control limit on an $\overline{X}$ chart).

   c. Construct a CUSUM chart, using $\overline{\overline{X}}$ for the target mean μ, and the estimate of $\sigma_{\overline{X}}$ found in part (b) for the standard deviation. Use the values $k = 0.5$ and $h = 4$.

   d. Is the process mean in control? If not, when is it first detected to be out of control?

   e. Construct an $\overline{X}$ chart, and use the Western Electric rules to determine whether the process mean is in control. (You did this already if you did Exercise 12 in Section 10.2.) Do the Western Electric rules give the same results as the CUSUM chart? If not, how are they different?

5. Concrete blocks to be used in a certain application are supposed to have a mean compressive strength of 1500 MPa. Samples of size 1 are used for quality control. The compressive strengths of the last 40 samples are given in the following table.

Sample	Strength	Sample	Strength
1	1487	21	1507
2	1463	22	1474
3	1499	23	1515
4	1502	24	1533
5	1473	25	1487
6	1520	26	1518
7	1520	27	1526
8	1495	28	1469
9	1503	29	1472
10	1499	30	1512
11	1497	31	1483
12	1516	32	1505
13	1489	33	1507
14	1545	34	1505
15	1498	35	1517
16	1503	36	1504
17	1522	37	1515
18	1502	38	1467
19	1499	39	1491
20	1484	40	1488

Previous results suggest that a value of $\sigma = 15$ is reasonable for this process.

a. Using the value 1500 for the target mean μ, and the values $k = 0.5$ and $h = 4$, construct a CUSUM chart.

b. Is the process mean in control? If not, when is it first detected to be out of control?

6. A quality-control apprentice is preparing a CUSUM chart. The values calculated for SL and SH are presented in the following table. Three of the values have been calculated incorrectly. Which are they?

Sample	SL	SH
1	0	0
2	0	0
3	0	0
4	−1.3280	0
5	−1.4364	0
6	−2.0464	0
7	−1.6370	0
8	−0.8234	0.2767
9	−0.4528	0.1106
10	0	0.7836
11	0.2371	0.0097
12	0.7104	0
13	0	0.2775
14	0	0.5842
15	0	0.3750
16	0	0.4658
17	0	0.1866
18	0	0.3277
19	−0.2036	0
20	0	−0.7345

10.5 Process Capability

Once a process is in a state of statistical control, it is important to evaluate its ability to produce output that conforms to design specifications. We consider variables data, and we assume that the quality characteristic of interest follows a normal distribution.

The first step in assessing process capability is to estimate the process mean and standard deviation. These estimates are denoted $\hat{\mu}$ and $\hat{\sigma}$, respectively. The data used to calculate $\hat{\mu}$ and $\hat{\sigma}$ are usually taken from control charts at a time when the process is in a state of control. The process mean is estimated with $\hat{\mu} = \overline{\overline{X}}$. The process standard deviation can be estimated by using either the average sample range $\overline{R}$ or the average sample standard deviation $\overline{s}$. Specifically, it has been shown that $\hat{\sigma}$ can be computed either by dividing $\overline{R}$ by a constant called d_2, or by dividing $\overline{s}$ by a constant called c_4. The values of the constants d_2 and c_4 depend on the sample size. Values are tabulated in Table A.8 (in Appendix A).

Summary

If a quality characteristic from a process in a state of control is normally distributed, then the process mean $\hat{\mu}$ and standard deviation $\hat{\sigma}$ can be estimated from control chart data as follows:

$$\hat{\mu} = \overline{\overline{X}}$$

$$\hat{\sigma} = \frac{\overline{R}}{d_2} \quad \text{or} \quad \hat{\sigma} = \frac{\overline{s}}{c_4}$$

The values of d_2 and c_4 depend on the sample size. Values are tabulated in Table A.8.

Note that the process standard deviation σ is not the same quantity that is used to compute the 3σ control limits on the $\overline{X}$ chart. The control limits are $\mu \pm 3\sigma_{\overline{X}}$, where $\sigma_{\overline{X}}$ is the standard deviation of the sample mean. The process standard deviation σ is the standard deviation of the quality characteristic of individual units. They are related by $\sigma_{\overline{X}} = \sigma/\sqrt{n}$, where n is the sample size.

To be fit for use, a quality characteristic must fall between a lower specification limit (LSL) and an upper specification limit (USL). Sometimes there is only one limit; this situation will be discussed at the end of this section. The specification limits are determined by design requirements. They are *not* the control limits found on control charts. We will assume that the process mean falls between the LSL and the USL.

We will discuss two indices of process capability, C_{pk} and C_p. The index C_{pk} describes the capability of the process as it is, while C_p describes the potential capability of the process. Note that the process capability index C_p has no relation to the quantity called Mallows' C_p that is used for linear model selection (discussed in Section 8.4). It is a coincidence that the two quantities have the same name.

The index C_{pk} is defined to be the distance from $\hat{\mu}$ to the nearest specification limit, divided by $3\hat{\sigma}$. Figure 10.17 presents an illustration where $\hat{\mu}$ is closer to the upper specification limit.

Definition

The index C_{pk} is equal either to

$$\frac{\hat{\mu} - \text{LSL}}{3\hat{\sigma}} \quad \text{or} \quad \frac{\text{USL} - \hat{\mu}}{3\hat{\sigma}}$$

whichever is less.

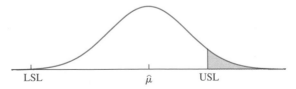

FIGURE 10.17 The normal curve represents the population of units produced by a process. The process mean is closer to the upper specification limit (USL) than to the lower specification limit (LSL). The index C_{pk} is therefore equal to $(\text{USL} - \hat{\mu})/3\hat{\sigma}$.

By convention, the minimum acceptable value for C_{pk} is 1. That is, a process is considered to be minimally capable if the process mean is three standard deviations from the nearest specification limit. A C_{pk} value of 1.33, indicating that the process mean is four standard deviations from the nearest specification limit, is generally considered good.

Example
10.11

The design specifications for a piston rod used in an automatic transmission call for the rod length to be between 71.4 and 72.8 mm. The process is monitored with an $\overline{X}$ chart and an S chart, using samples of size $n = 5$. These show the process to be in control. The values of $\overline{\overline{X}}$ and $\overline{s}$ are $\overline{\overline{X}} = 71.8$ mm and $\overline{s} = 0.20$ mm. Compute the value of C_{pk}. Is the process capability acceptable?

Solution

We estimate $\hat{\mu} = \overline{\overline{X}} = 71.8$. To compute $\hat{\sigma}$, we find from Table A.8 that $c_4 = 0.9400$ when the sample size is 5. Therefore $\hat{\sigma} = \overline{s}/c_4 = 0.20/0.9400 = 0.2128$. The specification limits are LSL = 71.4 mm and USL = 72.8 mm. The value $\hat{\mu}$ is closer to the LSL than to the USL. Therefore

$$C_{pk} = \frac{\hat{\mu} - \text{LSL}}{3\hat{\sigma}} = \frac{71.8 - 71.4}{(3)(0.2128)}$$
$$= 0.6266$$

Since $C_{pk} < 1$, the process capability is not acceptable.

Example
10.12

Refer to Example 10.11. Assume that it is possible to adjust the process mean to any desired value. To what value should it be set to maximize the value of C_{pk}? What will the value of C_{pk} be?

Solution

The specification limits are LSL = 71.4 and USL = 72.8. The value of C_{pk} will be maximized if the process mean is adjusted to the midpoint between the specification limits; that is, if $\mu = 72.1$. The process standard deviation is estimated with $\hat{\sigma} = 0.2128$. Therefore the maximum value of C_{pk} is $(72.1 - 71.4)/(3)(0.2128) = 1.0965$. The process capability would be acceptable.

The capability that can potentially be achieved by shifting the process mean to the midpoint between the upper and lower specification limits is called the **process capability index**, denoted C_p. If the process mean is at the midpoint between LSL and USL, then the distance from the mean to either specification limit is equal to one-half the distance between the specification limits, that is $\mu - \text{LSL} = \text{USL} - \mu = (\text{USL} - \text{LSL})/2$ (see Figure 10.18, page 510). It follows that

$$C_p = \frac{\text{USL} - \text{LSL}}{6\hat{\sigma}} \tag{10.5}$$

The process capability index C_p measures the potential capability of the process, that is the greatest capability that the process can achieve without reducing the process standard deviation.

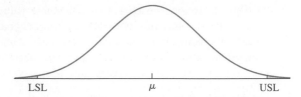

FIGURE 10.18 A process has maximum capability when the process mean is at the midpoint between the specification limits. In this case $\mu - \text{LSL} = \text{USL} - \mu = (\text{USL} - \text{LSL})/2$.

Example 10.13

Specifications for the output voltage of a certain electric circuit are 48 to 52 V. The process is in control with $\hat{\sigma} = 0.482$ V. Compute the process capability index C_p.

Solution

$$C_p = \frac{\text{USL} - \text{LSL}}{6\hat{\sigma}} = \frac{52 - 48}{(6)(0.482)}$$
$$= 1.38$$

The process capability is potentially good.

Estimating the Proportion of Nonconforming Units from Process Capability

Many people use the value of C_p to try to estimate the proportion of units that will be nonconforming. For example, if $C_p = 1$, then the specification limits are equal to $\hat{\mu} - 3\hat{\sigma}$ and $\hat{\mu} + 3\hat{\sigma}$, respectively. Therefore a unit will be nonconforming only if it is more than three standard deviations from the process mean. Now for a normal population, the proportion of items that are more than three standard deviations from the mean is equal to 0.0027. Therefore it is often stated that a process with $C_p = 1$ will produce 27 nonconforming parts per 10,000.

The problem with this is that the normality assumption is only approximate for real processes. The approximation may be very good near the middle of the curve, but it is often not good in the tails. Therefore the true proportion of nonconforming parts may be quite different from that predicted from the normal curve, especially when the proportion is very small. In general, estimates of small probabilities that are based on a normal approximation are extremely crude at best.

Six-Sigma Quality

The phrase "six-sigma quality" is now quite prevalent in discussions of quality control. A process is said to have six-sigma quality if the process capability index C_p has a value of 2.0 or greater. Equivalently, a process has six-sigma quality if the difference $\text{USL} - \text{LSL}$ is at least 12σ. When a process has six-sigma quality, then if the process mean is optimally adjusted, it is six standard deviations from each specification limit. In this case the proportion of nonconforming units will be virtually zero.

An important feature of a six-sigma process is that it can withstand moderate shifts in process mean without significant deterioration in capability. For example, even if the process mean shifts by 3σ in one direction or the other, it is still 3σ from the nearest specification limit, so the capability index will still be acceptable.

Example
10.14

Refer to Example 10.13. To what value must the process standard deviation be reduced in order for the process to attain six-sigma quality?

Solution

To attain six-sigma quality, the value of C_p must be at least 2.0. The value of σ for which this occurs is found by setting $C_p = 2.0$ and solving for σ. We obtain

$$2.0 = \frac{52 - 48}{6\sigma}$$

from which $\sigma = 0.33$.

In recent years, the term "six-sigma" has acquired a broader meaning and now refers to a business management strategy that promotes methods designed to identify and remove defects in manufacturing and business processes.

One-Sided Tolerances

Some characteristics have only one specification limit. For example, strengths usually have a lower specification limit but no upper limit, since for most applications a part cannot be too strong. The analog of C_{pk} when there is only a lower specification limit is the **lower capability index** C_{pl}; when there is only an upper limit, it is the **upper capability index** C_{pu}. Each of these quantities is defined to be the difference between the estimated process mean $\hat{\mu}$ and the specification limit, divided by $3\hat{\sigma}$.

Summary

If a process has only a lower specification limit (LSL), then the lower capability index is

$$C_{pl} = \frac{\hat{\mu} - \text{LSL}}{3\hat{\sigma}}$$

If a process has only an upper specification limit (USL), then the upper capability index is

$$C_{pu} = \frac{\text{USL} - \hat{\mu}}{3\hat{\sigma}}$$

There is no analog for C_p for processes with only one specification limit.

Exercises for Section 10.5

1. The design specifications for the intake valve in an internal combustion engine call for the valve clearance to be between 0.18 and 0.22 mm. Data from an $\overline{X}$ chart, based on samples of size 4, that shows that the process is in control, yield values of $\overline{\overline{X}} = 0.205$ mm and $\overline{s} = 0.002$ mm.

   a. Compute the value of C_{pk} for this process.

   b. Is the process capability acceptable? Explain.

2. The specifications for the fill volume of beverage cans is 11.95 to 12.10 oz. Data from an $\overline{X}$ chart, based on samples of size 5, that shows that the process is in control, yield values of $\overline{\overline{X}} = 12.01$ oz and $\overline{R} = 0.124$ oz.

   a. Compute the value of C_{pk} for this process.

   b. Is the process capability acceptable? Explain.

3. Refer to Exercise 1.

   a. To what value should the process mean be set to maximize the process capability?

   b. What will the process capability then be?

4. Refer to Exercise 2.

   a. To what value should the process mean be set to maximize the process capability?

   b. Is it possible to make the process capability acceptable simply by adjusting the process mean? Explain.

   c. When the process mean is adjusted to its optimum value, what value must be attained by the process standard deviation so that the process capability is acceptable?

   d. When the process mean is adjusted to its optimum value, what value must be attained by the process standard deviation so that the process has six-sigma quality?

5. A process has a process capability index of $C_p = 1.2$.

   a. Assume the process mean is set to its optimal value. Express the upper and lower specification limits in terms of the process mean and standard deviation.

   b. Using the normal curve, estimate the proportion of units that will be nonconforming.

   c. Is it likely or unlikely that the true proportion of nonconforming units will be quite different from the estimate in part (b)? Explain.

Supplementary Exercises for Chapter 10

1. A process is monitored for defective items by taking a sample of 250 items each day and calculating the proportion that are defective. Let p_i be the proportion of defective items in the ith sample. For the last 50 samples, the sum of the proportions is $\sum_{i=1}^{50} p_i = 2.98$. Calculate the center line and the 3σ upper and lower control limits for a p chart.

2. Someone constructs an $\overline{X}$ chart where the control limits are at $\pm 2.5\sigma_{\overline{X}}$ rather than at $\pm 3\sigma_{\overline{X}}$.

   a. If the process is in control, what is the ARL for this chart?

   b. If the process mean shifts by $1\sigma_{\overline{X}}$, what is the ARL for this chart?

   c. In units of $\sigma_{\overline{X}}$, how large an upward shift can be detected with an ARL of 10?

3. Samples of three resistors are taken periodically, and the resistances, in ohms, are measured. The following table presents the means, ranges, and standard deviations for 30 consecutive samples.

Sample	$\overline{X}$	R	s
1	5.114	0.146	0.077
2	5.144	0.158	0.085
3	5.220	0.057	0.031
4	5.196	0.158	0.081
5	5.176	0.172	0.099
6	5.222	0.030	0.017
7	5.209	0.118	0.059
8	5.212	0.099	0.053
9	5.238	0.157	0.085
10	5.152	0.104	0.054
11	5.163	0.051	0.026
12	5.221	0.105	0.055
13	5.144	0.132	0.071
14	5.098	0.123	0.062
15	5.070	0.083	0.042
16	5.029	0.073	0.038
17	5.045	0.161	0.087
18	5.008	0.138	0.071
19	5.029	0.082	0.042
20	5.038	0.109	0.055
21	4.962	0.066	0.034
22	5.033	0.078	0.041
23	4.993	0.085	0.044
24	4.961	0.126	0.066
25	4.976	0.094	0.047
26	5.005	0.135	0.068
27	5.022	0.120	0.062
28	5.077	0.140	0.074
29	5.033	0.049	0.026
30	5.068	0.146	0.076

The means are $\overline{\overline{X}} = 5.095$, $\overline{R} = 0.110$, and $\overline{s} = 0.058$.

a. Compute the 3σ limits for the R chart. Is the variance out of control at any point? If so, delete the samples that are out of control and recompute $\overline{\overline{X}}$ and $\overline{R}$.

b. Compute the 3σ limits for the $\overline{X}$ chart. On the basis of the 3σ limits, is the process mean in control? If not, at what point is it first detected to be out of control?

c. On the basis of the Western Electric rules, is the process mean in control? If not, when is it first detected to be out of control?

4. Repeat Exercise 3, using the S chart in place of the R chart.

5. Refer to Exercise 3.
   a. Delete any samples necessary to bring the process variation under control. (You did this already if you did Exercise 3.)
   b. Use $\overline{R}$ to estimate $\sigma_{\overline{X}}$ ($\sigma_{\overline{X}}$ is the difference between $\overline{\overline{X}}$ and the 1σ control limit on an $\overline{X}$ chart).
   c. Construct a CUSUM chart, using $\overline{\overline{X}}$ for the target mean μ, and the estimate of $\sigma_{\overline{X}}$ found in part (b) for the standard deviation. Use the values $k = 0.5$ and $h = 4$.
   d. Is the process mean in control? If not, when is it first detected to be out of control?
   e. Construct an $\overline{X}$ chart, and use the Western Electric rules to determine whether the process mean is in control. (You did this already if you did Exercise 3.) Do the Western Electric rules give the same results as the CUSUM chart? If not, how are they different?

6. A process is monitored for flaws by taking a sample of size 50 each hour and counting the total number of flaws in the sample items. The total number of flaws over the last 30 samples is 658.
   a. Compute the center line and upper and lower 3σ control limits.
   b. The tenth sample had three flaws. Was the process out of control at that time? Explain.

7. To set up a p chart to monitor a process that produces computer chips, samples of 500 chips are taken daily, and the number of defective chips in each sample is counted. The numbers of defective chips for each of the last 25 days are as follows:

   25 22 14 24 18 16 20 27 19 20 22 7 24 26
   11 14 18 29 21 32 29 34 34 30 24

   a. Compute the upper and lower 3σ limits for a p chart.
   b. At which sample is the process first detected to be out of control?
   c. Suppose that the special cause that resulted in the out-of-control condition is determined. Should this cause be remedied? Explain.

Appendix A

Tables

- **Table A.1: Cumulative Binomial Distribution**
- **Table A.2: Cumulative Normal Distribution**
- **Table A.3: Upper Percentage Points for the Student's t Distribution**
- **Table A.4: Tolerance Factors for the Normal Distribution**
- **Table A.5: Upper Percentage Points for the χ^2 Distribution**
- **Table A.6: Upper Percentage Points for the F Distribution**
- **Table A.7: Upper Percentage Points for the Studentized Range**
- **Table A.8: Control Chart Constants**

TABLE A.1 Cumulative binomial distribution

$$F(x) = P(X \le x) = \sum_{k=0}^{x} \frac{n!}{k!(n-k)!} p^k (1-p)^{(n-k)}$$

								p						
n	x	0.05	0.10	0.20	0.25	0.30	0.40	0.50	0.60	0.70	0.75	0.80	0.90	0.95
2	0	0.902	0.810	0.640	0.562	0.490	0.360	0.250	0.160	0.090	0.062	0.040	0.010	0.003
	1	0.997	0.990	0.960	0.938	0.910	0.840	0.750	0.640	0.510	0.438	0.360	0.190	0.098
	2	1.000	1.000	1.000	1.000	1.000	1.000	1.000	1.000	1.000	1.000	1.000	1.000	1.000
3	0	0.857	0.729	0.512	0.422	0.343	0.216	0.125	0.064	0.027	0.016	0.008	0.001	0.000
	1	0.993	0.972	0.896	0.844	0.784	0.648	0.500	0.352	0.216	0.156	0.104	0.028	0.007
	2	1.000	0.999	0.992	0.984	0.973	0.936	0.875	0.784	0.657	0.578	0.488	0.271	0.143
	3	1.000	1.000	1.000	1.000	1.000	1.000	1.000	1.000	1.000	1.000	1.000	1.000	1.000
4	0	0.815	0.656	0.410	0.316	0.240	0.130	0.062	0.026	0.008	0.004	0.002	0.000	0.000
	1	0.986	0.948	0.819	0.738	0.652	0.475	0.313	0.179	0.084	0.051	0.027	0.004	0.000
	2	1.000	0.996	0.973	0.949	0.916	0.821	0.688	0.525	0.348	0.262	0.181	0.052	0.014
	3	1.000	1.000	0.998	0.996	0.992	0.974	0.938	0.870	0.760	0.684	0.590	0.344	0.185
	4	1.000	1.000	1.000	1.000	1.000	1.000	1.000	1.000	1.000	1.000	1.000	1.000	1.000
5	0	0.774	0.590	0.328	0.237	0.168	0.078	0.031	0.010	0.002	0.001	0.000	0.000	0.000
	1	0.977	0.919	0.737	0.633	0.528	0.337	0.187	0.087	0.031	0.016	0.007	0.000	0.000
	2	0.999	0.991	0.942	0.896	0.837	0.683	0.500	0.317	0.163	0.104	0.058	0.009	0.001
	3	1.000	1.000	0.993	0.984	0.969	0.913	0.812	0.663	0.472	0.367	0.263	0.081	0.023
	4	1.000	1.000	1.000	0.999	0.998	0.990	0.969	0.922	0.832	0.763	0.672	0.410	0.226
	5	1.000	1.000	1.000	1.000	1.000	1.000	1.000	1.000	1.000	1.000	1.000	1.000	1.000
6	0	0.735	0.531	0.262	0.178	0.118	0.047	0.016	0.004	0.001	0.000	0.000	0.000	0.000
	1	0.967	0.886	0.655	0.534	0.420	0.233	0.109	0.041	0.011	0.005	0.002	0.000	0.000
	2	0.998	0.984	0.901	0.831	0.744	0.544	0.344	0.179	0.070	0.038	0.017	0.001	0.000
	3	1.000	0.999	0.983	0.962	0.930	0.821	0.656	0.456	0.256	0.169	0.099	0.016	0.002
	4	1.000	1.000	0.998	0.995	0.989	0.959	0.891	0.767	0.580	0.466	0.345	0.114	0.033
	5	1.000	1.000	1.000	1.000	0.999	0.996	0.984	0.953	0.882	0.822	0.738	0.469	0.265
	6	1.000	1.000	1.000	1.000	1.000	1.000	1.000	1.000	1.000	1.000	1.000	1.000	1.000
7	0	0.698	0.478	0.210	0.133	0.082	0.028	0.008	0.002	0.000	0.000	0.000	0.000	0.000
	1	0.956	0.850	0.577	0.445	0.329	0.159	0.063	0.019	0.004	0.001	0.000	0.000	0.000
	2	0.996	0.974	0.852	0.756	0.647	0.420	0.227	0.096	0.029	0.013	0.005	0.000	0.000
	3	1.000	0.997	0.967	0.929	0.874	0.710	0.500	0.290	0.126	0.071	0.033	0.003	0.000
	4	1.000	1.000	0.995	0.987	0.971	0.904	0.773	0.580	0.353	0.244	0.148	0.026	0.004
	5	1.000	1.000	1.000	0.999	0.996	0.981	0.938	0.841	0.671	0.555	0.423	0.150	0.044
	6	1.000	1.000	1.000	1.000	1.000	0.998	0.992	0.972	0.918	0.867	0.790	0.522	0.302
	7	1.000	1.000	1.000	1.000	1.000	1.000	1.000	1.000	1.000	1.000	1.000	1.000	1.000

Continued on page 516

TABLE A.1 Cumulative binomial distribution (continued)

								p						
n	x	0.05	0.10	0.20	0.25	0.30	0.40	0.50	0.60	0.70	0.75	0.80	0.90	0.95
8	0	0.663	0.430	0.168	0.100	0.058	0.017	0.004	0.001	0.000	0.000	0.000	0.000	0.000
	1	0.943	0.813	0.503	0.367	0.255	0.106	0.035	0.009	0.001	0.000	0.000	0.000	0.000
	2	0.994	0.962	0.797	0.679	0.552	0.315	0.145	0.050	0.011	0.004	0.001	0.000	0.000
	3	1.000	0.995	0.944	0.886	0.806	0.594	0.363	0.174	0.058	0.027	0.010	0.000	0.000
	4	1.000	1.000	0.990	0.973	0.942	0.826	0.637	0.406	0.194	0.114	0.056	0.005	0.000
	5	1.000	1.000	0.999	0.996	0.989	0.950	0.855	0.685	0.448	0.321	0.203	0.038	0.006
	6	1.000	1.000	1.000	1.000	0.999	0.991	0.965	0.894	0.745	0.633	0.497	0.187	0.057
	7	1.000	1.000	1.000	1.000	1.000	0.999	0.996	0.983	0.942	0.900	0.832	0.570	0.337
	8	1.000	1.000	1.000	1.000	1.000	1.000	1.000	1.000	1.000	1.000	1.000	1.000	1.000
9	0	0.630	0.387	0.134	0.075	0.040	0.010	0.002	0.000	0.000	0.000	0.000	0.000	0.000
	1	0.929	0.775	0.436	0.300	0.196	0.071	0.020	0.004	0.000	0.000	0.000	0.000	0.000
	2	0.992	0.947	0.738	0.601	0.463	0.232	0.090	0.025	0.004	0.001	0.000	0.000	0.000
	3	0.999	0.992	0.914	0.834	0.730	0.483	0.254	0.099	0.025	0.010	0.003	0.000	0.000
	4	1.000	0.999	0.980	0.951	0.901	0.733	0.500	0.267	0.099	0.049	0.020	0.001	0.000
	5	1.000	1.000	0.997	0.990	0.975	0.901	0.746	0.517	0.270	0.166	0.086	0.008	0.001
	6	1.000	1.000	1.000	0.999	0.996	0.975	0.910	0.768	0.537	0.399	0.262	0.053	0.008
	7	1.000	1.000	1.000	1.000	1.000	0.996	0.980	0.929	0.804	0.700	0.564	0.225	0.071
	8	1.000	1.000	1.000	1.000	1.000	1.000	0.998	0.990	0.960	0.925	0.866	0.613	0.370
	9	1.000	1.000	1.000	1.000	1.000	1.000	1.000	1.000	1.000	1.000	1.000	1.000	1.000
10	0	0.599	0.349	0.107	0.056	0.028	0.006	0.001	0.000	0.000	0.000	0.000	0.000	0.000
	1	0.914	0.736	0.376	0.244	0.149	0.046	0.011	0.002	0.000	0.000	0.000	0.000	0.000
	2	0.988	0.930	0.678	0.526	0.383	0.167	0.055	0.012	0.002	0.000	0.000	0.000	0.000
	3	0.999	0.987	0.879	0.776	0.650	0.382	0.172	0.055	0.011	0.004	0.001	0.000	0.000
	4	1.000	0.998	0.967	0.922	0.850	0.633	0.377	0.166	0.047	0.020	0.006	0.000	0.000
	5	1.000	1.000	0.994	0.980	0.953	0.834	0.623	0.367	0.150	0.078	0.033	0.002	0.000
	6	1.000	1.000	0.999	0.996	0.989	0.945	0.828	0.618	0.350	0.224	0.121	0.013	0.001
	7	1.000	1.000	1.000	1.000	0.998	0.988	0.945	0.833	0.617	0.474	0.322	0.070	0.012
	8	1.000	1.000	1.000	1.000	1.000	0.998	0.989	0.954	0.851	0.756	0.624	0.264	0.086
	9	1.000	1.000	1.000	1.000	1.000	1.000	0.999	0.994	0.972	0.944	0.893	0.651	0.401
	10	1.000	1.000	1.000	1.000	1.000	1.000	1.000	1.000	1.000	1.000	1.000	1.000	1.000
11	0	0.569	0.314	0.086	0.042	0.020	0.004	0.000	0.000	0.000	0.000	0.000	0.000	0.000
	1	0.898	0.697	0.322	0.197	0.113	0.030	0.006	0.001	0.000	0.000	0.000	0.000	0.000
	2	0.985	0.910	0.617	0.455	0.313	0.119	0.033	0.006	0.001	0.000	0.000	0.000	0.000
	3	0.998	0.981	0.839	0.713	0.570	0.296	0.113	0.029	0.004	0.001	0.000	0.000	0.000
	4	1.000	0.997	0.950	0.885	0.790	0.533	0.274	0.099	0.022	0.008	0.002	0.000	0.000
	5	1.000	1.000	0.988	0.966	0.922	0.753	0.500	0.247	0.078	0.034	0.012	0.000	0.000
	6	1.000	1.000	0.998	0.992	0.978	0.901	0.726	0.467	0.210	0.115	0.050	0.003	0.000
	7	1.000	1.000	1.000	0.999	0.996	0.971	0.887	0.704	0.430	0.287	0.161	0.019	0.002
	8	1.000	1.000	1.000	1.000	0.999	0.994	0.967	0.881	0.687	0.545	0.383	0.090	0.015
	9	1.000	1.000	1.000	1.000	1.000	0.999	0.994	0.970	0.887	0.803	0.678	0.303	0.102
	10	1.000	1.000	1.000	1.000	1.000	1.000	1.000	0.996	0.980	0.958	0.914	0.686	0.431
	11	1.000	1.000	1.000	1.000	1.000	1.000	1.000	1.000	1.000	1.000	1.000	1.000	1.000

Continued on page 517

TABLE A.1 Cumulative binomial distribution (continued)

								p						
n	x	0.05	0.10	0.20	0.25	0.30	0.40	0.50	0.60	0.70	0.75	0.80	0.90	0.95
12	0	0.540	0.282	0.069	0.032	0.014	0.002	0.000	0.000	0.000	0.000	0.000	0.000	0.000
	1	0.882	0.659	0.275	0.158	0.085	0.020	0.003	0.000	0.000	0.000	0.000	0.000	0.000
	2	0.980	0.889	0.558	0.391	0.253	0.083	0.019	0.003	0.000	0.000	0.000	0.000	0.000
	3	0.998	0.974	0.795	0.649	0.493	0.225	0.073	0.015	0.002	0.000	0.000	0.000	0.000
	4	1.000	0.996	0.927	0.842	0.724	0.438	0.194	0.057	0.009	0.003	0.001	0.000	0.000
	5	1.000	0.999	0.981	0.946	0.882	0.665	0.387	0.158	0.039	0.014	0.004	0.000	0.000
	6	1.000	1.000	0.996	0.986	0.961	0.842	0.613	0.335	0.118	0.054	0.019	0.001	0.000
	7	1.000	1.000	0.999	0.997	0.991	0.943	0.806	0.562	0.276	0.158	0.073	0.004	0.000
	8	1.000	1.000	1.000	1.000	0.998	0.985	0.927	0.775	0.507	0.351	0.205	0.026	0.002
	9	1.000	1.000	1.000	1.000	1.000	0.997	0.981	0.917	0.747	0.609	0.442	0.111	0.020
	10	1.000	1.000	1.000	1.000	1.000	1.000	0.997	0.980	0.915	0.842	0.725	0.341	0.118
	11	1.000	1.000	1.000	1.000	1.000	1.000	1.000	0.998	0.986	0.968	0.931	0.718	0.460
	12	1.000	1.000	1.000	1.000	1.000	1.000	1.000	1.000	1.000	1.000	1.000	1.000	1.000
13	0	0.513	0.254	0.055	0.024	0.010	0.001	0.000	0.000	0.000	0.000	0.000	0.000	0.000
	1	0.865	0.621	0.234	0.127	0.064	0.013	0.002	0.000	0.000	0.000	0.000	0.000	0.000
	2	0.975	0.866	0.502	0.333	0.202	0.058	0.011	0.001	0.000	0.000	0.000	0.000	0.000
	3	0.997	0.966	0.747	0.584	0.421	0.169	0.046	0.008	0.001	0.000	0.000	0.000	0.000
	4	1.000	0.994	0.901	0.794	0.654	0.353	0.133	0.032	0.004	0.001	0.000	0.000	0.000
	5	1.000	0.999	0.970	0.920	0.835	0.574	0.291	0.098	0.018	0.006	0.001	0.000	0.000
	6	1.000	1.000	0.993	0.976	0.938	0.771	0.500	0.229	0.062	0.024	0.007	0.000	0.000
	7	1.000	1.000	0.999	0.994	0.982	0.902	0.709	0.426	0.165	0.080	0.030	0.001	0.000
	8	1.000	1.000	1.000	0.999	0.996	0.968	0.867	0.647	0.346	0.206	0.099	0.006	0.000
	9	1.000	1.000	1.000	1.000	0.999	0.992	0.954	0.831	0.579	0.416	0.253	0.034	0.003
	10	1.000	1.000	1.000	1.000	1.000	0.999	0.989	0.942	0.798	0.667	0.498	0.134	0.025
	11	1.000	1.000	1.000	1.000	1.000	1.000	0.998	0.987	0.936	0.873	0.766	0.379	0.135
	12	1.000	1.000	1.000	1.000	1.000	1.000	1.000	0.999	0.990	0.976	0.945	0.746	0.487
	13	1.000	1.000	1.000	1.000	1.000	1.000	1.000	1.000	1.000	1.000	1.000	1.000	1.000
14	0	0.488	0.229	0.044	0.018	0.007	0.001	0.000	0.000	0.000	0.000	0.000	0.000	0.000
	1	0.847	0.585	0.198	0.101	0.047	0.008	0.001	0.000	0.000	0.000	0.000	0.000	0.000
	2	0.970	0.842	0.448	0.281	0.161	0.040	0.006	0.001	0.000	0.000	0.000	0.000	0.000
	3	0.996	0.956	0.698	0.521	0.355	0.124	0.029	0.004	0.000	0.000	0.000	0.000	0.000
	4	1.000	0.991	0.870	0.742	0.584	0.279	0.090	0.018	0.002	0.000	0.000	0.000	0.000
	5	1.000	0.999	0.956	0.888	0.781	0.486	0.212	0.058	0.008	0.002	0.000	0.000	0.000
	6	1.000	1.000	0.988	0.962	0.907	0.692	0.395	0.150	0.031	0.010	0.002	0.000	0.000
	7	1.000	1.000	0.998	0.990	0.969	0.850	0.605	0.308	0.093	0.038	0.012	0.000	0.000
	8	1.000	1.000	1.000	0.998	0.992	0.942	0.788	0.514	0.219	0.112	0.044	0.001	0.000
	9	1.000	1.000	1.000	1.000	0.998	0.982	0.910	0.721	0.416	0.258	0.130	0.009	0.000
	10	1.000	1.000	1.000	1.000	1.000	0.996	0.971	0.876	0.645	0.479	0.302	0.044	0.004
	11	1.000	1.000	1.000	1.000	1.000	0.999	0.994	0.960	0.839	0.719	0.552	0.158	0.030
	12	1.000	1.000	1.000	1.000	1.000	1.000	0.999	0.992	0.953	0.899	0.802	0.415	0.153
	13	1.000	1.000	1.000	1.000	1.000	1.000	1.000	0.999	0.993	0.982	0.956	0.771	0.512
	14	1.000	1.000	1.000	1.000	1.000	1.000	1.000	1.000	1.000	1.000	1.000	1.000	1.000

Continued on page 518

TABLE A.1 Cumulative binomial distribution (continued)

								p						
n	x	0.05	0.10	0.20	0.25	0.30	0.40	0.50	0.60	0.70	0.75	0.80	0.90	0.95
15	0	0.463	0.206	0.035	0.013	0.005	0.000	0.000	0.000	0.000	0.000	0.000	0.000	0.000
	1	0.829	0.549	0.167	0.080	0.035	0.005	0.000	0.000	0.000	0.000	0.000	0.000	0.000
	2	0.964	0.816	0.398	0.236	0.127	0.027	0.004	0.000	0.000	0.000	0.000	0.000	0.000
	3	0.995	0.944	0.648	0.461	0.297	0.091	0.018	0.002	0.000	0.000	0.000	0.000	0.000
	4	0.999	0.987	0.836	0.686	0.515	0.217	0.059	0.009	0.001	0.000	0.000	0.000	0.000
	5	1.000	0.998	0.939	0.852	0.722	0.403	0.151	0.034	0.004	0.001	0.000	0.000	0.000
	6	1.000	1.000	0.982	0.943	0.869	0.610	0.304	0.095	0.015	0.004	0.001	0.000	0.000
	7	1.000	1.000	0.996	0.983	0.950	0.787	0.500	0.213	0.050	0.017	0.004	0.000	0.000
	8	1.000	1.000	0.999	0.996	0.985	0.905	0.696	0.390	0.131	0.057	0.018	0.000	0.000
	9	1.000	1.000	1.000	0.999	0.996	0.966	0.849	0.597	0.278	0.148	0.061	0.002	0.000
	10	1.000	1.000	1.000	1.000	0.999	0.991	0.941	0.783	0.485	0.314	0.164	0.013	0.001
	11	1.000	1.000	1.000	1.000	1.000	0.998	0.982	0.909	0.703	0.539	0.352	0.056	0.005
	12	1.000	1.000	1.000	1.000	1.000	1.000	0.996	0.973	0.873	0.764	0.602	0.184	0.036
	13	1.000	1.000	1.000	1.000	1.000	1.000	1.000	0.995	0.965	0.920	0.833	0.451	0.171
	14	1.000	1.000	1.000	1.000	1.000	1.000	1.000	1.000	0.995	0.987	0.965	0.794	0.537
	15	1.000	1.000	1.000	1.000	1.000	1.000	1.000	1.000	1.000	1.000	1.000	1.000	1.000
16	0	0.440	0.185	0.028	0.010	0.003	0.000	0.000	0.000	0.000	0.000	0.000	0.000	0.000
	1	0.811	0.515	0.141	0.063	0.026	0.003	0.000	0.000	0.000	0.000	0.000	0.000	0.000
	2	0.957	0.789	0.352	0.197	0.099	0.018	0.002	0.000	0.000	0.000	0.000	0.000	0.000
	3	0.993	0.932	0.598	0.405	0.246	0.065	0.011	0.001	0.000	0.000	0.000	0.000	0.000
	4	0.999	0.983	0.798	0.630	0.450	0.167	0.038	0.005	0.000	0.000	0.000	0.000	0.000
	5	1.000	0.997	0.918	0.810	0.660	0.329	0.105	0.019	0.002	0.000	0.000	0.000	0.000
	6	1.000	0.999	0.973	0.920	0.825	0.527	0.227	0.058	0.007	0.002	0.000	0.000	0.000
	7	1.000	1.000	0.993	0.973	0.926	0.716	0.402	0.142	0.026	0.007	0.001	0.000	0.000
	8	1.000	1.000	0.999	0.993	0.974	0.858	0.598	0.284	0.074	0.027	0.007	0.000	0.000
	9	1.000	1.000	1.000	0.998	0.993	0.942	0.773	0.473	0.175	0.080	0.027	0.001	0.000
	10	1.000	1.000	1.000	1.000	0.998	0.981	0.895	0.671	0.340	0.190	0.082	0.003	0.000
	11	1.000	1.000	1.000	1.000	1.000	0.995	0.962	0.833	0.550	0.370	0.202	0.017	0.001
	12	1.000	1.000	1.000	1.000	1.000	0.999	0.989	0.935	0.754	0.595	0.402	0.068	0.007
	13	1.000	1.000	1.000	1.000	1.000	1.000	0.998	0.982	0.901	0.803	0.648	0.211	0.043
	14	1.000	1.000	1.000	1.000	1.000	1.000	1.000	0.997	0.974	0.937	0.859	0.485	0.189
	15	1.000	1.000	1.000	1.000	1.000	1.000	1.000	1.000	0.997	0.990	0.972	0.815	0.560
	16	1.000	1.000	1.000	1.000	1.000	1.000	1.000	1.000	1.000	1.000	1.000	1.000	1.000
17	0	0.418	0.167	0.023	0.008	0.002	0.000	0.000	0.000	0.000	0.000	0.000	0.000	0.000
	1	0.792	0.482	0.118	0.050	0.019	0.002	0.000	0.000	0.000	0.000	0.000	0.000	0.000
	2	0.950	0.762	0.310	0.164	0.077	0.012	0.001	0.000	0.000	0.000	0.000	0.000	0.000
	3	0.991	0.917	0.549	0.353	0.202	0.046	0.006	0.000	0.000	0.000	0.000	0.000	0.000
	4	0.999	0.978	0.758	0.574	0.389	0.126	0.025	0.003	0.000	0.000	0.000	0.000	0.000
	5	1.000	0.995	0.894	0.765	0.597	0.264	0.072	0.011	0.001	0.000	0.000	0.000	0.000
	6	1.000	0.999	0.962	0.893	0.775	0.448	0.166	0.035	0.003	0.001	0.000	0.000	0.000
	7	1.000	1.000	0.989	0.960	0.895	0.641	0.315	0.092	0.013	0.003	0.000	0.000	0.000
	8	1.000	1.000	0.997	0.988	0.960	0.801	0.500	0.199	0.040	0.012	0.003	0.000	0.000
	9	1.000	1.000	1.000	0.997	0.987	0.908	0.685	0.359	0.105	0.040	0.011	0.000	0.000

Continued on page 519

TABLE A.1 Cumulative binomial distribution (continued)

								p						
n	*x*	0.05	0.10	0.20	0.25	0.30	0.40	0.50	0.60	0.70	0.75	0.80	0.90	0.95
17	10	1.000	1.000	1.000	0.999	0.997	0.965	0.834	0.552	0.225	0.107	0.038	0.001	0.000
	11	1.000	1.000	1.000	1.000	0.999	0.989	0.928	0.736	0.403	0.235	0.106	0.005	0.000
	12	1.000	1.000	1.000	1.000	1.000	0.997	0.975	0.874	0.611	0.426	0.242	0.022	0.001
	13	1.000	1.000	1.000	1.000	1.000	1.000	0.994	0.954	0.798	0.647	0.451	0.083	0.009
	14	1.000	1.000	1.000	1.000	1.000	1.000	0.999	0.988	0.923	0.836	0.690	0.238	0.050
	15	1.000	1.000	1.000	1.000	1.000	1.000	1.000	0.998	0.981	0.950	0.882	0.518	0.208
	16	1.000	1.000	1.000	1.000	1.000	1.000	1.000	1.000	0.998	0.992	0.977	0.833	0.582
	17	1.000	1.000	1.000	1.000	1.000	1.000	1.000	1.000	1.000	1.000	1.000	1.000	1.000
18	0	0.397	0.150	0.018	0.006	0.002	0.000	0.000	0.000	0.000	0.000	0.000	0.000	0.000
	1	0.774	0.450	0.099	0.039	0.014	0.001	0.000	0.000	0.000	0.000	0.000	0.000	0.000
	2	0.942	0.734	0.271	0.135	0.060	0.008	0.001	0.000	0.000	0.000	0.000	0.000	0.000
	3	0.989	0.902	0.501	0.306	0.165	0.033	0.004	0.000	0.000	0.000	0.000	0.000	0.000
	4	0.998	0.972	0.716	0.519	0.333	0.094	0.015	0.001	0.000	0.000	0.000	0.000	0.000
	5	1.000	0.994	0.867	0.717	0.534	0.209	0.048	0.006	0.000	0.000	0.000	0.000	0.000
	6	1.000	0.999	0.949	0.861	0.722	0.374	0.119	0.020	0.001	0.000	0.000	0.000	0.000
	7	1.000	1.000	0.984	0.943	0.859	0.563	0.240	0.058	0.006	0.001	0.000	0.000	0.000
	8	1.000	1.000	0.996	0.981	0.940	0.737	0.407	0.135	0.021	0.005	0.001	0.000	0.000
	9	1.000	1.000	0.999	0.995	0.979	0.865	0.593	0.263	0.060	0.019	0.004	0.000	0.000
	10	1.000	1.000	1.000	0.999	0.994	0.942	0.760	0.437	0.141	0.057	0.016	0.000	0.000
	11	1.000	1.000	1.000	1.000	0.999	0.980	0.881	0.626	0.278	0.139	0.051	0.001	0.000
	12	1.000	1.000	1.000	1.000	1.000	0.994	0.952	0.791	0.466	0.283	0.133	0.006	0.000
	13	1.000	1.000	1.000	1.000	1.000	0.999	0.985	0.906	0.667	0.481	0.284	0.028	0.002
	14	1.000	1.000	1.000	1.000	1.000	1.000	0.996	0.967	0.835	0.694	0.499	0.098	0.011
	15	1.000	1.000	1.000	1.000	1.000	1.000	0.999	0.992	0.940	0.865	0.729	0.266	0.058
	16	1.000	1.000	1.000	1.000	1.000	1.000	1.000	0.999	0.986	0.961	0.901	0.550	0.226
	17	1.000	1.000	1.000	1.000	1.000	1.000	1.000	1.000	0.998	0.994	0.982	0.850	0.603
	18	1.000	1.000	1.000	1.000	1.000	1.000	1.000	1.000	1.000	1.000	1.000	1.000	1.000
19	0	0.377	0.135	0.014	0.004	0.001	0.000	0.000	0.000	0.000	0.000	0.000	0.000	0.000
	1	0.755	0.420	0.083	0.031	0.010	0.001	0.000	0.000	0.000	0.000	0.000	0.000	0.000
	2	0.933	0.705	0.237	0.111	0.046	0.005	0.000	0.000	0.000	0.000	0.000	0.000	0.000
	3	0.987	0.885	0.455	0.263	0.133	0.023	0.002	0.000	0.000	0.000	0.000	0.000	0.000
	4	0.998	0.965	0.673	0.465	0.282	0.070	0.010	0.001	0.000	0.000	0.000	0.000	0.000
	5	1.000	0.991	0.837	0.668	0.474	0.163	0.032	0.003	0.000	0.000	0.000	0.000	0.000
	6	1.000	0.998	0.932	0.825	0.666	0.308	0.084	0.012	0.001	0.000	0.000	0.000	0.000
	7	1.000	1.000	0.977	0.923	0.818	0.488	0.180	0.035	0.003	0.000	0.000	0.000	0.000
	8	1.000	1.000	0.993	0.971	0.916	0.667	0.324	0.088	0.011	0.002	0.000	0.000	0.000
	9	1.000	1.000	0.998	0.991	0.967	0.814	0.500	0.186	0.033	0.009	0.002	0.000	0.000
	10	1.000	1.000	1.000	0.998	0.989	0.912	0.676	0.333	0.084	0.029	0.007	0.000	0.000
	11	1.000	1.000	1.000	1.000	0.997	0.965	0.820	0.512	0.182	0.077	0.023	0.000	0.000
	12	1.000	1.000	1.000	1.000	0.999	0.988	0.916	0.692	0.334	0.175	0.068	0.002	0.000
	13	1.000	1.000	1.000	1.000	1.000	0.997	0.968	0.837	0.526	0.332	0.163	0.009	0.000
	14	1.000	1.000	1.000	1.000	1.000	0.999	0.990	0.930	0.718	0.535	0.327	0.035	0.002

Continued on page 520

TABLE A.1 Cumulative binomial distribution (continued)

								p						
n	*x*	0.05	0.10	0.20	0.25	0.30	0.40	0.50	0.60	0.70	0.75	0.80	0.90	0.95
19	15	1.000	1.000	1.000	1.000	1.000	1.000	0.998	0.977	0.867	0.737	0.545	0.115	0.013
	16	1.000	1.000	1.000	1.000	1.000	1.000	1.000	0.995	0.954	0.889	0.763	0.295	0.067
	17	1.000	1.000	1.000	1.000	1.000	1.000	1.000	0.999	0.990	0.969	0.917	0.580	0.245
	18	1.000	1.000	1.000	1.000	1.000	1.000	1.000	1.000	0.999	0.996	0.986	0.865	0.623
	19	1.000	1.000	1.000	1.000	1.000	1.000	1.000	1.000	1.000	1.000	1.000	1.000	1.000
20	0	0.358	0.122	0.012	0.003	0.001	0.000	0.000	0.000	0.000	0.000	0.000	0.000	0.000
	1	0.736	0.392	0.069	0.024	0.008	0.001	0.000	0.000	0.000	0.000	0.000	0.000	0.000
	2	0.925	0.677	0.206	0.091	0.035	0.004	0.000	0.000	0.000	0.000	0.000	0.000	0.000
	3	0.984	0.867	0.411	0.225	0.107	0.016	0.001	0.000	0.000	0.000	0.000	0.000	0.000
	4	0.997	0.957	0.630	0.415	0.238	0.051	0.006	0.000	0.000	0.000	0.000	0.000	0.000
	5	1.000	0.989	0.804	0.617	0.416	0.126	0.021	0.002	0.000	0.000	0.000	0.000	0.000
	6	1.000	0.998	0.913	0.786	0.608	0.250	0.058	0.006	0.000	0.000	0.000	0.000	0.000
	7	1.000	1.000	0.968	0.898	0.772	0.416	0.132	0.021	0.001	0.000	0.000	0.000	0.000
	8	1.000	1.000	0.990	0.959	0.887	0.596	0.252	0.057	0.005	0.001	0.000	0.000	0.000
	9	1.000	1.000	0.997	0.986	0.952	0.755	0.412	0.128	0.017	0.004	0.001	0.000	0.000
	10	1.000	1.000	0.999	0.996	0.983	0.872	0.588	0.245	0.048	0.014	0.003	0.000	0.000
	11	1.000	1.000	1.000	0.999	0.995	0.943	0.748	0.404	0.113	0.041	0.010	0.000	0.000
	12	1.000	1.000	1.000	1.000	0.999	0.979	0.868	0.584	0.228	0.102	0.032	0.000	0.000
	13	1.000	1.000	1.000	1.000	1.000	0.994	0.942	0.750	0.392	0.214	0.087	0.002	0.000
	14	1.000	1.000	1.000	1.000	1.000	0.998	0.979	0.874	0.584	0.383	0.196	0.011	0.000
	15	1.000	1.000	1.000	1.000	1.000	1.000	0.994	0.949	0.762	0.585	0.370	0.043	0.003
	16	1.000	1.000	1.000	1.000	1.000	1.000	0.999	0.984	0.893	0.775	0.589	0.133	0.016
	17	1.000	1.000	1.000	1.000	1.000	1.000	1.000	0.996	0.965	0.909	0.794	0.323	0.075
	18	1.000	1.000	1.000	1.000	1.000	1.000	1.000	0.999	0.992	0.976	0.931	0.608	0.264
	19	1.000	1.000	1.000	1.000	1.000	1.000	1.000	1.000	0.999	0.997	0.988	0.878	0.642
	20	1.000	1.000	1.000	1.000	1.000	1.000	1.000	1.000	1.000	1.000	1.000	1.000	1.000

TABLE A.2 Cumulative normal distribution

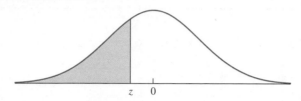

z	0.00	0.01	0.02	0.03	0.04	0.05	0.06	0.07	0.08	0.09
−3.6	.0002	.0002	.0001	.0001	.0001	.0001	.0001	.0001	.0001	.0001
−3.5	.0002	.0002	.0002	.0002	.0002	.0002	.0002	.0002	.0002	.0002
−3.4	.0003	.0003	.0003	.0003	.0003	.0003	.0003	.0003	.0003	.0002
−3.3	.0005	.0005	.0005	.0004	.0004	.0004	.0004	.0004	.0004	.0003
−3.2	.0007	.0007	.0006	.0006	.0006	.0006	.0006	.0005	.0005	.0005
−3.1	.0010	.0009	.0009	.0009	.0008	.0008	.0008	.0008	.0007	.0007
−3.0	.0013	.0013	.0013	.0012	.0012	.0011	.0011	.0011	.0010	.0010
−2.9	.0019	.0018	.0018	.0017	.0016	.0016	.0015	.0015	.0014	.0014
−2.8	.0026	.0025	.0024	.0023	.0023	.0022	.0021	.0021	.0020	.0019
−2.7	.0035	.0034	.0033	.0032	.0031	.0030	.0029	.0028	.0027	.0026
−2.6	.0047	.0045	.0044	.0043	.0041	.0040	.0039	.0038	.0037	.0036
−2.5	.0062	.0060	.0059	.0057	.0055	.0054	.0052	.0051	.0049	.0048
−2.4	.0082	.0080	.0078	.0075	.0073	.0071	.0069	.0068	.0066	.0064
−2.3	.0107	.0104	.0102	.0099	.0096	.0094	.0091	.0089	.0087	.0084
−2.2	.0139	.0136	.0132	.0129	.0125	.0122	.0119	.0116	.0113	.0110
−2.1	.0179	.0174	.0170	.0166	.0162	.0158	.0154	.0150	.0146	.0143
−2.0	.0228	.0222	.0217	.0212	.0207	.0202	.0197	.0192	.0188	.0183
−1.9	.0287	.0281	.0274	.0268	.0262	.0256	.0250	.0244	.0239	.0233
−1.8	.0359	.0351	.0344	.0336	.0329	.0322	.0314	.0307	.0301	.0294
−1.7	.0446	.0436	.0427	.0418	.0409	.0401	.0392	.0384	.0375	.0367
−1.6	.0548	.0537	.0526	.0516	.0505	.0495	.0485	.0475	.0465	.0455
−1.5	.0668	.0655	.0643	.0630	.0618	.0606	.0594	.0582	.0571	.0559
−1.4	.0808	.0793	.0778	.0764	.0749	.0735	.0721	.0708	.0694	.0681
−1.3	.0968	.0951	.0934	.0918	.0901	.0885	.0869	.0853	.0838	.0823
−1.2	.1151	.1131	.1112	.1093	.1075	.1056	.1038	.1020	.1003	.0985
−1.1	.1357	.1335	.1314	.1292	.1271	.1251	.1230	.1210	.1190	.1170
−1.0	.1587	.1562	.1539	.1515	.1492	.1469	.1446	.1423	.1401	.1379
−0.9	.1841	.1814	.1788	.1762	.1736	.1711	.1685	.1660	.1635	.1611
−0.8	.2119	.2090	.2061	.2033	.2005	.1977	.1949	.1922	.1894	.1867
−0.7	.2420	.2389	.2358	.2327	.2296	.2266	.2236	.2206	.2177	.2148
−0.6	.2743	.2709	.2676	.2643	.2611	.2578	.2546	.2514	.2483	.2451
−0.5	.3085	.3050	.3015	.2981	.2946	.2912	.2877	.2843	.2810	.2776
−0.4	.3446	.3409	.3372	.3336	.3300	.3264	.3228	.3192	.3156	.3121
−0.3	.3821	.3783	.3745	.3707	.3669	.3632	.3594	.3557	.3520	.3483
−0.2	.4207	.4168	.4129	.4090	.4052	.4013	.3974	.3936	.3897	.3859
−0.1	.4602	.4562	.4522	.4483	.4443	.4404	.4364	.4325	.4286	.4247
−0.0	.5000	.4960	.4920	.4880	.4840	.4801	.4761	.4721	.4681	.4641

Continued on page 522

TABLE A.2 Cumulative normal distribution (continued)

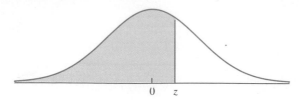

z	0.00	0.01	0.02	0.03	0.04	0.05	0.06	0.07	0.08	0.09
0.0	.5000	.5040	.5080	.5120	.5160	.5199	.5239	.5279	.5319	.5359
0.1	.5398	.5438	.5478	.5517	.5557	.5596	.5636	.5675	.5714	.5753
0.2	.5793	.5832	.5871	.5910	.5948	.5987	.6026	.6064	.6103	.6141
0.3	.6179	.6217	.6255	.6293	.6331	.6368	.6406	.6443	.6480	.6517
0.4	.6554	.6591	.6628	.6664	.6700	.6736	.6772	.6808	.6844	.6879
0.5	.6915	.6950	.6985	.7019	.7054	.7088	.7123	.7157	.7190	.7224
0.6	.7257	.7291	.7324	.7357	.7389	.7422	.7454	.7486	.7517	.7549
0.7	.7580	.7611	.7642	.7673	.7704	.7734	.7764	.7794	.7823	.7852
0.8	.7881	.7910	.7939	.7967	.7995	.8023	.8051	.8078	.8106	.8133
0.9	.8159	.8186	.8212	.8238	.8264	.8289	.8315	.8340	.8365	.8389
1.0	.8413	.8438	.8461	.8485	.8508	.8531	.8554	.8577	.8599	.8621
1.1	.8643	.8665	.8686	.8708	.8729	.8749	.8770	.8790	.8810	.8830
1.2	.8849	.8869	.8888	.8907	.8925	.8944	.8962	.8980	.8997	.9015
1.3	.9032	.9049	.9066	.9082	.9099	.9115	.9131	.9147	.9162	.9177
1.4	.9192	.9207	.9222	.9236	.9251	.9265	.9279	.9292	.9306	.9319
1.5	.9332	.9345	.9357	.9370	.9382	.9394	.9406	.9418	.9429	.9441
1.6	.9452	.9463	.9474	.9484	.9495	.9505	.9515	.9525	.9535	.9545
1.7	.9554	.9564	.9573	.9582	.9591	.9599	.9608	.9616	.9625	.9633
1.8	.9641	.9649	.9656	.9664	.9671	.9678	.9686	.9693	.9699	.9706
1.9	.9713	.9719	.9726	.9732	.9738	.9744	.9750	.9756	.9761	.9767
2.0	.9772	.9778	.9783	.9788	.9793	.9798	.9803	.9808	.9812	.9817
2.1	.9821	.9826	.9830	.9834	.9838	.9842	.9846	.9850	.9854	.9857
2.2	.9861	.9864	.9868	.9871	.9875	.9878	.9881	.9884	.9887	.9890
2.3	.9893	.9896	.9898	.9901	.9904	.9906	.9909	.9911	.9913	.9916
2.4	.9918	.9920	.9922	.9925	.9927	.9929	.9931	.9932	.9934	.9936
2.5	.9938	.9940	.9941	.9943	.9945	.9946	.9948	.9949	.9951	.9952
2.6	.9953	.9955	.9956	.9957	.9959	.9960	.9961	.9962	.9963	.9964
2.7	.9965	.9966	.9967	.9968	.9969	.9970	.9971	.9972	.9973	.9974
2.8	.9974	.9975	.9976	.9977	.9977	.9978	.9979	.9979	.9980	.9981
2.9	.9981	.9982	.9982	.9983	.9984	.9984	.9985	.9985	.9986	.9986
3.0	.9987	.9987	.9987	.9988	.9988	.9989	.9989	.9989	.9990	.9990
3.1	.9990	.9991	.9991	.9991	.9992	.9992	.9992	.9992	.9993	.9993
3.2	.9993	.9993	.9994	.9994	.9994	.9994	.9994	.9995	.9995	.9995
3.3	.9995	.9995	.9995	.9996	.9996	.9996	.9996	.9996	.9996	.9997
3.4	.9997	.9997	.9997	.9997	.9997	.9997	.9997	.9997	.9997	.9998
3.5	.9998	.9998	.9998	.9998	.9998	.9998	.9998	.9998	.9998	.9998
3.6	.9998	.9998	.9999	.9999	.9999	.9999	.9999	.9999	.9999	.9999

TABLE A.3 Upper percentage points for the Student's t distribution

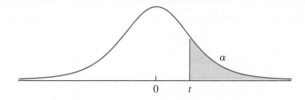

					α				
ν	.40	.25	.10	.05	.025	.01	.005	.001	.0005
1	0.325	1.000	3.078	6.314	12.706	31.821	63.657	318.309	636.619
2	0.289	0.816	1.886	2.920	4.303	6.965	9.925	22.327	31.599
3	0.277	0.765	1.638	2.353	3.182	4.541	5.841	10.215	12.924
4	0.271	0.741	1.533	2.132	2.776	3.747	4.604	7.173	8.610
5	0.267	0.727	1.476	2.015	2.571	3.365	4.032	5.893	6.869
6	0.265	0.718	1.440	1.943	2.447	3.143	3.707	5.208	5.959
7	0.263	0.711	1.415	1.895	2.365	2.998	3.499	4.785	5.408
8	0.262	0.706	1.397	1.860	2.306	2.896	3.355	4.501	5.041
9	0.261	0.703	1.383	1.833	2.262	2.821	3.250	4.297	4.781
10	0.260	0.700	1.372	1.812	2.228	2.764	3.169	4.144	4.587
11	0.260	0.697	1.363	1.796	2.201	2.718	3.106	4.025	4.437
12	0.259	0.695	1.356	1.782	2.179	2.681	3.055	3.930	4.318
13	0.259	0.694	1.350	1.771	2.160	2.650	3.012	3.852	4.221
14	0.258	0.692	1.345	1.761	2.145	2.624	2.977	3.787	4.140
15	0.258	0.691	1.341	1.753	2.131	2.602	2.947	3.733	4.073
16	0.258	0.690	1.337	1.746	2.120	2.583	2.921	3.686	4.015
17	0.257	0.689	1.333	1.740	2.110	2.567	2.898	3.646	3.965
18	0.257	0.688	1.330	1.734	2.101	2.552	2.878	3.610	3.922
19	0.257	0.688	1.328	1.729	2.093	2.539	2.861	3.579	3.883
20	0.257	0.687	1.325	1.725	2.086	2.528	2.845	3.552	3.850
21	0.257	0.686	1.323	1.721	2.080	2.518	2.831	3.527	3.819
22	0.256	0.686	1.321	1.717	2.074	2.508	2.819	3.505	3.792
23	0.256	0.685	1.319	1.714	2.069	2.500	2.807	3.485	3.768
24	0.256	0.685	1.318	1.711	2.064	2.492	2.797	3.467	3.745
25	0.256	0.684	1.316	1.708	2.060	2.485	2.787	3.450	3.725
26	0.256	0.684	1.315	1.706	2.056	2.479	2.779	3.435	3.707
27	0.256	0.684	1.314	1.703	2.052	2.473	2.771	3.421	3.690
28	0.256	0.683	1.313	1.701	2.048	2.467	2.763	3.408	3.674
29	0.256	0.683	1.311	1.699	2.045	2.462	2.756	3.396	3.659
30	0.256	0.683	1.310	1.697	2.042	2.457	2.750	3.385	3.646
35	0.255	0.682	1.306	1.690	2.030	2.438	2.724	3.340	3.591
40	0.255	0.681	1.303	1.684	2.021	2.423	2.704	3.307	3.551
60	0.254	0.679	1.296	1.671	2.000	2.390	2.660	3.232	3.460
120	0.254	0.677	1.289	1.658	1.980	2.358	2.617	3.160	3.373
∞	0.253	0.674	1.282	1.645	1.960	2.326	2.576	3.090	3.291

TABLE A.4 Tolerance factors for the normal distribution

Sample Size n	Confidence Level 95% Percent of Population Contained			Confidence Level 99% Percent of Population Contained		
	90%	95%	99%	90%	95%	99%
2	32.0187	37.6746	48.4296	160.1940	188.4915	242.3004
3	8.3795	9.9158	12.8613	18.9304	22.4009	29.0553
4	5.3692	6.3699	8.2993	9.3984	11.1501	14.5274
5	4.2749	5.0787	6.6338	6.6118	7.8550	10.2602
6	3.7123	4.4140	5.7746	5.3366	6.3453	8.3013
7	3.3686	4.0074	5.2481	4.6129	5.4877	7.1868
8	3.1358	3.7317	4.8907	4.1473	4.9355	6.4683
9	2.9670	3.5317	4.6310	3.8223	4.5499	5.9660
10	2.8385	3.3794	4.4330	3.5821	4.2647	5.5943
11	2.7372	3.2592	4.2766	3.3970	4.0449	5.3075
12	2.6550	3.1617	4.1496	3.2497	3.8700	5.0792
13	2.5868	3.0808	4.0441	3.1295	3.7271	4.8926
14	2.5292	3.0124	3.9549	3.0294	3.6081	4.7371
15	2.4799	2.9538	3.8785	2.9446	3.5073	4.6053
16	2.4371	2.9029	3.8121	2.8717	3.4207	4.4920
17	2.3995	2.8583	3.7538	2.8084	3.3453	4.3934
18	2.3662	2.8188	3.7022	2.7527	3.2792	4.3068
19	2.3366	2.7835	3.6560	2.7034	3.2205	4.2300
20	2.3099	2.7518	3.6146	2.6594	3.1681	4.1614
25	2.2083	2.6310	3.4565	2.4941	2.9715	3.9039
30	2.1398	2.5494	3.3497	2.3848	2.8414	3.7333
35	2.0899	2.4900	3.2719	2.3063	2.7479	3.6107
40	2.0516	2.4445	3.2122	2.2468	2.6770	3.5177
45	2.0212	2.4083	3.1647	2.1998	2.6211	3.4443
50	1.9964	2.3787	3.1259	2.1616	2.5756	3.3846
60	1.9578	2.3328	3.0657	2.1029	2.5057	3.2929
70	1.9291	2.2987	3.0208	2.0596	2.4541	3.2251
80	1.9068	2.2720	2.9859	2.0260	2.4141	3.1725
90	1.8887	2.2506	2.9577	1.9990	2.3819	3.1303
100	1.8738	2.2328	2.9343	1.9768	2.3555	3.0955
200	1.7981	2.1425	2.8158	1.8651	2.2224	2.9207
300	1.7670	2.1055	2.7671	1.8199	2.1685	2.8499
400	1.7492	2.0843	2.7392	1.7940	2.1377	2.8094
500	1.7373	2.0701	2.7206	1.7769	2.1173	2.7826
600	1.7287	2.0598	2.7071	1.7644	2.1024	2.7631
700	1.7220	2.0519	2.6967	1.7549	2.0911	2.7481
800	1.7167	2.0456	2.6884	1.7473	2.0820	2.7362
900	1.7124	2.0404	2.6816	1.7410	2.0746	2.7264
1000	1.7087	2.0361	2.6759	1.7358	2.0683	2.7182

TABLE A.5 Upper percentage points for the χ^2 distribution

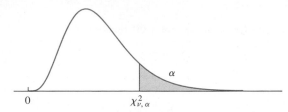

ν	.995	.99	.975	.95	.90	.10	.05	.025	.01	.005
1	0.000	0.000	0.001	0.004	0.016	2.706	3.841	5.024	6.635	7.879
2	0.010	0.020	0.051	0.103	0.211	4.605	5.991	7.378	9.210	10.597
3	0.072	0.115	0.216	0.352	0.584	6.251	7.815	9.348	11.345	12.838
4	0.207	0.297	0.484	0.711	1.064	7.779	9.488	11.143	13.277	14.860
5	0.412	0.554	0.831	1.145	1.610	9.236	11.070	12.833	15.086	16.750
6	0.676	0.872	1.237	1.635	2.204	10.645	12.592	14.449	16.812	18.548
7	0.989	1.239	1.690	2.167	2.833	12.017	14.067	16.013	18.475	20.278
8	1.344	1.646	2.180	2.733	3.490	13.362	15.507	17.535	20.090	21.955
9	1.735	2.088	2.700	3.325	4.168	14.684	16.919	19.023	21.666	23.589
10	2.156	2.558	3.247	3.940	4.865	15.987	18.307	20.483	23.209	25.188
11	2.603	3.053	3.816	4.575	5.578	17.275	19.675	21.920	24.725	26.757
12	3.074	3.571	4.404	5.226	6.304	18.549	21.026	23.337	26.217	28.300
13	3.565	4.107	5.009	5.892	7.042	19.812	22.362	24.736	27.688	29.819
14	4.075	4.660	5.629	6.571	7.790	21.064	23.685	26.119	29.141	31.319
15	4.601	5.229	6.262	7.261	8.547	22.307	24.996	27.488	30.578	32.801
16	5.142	5.812	6.908	7.962	9.312	23.542	26.296	28.845	32.000	34.267
17	5.697	6.408	7.564	8.672	10.085	24.769	27.587	30.191	33.409	35.718
18	6.265	7.015	8.231	9.390	10.865	25.989	28.869	31.526	34.805	37.156
19	6.844	7.633	8.907	10.117	11.651	27.204	30.144	32.852	36.191	38.582
20	7.434	8.260	9.591	10.851	12.443	28.412	31.410	34.170	37.566	39.997
21	8.034	8.897	10.283	11.591	13.240	29.615	32.671	35.479	38.932	41.401
22	8.643	9.542	10.982	12.338	14.041	30.813	33.924	36.781	40.289	42.796
23	9.260	10.196	11.689	13.091	14.848	32.007	35.172	38.076	41.638	44.181
24	9.886	10.856	12.401	13.848	15.659	33.196	36.415	39.364	42.980	45.559
25	10.520	11.524	13.120	14.611	16.473	34.382	37.652	40.646	44.314	46.928
26	11.160	12.198	13.844	15.379	17.292	35.563	38.885	41.923	45.642	48.290
27	11.808	12.879	14.573	16.151	18.114	36.741	40.113	43.195	46.963	49.645
28	12.461	13.565	15.308	16.928	18.939	37.916	41.337	44.461	48.278	50.993
29	13.121	14.256	16.047	17.708	19.768	39.087	42.557	45.722	49.588	52.336
30	13.787	14.953	16.791	18.493	20.599	40.256	43.773	46.979	50.892	53.672
31	14.458	15.655	17.539	19.281	21.434	41.422	44.985	48.232	52.191	55.003
32	15.134	16.362	18.291	20.072	22.271	42.585	46.194	49.480	53.486	56.328
33	15.815	17.074	19.047	20.867	23.110	43.745	47.400	50.725	54.776	57.648
34	16.501	17.789	19.806	21.664	23.952	44.903	48.602	51.966	56.061	58.964
35	17.192	18.509	20.569	22.465	24.797	46.059	49.802	53.203	57.342	60.275
36	17.887	19.233	21.336	23.269	25.643	47.212	50.998	54.437	58.619	61.581
37	18.586	19.960	22.106	24.075	26.492	48.363	52.192	55.668	59.893	62.883
38	19.289	20.691	22.878	24.884	27.343	49.513	53.384	56.896	61.162	64.181
39	19.996	21.426	23.654	25.695	28.196	50.660	54.572	58.120	62.428	65.476
40	20.707	22.164	24.433	26.509	29.051	51.805	55.758	59.342	63.691	66.766

For $\nu > 40$, $\chi^2_{\nu,\alpha} \approx 0.5(z_\alpha + \sqrt{2\nu - 1})^2$.

TABLE A.6 Upper percentage points for the F distribution

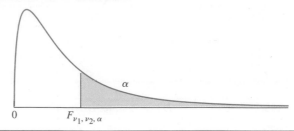

		ν_1								
ν_2	α	1	2	3	4	5	6	7	8	9
1	0.100	39.86	49.50	53.59	55.83	57.24	58.20	58.91	59.44	59.86
	0.050	161.45	199.50	215.71	224.58	230.16	233.99	236.77	238.88	240.54
	0.010	4052.18	4999.50	5403.35	5624.58	5763.65	5858.99	5928.36	5981.07	6022.47
	0.001	405284	500012	540382	562501	576405	585938	592874	598144	603040
2	0.100	8.53	9.00	9.16	9.24	9.29	9.33	9.35	9.37	9.38
	0.050	18.51	19.00	19.16	19.25	19.30	19.33	19.35	19.37	19.38
	0.010	98.50	99.00	99.17	99.25	99.30	99.33	99.36	99.37	99.39
	0.001	998.50	999.00	999.17	999.25	999.30	999.33	999.36	999.37	999.39
3	0.100	5.54	5.46	5.39	5.34	5.31	5.28	5.27	5.25	5.24
	0.050	10.13	9.55	9.28	9.12	9.01	8.94	8.89	8.85	8.81
	0.010	34.12	30.82	29.46	28.71	28.24	27.91	27.67	27.49	27.35
	0.001	167.03	148.50	141.11	137.10	134.58	132.85	131.58	130.62	129.86
4	0.100	4.54	4.32	4.19	4.11	4.05	4.01	3.98	3.95	3.94
	0.050	7.71	6.94	6.59	6.39	6.26	6.16	6.09	6.04	6.00
	0.010	21.20	18.00	16.69	15.98	15.52	15.21	14.98	14.80	14.66
	0.001	74.14	61.25	56.18	53.44	51.71	50.53	49.66	49.00	48.47
5	0.100	4.06	3.78	3.62	3.52	3.45	3.40	3.37	3.34	3.32
	0.050	6.61	5.79	5.41	5.19	5.05	4.95	4.88	4.82	4.77
	0.010	16.26	13.27	12.06	11.39	10.97	10.67	10.46	10.29	10.16
	0.001	47.18	37.12	33.20	31.09	29.75	28.83	28.16	27.65	27.24
6	0.100	3.78	3.46	3.29	3.18	3.11	3.05	3.01	2.98	2.96
	0.050	5.99	5.14	4.76	4.53	4.39	4.28	4.21	4.15	4.10
	0.010	13.75	10.92	9.78	9.15	8.75	8.47	8.26	8.10	7.98
	0.001	35.51	27.00	23.70	21.92	20.80	20.03	19.46	19.03	18.69
7	0.100	3.59	3.26	3.07	2.96	2.88	2.83	2.78	2.75	2.72
	0.050	5.59	4.74	4.35	4.12	3.97	3.87	3.79	3.73	3.68
	0.010	12.25	9.55	8.45	7.85	7.46	7.19	6.99	6.84	6.72
	0.001	29.25	21.69	18.77	17.20	16.21	15.52	15.02	14.63	14.33
8	0.100	3.46	3.11	2.92	2.81	2.73	2.67	2.62	2.59	2.56
	0.050	5.32	4.46	4.07	3.84	3.69	3.58	3.50	3.44	3.39
	0.010	11.26	8.65	7.59	7.01	6.63	6.37	6.18	6.03	5.91
	0.001	25.41	18.49	15.83	14.39	13.48	12.86	12.40	12.05	11.77
9	0.100	3.36	3.01	2.81	2.69	2.61	2.55	2.51	2.47	2.44
	0.050	5.12	4.26	3.86	3.63	3.48	3.37	3.29	3.23	3.18
	0.010	10.56	8.02	6.99	6.42	6.06	5.80	5.61	5.47	5.35
	0.001	22.86	16.39	13.90	12.56	11.71	11.13	10.70	10.37	10.11

Continued on page 527

TABLE A.6 Upper percentage points for the *F* distribution (continued)

ν_2	α	ν_1 10	12	15	20	25	30	40	50	60
1	0.100	60.19	60.71	61.22	61.74	62.05	62.26	62.53	62.69	62.79
	0.050	241.88	243.91	245.95	248.01	249.26	250.10	251.14	251.77	252.20
	0.010	6055.85	6106.32	6157.29	6208.73	6239.83	6260.65	6286.78	6302.52	6313.03
	0.001	606316	611276	616292	621362	624430	626486	659725	660511	6610390
2	0.100	9.39	9.41	9.42	9.44	9.45	9.46	9.47	9.47	9.47
	0.050	19.40	19.41	19.43	19.45	19.46	19.46	19.47	19.48	19.48
	0.010	99.40	99.42	99.43	99.45	99.46	99.47	99.47	99.48	99.48
	0.001	999.40	999.42	999.43	999.45	999.46	999.47	999.47	999.48	999.48
3	0.100	5.23	5.22	5.20	5.18	5.17	5.17	5.16	5.15	5.15
	0.050	8.79	8.74	8.70	8.66	8.63	8.62	8.59	8.58	8.57
	0.010	27.23	27.05	26.87	26.69	26.58	26.50	26.41	26.35	26.32
	0.001	129.25	128.32	127.37	126.42	125.84	125.45	124.96	124.66	124.47
4	0.100	3.92	3.90	3.87	3.84	3.83	3.82	3.80	3.80	3.79
	0.050	5.96	5.91	5.86	5.80	5.77	5.75	5.72	5.70	5.69
	0.010	14.55	14.37	14.20	14.02	13.91	13.84	13.75	13.69	13.65
	0.001	48.05	47.41	46.76	46.10	45.70	45.43	45.09	44.88	44.75
5	0.100	3.30	3.27	3.24	3.21	3.19	3.17	3.16	3.15	3.14
	0.050	4.74	4.68	4.62	4.56	4.52	4.50	4.46	4.44	4.43
	0.010	10.05	9.89	9.72	9.55	9.45	9.38	9.29	9.24	9.20
	0.001	26.92	26.42	25.91	25.39	25.08	24.87	24.60	24.44	24.33
6	0.100	2.94	2.90	2.87	2.84	2.81	2.80	2.78	2.77	2.76
	0.050	4.06	4.00	3.94	3.87	3.83	3.81	3.77	3.75	3.74
	0.010	7.87	7.72	7.56	7.40	7.30	7.23	7.14	7.09	7.06
	0.001	18.41	17.99	17.56	17.12	16.85	16.67	16.44	16.31	16.21
7	0.100	2.70	2.67	2.63	2.59	2.57	2.56	2.54	2.52	2.51
	0.050	3.64	3.57	3.51	3.44	3.40	3.38	3.34	3.32	3.30
	0.010	6.62	6.47	6.31	6.16	6.06	5.99	5.91	5.86	5.82
	0.001	14.08	13.71	13.32	12.93	12.69	12.53	12.33	12.20	12.12
8	0.100	2.54	2.50	2.46	2.42	2.40	2.38	2.36	2.35	2.34
	0.050	3.35	3.28	3.22	3.15	3.11	3.08	3.04	3.02	3.01
	0.010	5.81	5.67	5.52	5.36	5.26	5.20	5.12	5.07	5.03
	0.001	11.54	11.19	10.84	10.48	10.26	10.11	9.92	9.80	9.73
9	0.100	2.42	2.38	2.34	2.30	2.27	2.25	2.23	2.22	2.21
	0.050	3.14	3.07	3.01	2.94	2.89	2.86	2.83	2.80	2.79
	0.010	5.26	5.11	4.96	4.81	4.71	4.65	4.57	4.52	4.48
	0.001	9.89	9.57	9.24	8.90	8.69	8.55	8.37	8.26	8.19

Continued on page 528

TABLE A.6 Upper percentage points for the F distribution (continued)

ν_2	α	ν_1 1	2	3	4	5	6	7	8	9
10	0.100	3.29	2.92	2.73	2.61	2.52	2.46	2.41	2.38	2.35
	0.050	4.96	4.10	3.71	3.48	3.33	3.22	3.14	3.07	3.02
	0.010	10.04	7.56	6.55	5.99	5.64	5.39	5.20	5.06	4.94
	0.001	21.04	14.91	12.55	11.28	10.48	9.93	9.52	9.20	8.96
11	0.100	3.23	2.86	2.66	2.54	2.45	2.39	2.34	2.30	2.27
	0.050	4.84	3.98	3.59	3.36	3.20	3.09	3.01	2.95	2.90
	0.010	9.65	7.21	6.22	5.67	5.32	5.07	4.89	4.74	4.63
	0.001	19.69	13.81	11.56	10.35	9.58	9.05	8.66	8.35	8.12
12	0.100	3.18	2.81	2.61	2.48	2.39	2.33	2.28	2.24	2.21
	0.050	4.75	3.89	3.49	3.26	3.11	3.00	2.91	2.85	2.80
	0.010	9.33	6.93	5.95	5.41	5.06	4.82	4.64	4.50	4.39
	0.001	18.64	12.97	10.80	9.63	8.89	8.38	8.00	7.71	7.48
13	0.100	3.14	2.76	2.56	2.43	2.35	2.28	2.23	2.20	2.16
	0.050	4.67	3.81	3.41	3.18	3.03	2.92	2.83	2.77	2.71
	0.010	9.07	6.70	5.74	5.21	4.86	4.62	4.44	4.30	4.19
	0.001	17.82	12.31	10.21	9.07	8.35	7.86	7.49	7.21	6.98
14	0.100	3.10	2.73	2.52	2.39	2.31	2.24	2.19	2.15	2.12
	0.050	4.60	3.74	3.34	3.11	2.96	2.85	2.76	2.70	2.65
	0.010	8.86	6.51	5.56	5.04	4.69	4.46	4.28	4.14	4.03
	0.001	17.14	11.78	9.73	8.62	7.92	7.44	7.08	6.80	6.58
15	0.100	3.07	2.70	2.49	2.36	2.27	2.21	2.16	2.12	2.09
	0.050	4.54	3.68	3.29	3.06	2.90	2.79	2.71	2.64	2.59
	0.010	8.68	6.36	5.42	4.89	4.56	4.32	4.14	4.00	3.89
	0.001	16.59	11.34	9.34	8.25	7.57	7.09	6.74	6.47	6.26
16	0.100	3.05	2.67	2.46	2.33	2.24	2.18	2.13	2.09	2.06
	0.050	4.49	3.63	3.24	3.01	2.85	2.74	2.66	2.59	2.54
	0.010	8.53	6.23	5.29	4.77	4.44	4.20	4.03	3.89	3.78
	0.001	16.12	10.97	9.01	7.94	7.27	6.80	6.46	6.19	5.98
17	0.100	3.03	2.64	2.44	2.31	2.22	2.15	2.10	2.06	2.03
	0.050	4.45	3.59	3.20	2.96	2.81	2.70	2.61	2.55	2.49
	0.010	8.40	6.11	5.18	4.67	4.34	4.10	3.93	3.79	3.68
	0.001	15.72	10.66	8.73	7.68	7.02	6.56	6.22	5.96	5.75
18	0.100	3.01	2.62	2.42	2.29	2.20	2.13	2.08	2.04	2.00
	0.050	4.41	3.55	3.16	2.93	2.77	2.66	2.58	2.51	2.46
	0.010	8.29	6.01	5.09	4.58	4.25	4.01	3.84	3.71	3.60
	0.001	15.38	10.39	8.49	7.46	6.81	6.35	6.02	5.76	5.56
19	0.100	2.99	2.61	2.40	2.27	2.18	2.11	2.06	2.02	1.98
	0.050	4.38	3.52	3.13	2.90	2.74	2.63	2.54	2.48	2.42
	0.010	8.18	5.93	5.01	4.50	4.17	3.94	3.77	3.63	3.52
	0.001	15.08	10.16	8.28	7.27	6.62	6.18	5.85	5.59	5.39
20	0.100	2.97	2.59	2.38	2.25	2.16	2.09	2.04	2.00	1.96
	0.050	4.35	3.49	3.10	2.87	2.71	2.60	2.51	2.45	2.39
	0.010	8.10	5.85	4.94	4.43	4.10	3.87	3.70	3.56	3.46
	0.001	14.82	9.95	8.10	7.10	6.46	6.02	5.69	5.44	5.24

Continued on page 529

TABLE A.6 Upper percentage points for the F distribution (continued)

ν_2	α	ν_1								
		10	12	15	20	25	30	40	50	60
10	0.100	2.32	2.28	2.24	2.20	2.17	2.16	2.13	2.12	2.11
	0.050	2.98	2.91	2.85	2.77	2.73	2.70	2.66	2.64	2.62
	0.010	4.85	4.71	4.56	4.41	4.31	4.25	4.17	4.12	4.08
	0.001	8.75	8.45	8.13	7.80	7.60	7.47	7.30	7.19	7.12
11	0.100	2.25	2.21	2.17	2.12	2.10	2.08	2.05	2.04	2.03
	0.050	2.85	2.79	2.72	2.65	2.60	2.57	2.53	2.51	2.49
	0.010	4.54	4.40	4.25	4.10	4.01	3.94	3.86	3.81	3.78
	0.001	7.92	7.63	7.32	7.01	6.81	6.68	6.52	6.42	6.35
12	0.100	2.19	2.15	2.10	2.06	2.03	2.01	1.99	1.97	1.96
	0.050	2.75	2.69	2.62	2.54	2.50	2.47	2.43	2.40	2.38
	0.010	4.30	4.16	4.01	3.86	3.76	3.70	3.62	3.57	3.54
	0.001	7.29	7.00	6.71	6.40	6.22	6.09	5.93	5.83	5.76
13	0.100	2.14	2.10	2.05	2.01	1.98	1.96	1.93	1.92	1.90
	0.050	2.67	2.60	2.53	2.46	2.41	2.38	2.34	2.31	2.30
	0.010	4.10	3.96	3.82	3.66	3.57	3.51	3.43	3.38	3.34
	0.001	6.80	6.52	6.23	5.93	5.75	5.63	5.47	5.37	5.30
14	0.100	2.10	2.05	2.01	1.96	1.93	1.91	1.89	1.87	1.86
	0.050	2.60	2.53	2.46	2.39	2.34	2.31	2.27	2.24	2.22
	0.010	3.94	3.80	3.66	3.51	3.41	3.35	3.27	3.22	3.18
	0.001	6.40	6.13	5.85	5.56	5.38	5.25	5.10	5.00	4.94
15	0.100	2.06	2.02	1.97	1.92	1.89	1.87	1.85	1.83	1.82
	0.050	2.54	2.48	2.40	2.33	2.28	2.25	2.20	2.18	2.16
	0.010	3.80	3.67	3.52	3.37	3.28	3.21	3.13	3.08	3.05
	0.001	6.08	5.81	5.54	5.25	5.07	4.95	4.80	4.70	4.64
16	0.100	2.03	1.99	1.94	1.89	1.86	1.84	1.81	1.79	1.78
	0.050	2.49	2.42	2.35	2.28	2.23	2.19	2.15	2.12	2.11
	0.010	3.69	3.55	3.41	3.26	3.16	3.10	3.02	2.97	2.93
	0.001	5.81	5.55	5.27	4.99	4.82	4.70	4.54	4.45	4.39
17	0.100	2.00	1.96	1.91	1.86	1.83	1.81	1.78	1.76	1.75
	0.050	2.45	2.38	2.31	2.23	2.18	2.15	2.10	2.08	2.06
	0.010	3.59	3.46	3.31	3.16	3.07	3.00	2.92	2.87	2.83
	0.001	5.58	5.32	5.05	4.78	4.60	4.48	4.33	4.24	4.18
18	0.100	1.98	1.93	1.89	1.84	1.80	1.78	1.75	1.74	1.72
	0.050	2.41	2.34	2.27	2.19	2.14	2.11	2.06	2.04	2.02
	0.010	3.51	3.37	3.23	3.08	2.98	2.92	2.84	2.78	2.75
	0.001	5.39	5.13	4.87	4.59	4.42	4.30	4.15	4.06	4.00
19	0.100	1.96	1.91	1.86	1.81	1.78	1.76	1.73	1.71	1.70
	0.050	2.38	2.31	2.23	2.16	2.11	2.07	2.03	2.00	1.98
	0.010	3.43	3.30	3.15	3.00	2.91	2.84	2.76	2.71	2.67
	0.001	5.22	4.97	4.70	4.43	4.26	4.14	3.99	3.90	3.84
20	0.100	1.94	1.89	1.84	1.79	1.76	1.74	1.71	1.69	1.68
	0.050	2.35	2.28	2.20	2.12	2.07	2.04	1.99	1.97	1.95
	0.010	3.37	3.23	3.09	2.94	2.84	2.78	2.69	2.64	2.61
	0.001	5.08	4.82	4.56	4.29	4.12	4.00	3.86	3.77	3.70

Continued on page 530

TABLE A.6 Upper percentage points for the F distribution (continued)

ν_2	α	1	2	3	4	5	6	7	8	9
21	0.100	2.96	2.57	2.36	2.23	2.14	2.08	2.02	1.98	1.95
	0.050	4.32	3.47	3.07	2.84	2.68	2.57	2.49	2.42	2.37
	0.010	8.02	5.78	4.87	4.37	4.04	3.81	3.64	3.51	3.40
	0.001	14.59	9.77	7.94	6.95	6.32	5.88	5.56	5.31	5.11
22	0.100	2.95	2.56	2.35	2.22	2.13	2.06	2.01	1.97	1.93
	0.050	4.30	3.44	3.05	2.82	2.66	2.55	2.46	2.40	2.34
	0.010	7.95	5.72	4.82	4.31	3.99	3.76	3.59	3.45	3.35
	0.001	14.38	9.61	7.80	6.81	6.19	5.76	5.44	5.19	4.99
23	0.100	2.94	2.55	2.34	2.21	2.11	2.05	1.99	1.95	1.92
	0.050	4.28	3.42	3.03	2.80	2.64	2.53	2.44	2.37	2.32
	0.010	7.88	5.66	4.76	4.26	3.94	3.71	3.54	3.41	3.30
	0.001	14.20	9.47	7.67	6.70	6.08	5.65	5.33	5.09	4.89
24	0.100	2.93	2.54	2.33	2.19	2.10	2.04	1.98	1.94	1.91
	0.050	4.26	3.40	3.01	2.78	2.62	2.51	2.42	2.36	2.30
	0.010	7.82	5.61	4.72	4.22	3.90	3.67	3.50	3.36	3.26
	0.001	14.03	9.34	7.55	6.59	5.98	5.55	5.23	4.99	4.80
25	0.100	2.92	2.53	2.32	2.18	2.09	2.02	1.97	1.93	1.89
	0.050	4.24	3.39	2.99	2.76	2.60	2.49	2.40	2.34	2.28
	0.010	7.77	5.57	4.68	4.18	3.85	3.63	3.46	3.32	3.22
	0.001	13.88	9.22	7.45	6.49	5.89	5.46	5.15	4.91	4.71
26	0.100	2.91	2.52	2.31	2.17	2.08	2.01	1.96	1.92	1.88
	0.050	4.23	3.37	2.98	2.74	2.59	2.47	2.39	2.32	2.27
	0.010	7.72	5.53	4.64	4.14	3.82	3.59	3.42	3.29	3.18
	0.001	13.74	9.12	7.36	6.41	5.80	5.38	5.07	4.83	4.64
27	0.100	2.90	2.51	2.30	2.17	2.07	2.00	1.95	1.91	1.87
	0.050	4.21	3.35	2.96	2.73	2.57	2.46	2.37	2.31	2.25
	0.010	7.68	5.49	4.60	4.11	3.78	3.56	3.39	3.26	3.15
	0.001	13.61	9.02	7.27	6.33	5.73	5.31	5.00	4.76	4.57
28	0.100	2.89	2.50	2.29	2.16	2.06	2.00	1.94	1.90	1.87
	0.050	4.20	3.34	2.95	2.71	2.56	2.45	2.36	2.29	2.24
	0.010	7.64	5.45	4.57	4.07	3.75	3.53	3.36	3.23	3.12
	0.001	13.50	8.93	7.19	6.25	5.66	5.24	4.93	4.69	4.50
29	0.100	2.89	2.50	2.28	2.15	2.06	1.99	1.93	1.89	1.86
	0.050	4.18	3.33	2.93	2.70	2.55	2.43	2.35	2.28	2.22
	0.010	7.60	5.42	4.54	4.04	3.73	3.50	3.33	3.20	3.09
	0.001	13.39	8.85	7.12	6.19	5.59	5.18	4.87	4.64	4.45
30	0.100	2.88	2.49	2.28	2.14	2.05	1.98	1.93	1.88	1.85
	0.050	4.17	3.32	2.92	2.69	2.53	2.42	2.33	2.27	2.21
	0.010	7.56	5.39	4.51	4.02	3.70	3.47	3.30	3.17	3.07
	0.001	13.29	8.77	7.05	6.12	5.53	5.12	4.82	4.58	4.39
31	0.100	2.87	2.48	2.27	2.14	2.04	1.97	1.92	1.88	1.84
	0.050	4.16	3.30	2.91	2.68	2.52	2.41	2.32	2.25	2.20
	0.010	7.53	5.36	4.48	3.99	3.67	3.45	3.28	3.15	3.04
	0.001	13.20	8.70	6.99	6.07	5.48	5.07	4.77	4.53	4.34

Continued on page 531

TABLE A.6 Upper percentage points for the F distribution (continued)

ν_2	α	10	12	15	20	25	30	40	50	60
21	0.100	1.92	1.87	1.83	1.78	1.74	1.72	1.69	1.67	1.66
	0.050	2.32	2.25	2.18	2.10	2.05	2.01	1.96	1.94	1.92
	0.010	3.31	3.17	3.03	2.88	2.79	2.72	2.64	2.58	2.55
	0.001	4.95	4.70	4.44	4.17	4.00	3.88	3.74	3.64	3.58
22	0.100	1.90	1.86	1.81	1.76	1.73	1.70	1.67	1.65	1.64
	0.050	2.30	2.23	2.15	2.07	2.02	1.98	1.94	1.91	1.89
	0.010	3.26	3.12	2.98	2.83	2.73	2.67	2.58	2.53	2.50
	0.001	4.83	4.58	4.33	4.06	3.89	3.78	3.63	3.54	3.48
23	0.100	1.89	1.84	1.80	1.74	1.71	1.69	1.66	1.64	1.62
	0.050	2.27	2.20	2.13	2.05	2.00	1.96	1.91	1.88	1.86
	0.010	3.21	3.07	2.93	2.78	2.69	2.62	2.54	2.48	2.45
	0.001	4.73	4.48	4.23	3.96	3.79	3.68	3.53	3.44	3.38
24	0.100	1.88	1.83	1.78	1.73	1.70	1.67	1.64	1.62	1.61
	0.050	2.25	2.18	2.11	2.03	1.97	1.94	1.89	1.86	1.84
	0.010	3.17	3.03	2.89	2.74	2.64	2.58	2.49	2.44	2.40
	0.001	4.64	4.39	4.14	3.87	3.71	3.59	3.45	3.36	3.29
25	0.100	1.87	1.82	1.77	1.72	1.68	1.66	1.63	1.61	1.59
	0.050	2.24	2.16	2.09	2.01	1.96	1.92	1.87	1.84	1.82
	0.010	3.13	2.99	2.85	2.70	2.60	2.54	2.45	2.40	2.36
	0.001	4.56	4.31	4.06	3.79	3.63	3.52	3.37	3.28	3.22
26	0.100	1.86	1.81	1.76	1.71	1.67	1.65	1.61	1.59	1.58
	0.050	2.22	2.15	2.07	1.99	1.94	1.90	1.85	1.82	1.80
	0.010	3.09	2.96	2.81	2.66	2.57	2.50	2.42	2.36	2.33
	0.001	4.48	4.24	3.99	3.72	3.56	3.44	3.30	3.21	3.15
27	0.100	1.85	1.80	1.75	1.70	1.66	1.64	1.60	1.58	1.57
	0.050	2.20	2.13	2.06	1.97	1.92	1.88	1.84	1.81	1.79
	0.010	3.06	2.93	2.78	2.63	2.54	2.47	2.38	2.33	2.29
	0.001	4.41	4.17	3.92	3.66	3.49	3.38	3.23	3.14	3.08
28	0.100	1.84	1.79	1.74	1.69	1.65	1.63	1.59	1.57	1.56
	0.050	2.19	2.12	2.04	1.96	1.91	1.87	1.82	1.79	1.77
	0.010	3.03	2.90	2.75	2.60	2.51	2.44	2.35	2.30	2.26
	0.001	4.35	4.11	3.86	3.60	3.43	3.32	3.18	3.09	3.02
29	0.100	1.83	1.78	1.73	1.68	1.64	1.62	1.58	1.56	1.55
	0.050	2.18	2.10	2.03	1.94	1.89	1.85	1.81	1.77	1.75
	0.010	3.00	2.87	2.73	2.57	2.48	2.41	2.33	2.27	2.23
	0.001	4.29	4.05	3.80	3.54	3.38	3.27	3.12	3.03	2.97
30	0.100	1.82	1.77	1.72	1.67	1.63	1.61	1.57	1.55	1.54
	0.050	2.16	2.09	2.01	1.93	1.88	1.84	1.79	1.76	1.74
	0.010	2.98	2.84	2.70	2.55	2.45	2.39	2.30	2.25	2.21
	0.001	4.24	4.00	3.75	3.49	3.33	3.22	3.07	2.98	2.92
31	0.100	1.81	1.77	1.71	1.66	1.62	1.60	1.56	1.54	1.53
	0.050	2.15	2.08	2.00	1.92	1.87	1.83	1.78	1.75	1.73
	0.010	2.96	2.82	2.68	2.52	2.43	2.36	2.27	2.22	2.18
	0.001	4.19	3.95	3.71	3.45	3.28	3.17	3.03	2.94	2.87

Continued on page 532

		ν_1								
ν_2	α	1	2	3	4	5	6	7	8	9
32	0.100	2.87	2.48	2.26	2.13	2.04	1.97	1.91	1.87	1.83
	0.050	4.15	3.29	2.90	2.67	2.51	2.40	2.31	2.24	2.19
	0.010	7.50	5.34	4.46	3.97	3.65	3.43	3.26	3.13	3.02
	0.001	13.12	8.64	6.94	6.01	5.43	5.02	4.72	4.48	4.30
33	0.100	2.86	2.47	2.26	2.12	2.03	1.96	1.91	1.86	1.83
	0.050	4.14	3.28	2.89	2.66	2.50	2.39	2.30	2.23	2.18
	0.010	7.47	5.31	4.44	3.95	3.63	3.41	3.24	3.11	3.00
	0.001	13.04	8.58	6.88	5.97	5.38	4.98	4.67	4.44	4.26
34	0.100	2.86	2.47	2.25	2.12	2.02	1.96	1.90	1.86	1.82
	0.050	4.13	3.28	2.88	2.65	2.49	2.38	2.29	2.23	2.17
	0.010	7.44	5.29	4.42	3.93	3.61	3.39	3.22	3.09	2.98
	0.001	12.97	8.52	6.83	5.92	5.34	4.93	4.63	4.40	4.22
35	0.100	2.85	2.46	2.25	2.11	2.02	1.95	1.90	1.85	1.82
	0.050	4.12	3.27	2.87	2.64	2.49	2.37	2.29	2.22	2.16
	0.010	7.42	5.27	4.40	3.91	3.59	3.37	3.20	3.07	2.96
	0.001	12.90	8.47	6.79	5.88	5.30	4.89	4.59	4.36	4.18
36	0.100	2.85	2.46	2.24	2.11	2.01	1.94	1.89	1.85	1.81
	0.050	4.11	3.26	2.87	2.63	2.48	2.36	2.28	2.21	2.15
	0.010	7.40	5.25	4.38	3.89	3.57	3.35	3.18	3.05	2.95
	0.001	12.83	8.42	6.74	5.84	5.26	4.86	4.56	4.33	4.14
37	0.100	2.85	2.45	2.24	2.10	2.01	1.94	1.89	1.84	1.81
	0.050	4.11	3.25	2.86	2.63	2.47	2.36	2.27	2.20	2.14
	0.010	7.37	5.23	4.36	3.87	3.56	3.33	3.17	3.04	2.93
	0.001	12.77	8.37	6.70	5.80	5.22	4.82	4.53	4.30	4.11
38	0.100	2.84	2.45	2.23	2.10	2.01	1.94	1.88	1.84	1.80
	0.050	4.10	3.24	2.85	2.62	2.46	2.35	2.26	2.19	2.14
	0.010	7.35	5.21	4.34	3.86	3.54	3.32	3.15	3.02	2.92
	0.001	12.71	8.33	6.66	5.76	5.19	4.79	4.49	4.26	4.08
39	0.100	2.84	2.44	2.23	2.09	2.00	1.93	1.88	1.83	1.80
	0.050	4.09	3.24	2.85	2.61	2.46	2.34	2.26	2.19	2.13
	0.010	7.33	5.19	4.33	3.84	3.53	3.30	3.14	3.01	2.90
	0.001	12.66	8.29	6.63	5.73	5.16	4.76	4.46	4.23	4.05
40	0.100	2.84	2.44	2.23	2.09	2.00	1.93	1.87	1.83	1.79
	0.050	4.08	3.23	2.84	2.61	2.45	2.34	2.25	2.18	2.12
	0.010	7.31	5.18	4.31	3.83	3.51	3.29	3.12	2.99	2.89
	0.001	12.61	8.25	6.59	5.70	5.13	4.73	4.44	4.21	4.02
50	0.100	2.81	2.41	2.20	2.06	1.97	1.90	1.84	1.80	1.76
	0.050	4.03	3.18	2.79	2.56	2.40	2.29	2.20	2.13	2.07
	0.010	7.17	5.06	4.20	3.72	3.41	3.19	3.02	2.89	2.78
	0.001	12.22	7.96	6.34	5.46	4.90	4.51	4.22	4.00	3.82
60	0.100	2.79	2.39	2.18	2.04	1.95	1.87	1.82	1.77	1.74
	0.050	4.00	3.15	2.76	2.53	2.37	2.25	2.17	2.10	2.04
	0.010	7.08	4.98	4.13	3.65	3.34	3.12	2.95	2.82	2.72
	0.001	11.97	7.77	6.17	5.31	4.76	4.37	4.09	3.86	3.69
120	0.100	2.75	2.35	2.13	1.99	1.90	1.82	1.77	1.72	1.68
	0.050	3.92	3.07	2.68	2.45	2.29	2.18	2.09	2.02	1.96
	0.010	6.85	4.79	3.95	3.48	3.17	2.96	2.79	2.66	2.56
	0.001	11.38	7.32	5.78	4.95	4.42	4.04	3.77	3.55	3.38

Continued on page 533

ν_2	α	ν_1								
		10	12	15	20	25	30	40	50	60
32	0.100	1.81	1.76	1.71	1.65	1.62	1.59	1.56	1.53	1.52
	0.050	2.14	2.07	1.99	1.91	1.85	1.82	1.77	1.74	1.71
	0.010	2.93	2.80	2.65	2.50	2.41	2.34	2.25	2.20	2.16
	0.001	4.14	3.91	3.66	3.40	3.24	3.13	2.98	2.89	2.83
33	0.100	1.80	1.75	1.70	1.64	1.61	1.58	1.55	1.53	1.51
	0.050	2.13	2.06	1.98	1.90	1.84	1.81	1.76	1.72	1.70
	0.010	2.91	2.78	2.63	2.48	2.39	2.32	2.23	2.18	2.14
	0.001	4.10	3.87	3.62	3.36	3.20	3.09	2.94	2.85	2.79
34	0.100	1.79	1.75	1.69	1.64	1.60	1.58	1.54	1.52	1.50
	0.050	2.12	2.05	1.97	1.89	1.83	1.80	1.75	1.71	1.69
	0.010	2.89	2.76	2.61	2.46	2.37	2.30	2.21	2.16	2.12
	0.001	4.06	3.83	3.58	3.33	3.16	3.05	2.91	2.82	2.75
35	0.100	1.79	1.74	1.69	1.63	1.60	1.57	1.53	1.51	1.50
	0.050	2.11	2.04	1.96	1.88	1.82	1.79	1.74	1.70	1.68
	0.010	2.88	2.74	2.60	2.44	2.35	2.28	2.19	2.14	2.10
	0.001	4.03	3.79	3.55	3.29	3.13	3.02	2.87	2.78	2.72
36	0.100	1.78	1.73	1.68	1.63	1.59	1.56	1.53	1.51	1.49
	0.050	2.11	2.03	1.95	1.87	1.81	1.78	1.73	1.69	1.67
	0.010	2.86	2.72	2.58	2.43	2.33	2.26	2.18	2.12	2.08
	0.001	3.99	3.76	3.51	3.26	3.10	2.98	2.84	2.75	2.69
37	0.100	1.78	1.73	1.68	1.62	1.58	1.56	1.52	1.50	1.48
	0.050	2.10	2.02	1.95	1.86	1.81	1.77	1.72	1.68	1.66
	0.010	2.84	2.71	2.56	2.41	2.31	2.25	2.16	2.10	2.06
	0.001	3.96	3.73	3.48	3.23	3.07	2.95	2.81	2.72	2.66
38	0.100	1.77	1.72	1.67	1.61	1.58	1.55	1.52	1.49	1.48
	0.050	2.09	2.02	1.94	1.85	1.80	1.76	1.71	1.68	1.65
	0.010	2.83	2.69	2.55	2.40	2.30	2.23	2.14	2.09	2.05
	0.001	3.93	3.70	3.45	3.20	3.04	2.92	2.78	2.69	2.63
39	0.100	1.77	1.72	1.67	1.61	1.57	1.55	1.51	1.49	1.47
	0.050	2.08	2.01	1.93	1.85	1.79	1.75	1.70	1.67	1.65
	0.010	2.81	2.68	2.54	2.38	2.29	2.22	2.13	2.07	2.03
	0.001	3.90	3.67	3.43	3.17	3.01	2.90	2.75	2.66	2.60
40	0.100	1.76	1.71	1.66	1.61	1.57	1.54	1.51	1.48	1.47
	0.050	2.08	2.00	1.92	1.84	1.78	1.74	1.69	1.66	1.64
	0.010	2.80	2.66	2.52	2.37	2.27	2.20	2.11	2.06	2.02
	0.001	3.87	3.64	3.40	3.14	2.98	2.87	2.73	2.64	2.57
50	0.100	1.73	1.68	1.63	1.57	1.53	1.50	1.46	1.44	1.42
	0.050	2.03	1.95	1.87	1.78	1.73	1.69	1.63	1.60	1.58
	0.010	2.70	2.56	2.42	2.27	2.17	2.10	2.01	1.95	1.91
	0.001	3.67	3.44	3.20	2.95	2.79	2.68	2.53	2.44	2.38
60	0.100	1.71	1.66	1.60	1.54	1.50	1.48	1.44	1.41	1.40
	0.050	1.99	1.92	1.84	1.75	1.69	1.65	1.59	1.56	1.53
	0.010	2.63	2.50	2.35	2.20	2.10	2.03	1.94	1.88	1.84
	0.001	3.54	3.32	3.08	2.83	2.67	2.55	2.41	2.32	2.25
120	0.100	1.65	1.60	1.55	1.48	1.44	1.41	1.37	1.34	1.32
	0.050	1.91	1.83	1.75	1.66	1.60	1.55	1.50	1.46	1.43
	0.010	2.47	2.34	2.19	2.03	1.93	1.86	1.76	1.70	1.66
	0.001	3.24	3.02	2.78	2.53	2.37	2.26	2.11	2.02	1.95

TABLE A.7 Upper percentage points for the Studentized range q_{v_1, v_2}

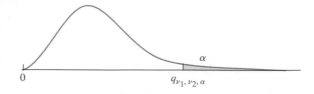

v_2	α	2	3	4	5	6	7	8	9	10	11	12	13	14	15
1	0.10	8.93	13.44	16.36	18.49	20.15	21.51	22.64	23.62	24.48	25.24	25.92	26.54	27.10	27.62
	0.05	17.97	26.98	32.82	37.08	40.41	43.12	45.40	47.36	49.07	50.59	51.96	53.20	54.33	55.36
	0.01	90.03	135.0	164.3	185.6	202.2	215.8	227.2	237.0	245.6	253.2	260.0	266.2	271.8	277.0
2	0.10	4.13	5.73	6.77	7.54	8.14	8.63	9.05	9.41	9.72	10.01	10.26	10.49	10.70	10.89
	0.05	6.08	8.33	9.80	10.88	11.74	12.44	13.03	13.54	13.99	14.39	14.75	15.08	15.38	15.65
	0.01	14.04	19.02	22.29	24.72	26.63	28.20	29.53	30.68	31.69	32.59	33.40	34.13	34.81	35.43
3	0.10	3.33	4.47	5.20	5.74	6.16	6.51	6.81	7.06	7.29	7.49	7.67	7.83	7.98	8.12
	0.05	4.50	5.91	6.82	7.50	8.04	8.48	8.85	9.18	9.46	9.72	9.95	10.15	10.35	10.52
	0.01	8.26	10.62	12.17	13.33	14.24	15.00	15.64	16.20	16.69	17.13	17.53	17.89	18.22	18.52
4	0.10	3.01	3.98	4.59	5.03	5.39	5.68	5.93	6.14	6.33	6.49	6.65	6.78	6.91	7.02
	0.05	3.93	5.04	5.76	6.29	6.71	7.05	7.35	7.60	7.83	8.03	8.21	8.37	8.52	8.66
	0.01	6.51	8.12	9.17	9.96	10.58	11.10	11.55	11.93	12.27	12.57	12.84	13.09	13.32	13.53
5	0.10	2.85	3.72	4.26	4.66	4.98	5.24	5.46	5.65	5.82	5.97	6.10	6.22	6.34	6.44
	0.05	3.64	4.60	5.22	5.67	6.03	6.33	6.58	6.80	6.99	7.17	7.32	7.47	7.60	7.72
	0.01	5.70	6.98	7.80	8.42	8.91	9.32	9.67	9.97	10.24	10.48	10.70	10.89	11.08	11.24
6	0.10	2.75	3.56	4.07	4.44	4.73	4.97	5.17	5.34	5.50	5.64	5.76	5.87	5.98	6.07
	0.05	3.46	4.34	4.90	5.30	5.63	5.90	6.12	6.32	6.49	6.65	6.79	6.92	7.03	7.14
	0.01	5.24	6.33	7.03	7.56	7.97	8.32	8.61	8.87	9.10	9.30	9.48	9.65	9.81	9.95
7	0.10	2.68	3.45	3.93	4.28	4.55	4.78	4.97	5.14	5.28	5.41	5.53	5.64	5.74	5.83
	0.05	3.34	4.16	4.68	5.06	5.36	5.61	5.82	6.00	6.16	6.30	6.43	6.55	6.66	6.76
	0.01	4.95	5.92	6.54	7.01	7.37	7.68	7.94	8.17	8.37	8.55	8.71	8.86	9.00	9.12
8	0.10	2.63	3.37	3.83	4.17	4.43	4.65	4.83	4.99	5.13	5.25	5.36	5.46	5.56	5.64
	0.05	3.26	4.04	4.53	4.89	5.17	5.40	5.60	5.77	5.92	6.05	6.18	6.29	6.39	6.48
	0.01	4.75	5.64	6.20	6.62	6.96	7.24	7.47	7.68	7.86	8.03	8.18	8.31	8.44	8.55
9	0.10	2.59	3.32	3.76	4.08	4.34	4.54	4.72	4.87	5.01	5.13	5.23	5.33	5.42	5.51
	0.05	3.20	3.95	4.41	4.76	5.02	5.24	5.43	5.59	5.74	5.87	5.98	6.09	6.19	6.28
	0.01	4.60	5.43	5.96	6.35	6.66	6.91	7.13	7.33	7.49	7.65	7.78	7.91	8.03	8.13
10	0.10	2.56	3.27	3.70	4.02	4.26	4.47	4.64	4.78	4.91	5.03	5.13	5.23	5.32	5.40
	0.05	3.15	3.88	4.33	4.65	4.91	5.12	5.30	5.46	5.60	5.72	5.83	5.93	6.03	6.11
	0.01	4.48	5.27	5.77	6.14	6.43	6.67	6.87	7.05	7.21	7.36	7.49	7.60	7.71	7.81
11	0.10	2.54	3.23	3.66	3.96	4.20	4.40	4.57	4.71	4.84	4.95	5.05	5.15	5.23	5.31
	0.05	3.11	3.82	4.26	4.57	4.82	5.03	5.20	5.35	5.49	5.61	5.71	5.81	5.90	5.98
	0.01	4.39	5.15	5.62	5.97	6.25	6.48	6.67	6.84	6.99	7.13	7.25	7.36	7.46	7.56
12	0.10	2.52	3.20	3.62	3.92	4.16	4.35	4.51	4.65	4.78	4.89	4.99	5.08	5.16	5.24
	0.05	3.08	3.77	4.20	4.51	4.75	4.95	5.12	5.27	5.39	5.51	5.61	5.71	5.80	5.88
	0.01	4.32	5.05	5.50	5.84	6.10	6.32	6.51	6.67	6.81	6.94	7.06	7.17	7.26	7.36

Continued on page 535

TABLE A.7 Upper percentage points for the Studentized range q_{ν_1, ν_2} (continued)

ν_2	α	2	3	4	5	6	7	8	9	10	11	12	13	14	15
								ν_1							
13	0.10	2.50	3.18	3.59	3.88	4.12	4.30	4.46	4.60	4.72	4.83	4.93	5.02	5.10	5.18
	0.05	3.06	3.73	4.15	4.45	4.69	4.88	5.05	5.19	5.32	5.43	5.53	5.63	5.71	5.79
	0.01	4.26	4.96	5.40	5.73	5.98	6.19	6.37	6.53	6.67	6.79	6.90	7.01	7.10	7.19
14	0.10	2.49	3.16	3.56	3.85	4.08	4.27	4.42	4.56	4.68	4.79	4.88	4.97	5.05	5.12
	0.05	3.03	3.70	4.11	4.41	4.64	4.83	4.99	5.13	5.25	5.36	5.46	5.55	5.64	5.71
	0.01	4.21	4.89	5.32	5.63	5.88	6.08	6.26	6.41	6.54	6.66	6.77	6.87	6.96	7.05
15	0.10	2.48	3.14	3.54	3.83	4.05	4.23	4.39	4.52	4.64	4.75	4.84	4.93	5.01	5.08
	0.05	3.01	3.67	4.08	4.37	4.59	4.78	4.94	5.08	5.20	5.31	5.40	5.49	5.57	5.65
	0.01	4.17	4.84	5.25	5.56	5.80	5.99	6.16	6.31	6.44	6.55	6.66	6.76	6.84	6.93
16	0.10	2.47	3.12	3.52	3.80	4.03	4.21	4.36	4.49	4.61	4.71	4.81	4.89	4.97	5.04
	0.05	3.00	3.65	4.05	4.33	4.56	4.74	4.90	5.03	5.15	5.26	5.35	5.44	5.52	5.59
	0.01	4.13	4.79	5.19	5.49	5.72	5.92	6.08	6.22	6.35	6.46	6.56	6.66	6.74	6.82
17	0.10	2.46	3.11	3.50	3.78	4.00	4.18	4.33	4.46	4.58	4.68	4.77	4.86	4.93	5.01
	0.05	2.98	3.63	4.02	4.30	4.52	4.70	4.86	4.99	5.11	5.21	5.31	5.39	5.47	5.54
	0.01	4.10	4.74	5.14	5.43	5.66	5.85	6.01	6.15	6.27	6.38	6.48	6.57	6.66	6.73
18	0.10	2.45	3.10	3.49	3.77	3.98	4.16	4.31	4.44	4.55	4.65	4.75	4.83	4.90	4.98
	0.05	2.97	3.61	4.00	4.28	4.49	4.67	4.82	4.96	5.07	5.17	5.27	5.35	5.43	5.50
	0.01	4.07	4.70	5.09	5.38	5.60	5.79	5.94	6.08	6.20	6.31	6.41	6.50	6.58	6.65
19	0.10	2.45	3.09	3.47	3.75	3.97	4.14	4.29	4.42	4.53	4.63	4.72	4.80	4.88	4.95
	0.05	2.96	3.59	3.98	4.25	4.47	4.65	4.79	4.92	5.04	5.14	5.23	5.31	5.39	5.46
	0.01	4.05	4.67	5.05	5.33	5.55	5.73	5.89	6.02	6.14	6.25	6.34	6.43	6.51	6.58
20	0.10	2.44	3.08	3.46	3.74	3.95	4.12	4.27	4.40	4.51	4.61	4.70	4.78	4.85	4.92
	0.05	2.95	3.58	3.96	4.23	4.45	4.62	4.77	4.90	5.01	5.11	5.20	5.28	5.36	5.43
	0.01	4.02	4.64	5.02	5.29	5.51	5.69	5.84	5.97	6.09	6.19	6.28	6.37	6.45	6.52
24	0.10	2.42	3.05	3.42	3.69	3.90	4.07	4.21	4.34	4.44	4.54	4.63	4.71	4.78	4.85
	0.05	2.92	3.53	3.90	4.17	4.37	4.54	4.68	4.81	4.92	5.01	5.10	5.18	5.25	5.32
	0.01	3.96	4.55	4.91	5.17	5.37	5.54	5.69	5.81	5.92	6.02	6.11	6.19	6.26	6.33
30	0.10	2.40	3.02	3.39	3.65	3.85	4.02	4.16	4.28	4.38	4.47	4.56	4.64	4.71	4.77
	0.05	2.89	3.49	3.85	4.10	4.30	4.46	4.60	4.72	4.82	4.92	5.00	5.08	5.15	5.21
	0.01	3.89	4.45	4.80	5.05	5.24	5.40	5.54	5.65	5.76	5.85	5.93	6.01	6.08	6.14
40	0.10	2.38	2.99	3.35	3.60	3.80	3.96	4.10	4.21	4.32	4.41	4.49	4.56	4.63	4.69
	0.05	2.86	3.44	3.79	4.04	4.23	4.39	4.52	4.63	4.73	4.82	4.90	4.98	5.04	5.11
	0.01	3.82	4.37	4.70	4.93	5.11	5.26	5.39	5.50	5.60	5.69	5.76	5.83	5.90	5.96
60	0.10	2.36	2.96	3.31	3.56	3.75	3.91	4.04	4.16	4.25	4.34	4.42	4.49	4.56	4.62
	0.05	2.83	3.40	3.74	3.98	4.16	4.31	4.44	4.55	4.65	4.73	4.81	4.88	4.94	5.00
	0.01	3.76	4.28	4.59	4.82	4.99	5.13	5.25	5.36	5.45	5.53	5.60	5.67	5.73	5.78
120	0.10	2.34	2.93	3.28	3.52	3.71	3.86	3.99	4.10	4.19	4.28	4.35	4.42	4.48	4.54
	0.05	2.80	3.36	3.68	3.92	4.10	4.24	4.36	4.47	4.56	4.64	4.71	4.78	4.84	4.90
	0.01	3.70	4.20	4.50	4.71	4.87	5.01	5.12	5.21	5.30	5.37	5.44	5.50	5.56	5.61
∞	0.10	2.33	2.90	3.24	3.48	3.66	3.81	3.93	4.04	4.13	4.21	4.28	4.35	4.41	4.47
	0.05	2.77	3.31	3.63	3.86	4.03	4.17	4.29	4.39	4.47	4.55	4.62	4.68	4.74	4.80
	0.01	3.64	4.12	4.40	4.60	4.76	4.88	4.99	5.08	5.16	5.23	5.29	5.35	5.40	5.45

TABLE A.8 Control chart constants

Subgroup Size n	A_2	A_3	B_3	B_4	D_3	D_4	c_4	d_2
2	1.880	2.659	0.000	3.267	0.000	3.267	0.7979	1.128
3	1.023	1.954	0.000	2.568	0.000	2.575	0.8862	1.693
4	0.729	1.628	0.000	2.266	0.000	2.282	0.9213	2.059
5	0.577	1.427	0.000	2.089	0.000	2.114	0.9400	2.326
6	0.483	1.287	0.030	1.970	0.000	2.004	0.9515	2.534
7	0.419	1.182	0.118	1.882	0.076	1.924	0.9594	2.704
8	0.373	1.099	0.185	1.815	0.136	1.864	0.9650	2.847
9	0.337	1.032	0.239	1.761	0.184	1.816	0.9693	2.970
10	0.308	0.975	0.284	1.716	0.223	1.777	0.9727	3.078
11	0.285	0.927	0.321	1.679	0.256	1.744	0.9754	3.173
12	0.266	0.866	0.354	1.646	0.283	1.717	0.9776	3.258
13	0.249	0.850	0.382	1.618	0.307	1.693	0.9794	3.336
14	0.235	0.817	0.406	1.594	0.328	1.672	0.9810	3.407
15	0.223	0.789	0.428	1.572	0.347	1.653	0.9823	3.472
16	0.212	0.763	0.448	1.552	0.363	1.637	0.9835	3.532
17	0.203	0.739	0.466	1.534	0.378	1.622	0.9845	3.588
18	0.194	0.718	0.482	1.518	0.391	1.609	0.9854	3.640
19	0.187	0.698	0.497	1.503	0.403	1.597	0.9862	3.689
20	0.180	0.680	0.510	1.490	0.415	1.585	0.9869	3.735
21	0.173	0.663	0.523	1.477	0.425	1.575	0.9876	3.778
22	0.167	0.647	0.534	1.466	0.434	1.566	0.9882	3.819
23	0.162	0.633	0.545	1.455	0.443	1.557	0.9887	3.858
24	0.157	0.619	0.555	1.445	0.452	1.548	0.9892	3.895
25	0.153	0.606	0.565	1.435	0.459	1.541	0.9896	3.931

For $n > 25$: $A_3 \approx 3/\sqrt{n}$, $B_3 \approx 1 - 3/\sqrt{2n}$, and $B_4 \approx 1 + 3/\sqrt{2n}$.

Appendix B

Bibliography

Agresti, A. (2007). *An Introduction to Categorical Data Analysis*, 2nd ed. John Wiley & Sons, New York. An authoritative and comprehensive treatment of the subject, suitable for a broad audience.

Belsey, D., Kuh, E., and Welsch, R. (2004). *Regression Diagnostics: Identifying Influential Data and Sources of Collinearity*. John Wiley & Sons, New York. A presentation of methods for evaluating the reliability of regression estimates.

Bevington, P., and Robinson, D. (2003). *Data Reduction and Error Analysis for the Physical Sciences*, 3rd ed. McGraw-Hill, Boston. An introduction to data analysis, with emphasis on propagation of error and calculation.

Bickel, P., and Doksum, K. (2007). *Mathematical Statistics: Basic Ideas and Selected Topics*, Vol. I, 2nd ed. Prentice-Hall, Upper Saddle River, NJ. A thorough treatment of the mathematical principles of statistics, at a fairly advanced level.

Box, G., and Draper, N. (1987). *Empirical Model-Building and Response Surfaces*. John Wiley & Sons, New York. An excellent practical introduction to the fitting of curves and higher-order surfaces to data.

Box, G., Hunter, W., and Hunter, J. (2005). *Statistics for Experimenters*, 2nd ed. John Wiley & Sons, New York. A very intuitive and practical introduction to the basic principles of data analysis and experimental design.

Brockwell, R., and Davis, R. (2002). *Introduction to Time Series and Forecasting*, 2nd ed. Springer-Verlag, New York. An excellent introductory text at the undergraduate level, more rigorous than Chatfield (2003).

Casella, G., and Berger, R. (2002). *Statistical Inference*, 2nd ed. Duxbury, Pacific Grove, CA. A fairly rigorous development of the theory of statistics.

Chatfield, C. (2003). *An Analysis of Time Series: An Introduction*, 6th ed. CRC Press, Boca Raton, FL. An intuitive presentation, at a somewhat less advanced level than Brockwell and Davis (2002).

Chatfield, C. (1983). *Statistics for Technology*, 3rd ed., revised. Chapman and Hall/CRC, Boca Raton, FL. A clear and concise introduction to basic principles of statistics, oriented toward engineers and scientists.

Cochran, W. (1977). *Sampling Techniques*, 3rd ed. John Wiley & Sons, New York. A comprehensive account of sampling theory.

Cook, D., and Weisberg, S. (1994). *Applied Regression Including Computing and Graphics*. John Wiley & Sons, New York. A presentation of graphical methods for analyzing data with linear models, with an emphasis on graphical methods.

DeGroot, M., and Schervish, M. (2002). *Probability and Statistics*, 3rd ed. Addison-Wesley, Reading, MA. A very readable introduction at a somewhat higher mathematical level than this book.

Draper, N., and Smith, H. (1998). *Applied Regression Analysis*, 3rd ed. John Wiley & Sons, New York. An extensive and authoritative treatment of linear regression.

Efron, B., and Tibshirani, R. (1993). *An Introduction to the Bootstrap*. Chapman and Hall, New York. A clear and comprehensive introduction to bootstrap methods.

Freedman, D., Pisani, R., and Purves, R. (2007). *Statistics*, 4th ed. Norton, New York. An excellent intuitive introduction to the fundamental principles of statistics.

Hocking, R. (2003). *Methods and Applications of Linear Models: Regression and the Analysis of Variance*, 2nd ed. John Wiley & Sons, New York. A thorough treatment of the theory and applications of regression and analysis of variance.

Kenett, R., and Zacks, S. (1998). *Modern Industrial Statistics*. Brooks/Cole, Pacific Grove, CA. An up-to-date treatment of the subject with emphasis on industrial engineering.

Larsen, R., and Marx, M. (2006). *An Introduction to Mathematical Statistics and Its Applications*, 4th ed. Prentice-Hall, Upper Saddle River, NJ. An introduction to statistics at a higher mathematical level than that of this book. Contains many good examples.

Lee, P. (1997). *Bayesian Statistics: An Introduction*, 3rd ed. Hodder Arnold, London. A clear and basic introduction to statistical methods that are based on the subjective view of probability.

Lehmann, E., and D'Abrera, H. (2006). *Nonparametrics: Statistical Methods Based on Ranks*, 2nd ed. Springer, New York. Thorough presentation of basic distribution-free methods.

Miller, A. (2002). *Subset Selection in Regression*, 2nd ed. Chapman and Hall, London. A strong and concise treatment of the basic principles of model selection.

Miller, R. (1997). *Beyond ANOVA: The Basics of Applied Statistics*. Chapman and Hall/CRC, Boca Raton, FL. A very practical and intuitive treatment of methods useful in analyzing real data, when standard assumptions may not be satisfied.

Montgomery, D. (2009a). *Design and Analysis of Experiments*, 7th ed. John Wiley & Sons, New York. A thorough exposition of the methods of factorial experiments, focusing on engineering applications.

Montgomery, D. (2009b). *Introduction to Statistical Quality Control*, 6th ed. John Wiley & Sons, New York. A comprehensive and readable introduction to the subject.

Mood, A., Graybill, F., and Boes, D. (1974). *Introduction to the Theory of Statistics*, 3rd ed. McGraw-Hill, Boston. A classic introduction to mathematical statistics and an excellent reference.

Mosteller, F., and Tukey, J. (1977). *Data Analysis and Regression*. Addison-Wesley, Reading, MA. An intuitive and philosophical presentation of some very practical ideas.

Rice, J. (2006). *Mathematical Statistics and Data Analysis*, 3rd ed. Wadsworth, Belmont, CA. A good blend of theory and practice, at a somewhat higher level than this book.

Ross, S. (2005). *A First Course in Probability*, 7th ed. Prentice-Hall, Upper Saddle River, NJ. A mathematically sophisticated introduction to probability.

Ross, S. (2004). *Introduction to Probability and Statistics for Engineers and Scientists*, 3rd ed. Harcourt/Academic Press, San Diego. An introduction at a somewhat higher mathematical level than this book.

Salsburg, D. (2001). *The Lady Tasting Tea*. W. H. Freeman and Company, New York. An insightful discussion of the influence of statistics on 20th-century science, with many fascinating anecdotes about famous statisticians.

Tanur, J., Pieters, R., and Mosteller, F. (eds.) (1989). *Statistics: A Guide to the Unknown*, 3rd ed. Wadsworth/Brooks-Cole, Pacific Grove, CA. A collection of case studies illustrating a variety of statistical applications.

Taylor, J. (1997). *An Introduction to Error Analysis*, 2nd ed. University Science Books, Sausalito, CA. A thorough treatment of propagation of error, along with a selection of other topics in data analysis.

Tufte, E. (2001). *The Visual Display of Quantitative Information*, 2nd ed. Graphics Press, Cheshire, CT. A clear and compelling demonstration of the principles of effective statistical graphics, containing numerous examples.

Tukey, J. (1977). *Exploratory Data Analysis*. Addison-Wesley, Reading, MA. A wealth of techniques for summarizing and describing data.

Wackerly, D., Mendenhall, W., and Scheaffer, R. (2007). *Mathematical Statistics with Applications*, 7th ed. Duxbury, Pacific Grove, CA. An introduction to statistics at a somewhat higher mathematical level than this book.

Weisberg, S. (2005). *Applied Linear Regression*, 3rd ed. John Wiley & Sons, New York. A concise introduction to the application of linear regression models, including diagnostics, model building, and interpretation of output.

Answers to Selected Exercises

Section 1.1

1. (a) The population consists of all the bolts in the shipment. It is tangible.
    (b) The population consists of all measurements that could be made on that resistor with that ohmmeter. It is conceptual.
    (c) The population consists of all residents of the town. It is tangible.
    (d) The population consists of all welds that could be made by that process. It is conceptual.
    (e) The population consists of all parts manufactured that day. It is tangible.

3. (a) False   (b) True

5. (a) No.   What is important is the population proportion of defectives; the sample proportion is only an approximation. The population proportion for the new process may in fact be greater or less than that of the old process.
    (b) No.   The population proportion for the new process may be 12% or more, even though the sample proportion was only 11%.
    (c) Finding 2 defective circuits in the sample.

7. A good knowledge of the process that generated the data.

Section 1.2

1. (a) The mean will be divided by 2.2.
    (b) The standard deviation will be divided by 2.2.

3. False

5. No.   In the sample 1,2,4 the mean is 7/3, which does not appear at all.

7. The sample size can be any odd number.

9. Yes.   If all the numbers in the list are the same, the standard deviation will equal 0.

11. 169.6 cm.

13. (a) All would be multiplied by 2.54.
     (b) Not exactly the same, since the measurements would be a little different the second time.

15. (a) The tertiles are 45 and 77.5.   (b) The quintiles are 32, 47.5, 75, and 85.5.

Section 1.3

1. (a)

Stem	Leaf
11	6
12	678
13	13678
14	13368
15	126678899
16	122345556
17	013344467
18	1333558
19	2
20	3

(d) The boxplot shows no outliers.

3.

Stem	Leaf
1	1588
2	00003468
3	0234588
4	0346
5	2235666689
6	00233459
7	113558
8	568
9	1225
10	1
11	
12	2
13	06
14	
15	
16	
17	1
18	6
19	9
20	
21	
22	
23	3

There are 23 stems in this plot. An advantage of this plot over the one in Figure 1.6 is that the values are given to the tenths digit instead of to the ones digit. A disadvantage is that there are too many stems, and many of them are empty.

5. (c) The yields for catalyst B are considerably more spread out than those for catalyst A. The median yield for catalyst A is greater than the median for catalyst B. The median yield for B is closer to the first quartile than the third, but the lower whisker is longer than the upper one, so the median is approximately equidistant from the extremes of the data. Thus the yields for catalyst B are approximately symmetric. The largest yield for catalyst A is an outlier; the remaining yields for catalyst A are approximately symmetric.

7. (a) Closest to 25%   (b) 130–135 mm

9. 82 is the only outlier.

11. (ii)

13. (a) A: 4.60, B: 3.86   (b) Yes.
  (c) No. The minimum value of -2.235 is an "outlier," since it is more than 1.5 times the interquartile range below the first quartile. The lower whisker should extend to the smallest point that is not an outlier, but the value of this point is not given.

15. (b) The boxplot indicates that the value 470 is an outlier.
  (d) The dotplot indicates that the value 384 is detached from the bulk of the data, and thus could be considered to be an outlier.

Supplementary Exercises for Chapter 1

1. The mean and standard deviation both increase by 5%.

3. (a) False   (b) True   (c) False   (d) True

5. (a) It is not possible to tell by how much the mean changes.
  (b) If there are more than two numbers on the list, the median is unchanged. If there are only two numbers on the list, the median is changed, but we cannot tell by how much.
  (c) It is not possible to tell by how much the standard deviation changes.

7. (a) The mean decreases by 0.774.   (b) The mean changes to 24.226.   (c) The median is unchanged.
  (d) It is not possible to tell by how much the standard deviation changes.

9. Statement (i) is true.

11. (a) Skewed to the left. The 85th percentile is much closer to the median (50th percentile) than the 15th percentile is. Therefore the histogram is likely to have a longer left-hand tail than right-hand tail.
  (b) Skewed to the right. The 15th percentile is much closer to the median (50th percentile) than the 85th percentile is. Therefore the histogram is likely to have a longer right-hand tail than left-hand tail.

13. (b) Each sample contains one outlier.
  (c) In the Sacaton boxplot, the median is about midway between the first and third quartiles, suggesting that the data between these quartiles are fairly symmetric. The upper whisker of the box is much longer than the lower whisker, and there is an outlier on the upper side. This indicates that the data as a whole are skewed to the right. In the Gila Plain boxplot data, the median is about midway between the first and third quartiles, suggesting that the data between these quartiles are fairly symmetric. The upper whisker is slightly longer than the lower whisker, and there is an outlier on the upper side. This suggest that the data as a whole are somewhat skewed to the right. In the Casa Grande boxplot, the median is very close to the first quartile. This suggests that there are several values very close to each other about one-fourth of the way through the data. The two whiskers are of about equal length, which suggests that the tails are about equal, except for the outlier on the upper side.

Section 2.1

1. 0.8242

3. (a) The correlation coefficient is appropriate. The points are approximately clustered around a line.

(b) The correlation coefficient is not appropriate. The relationship is curved, not linear.

(c) The correlation coefficient is not appropriate. The plot contains outliers.

5. More than 0.6

7. (a) Between temperature and yield, $r = 0.7323$; between stirring rate and yield, $r = 0.7513$; between temperature and stirring rate, $r = 0.9064$.

(b) No, the result might be due to confounding, since the correlation between temperature and stirring rate is far from 0.

(c) No, the result might be due to confounding, since the correlation between temperature and stirring rate is far from 0.

Section 2.2

1. (a) 144.89 kg   (b) 2.55 kg

3. (a) 18.869 inches   (b) 70.477 inches

    (c) No, some of the men whose points lie below the least-squares line will have shorter arms.

5. (b) $y = 8.5593 - 0.15513x$.

(c) By 0.776 miles per gallon.

(d) 6.23 miles per gallon.

(e) miles per gallon per ton

(f) miles per gallon

7. (b) $y = 12.193 - 0.833x$

(c) $(8.86, -0.16), (8.69, 0.11), (8.53, -0.23), (8.36, 0.34), (8.19, -0.09), (8.03, -0.03), (7.86, 0.24), (7.69, 0.01),$
$(7.53, -0.03), (7.36, -0.16)$

(d) Decrease by 0.0833 hours.   (e) 8.53 hours   (f) 4.79%

9. (a) $\hat{\beta}_0 = -0.0390$, $\hat{\beta}_1 = 1.017$ (b) 1.283

11. (b) $y = 2.9073 + 0.88824x$ (c) 10.659   (d) 47.319   (e) $109,310

Section 2.3

1. (b) It is appropriate for Type 1, as the scatterplot shows a clear linear trend. It is not appropriate for Type 2, since the scatterplot has a curved pattern. It is not appropriate for Type 3, as the scatterplot contains an outlier.

3. (a) $y = -3.26 + 2.0229x$ (b) No, this requires extrapolation.   (c) Yes, the prediction is 6.8545.
(d) No, this requires extrapolation.

5. 0.8492

7. $y = 19.499 + 1.3458x$

9. $y = 20 + 10x$

Supplementary Exercises for Chapter 2

1. (iii) equal to $47,500

3. Closest to -1. If two people differ in age by x years, the graduation year of the older one will be approximately x years less than that of the younger one. Therefore the points on a scatterplot of age versus graduation year would lie very close to a straight line with negative slope.

5. (a) $y = 2$ (b) $y = -4/3$

(c) For the correlation to be equal to -1, the points would have to lie on a straight line with negative slope. There is no value for y for which this is the case.

7. (a) $y = 337.1334 + 0.0980x$ (b) 827

(c) $\ln y = -0.4658 + 0.8198 \ln x$ (d) 676

(g) $\ln y$ versus $\ln x$.   (h) The one in part (d), since it is based on a scatterplot with a more linear relationship.

Section 3.1

1. 0.88

3. (a) 0.15   (b) 0.6667

5. 0.94

7. (a) 0.6   (b) 0.9

Section 3.2

1. (a) 0.03   (b) 0.68   (c) 0.32

3. (a) 0.88   (b) 0.1715   (c) 0.4932   (d) 0.8433

5. (a) 0.8   (b) 0.7   (c) 0.7   (d) Yes

7. 0.9997

9. (a) 0.9904   (b) 0.1   (c) 0.2154   (d) 7

11. 0.82

Section 3.3

1. (a) Discrete   (b) Continuous   (c) Discrete   (d) Continuous   (e) Discrete

3. (a) 2.3   (b) 1.81   (c) 1.345   (d)

y	10	20	30	40	50
$p(y)$	0.4	0.2	0.2	0.1	0.1

(e) 23   (f) 181

(g) 13.45

5. (a) $c = 0.1$ (b) 0.2   (c) 3   (d) 1   (e) 1

7. (a) 1/16   (b) 106.67 Ω (c) 9.4281 Ω (d) $F(x) = \begin{cases} 0 & x < 80 \\ x^2/1600 - x/10 + 4 & 80 \le x < 120 \\ 1 & x \ge 120 \end{cases}$

9. (a) 10 months   (b) 10 months   (c) $F(t) = \begin{cases} 0 & t < 0 \\ 1 - e^{-0.1t} & t \ge 0 \end{cases}$ (d) 0.6988

11. (a) 0.8   (b) 0.7   (c) 0.2864   (d) 0.2236   (e) 0.9552   (f) $F(x) = \begin{cases} 0 & x < 0 \\ 0.6x^2 + 0.4x^3 & 0 \le x < 1 \\ 1 & x \ge 1 \end{cases}$

13. (a) 0.2428   (b) 0.5144   (c) 3   (d) 0.5684   (e) 0.5832   (f) $F(x) = \begin{cases} 0 & x < 2 \\ (-x^3 + 9x^2 - 28)/52 & 2 \le x < 4 \\ 1 & x \ge 4 \end{cases}$

15. (a) 67/256   (b) 109/256   (c) 2.4%   (d) 0.64   (e) $F(x) = \begin{cases} 0 & x < 0 \\ (x^3/16 - 3x^4/256) & 0 \le x < 4 \\ 1 & x \ge 4 \end{cases}$

Section 3.4

1. (a) $\mu = 28.5$, $\sigma = 1.2$ (b) $\mu = -2.7$, $\sigma = 0.412$ (c) $\mu = 36.7$, $\sigma = 0.566$

3. $\mu = 2.013$ L, $\sigma = 0.00102$ L

5. (a) 17.5 mm   (b) 0.224 mm

7. (a) 0.650   (b) 0.158

9. (a) 150 cm   (b) 0.447 cm

11. (a) 500   (b) 8.944   (c) 25   (d) 0.4472

13. 9.80 ± 0.39 m/sec²

Supplementary Exercises for Chapter 3

1. (a) 0.20   (b) 0.95

3. (a) That the gauges fail independently.
(b) One cause of failure, a fire, will cause both gauges to fail. Therefore, they do not fail independently.
(c) Too low.   The correct calculation would use P(second gauge fails|first gauge fails) in place of P(second gauge fails). Because there is a chance that both gauges fail together in a fire, the condition that the first gauge fails makes it more likely that the second gauge fails as well.
Therefore P(second gauge fails|first gauge fails) $> P$(second gauge fails).

5. 0.9125

7. (a) $\mu = 6$, $\sigma^2 = 9$ (b) $\mu = 4$, $\sigma^2 = 10$ (c) $\mu = 0$, $\sigma^2 = 10$ (d) $\mu = 16$, $\sigma^2 = 328$

9. 1.7289 ± 0.0058 sec

11. (a) $1 - 3e^{-2} = 0.5940$ (b) 0.3587   (c) 2   (d) $F(x) = \begin{cases} 0 & x < 0 \\ 1 - (x+1)e^{-x} & x > 0 \end{cases}$

13. With this process, the probability that a ring meets the specification is 0.641. With the process in Exercise 12, the probability is 0.568. Therefore this process is better than the one in Exercise 12.

15. (a) 0.2993   (b) 0.00288

Section 4.1

1. (a) 0.0425   (b) 0.2508   (c) 0.1662   (d) 0.0464   (e) 6   (f) 2.4

3. (a) 0.1312   (b) 0.8936   (c) 0.5798   (d) 0.1495

5. (a) 0.1028   (b) 0.6477   (c) 0.0388   (d) 3   (e) 1.597

7. (a) 0.21875   (b) 4   (c) 2   (d) 1.414

9. (a) 0.95   (b) 0.0861

11. (a) 0.9914   (b) 4

13. (a) 1.346×10^{-4}
(b) Yes, only about 13 or 14 out of every 100,000 samples of size 10 would have 7 or more defective items.
(c) Yes, because 7 defectives in a sample of size 10 is an unusually large number for a good shipment.
(d) 0.4557
(e) No, in about 45% of the samples of size 10, 2 or more items would be defective.
(f) No, because 2 defectives in a sample of size 10 is not an unusually large number for a good shipment.

15. (a) 0.8369   (b) 9

Section 4.2

1. (a) 0.2240   (b) 0.0498   (c) 0.4232   (d) 0.5768   (e) 3   (f) 1.732

3. (a) 0.0902   (b) 0.4060   (c) 0.7218   (d) 2   (e) 1.414

5. (a) 0.1563   (b) 0.0688   (c) 0.6767

7. (a) 0.1404   (b) 0.2378   (c) 5   (d) 2.2361

9. (ii)

11. (a) 7.295×10^{-3}
(b) Yes. If the mean concentration is 7 particles per mL, then only about 7 in every thousand 1 mL samples will contain 1 or fewer particles.
(c) Yes, because 1 particle in a 1 mL sample is an unusually small number if the mean concentration is 7 particles per mL.
(d) 0.4497
(e) No. If the mean concentration is 7 particles per mL, then about 45% of all 1 mL samples will contain 6 or fewer particles.
(f) No, because 6 particles in a 1 mL sample is not an unusually small number if the mean concentration is 7 particles per mL.

Section 4.3

1. (a) 0.7734   (b) 0.2195   (c) 0.4837   (d) 0.8702

3. (a) 1.00   (b) 0.86   (c) 1.50   (d) −1.70   (e) 1.45

5. (a) 0.1292   (b) ≈ 429   (c) 96th percentile   (d) 0.4649

7. (a) 0.0228   (b) 1144 h   (c) 89th percentile   (d) 0.3721

9. (a) 0.0764   (b) 9.062 GPa   (c) 12.303 GPa

11. (a) 0.0336   (b) Yes, the proportion of days shut down in this case would be only 0.0228.

Section 4.4

1. (a) 3.5966   (b) 0.5293   (c) 3.3201   (d) 5.5400

3. (a) 25.212   (b) 3.9828   (c) 24.903   (d) 0.2148   (e) 27.666

5. (a) $1.0565   (b) 0.0934   (c) $1.0408   (d) 0.2090

Section 4.5

1. (a) 2   (b) 4   (c) 0.0821   (d) 1.3863

3. (a) 4 microns   (b) 4 microns   (c) 0.5276   (d) 0.0639   (e) 2.7726 microns   (f) 5.5452 microns
(g) 18.4207 microns

5. No. If the lifetimes were exponentially distributed, the proportion of used components lasting longer than 5 years would be the same as the proportion of new components lasting longer than 5 years, because of the lack of memory property.

Section 4.6

1. (a) 5   (b) 2.8868

3. (a) 3   (b) 1.2247

5. (a) 1   (b) 5   (c) 0.8647   (d) 0.0863   (e) 0.1078

7. (a) 0.8490   (b) 0.5410   (c) 1899.2 hours   (d) 8.761×10^{-4}

9. (a) 0.3679   (b) 0.2978   (c) 0.4227

Section 4.7

1. (a) No   (b) No   (c) Yes

3. These data do not appear to come from an approximately normal distribution.

5. The PM data do not appear to come from an approximately normal distribution.

7. Yes. If the logs of the PM data come from a normal population, then the PM data come from a lognormal population, and vice versa.

Section 4.8

1. (a) 0.2743   (b) 0.0359

3. (a) 0.0170   (b) 126.8 lb   (c) ≈ 105

5. 0.9222

7. (a) 0.0233   (b) 1.8854   (c) ≈ 110

9. 0.5793

11. (a) 0.0002
(b) Yes. Only about 2 in 10,000 samples of size 1000 will have 75 or more nonconforming tiles if the goal has been reached.
(c) No, because 75 nonconforming tiles in a sample of 1000 is an unusually large number if the goal has been reached.
(d) 0.3594
(e) No. More than 1/3 of the samples of size 1000 will have 53 or more nonconforming tiles if the goal has been reached.
(f) Yes, because 53 nonconforming tiles in a sample of 1000 is an not unusually large number if the goal has been reached.

Supplementary Exercises for Chapter 4

1. 0.9744

3. (a) 0.2503   (b) 0.4744   (c) 0.1020   (d) 0.1414   (e) 0.8508

5. (a) 0.9044   (b) 0.00427   (c) 0.00512

7. (a) 0.6826   (b) $z = 1.28$ (c) 0.0010

9. (a) 0.0668   (b) 0.6687   (c) 0.0508

11. (a) 0.8830   (b) 0.4013   (c) 0.0390 (0.1065 is a spurious root.)

13. (a) 0.4889   (b) 0.8679

15. (a) 0.4090
(b) No. More than 40% of the samples will have a total weight of 914.8 ounces or less if the claim is true.
(c) No, because a total weight of 914.8 ounces is not unusually small if the claim is true.
(d) ≈ 0
(e) Yes. Almost none of the samples will have a total weight of 910.3 ounces or less if the claim is true.
(f) Yes, because a total weight of 910.3 ounces is unusually small if the claim is true.

Section 5.1

1. iii.

3. (a) Bias $= 0$, Variance $= 1/2$, MSE $= 1/2$ (b) Bias $= 0$, Variance $= 5/9$, MSE $= 5/9$
(c) Bias $= \mu/2$, Variance $= 1/8$, MSE $= \mu^2/4 + 1/8$

5. For $-1.3123 < \mu < 1.3123$

Section 5.2

1. (a) 1.645   (b) 1.37   (c) 2.81   (d) 1.15

3. Up, down

5. (a) (175.50, 180.50)   (b) (174.70, 181.30)   (c) 88.12%   (d) 189   (e) 327

7. (a) (6186.7, 6273.3)   (b) (6173.0, 6287.0)   (c) 74.16%   (d) 301   (e) 521

9. (a) (1.538, 1.582)   (b) (1.534, 1.586)   (c) 92.66%   (d) 385   (e) 543

11. (a) (27.04, 30.96)   (b) (26.42, 31.58)   (c) 86.64%   (d) 312   (e) 540

13. (a) 17.03   (b) 98.78%

15. (a) 1.7634   (b) 97.13%

17. (a) 70.33   (b) 99.29%

Section 5.3

1. (a) 0.40   (b) (0.294, 0.517)   (c) (0.272, 0.538)   (d) 89   (e) 127

3. (a) (0.0886, 0.241)   (b) (0.101, 0.229)   (c) 327   (d) 229

5. (a) (0.07672, 0.08721)   (b) (0.07506, 0.08887)   (c) 87.29%

7. 0.811

9. (a) (0.490, 0.727)   (b) (0.863, 1)   (c) (0.572, 0.906)

11. (a) 381   (b) (0.133, 0.290)   (c) 253

13. (a) (0.271, 0.382)   (b) 658   (c) 748

Section 5.4

1. (a) 1.860   (b) 2.776   (c) 2.763   (d) 12.706

3. (a) 95%   (b) 98%   (c) 90%   (d) 99%   (e) 99.9%

5. (12.318, 13.762)

7. (a)   (b) Yes, the 99% confidence interval is (3.225, 3.253).

(c)   (d) No, the data set contains an outlier.

9. (5.303, 6.497)

11. Yes, there are no outliers. A 95% confidence interval is (203.81, 206.45).

13. (0.515, 1.985)

16. (a) 2.3541   (b) 0.888   (c) 3.900

17. (a) (10.030, 33.370)
(b) No. The minimum possible value is 0, which is less than two sample standard deviations below the sample mean. Therefore it is impossible to observe a value that is two or more sample standard deviations below the sample mean. This suggests that the sample may not come from a normal population.

Section 5.5

1. (a) (96.559, 106.241)   (b) (96.321, 106.479)

3. (a) (3.8311, 7.9689)   (b) (3.3875, 8.4125)

5. (a) (83.454, 89.666)   (b) (79.808, 93.312)

Supplementary Exercises for Chapter 5

1. (a) 0.74   (b) (0.603, 0.842)   (c) 74   (d) (0.565, 0.879)   (e) 130

3. The narrowest interval, (4.20, 5.83), is the 90% confidence interval; the widest interval, (3.57, 6.46), is the 99% confidence interval; and (4.01, 6.02) is the 95% confidence interval.

5. 845

7. 441

9. (a) (0.0529, 0.1055)   (b) 697   (c) (0.0008, 0.556)

11. (a) False   (b) False   (c) True   (d) False

13. 280

15. (0.21525, 0.23875)

17. (a) False   (b) True   (c) False

19. (a) (36.804, 37.196)   (b) 68%   (c) The measurements come from a normal population.
    (d) (36.774, 37.226)

21. (a) Since X is normally distributed with mean $n\lambda$, it follows that for a proportion $1 - \alpha$ of all possible samples, $-z_{\alpha/2}\sigma_X < X - n\lambda < z_{\alpha/2}\sigma_X$. Multiplying by -1 and adding X across the inequality yields $X - z_{\alpha/2}\sigma_X < n\lambda < X + z_{\alpha/2}\sigma_X$, which is the desired result.
    (b) Since n is a constant, $\sigma_{X/n} = \sigma_X/n = \sqrt{n\lambda}/n = \sqrt{\lambda/n}$. Therefore $\sigma_{\hat{\lambda}} = \sigma_X/n$.
    (c) Divide the inequality in part (a) by n.
    (d) Substitute $\sqrt{\hat{\lambda}/n}$ for $\sigma_{\hat{\lambda}}$ in part (c). It can then be seen that for a proportion $1 - \alpha$ of all possible samples, $\hat{\lambda} - z_{\alpha/2}\sqrt{\hat{\lambda}/n} < \lambda < \hat{\lambda} + z_{\alpha/2}\sqrt{\hat{\lambda}/n}$. The interval $\hat{\lambda} \pm z_{\alpha/2}\sqrt{\hat{\lambda}/n}$ is therefore a level $1 - \alpha$ confidence interval for λ.
    (e) (53.210, 66.790)

Section 6.1

1. (a) 0.0014   (b) 0.14%

3. (a) 0.2584   (b) 25.84%

5. (a) 0.0011
   (b) If the mean level were no greater than 5 mg/L, the chance of observing a sample mean as large as 5.4 would be only 0.0011. Therefore we are convinced that the level is greater than 5 mg/L.

7. (a) ≈ 0
   (b) If the mean daily output were 740 tons or more, the chance of observing a sample mean as small as the value of 715 that was actually observed would be nearly 0. Therefore we are convinced that the mean daily output is not 740 tons or more, but instead is less than 740 tons.

9. (a) 0.1867
   (b) If the mean lifetime were 750 kilocycles, the chance of observing a sample mean as large as the value of 763 that was actually observed would be 0.1867. Since this is not a small probability, it is plausible that the mean lifetime is 750 kilocycles.

11. (ii)

13. $P = 0.3174$

15. (a) 0.2153   (b) 2.65   (c) 0.0040

Section 6.2

1. $P = 0.10$

3. (iv)

5. (a) True   (b) True   (c) False

7. (a) No. The P-value is 0.177, which is greater than 0.05.

(b) The value 36 is contained in the 95% confidence interval for μ. Therefore the hypothesis $H_0 : \mu = 36$ versus $H_1 : \mu \neq 36$ cannot be rejected at the 5% level.

9. (a) $H_0 : \mu \leq 10$ (b) $H_0 : \mu = 20$ (c) $H_0 : \mu \leq 8$

11. (a) (ii) The scale is out of calibration.
    (b) (iii) The scale might be in calibration.
    (c) No. The scale is in calibration only if $\mu = 10$. The strongest evidence in favor of this hypothesis would occur if $\overline{X} = 10$. But since there is random variation in $\overline{X}$, we cannot be sure even then that $\mu = 10$.

13. No, she cannot conclude that the null hypothesis is true.

15. (i)

17. (a) Yes. Quantities greater than the upper confidence bound will have P-values less than 0.05. Therefore $P < 0.05$.
    (b) No, we would need to know the 99% upper confidence bound to determine whether $P < 0.01$.

19. Yes, we can compute the P-value exactly. Since the 95% upper confidence bound is 3.45, we know that $3.40 + 1.645s/\sqrt{n} = 3.45$. Therefore $s/\sqrt{n} = 0.0304$. The z-score is $(3.40 - 3.50)/0.0304 = -3.29$. The P-value is 0.0005, which is less than 0.01.

Section 6.3

1. Yes, $P = 0.0062$.

3. No, $P = 0.1292$.

5. Yes, $P = 0.0158$.

7. No, $P = 0.1867$.

9. No, $P = 0.1515$.

11. Yes, $P = 0.0040$.

13. (a) 0.69   (b) -0.49 (c) 0.3121

Section 6.4

1. (a) $t_2 = 0.6547$, $0.50 < P < 0.80$. The instrument may well be calibrated correctly.
   (b) The t-test cannot be performed, because the sample standard deviation cannot be computed from a sample of size 1.

3. (a) $H_0 : \mu \geq 16$ versus $H_1 : \mu < 16$ (b) $t_9 = -2.7388$ (c) $0.01 < P < 0.025$, reject H_0.

5. (a) No, $0.10 < P < 0.25$ (b) Yes, $0.001 < P < 0.005$ (c) No, $0.20 < P < 0.50$

7. (a)   (b) Yes, $t_6 = 1.4194$, $0.20 < P < 0.50$, do not reject H_0.

   (c)   (d) No, the sample contains an outlier.

9. (a) Yes, $0.001 < P < 0.005$ (b) No, $0.10 < P < 0.20$

11. Yes, $t_3 = -4.0032$, $0.01 < P < 0.025$.

13. (a) 6.0989   (b) 9.190   (c) 17.384   (d) -1.48

Section 6.5

1. (a) $H_0: p_1 = 0.85, \ p_2 = 0.10, \ p_3 = 0.05$
(b) $425, 50, 25$ (c) $\chi_2^2 = 10.4412$
(d) $0.005 < P < 0.01$. The true percentages differ from 85%, 10%, and 5%.

3. The expected values are

Net Excess Capacity	Small	Large
$< 0\%$	81.7572	99.2428
$0 - 10\%$	44.7180	54.2820
$11 - 20\%$	14.0026	16.9974
$21 - 30\%$	4.9687	6.0313
$> 30\%$	27.5535	33.4465

$\chi_4^2 = 12.9451, 0.01 < P < 0.05$. It is reasonable to conclude that the distributions differ.

5. Yes, $\chi_4^2 = 10.829, 0.025 < P < 0.05$

7. (a)   10.30   13.35   13.35
         6.96   9.02   9.02
         9.74   12.62   12.62

(b) $\chi_4^2 = 6.4808, P > 0.10$. There is no evidence that the rows and columns are not independent.

9. (iii)

11. $\chi_3^2 = 2.133, P > 0.10$. There is no evidence that the engineer's claim is incorrect.

13. Yes, $\chi_{11}^2 = 41.3289, P < 0.05$.

Section 6.6

1. (a) False   (b) True   (c) True

3. The 1% level

5. (a) Type I error   (b) Correct decision   (c) Correct decision   (d) Type II error

7. (a) Reject H_0 if $\overline{X} \geq 100.0196$ or if $\overline{X} \leq 99.9804$.   (b) Reject H_0 if $\overline{X} \geq 100.01645$ or if $\overline{X} \leq 99.98355$.
(c) Yes   (d) No   (e) 13.36%

Section 6.7

1. (a) True   (b) True   (c) False   (d) False

3. increase

5. (ii)

7. (a) $H_0: \mu \geq 50,000$ versus $H_1: \mu < 50,000$. H_1 is true.   (b) The level is 0.1151; the power is 0.4207.
(c) 0.2578   (d) 0.4364   (e) 618

9. (a) Two-tailed   (b) $p = 0.5$ (c) $p = 0.4$
   (d) Less than 0.7. The power for a sample size of 150 is 0.691332, and the power for a smaller sample size of 100 would be less than this.
   (e) Greater than 0.6. The power for a sample size of 150 is 0.691332, and the power for a larger sample size of 200 would be greater than this.
   (f) Greater than 0.65. The power against the alternative $p = 0.4$ is 0.691332, and the alternative $p = 0.3$ is farther from the null than $p = 0.4$. So the power against the alternative $p = 0.3$ is greater than 0.691332.
   (g) It's impossible to tell from the output. The power against the alternative $p = 0.45$ will be less than the power against $p = 0.4$, which is 0.691332. But we cannot tell without calculating whether it will be less than 0.65.

11. (a) Two-tailed
   (b) Less than 0.9. The sample size of 60 is the smallest that will produce power greater than or equal to the target power of 0.9.
   (c) Greater than 0.9. The power is greater than 0.9 against a difference of 3, so it will be greater than 0.9 against any difference greater than 3.

Section 6.8

1. (a) The Bonferroni-adjusted P-value is 0.012. Since this value is small, we can conclude that this setting reduces the proportion of defective parts.
   (b) The Bonferroni-adjusted P-value is 0.18. Since this value is not so small, we cannot conclude that this setting reduces the proportion of defective parts.

3. 0.0025

5. (a) No. If the mean burnout amperage is equal to 15 amps every day, the probability of rejecting H_0 is 0.05 each day. The number of times in 200 days that H_0 is rejected is then a Binomial random variable with $n = 200$, $p = 0.05$. The probability of rejecting H_0 10 or more times in 200 days is then approximately equal to 0.5636. So it would not be unusual to reject H_0 10 times in 200 trials if H_0 is always true.
   (b) Yes. If the mean burnout amperage is equal to 15 amps every day, the probability of rejecting H_0 is 0.05 each day. The number of times in 200 days that H_0 is rejected is then a Binomial random variable with $n - 200$, $p = 0.05$. The probability of rejecting H_0 20 or more times in 200 days is then approximately equal to 0.0010. So it would be quite unusual to reject H_0 20 times in 200 trials if H_0 is always true.

Supplementary Exercises for Chapter 6

1. Yes, $P = 0.0008$.

3. (a) $H_0 : \mu \geq 90$ versus $H_1 : \mu < 90$ (b) $\overline{X} < 89.3284$
   (c) This is not an appropriate rejection region. The rejection region should consist of values for $\overline{X}$ that will make the P-value of the test less than a chosen threshold level. This rejection region consists of values for which the P-value will be greater than some level.
   (d) This is an appropriate rejection region. The level of the test is 0.0708.
   (e) This is not an appropriate rejection region. The rejection region should consist of values for $\overline{X}$ that will make the P-value of the test less than a chosen threshold level. This rejection region contains values of $\overline{X}$ for which the P-value will be large.

5. (a) 0.05   (b) 0.1094

7. The Bonferroni-adjusted P-value is 0.1228. We cannot conclude that the failure rate on line 3 is less than 0.10.

9. No. $\chi_2^2 = 2.1228$, $P > 0.10$.

Section 7.1

1. (9.683, 11.597)

3. (5.589, 8.211)

5. (2.021, 4.779)

7. It is not possible. The amounts of time spent in bed and spent asleep in bed are not independent.

9. (0.0613, 0.1987)

11. Test $H_0: \mu_2 - \mu_1 \leq 50$ versus $H_1: \mu_2 - \mu_1 > 50$, where μ_1 is the mean for the more expensive product and μ_2 is the mean for the less expensive product. $P = 0.0188$. The more expensive inhibitor should be used.

13. Yes, $P = 0.0122$.

15. No, $P = 0.1292$.

17. (a) $H_0: \mu_1 - \mu_2 \leq 0$ versus $H_1: \mu_1 - \mu_2 > 0$, $P = 0.2119$. We cannot conclude that the mean score on one-tailed questions is greater.
(b) $H_0: \mu_1 - \mu_2 = 0$ versus $H_1: \mu_1 - \mu_2 \neq 0$, $P = 0.4238$. We cannot conclude that the mean score on one-tailed questions differs from the mean score on two-tailed questions.

19. (a) Yes, $P = 0.0035$. (b) Yes, $P = 0.0217$.

21. (a) (i) 11.128, (ii) 0.380484   (b) 0.0424, similar to the P-value computed with the t statistic.
(c) $(-0.3967, 5.7367)$

Section 7.2

1. (0.0374, 0.0667)

3. (0.00346, 0.03047).

5. $(-0.005507, 0.1131)$

7. No. The sample proportions come from the same sample rather than from two independent samples.

9. (a) $H_0: p_1 - p_2 \geq 0$ vs. $H_1: p_1 - p_2 < 0$ (b) $P = 0.1492$ (c) Machine 1

11. Yes, $P = 0.0018$.

13. Yes, $P = 0.0207$.

15. Yes, $P = 0.0179$.

17. No, $P = 0.7114$.

19. No, these are not simple random samples.

21. (a) 0.660131   (b) 49   (c) 1.79   (d) 0.073

Section 7.3

1. (1.117, 16.026)

3. (7.798, 30.602)

5. (20.278, 25.922)

7. (0.0447, 0.173)

9. (38.931, 132.244)

11. Yes, $t_{16} = 2.9973$, $0.001 < P < 0.005$.

13. No, $t_{14} = 1.0236$, $0.20 < P < 0.50$.

15. Yes, $t_{11} = 2.6441$, $0.02 < P < 0.05$.

17. No, $t_7 = 0.3444$, $0.50 < P < 0.80$.

19. No, $t_{15} = 1.0024$, $0.10 < P < 0.25$.

21. (a) Yes, $t_6 = 3.2832$, $0.005 < P < 0.01$.   (b) No, $t_6 = 1.2649$, $0.10 < P < 0.25$.

23. (a) 0.197   (b) 0.339   (c) -1.484 (d) -6.805

Section 7.4

1. (0.3077, 0.5109)

3. (2.712, 4.030)

5. (2.090, 11.384)

7. (a) (0.747, 2.742)   (b) 80%

9. (a) $H_0: \mu_1 - \mu_2 = 0$ versus $H_1: \mu_1 - \mu_2 \neq 0$ (b) $t_6 = 2.1187$ (c) $0.05 < P < 0.10$, H_0 is suspect.

11. Yes, $t_5 = 6.0823$, $0.001 < P < 0.002$.

13. Yes, $t_6 = 7.0711$, $P < 0.001$.

15. No, $t_7 = 1.0293$, $0.20 < P < 0.50$.

17. (a) Let μ_R be the mean number of miles per gallon for taxis using radial tires, and let μ_B be the mean number of miles per gallon for taxis using bias tires. The appropriate null and alternate hypotheses are $H_0: \mu_R - \mu_B \leq 0$ versus $H_1: \mu_R - \mu_B > 0$. The value of the test statistic is $t_9 = 8.9532$, so $P < 0.0005$.
(b) The appropriate null and alternate hypotheses are $H_0: \mu_R - \mu_B \leq 2$ versus $H_1: \mu_R - \mu_B > 2$. The value of the test statistic is $t_9 = 3.3749$, so $0.001 < P < 0.005$.

19. (a) 1.1050   (b) 2.8479   (c) 4.0665   (d) 3.40

Section 7.5

1. 2.51

3. (a) 0.01   (b) 0.02

5. No. $F_{8, 13} = 1.0429$, $P > 0.20$.

Supplementary Exercises for Chapter 7

1. No, $P = 0.1635$.

3. Yes, $P = 0.0006$.

5. $(0.0591, 0.208)$

7. (a) $(-0.0446, 0.103)$
   (b) If 100 additional chips were sampled from the less expensive process, the width of the confidence interval would be approximately ± 0.0709. If 50 additional chips were sampled from the more expensive process, the width of the confidence interval would be approximately ± 0.0569. If 50 additional chips were sampled from the less expensive process and 25 additional chips were sampled from the more expensive process, the width of the confidence interval would be approximately ± 0.0616. Therefore the greatest increase in precision would be achieved by sampling 50 additional chips from the more expensive process.

9. No, because the two samples are not independent.

11. Yes, $P = 0.0367$.

13. $(1.942, 19.725)$

15. $(-0.420, 0.238)$

17. This requires a test for the difference between two means. The data are unpaired. Let μ_1 represent the population mean annual cost for cars using regular fuel, and let μ_2 represent the population mean annual cost for cars using premium fuel. Then the appropriate null and alternate hypotheses are $H_0 : \mu_1 - \mu_2 \geq 0$ versus $H_1 : \mu_1 - \mu_2 < 0$. The test statistic is the difference in the sample mean costs between the two groups. The z table should be used to find the P-value.

19. $(0.1234, 0.8766)$

21. Yes, $t_7 = -3.0151, 0.01 < P < 0.02$

23. (a) Let μ_A be the mean thrust/weight ratio for Fuel A, and let μ_B be the mean thrust/weight ratio for Fuel B. The appropriate null and alternate hypotheses are $H_0 : \mu_A - \mu_B \leq 0$ vs. $H_1 : \mu_A - \mu_B > 0$.
   (b) Yes. $t_{29} = 2.0339, 0.025 < P < 0.05$.

Section 8.1

1. (a) $\hat{\beta}_0 = 7.6233, \hat{\beta}_1 = 0.32964$ (b) 17.996   (c) For β_0: $(-0.744, 15.991)$, for β_1: $(0.208, 0.451)$
   (d) Yes. $t_{10} = -3.119, 0.005 < P < 0.01$.   (e) $(16.722, 24.896)$ (f) $(10.512, 31.106)$

3. (a) The slope is -0.7524; the intercept is 88.761.
   (b) Yes, the P-value for the slope is ≈ 0, so humidity is related to ozone level.
   (c) 51.14 ppb   (d) -0.469 (e) $(41.6, 45.6)$
   (f) No. A reasonable range of predicted values is given by the 95% prediction interval, which is $(20.86, 66.37)$.

5. (a) $H_0 : \beta_A - \beta_B = 0$ (b) Yes. $z = -4.55, P \approx 0$.

7. (a) $y = 73.266 - 0.49679x$ (b) For β_0: $(62.57, 83.96)$. For β_1: $(-0.952, -0.042)$.
   (c) $(45.28, 76.41)$ (d) The one where the horizontal expansion is 30 will be wider.

9. (a) $y = -0.32584 + 0.22345x$.   (b) For β_0, $(-2.031, 1.379)$, for β_1, $(0.146, 0.301)$.   (c) 4.14
   (d) $(3.727, 4.559)$ (e) $(1.585, 6.701)$

11. The confidence interval at 1.5 would be the shortest. The confidence interval at 1.8 would be the longest.

13. (a) 0.256   (b) 0.80   (c) 1.13448   (d) 0.001

15. (a) 553.71   (b) 162.06   (c) Below
(d) There is a greater amount of vertical spread on the right side of the plot than on the left.

17. $t_{21} = -2.710$, $0.01 < P < 0.02$, we can conclude that $\rho \neq 0$.

Section 8.2

1. (a) $\ln y = -0.4442 + 0.79833 \ln x$ (b) 330.95   (c) 231.76   (d) (53.19, 1009.89)

3. (a) The least-squares line is $y = 1.0014 - 0.015071t$.   (b) The least-squares line is $\ln y = 0.0014576 - 0.015221t$.
(c) The least-squares line is $y = 0.99922 - 0.012385t^{1.5}$.
(d) The model $\ln y = 0.99922 - 0.012385t^{1.5}$ fits best. Its residual plot shows the least pattern.
(e) 0.991

5. (a) $y = 20.162 + 1.269x$
(b) There is no apparent pattern to the residual plot.
(c) The residuals increase over time. The linear model is not appropriate as is. Time, or other variables, must be included in the model.

7. $\ln W = \beta_0 + \beta_1 \ln L + \varepsilon$, where $\beta_0 = \ln a$ and $\beta_1 = b$.

9. (a) A physical law.
(b) It would be better to redo the experiment. If the results of an experiment violate a physical law, then something was wrong with the experiment, and you can't fix it by transforming variables.

Section 8.3

1. (a) 49.617 kg/mm²   (b) 33.201 kg/mm²   (c) 2.1245 kg/mm²

3. There is no obvious pattern to the residual plot, so the linear model appears to fit well.

5. (a) 25.465
(b) No, the predicted change depends on the values of the other independent variables, because of the interaction terms.
(c) 0.9691   (d) $F_{9,17} = 59.204$. Yes, the null hypothesis can be rejected.

7. (a) 2.3411 liters   (b) 0.06768 liters
(c) Nothing is wrong. In theory, the constant estimates FEV_1 for an individual whose values for the other variables are all equal to zero. Since these values are outside the range of the data (e.g., no one has zero height), the constant need not represent a realistic value for an actual person.

9. (a) 3.572   (b) 0.098184
(c) Nothing is wrong. The constant estimates the pH for a pulp whose values for the other variables are all equal to zero. Since these values are outside the range of the data (e.g., no pulp has zero density), the constant need not represent a realistic value for an actual pulp.
(d) (3.4207, 4.0496)   (e) (2.2333, 3.9416)
(f) Pulp B. The standard deviation of its predicted pH (SE Fit) is smaller than that of Pulp A (0.1351 versus 0.2510).

11. (a) -2.05 (b) 0.3521   (c) -0.2445 (d) 4.72   (e) 13.92   (f) 18.316   (g) 4.54   (h) 9

13. (a) $135.92°\text{F}$
    (b) No. The change in the predicted flash point due to a change in acetic acid concentration depends on the butyric acid concentration as well, because of the interaction between these two variables.
    (c) Yes. The predicted flash point will change by $-13.897°\text{F}$.

15. (a) $0.2286, -0.5743, 0.3514, 0.1057, -0.1114, 0.0000$
    (b) $SSE = 0.5291, SST = 16.7083$
    (c) $s^2 = 0.1764$ (d) $R^2 = 0.9683$
    (e) $F = 45.864$. There are 2 and 3 degrees of freedom.
    (f) Yes, the P-value corresponding to the F statistic with 2 and 3 degrees of freedom is between 0.001 and 0.01, so it is less than 0.05.

17. (a) 2.0711   (b) 0.17918
    (c) PP is more useful, because its P-value is small, while the P-value of CP is fairly large.
    (d) The percent change in GDP would be expected to be larger in Sweden, because the coefficient of PP is negative.

19. (a) $y = -0.012167 + 0.043258x + 2.9205x^2$ (b) (2.830, 3.011)   (c) (5.660, 6.022)
    (d) $\hat{\beta}_0$: $t_7 = -1.1766$, $P = 0.278$, $\hat{\beta}_1$: $t_7 = 1.0017$, $P = 0.350$, $\hat{\beta}_2$: $t_7 = 76.33$, $P = 0.000$.
    (e) No, the P-value of 0.278 is not small enough to reject the null hypothesis that $\beta_0 = 0$.
    (f) No, the P-value of 0.350 is not small enough to reject the null hypothesis that $\beta_1 = 0$.

Section 8.4

1. (a) False   (b) True   (c) False   (d) True

3. (iv)

5. The four-variable model with the highest value of R^2 has a lower R^2 than the three-variable model with the highest value of R^2. This is impossible.

7. (a) 0.2803
    (b) 3 degrees of freedom in the numerator and 157 in the denominator.
    (c) $P > 0.10$. The reduced model is plausible.
    (d) This is not correct. It is possible for a group of variables to be fairly strongly related to an independent variable, even though none of the variables individually is strongly related.
    (e) No mistake. If y is the dependent variable, then the total sum of squares is $\sum (y_i - \bar{y})^2$. This quantity does not involve the independent variables.

9. (a) β_0: $P = 0.044$, β_1: $P = 0.182$, β_2: $P = 0.006$
    (b) β_0: $P = 0.414$, β_1: $P = 0.425$
    (c) β_0: $P = 0.000$, β_2: $P = 0.007$
    (d) The model containing x_2 as the only independent variable is best. There is no evidence that the coefficient of x_1 differs from 0.

11. The model $y = \beta_0 + \beta_1 x_2 + \varepsilon$ is a good one. One way to see this is to compare the fit of this model to the full quadratic model. The F statistic for testing the plausibility of the reduced model is 0.7667. $P > 0.10$. The large P-value indicates that the reduced model is plausible.

Supplementary Exercises for Chapter 8

1. (a) $y = 0.041496 + 0.0073664x$ (b) $(-0.00018, 0.01492)$ (c) $(0.145, 0.232)$ (d) $(0.0576, 0.320)$

3. (a) $\ln y = \beta_0 + \beta_1 \ln x$, where $\beta_0 = \ln k$ and $\beta_1 = r$.
(b) The least-squares line is $\ln y = -1.7058 + 0.65033 \ln x$. Therefore $\hat{r} = 0.65033$ and $\hat{k} = e^{-1.7058} = 0.18162$.
(c) $t_3 = 4.660$, $P = 0.019$. No, it is not plausible.

5. (b) $T_{i+1} = 120.18 - 0.696T_i$.   (c) $(-0.888, -0.503)$ (d) 71.48 minutes.   (e) $(68.40, 74.56)$
(f) $(45.00, 97.95)$

7. (a) $\hat{\beta}_0 = 0.8182$, $\hat{\beta}_1 = 0.9418$
(b) No. $t_9 = 1.274$, $0.20 < P < 0.50$.
(c) Yes. $t_9 = -5.359$, $P < 0.001$.
(d) Yes, since we can conclude that $\beta_1 \neq 1$, we can conclude that the machine is out of calibration.
(e) $(18.58, 20.73)$ (f) $(75.09, 77.23)$
(g) No, when the true value is 20, the result of part (e) shows that a 95% confidence interval for the mean of the measured values is $(18.58, 20.73)$. Therefore it is plausible that the mean measurement will be 20, so that the machine is in calibration.

9. (ii)

11. (a) 145.63   (b) Yes. $r = -\sqrt{\text{R-sq}} = -0.988$. Note that r is negative because the slope of the least-squares line is negative.   (c) 145.68.

13. (a) 24.6%   (b) 5.43%   (c) No, we need to know the oxygen content.

15. (a) 0.207   (b) 0.8015   (c) 3.82   (d) 1.200   (e) 2   (f) 86.81   (g) 43.405   (h) 30.14   (i) 14

17. (a) Neighbors $= 10.84 - 0.0739\text{Speed} - 0.127\text{Pause} + 0.00111\text{Speed}^2 + 0.00167\text{Pause}^2 - 0.00024\text{Speed}\cdot\text{Pause}$
(b) Drop the interaction term Speed·Pause. Neighbors $= 10.97 - 0.0799\text{Speed} - 0.133\text{Pause} + 0.00111\text{Speed}^2 + 0.00167\text{Pause}^2$
Comparing this model with the one in part (a), $F_{1,24} = 0.7664$, $P > 0.10$.
(c) There is a some suggestion of heteroscedasticity, but it is hard to be sure without more data.
(d) No, comparing with the full model containing Speed, Pause, Speed^2, and Pause^2, and Speed·Pause, the F statistic is $F_{3,24} = 15.70$, and $P < 0.001$.
(e)

Vars	R-Sq	C-p	Speed	Pause	Speed2	Pause2	Speed*Pause
1	61.5	92.5	X				
1	60.0	97.0					X
2	76.9	47.1	X	X			
2	74.9	53.3	X	X			
3	90.3	7.9	X	X		X	
3	87.8	15.5		X		X	X
4	92.0	4.8	X	X	X	X	
4	90.5	9.2	X	X		X	X
5	92.2	6.0	X	X	X	X	X

(f) The model containing the dependent variables Speed, Pause, Speed^2 and Pause^2 has both the lowest value of C_p and the largest value of adjusted R^2.

19. (a) 3   (b) Maximum occurs when tensile strength is 10.188.

21. (a)

Source	DF	SS	MS	F	P
Regression	5	20.35	4.07	894.19	0.000
Residual Error	10	0.046	0.0046		
Total	15	20.39			

(b) The model containing the variables x_1, x_2, and x_2^2 is one good model.
(c) The model with the best adjusted R^2 (0.99716) contains the variables x_2, x_1^2, and x_2^2. This model is also the model with the smallest value of Mallows' C_p (2.2). This is not the best model, since it contains x_1^2 but not x_1. The model containing x_1, x_2, and x_2^2, suggested in the answer to part (b), is better. Note that the adjusted R^2 for the model in part (b) is 0.99704, which differs negligibly from that of the model with the largest adjusted R^2 value.

23. (a)
Predictor	Coef	StDev
Constant	1.1623	0.17042
t	0.059718	0.0088901
t^2	−0.00027482	0.000069662

(b) 17.68 minutes   (c) (0.03143, 0.08801)
(d) The reaction rate is decreasing with time if $\beta_2 < 0$. We therefore test $H_0 : \beta_2 \geq 0$ vs. $H_1 : \beta_2 < 0$. The test statistic is $t_3 = 3.945$, $P = 0.029/2 = 0.0145$. It is reasonable to conclude that the reaction rate decreases with time.

25. (a) The 17-variable model containing the independent variables x_1, x_2, x_3, x_6, x_7, x_8, x_9, x_{11}, x_{13}, x_{14}, x_{16}, x_{18}, x_{19}, x_{20}, x_{21}, x_{22}, and x_{23} has adjusted R^2 equal to 0.98446. The fitted model is

$$\begin{aligned} y = {} & - 1569.8 - 24.909x_1 + 196.95x_2 + 8.8669x_3 \\ & - 2.2359x_6 - 0.077581x_7 + 0.057329x_8 \\ & - 1.3057x_9 - 12.227x_{11} + 44.143x_{13} \\ & + 4.1883x_{14} + 0.97071x_{16} + 74.775x_{18} \\ & + 21.656x_{19} - 18.253x_{20} + 82.591x_{21} \\ & - 37.553x_{22} + 329.8x_{23} \end{aligned}$$

(b) The 8-variable model containing the independent variables x_1, x_2, x_5, x_8, x_{10}, x_{11}, x_{14}, and x_{21} has Mallows' C_p equal to 1.7. The fitted model is

$$\begin{aligned} y = {} & - 665.98 - 24.782x_1 + 76.499x_2 + 121.96x_5 \\ & + 0.024247x_8 + 20.4x_{10} - 7.1313x_{11} \\ & + 2.4466x_{14} + 47.85x_{21} \end{aligned}$$

(c) Using a value of 0.15 for both α-to-enter and α-to-remove, the equation chosen by stepwise regression is
$y = -927.72 + 142.40x_5 + 0.081701x_7 + 21.698x_{10} + 0.41270x_{16} + 45.672x_{21}$.

(d) The 13-variable model below has adjusted R^2 equal to 0.95402. (There are also two 12-variable models whose adjusted R^2 is only very slightly lower.)

$$z = 8663.2 - 313.31x_3 - 14.46x_6 + 0.358x_7$$
$$- 0.078746x_8 + 13.998x_9 + 230.24x_{10}$$
$$- 188.16x_{13} + 5.4133x_{14} + 1928.2x_{15}$$
$$- 8.2533x_{16} + 294.94x_{19} + 129.79x_{22}$$
$$- 3020.7x_{23}$$

(e) The 2-variable model $z = -1660.9 + 0.67152x_7 + 134.28x_{10}$ has Mallows' C_p equal to -4.0.

(f) Using a value of 0.15 for both α-to-enter and α-to-remove, the equation chosen by stepwise regression is $z = -1660.9 + 0.67152x_7 + 134.28x_{10}$

(g) The 17-variable model below has adjusted R^2 equal to 0.97783.

$$w = 700.56 - 21.701x_2 - 20.000x_3 + 21.813x_4$$
$$+ 62.599x_5 + 0.016156x_7 - 0.012689x_8$$
$$+ 1.1315x_9 + 15.245x_{10} + 1.1103x_{11}$$
$$- 20.523x_{13} - 90.189x_{15} - 0.77442x_{16}$$
$$+ 7.5559x_{19} + 5.9163x_{20} - 7.5497x_{21}$$
$$+ 12.994x_{22} - 271.32x_{23}$$

(h) The 13-variable model below has Mallows' C_p equal to 8.0.

$$w = 567.06 - 23.582x_2 - 16.766x_3 + 90.482x_5$$
$$+ 0.0082274x_7 - 0.011004x_8 + 0.89554x_9$$
$$+ 12.131x_{10} - 11.984x_{13} - 0.67302x_{16}$$
$$+ 11.097x_{19} + 4.6448x_{20} + 11.108x_{22}$$
$$- 217.82x_{23}$$

(i) Using a value of 0.15 for both α-to-enter and α-to-remove, the equation chosen by stepwise regression is $w = 130.92 - 28.085x_2 + 113.49x_5 + 0.16802x_9 - 0.20216x_{16} + 11.417x_{19} + 12.068x_{21} - 78.371x_{23}$.

Section 9.1

1. (a)

Source	DF	SS	MS	F	P
Temperature	3	202.44	67.481	59.731	0.000
Error	16	18.076	1.1297		
Total	19	220.52			

(b) Yes. $F_{3,16} = 59.731$, $P < 0.001$ ($P \approx 0$).

3. (a)

Source	DF	SS	MS	F	P
Treatment	4	19.009	4.7522	2.3604	0.117
Error	11	22.147	2.0133		
Total	15	41.155			

(b) No. $F_{4, 11} = 2.3604$, $P > 0.10$ $(P = 0.117)$.

5. (a)

Source	DF	SS	MS	F	P
Site	3	1.4498	0.48327	2.1183	0.111
Error	47	10.723	0.22815		
Total	50	12.173			

(b) No. $F_{3, 47} = 2.1183$, $P > 0.10$ $(P = 0.111)$.

7. (a)

Source	DF	SS	MS	F	P
Group	3	0.19218	0.064062	1.8795	0.142
Error	62	2.1133	0.034085		
Total	65	2.3055			

(b) No. $F_{3, 62} = 1.8795$, $P > 0.10$ $(P = 0.142)$

9. (a)

Source	DF	SS	MS	F	P
Temperature	2	148.56	74.281	10.530	0.011
Error	6	42.327	7.0544		
Total	8	190.89			

(b) Yes. $F_{2, 6} = 10.530$, $0.01 < P < 0.05$ $(P = 0.011)$.

11. No, $F_{3, 16} = 15.8255$, $P < 0.001$ $(P \approx 4.8 \times 10^{-5})$.

13. (a)

Source	DF	SS	MS	F	P
Temperature	3	58.650	19.550	8.4914	0.001
Error	16	36.837	2.3023		
Total	19	95.487			

(b) Yes, $F_{3, 16} = 8.4914$, $0.001 < P < 0.01$ $(P = 0.0013)$.

15. (a)

Source	DF	SS	MS	F	P
Grade	3	1721.4	573.81	9.4431	0.000
Error	96	5833.4	60.765		
Total	99	7554.9			

(b) Yes, $F_{3,96} = 9.4431$, $P < 0.001$ ($P \approx 0$).

17. (a)

Source	DF	SS	MS	F	P
Soil	2	2.1615	1.0808	5.6099	0.0104
Error	23	4.4309	0.19265		
Total	25	6.5924			

(b) Yes. $F_{2,23} = 5.6099$, $0.01 < P < 0.05$ ($P = 0.0104$)

Section 9.2

1. (a) Yes, $F_{5,6} = 46.64$, $P \approx 0$.
(b) A and B, A and C, A and D, A and E, B and C, B and D, B and E, B and F, D and F.

3. We cannot conclude at the 5% level that any of the treatment means differ.

5. Means 1 and 3 differ at the 5% level.

7. (a) $F_{3,16} = 5.08$, $0.01 < P < 0.05$. The null hypothesis of no difference is rejected at the 5% level.
(b) Catalyst 1 and catalyst 2 both differ significantly from catalyst 4.

9. Any value of *MSE* satisfying $5.099 < MSE < 6.035$.

Section 9.3

1. (a) 3   (b) 2   (c) 6   (d) 24
(e)

Source	DF	SS	MS	F	P
Oil	3	1.0926	0.36420	5.1314	0.007
Ring	2	0.9340	0.46700	6.5798	0.005
Interaction	6	0.2485	0.041417	0.58354	0.740
Error	24	1.7034	0.070975		
Total	35	3.9785			

(f) Yes. $F_{6,24} = 0.58354$, $P > 0.10$ ($P = 0.740$).
(g) No, some of the main effects of oil type are nonzero. $F_{3,24} = 5.1314$, $0.001 < P < 0.01$ ($P = 0.007$).
(h) No, some of the main effects of piston ring type are nonzero. $F_{2,24} = 6.5798$, $0.001 < P < 0.01$ ($P = 0.005$).

3. (a)

Source	DF	SS	MS	F	P
Mold Temp.	4	69738	17434.5	6.7724	0.000
Alloy	2	8958	4479.0	1.7399	0.187
Interaction	8	7275	909.38	0.35325	0.939
Error	45	115845	2574.3		
Total	59	201816			

(b) Yes. $F_{8,45} = 0.35325$, $P > 0.10$ ($P = 0.939$).
(c) No, some of the main effects of mold temperature are nonzero. $F_{4,45} = 6.7724$, $P < 0.001$ ($P \approx 0$).
(d) Yes. $F_{3,45} = 1.7399$, $P > 0.10$, ($P = 0.187$).

5. (a)

Source	DF	SS	MS	F	P
Solution	1	1993.9	1993.9	5.1983	0.034
Temperature	1	78.634	78.634	0.20500	0.656
Interaction	1	5.9960	5.9960	0.015632	0.902
Error	20	7671.4	383.57		
Total	23	9750.0			

(b) Yes, $F_{1,20} = 0.015632$, $P > 0.10$ ($P = 0.902$).
(c) Yes, since the additive model is plausible. The mean yield stress differs between Na_2HPO_4 and NaCl: $F_{1,20} = 5.1983$, $0.01 < P < 0.05$ ($P = 0.034$).
(d) There is no evidence that the temperature affects yield stress: $F_{1,20} = 0.20500$, $P > 0.10$ ($P = 0.656$).

7. (a)

Source	DF	SS	MS	F	P
Adhesive	1	17.014	17.014	6.7219	0.024
Curing Pressure	2	35.663	17.832	7.0450	0.009
Interaction	2	39.674	19.837	7.8374	0.007
Error	12	30.373	2.5311		
Total	17	122.73			

(b) No. $F_{2,12} = 7.8374$, $0.001 < P < 0.01$ ($P = 0.007$).
(c) No, because the additive model is rejected.
(d) No, because the additive model is rejected.

9. (a)

Source	DF	SS	MS	F	P
Taper Material	1	0.0591	0.0591	23.630	0.000
Neck Length	2	0.0284	0.0142	5.6840	0.010
Interaction	2	0.0090	0.0045	1.8185	0.184
Error	24	0.0600	0.0025		
Total	29	0.1565			

(b) The additive model is plausible. The value of the test statistic is 1.8185, its null distribution is $F_{2,24}$, and $P > 0.10$ ($P = 0.184$).

(c) Yes, since the additive model is plausible. The mean coefficient of friction differs between CPTi-ZrO$_2$ and TiAlloy-ZrO$_2$: $F_{1,24} = 23.630$, $P < 0.001$.

11. (a)

Source	DF	SS	MS	F	P
Concentration	2	0.37936	0.18968	3.8736	0.040
Delivery Ratio	2	7.34	3.67	74.949	0.000
Interaction	4	3.4447	0.86118	17.587	0.000
Error	18	0.8814	0.048967		
Total	26	12.045			

(b) No. The The value of the test statistic is 17.587, its null distribution is $F_{4,18}$, and $P \approx 0$.

(c) The slopes of the line segments are quite different from one another, indicating a high degree of interaction.

13. (a)

Source	DF	SS	MS	F	P
Wafer	2	114661.4	57330.7	11340.1	0.000
Operator	2	136.78	68.389	13.53	0.002
Interaction	4	6.5556	1.6389	0.32	0.855
Error	9	45.500	5.0556		
Total	17	114850.3			

(b) There are differences among the operators. $F_{2,9} = 13.53$, $0.01 < P < 0.001$ ($P = 0.002$).

15. (a)

Source	DF	SS	MS	F	P
PVAL	2	125.41	62.704	8.2424	0.003
DCM	2	1647.9	823.94	108.31	0.000
Interaction	4	159.96	39.990	5.2567	0.006
Error	18	136.94	7.6075		
Total	26	2070.2			

(b) Since the interaction terms are not equal to 0, ($F_{4,18} = 5.2562$, $P = 0.006$), we cannot interpret the main effects. Therefore we compute the cell means. These are

	DCM (mL)		
PVAL	50	40	30
0.5	97.8	92.7	74.2
1.0	93.5	80.8	75.4
2.0	94.2	88.6	78.8

We conclude that a DCM level of 50 mL produces greater encapsulation efficiency than either of the other levels. If DCM = 50, the PVAL concentration does not have much effect. Note that for DCM = 50, encapsulation efficiency is maximized at the lowest PVAL concentration, but for DCM = 30 it is maximized at the highest PVAL concentration. This is the source of the significant interaction.

Section 9.4

1. (a) Liming is the blocking factor, soil is the treatment factor.
   (b)

Source	DF	SS	MS	F	P
Soil	3	1.178	0.39267	18.335	0.000
Block	4	5.047	1.2617	58.914	0.000
Error	12	0.257	0.021417		
Total	19	6.482			

   (c) Yes, $F_{3,12} = 18.335$, $P \approx 0$.

3. (a)

Source	DF	SS	MS	F	P
Lighting	3	9943	3314.3	3.3329	0.036
Block	2	11432	5716.0	5.7481	0.009
Interaction	6	6135	1022.5	1.0282	0.431
Error	24	23866	994.42		
Total	35	51376			

   (b) Yes. The P-value for interactions is large (0.431).
   (c) Yes. The P-value for lighting is small (0.036).

5. (a)

Source	DF	SS	MS	F	P
Machine	9	339032	37670	2.5677	0.018
Block	5	1860838	372168	25.367	0.000
Error	45	660198	14671		
Total	59	2860069			

   (b) Yes, $F_{9,45} = 2.5677$, $P = 0.018$.

7. (a) One motor of each type should be tested on each day. The order in which the motors are tested on any given day should be chosen at random. This is a randomized block design, in which the days are the blocks. It is not a completely randomized design, since randomization occurs only within blocks.
   (b) The test statistic is

$$\frac{\sum_{i=1}^{5}(\overline{X}_{i.} - \overline{X}_{..})^2}{\sum_{j=1}^{4}\sum_{i=1}^{5}(X_{ij} - \overline{X}_{i.} - \overline{X}_{.j} - \overline{X}_{..})^2/12}.$$

Section 9.5

1.

	A	B	C	D
1	−	−	−	−
ad	+	−	−	+
bd	−	+	−	+
ab	+	+	−	−
cd	−	−	+	+
ac	+	−	+	−
bc	−	+	+	−
abcd	+	+	+	+

The alias pairs are $\{A, BCD\}$, $\{B, ACD\}$, $\{C, ABD\}$, $\{D, ABC\}$, $\{AB, CD\}$, $\{AC, BD\}$, and $\{AD, BC\}$

3. (a)

Term	Effect	DF	SS	MS	F	P
A	6.75	1	182.25	182.25	11.9508	0.009
B	9.50	1	361.00	361.00	23.6721	0.001
C	1.00	1	4.00	4.00	0.2623	0.622
AB	2.50	1	25.00	25.00	1.6393	0.236
AC	0.50	1	1.00	1.00	0.0656	0.804
BC	0.75	1	2.25	2.25	0.1475	0.711
ABC	−2.75	1	30.25	30.25	1.9836	0.197
Error		8	122.00	15.25		
Total		15	727.75			

(b) Factors A and B (temperature and concentration) seem to have an effect on yield. There is no evidence that pH has an effect. None of the interactions appears to be significant. Their P-values are all greater than 0.19.

(c) Since the effect of temperature is positive and statistically significant, we can conclude that the mean yield is higher when temperature is high.

5. (a)

Term	Effect
A	3.3750
B	23.625
C	1.1250
AB	−2.8750
AC	−1.3750
BC	−1.6250
ABC	1.8750

(b) No, since the design is unreplicated, there is no error sum of squares.

(c) No, none of the interaction terms are nearly as large as the main effect of factor B.

(d) If the additive model is known to hold, then the ANOVA table below shows that the main effect of B is not equal to 0, while the main effects of A and C may be equal to 0.

Term	Effect	DF	SS	MS	F	P
A	3.3750	1	22.781	22.781	2.7931	0.170
B	23.625	1	1116.3	1116.3	136.86	0.000
C	1.1250	1	2.5312	2.5312	0.31034	0.607
Error		4	32.625	8.1562		
Total		7	1174.2			

7. (a)

Term	Effect
A	2.445
B	0.140
C	−0.250
AB	1.450
AC	0.610
BC	0.645
ABC	−0.935

(b) No, since the design is unreplicated, there is no error sum of squares.

(c) The estimates lie nearly on a straight line, so none of the factors can clearly be said to influence the resistance.

9. (a)

Term	Effect
A	1.2
B	3.25
C	−16.05
D	−2.55
AB	2
AC	2.9
AD	−1.2
BC	1.05
BD	−1.45
CD	−1.6
ABC	−0.8
ABD	−1.9
ACD	−0.15
BCD	0.8
ABCD	0.65

(b) Factor C is the only one that really stands out.

11. (a)

Term	Effect	DF	SS	MS	F	P
A	14.245	1	811.68	811.68	691.2	0.000
B	8.0275	1	257.76	257.76	219.5	0.000
C	−6.385	1	163.07	163.07	138.87	0.000
AB	−1.68	1	11.29	11.29	9.6139	0.015
AC	−1.1175	1	4.9952	4.9952	4.2538	0.073
BC	−0.535	1	1.1449	1.1449	0.97496	0.352
ABC	−1.2175	1	5.9292	5.9292	5.0492	0.055
Error		8	9.3944	1.1743		
Total		15	1265.3			

(b) All main effects are significant, as is the AB interaction. Only the BC interaction has a P value that is reasonably large. All three factors appear to be important, and they seem to interact considerably with each other.

13. (ii) The sum of the main effect of A and the $BCDE$ interaction.

Supplementary Exercises for Chapter 9

1.

Source	DF	SS	MS	F	P
Gypsum	3	0.013092	0.0043639	0.289	0.832
Error	8	0.12073	0.015092		
Total	11	0.13383			

The value of the test statistic is $F_{3,8} = 0.289$; $P > 0.10$ ($P = 0.832$). There is no evidence that the pH differs with the amount of gypsum added.

3.

Source	DF	SS	MS	F	P
Day	2	1.0908	0.54538	22.35	0.000
Error	36	0.87846	0.024402		
Total	38	1.9692			

We conclude that the mean sugar content differs among the three days ($F_{2,36} = 22.35$, $P \approx 0$).

5. (a) No. The variances are not constant across groups. In particular, there is an outlier in group 1.
(b) No, for the same reasons as in part (a).
(c)

Source	DF	SS	MS	F	P
Group	4	5.2029	1.3007	8.9126	0.000
Error	35	5.1080	0.14594		
Total	39	10.311			

We conclude that the mean dissolve time differs among the groups ($F_{4,35} = 8.9126$, $P \approx 0$).

7. The recommendation is not a good one. The engineer is trying to interpret the main effects without looking at the interactions. The small P-value for the interactions indicates that they must be taken into account. Looking at the cell means, it is clear that if design 2 is used, then the less expensive material performs just as well as the more expensive material. The best recommendation, therefore, is to use design 2 with the less expensive material.

9. (a)

Source	DF	SS	MS	F	P
Base	3	13495	4498.3	7.5307	0.000
Instrument	2	90990	45495	76.164	0.000
Interaction	6	12050	2008.3	3.3622	0.003
Error	708	422912	597.33		
Total	719	539447			

(b) No, it is not appropriate because there are interactions between the row and column effects ($F_{6,708} = 3.3622$, $P = 0.003$).

11. (a) Yes. $F_{4,15} = 8.7139$, $P = 0.001$.
   (b) $q_{5,20} = 4.23$, $MSE = 29.026$, $J = 4$. The 5% critical value is therefore $4.23\sqrt{29.026/4} = 11.39$. The sample means for the five channels are $\overline{X}_1 = 44.000$, $\overline{X}_2 = 44.100$, $\overline{X}_3 = 30.900$, $\overline{X}_4 = 28.575$, $\overline{X}_5 = 44.425$. We can therefore conclude that channels 3 and 4 differ from channels 1, 2, and 5.

13. No. $F_{4,289} = 1.5974$, $P > 0.10$ ($P = 0.175$).

15. (a)

Term	Effect	Term	Effect
A	3.9875	BD	−0.0875
B	2.0375	CD	0.6375
C	1.7125	ABC	−0.2375
D	3.7125	ABD	0.5125
AB	−0.1125	ACD	0.4875
AC	0.0125	BCD	−0.3125
AD	−0.9375	$ABCD$	−0.7125
BC	0.7125		

(b) The main effects are noticeably larger than the interactions, and the main effects for A and D are noticeably larger than those for B and C.

(c)

	Effect	DF	SS	MS	F	P
A	3.9875	1	63.601	63.601	68.415	0.000
B	2.0375	1	16.606	16.606	17.863	0.008
C	1.7125	1	11.731	11.731	12.619	0.016
D	3.7125	1	55.131	55.131	59.304	0.001
AB	−0.1125	1	0.0506	0.0506	0.05446	0.825
AC	0.0125	1	0.0006	0.0006	0.00067	0.980
AD	−0.9375	1	3.5156	3.5156	3.7818	0.109
BC	0.7125	1	2.0306	2.0306	2.1843	0.199
BD	−0.0875	1	0.0306	0.0306	0.0329	0.863
CD	0.6375	1	1.6256	1.6256	1.7487	0.243
Interaction		5	4.6481	0.92963		
Total		15	158.97			

We can conclude that each of the factors A, B, C, and D has an effect on the outcome.

(d) The F statistics are computed by dividing the mean square for each effect (equal to its sum of squares) by the error mean square 1.04. The degrees of freedom for each F statistic are 1 and 4. The results are summarized in the following table.

	Effect	DF	SS	MS	F	P
A	3.9875	1	63.601	63.601	61.154	0.001
B	2.0375	1	16.606	16.606	15.967	0.016
C	1.7125	1	11.731	11.731	11.279	0.028
D	3.7125	1	55.131	55.131	53.01	0.002
AB	−0.1125	1	0.0506	0.0506	0.04868	0.836
AC	0.0125	1	0.0006	0.0006	0.00060	0.982
AD	−0.9375	1	3.5156	3.5156	3.3804	0.140
BC	0.7125	1	2.0306	2.0306	1.9525	0.235
BD	−0.0875	1	0.0306	0.0306	0.02945	0.872
CD	0.6375	1	1.6256	1.6256	1.5631	0.279
ABC	−0.2375	1	0.2256	0.2256	0.21695	0.666
ABD	0.5125	1	1.0506	1.0506	1.0102	0.372
ACD	0.4875	1	0.9506	0.9506	0.91406	0.393
BCD	−0.3125	1	0.3906	0.3906	0.3756	0.573
$ABCD$	−0.7125	1	2.0306	2.0306	1.9525	0.235

(e) Yes. None of the P-values for the third- or higher-order interactions are small.

(f) We can conclude that each of the factors A, B, C, and D has an effect on the outcome.

17. (a)

Source	DF	SS	MS	F	P
H_2SO_4	2	457.65	228.83	8.8447	0.008
$CaCl_2$	2	38783	19391	749.53	0.000
Interaction	4	279.78	69.946	2.7036	0.099
Error	9	232.85	25.872		
Total	17	39753			

(b) The P-value for interactions is 0.099. One cannot rule out the additive model.

(c) Yes, $F_{2,9} = 8.8447, 0.001 < P < 0.01$ ($P = 0.008$). (d) Yes, $F_{2,9} = 749.53, P \approx 0.000$.

Section 10.1

1. (a) Count   (b) Continuous   (c) Binary   (d) Continuous

3. (a) is in control   (b) has high capability

5. (a) False   (b) False   (c) True   (d) True

Section 10.2

1. (a) LCL $= 0$, UCL $= 10.931$ (b) LCL $= 0$, UCL $= 4.721$ (c) LCL $= 20.258$, UCL $= 27.242$
(d) LCL $= 20.358$, UCL $= 27.142$

3. (a) LCL = 0, UCL = 0.2949, the variance is in control.
   (b) LCL = 2.4245, UCL = 2.5855. The process is out of control for the first time on sample 8.
   (c) 1σ limits are 2.4782, 2.5318; 2σ limits are 2.4513, 2.5587. The process is out of control for the first time on sample 7, where two out of the last three samples are below the lower 2σ control limit.

5. (a) 15.27   (b) 15.13   (c) 1.92   (d) 13

7. (a) 0.126   (b) 0.237   (c) 0.582   (d) 256

9. (a) LCL = 0.0163, UCL = 0.1597. The variance is in control.
   (b) LCL = 9.8925, UCL = 10.0859. The process is out of control for the first time on sample 3.
   (c) 1σ limits are 9.9570, 10.0214; 2σ limits are 9.9247, 10.0537. The process is out of control for the first time on sample 3, where one sample is above the upper 3σ control limit.

11. (a) LCL = 0, UCL = 0.971. The variance is in control.
    (b) LCL = 9.147, UCL = 10.473. The process is in control.
    (c) 1σ limits are 9.589, 10.031; 2σ limits are 9.368, 10.252. The process is out of control for the first time on sample 9, where 2 of the last three sample means are below the lower 2σ control limit.

13. (a) LCL = 0, UCL = 6.984. The variance is out of control on sample 8. After deleting this sample, $\overline{\overline{X}} = 150.166$, $\overline{R} = 6.538$, $\overline{s} = 2.911$. The new limits for the S chart are 0 and 6.596. The variance is now in control.
    (b) LCL = 145.427, UCL = 154.905. The process is in control.
    (c) 1σ limits are 148.586, 151.746; 2σ limits are 147.007, 153.325. The process is out of control for the first time on sample 10, where four of the last five sample means are below the lower 1σ control limit.

Section 10.3

1. Centerline is 0.0355, LCL is 0.00345, UCL is 0.06755

3. Yes, the 3σ control limits are 0.0317 and 0.1553.

5. (iv). The sample size must be large enough so the mean number of defectives per sample is at least 10.

7. It was out of control. The UCL is 23.13.

Section 10.4

1. (a) No samples need be deleted.   (b) $\sigma_{\overline{X}} = (0.577)(0.1395)/3 = 0.0268$
   (c) UCL = 0.107, LCL = −0.107   (d) The process is out of control for the first time on sample 8.
   (e) The Western Electric rules specify that the process is out of control for the first time on sample 7.

3. (a) No samples need be deleted.   (b) $\sigma_{\overline{X}} = (0.577)(1.14)/3 = 0.219$ (c) UCL = 0.877, LCL = −0.877
   (d) The process is out of control for the first time on sample 9.
   (e) The Western Electric rules specify that the process is out of control for the first time on sample 9.

5. (a) UCL = 60, LCL = −60   (b) The process is in control.

Section 10.5

1. (a) $C_{pk} = 2.303$ (b) Yes. Since $C_{pk} > 1$, the process capability is acceptable.

3. (a) 0.20   (b) 3.071

5. (a) $\mu \pm 3.6\sigma$ (b) 0.0004   (c) Likely. The normal approximation is likely to be inaccurate in the tails.

Supplementary Exercises for Chapter 10

1. Centerline is 0.0596, LCL is 0.0147, UCL is 0.1045.

3. (a) LCL = 0, UCL = 0.283. The variance is in control.
(b) LCL = 4.982, UCL = 5.208. The process is out of control on sample 3.
(c) 1σ limits are 5.057, 5.133; 2σ limits are 5.020, 5.170. The process is out of control for the first time on sample 3, where a sample mean is above the upper 3σ control limit.

5. (a) No samples need be deleted.   (b) $\sigma_{\overline{X}} = (1.023)(0.110)/3 = 0.0375$ (c) UCL = 0.15, LCL = −0.15
(d) The process is out of control on sample 4.
(e) The Western Electric rules specify that the process is out of control on sample 3.

7. LCL = 0.0170, UCL = 0.0726   (b) Sample 12
(c) No, this special cause improves the process. It should be preserved rather than eliminated.

INDEX

2^3 factorial experiment
  analysis of variance table, 452
  effect estimates, 450
  effect sum of squares, 451
  error sum of squares, 451
  estimating effects, 448
  F test, 452
  hypothesis test, 452
  notation, 448
  sign table, 449
2^p factorial experiment
  effect estimates, 455
  effect sum of squares, 455
  error sum of squares, 455
  F test, 455
  sign table, 456
  without replication, 455

A

Addition rule for probabilities, 73
Additive model, 424
Adjusted R^2, 351, 370
Aliasing, 461
Alternate hypothesis, 213
Analysis of variance
  one-way, *see* One-way analysis of variance
  two-way, *see* Two-way analysis of variance
Analysis of variance identity
  for multiple regression, 347
  for one-way analysis of variance, 408
Analysis of variance table
  in 2^3 factorial experiment, 452
  in multiple regression, 350–351
  in one-way analysis of variance, 404
  in simple linear regression, 327–328
  in two-way analysis of variance, 426
ARL, *see* Average run length
Assignable cause, 478
Average, 11
Average run length, 487

B

Backward elimination, 371
Balanced design, 407, 422
Bayesian statistics, 223
Bernoulli trial, 119
Best subsets regression, 368
Bias, 174
Binomial distribution, 119–124
  mean, 124
  normal approximation to, 164
  probability mass function, 121
  variance, 124
Bonferroni method, 263
Boxplot, 26–30
  comparative, 27, 28
  representing outliers in, 26

C

c chart, 501
  control limits for, 502
C_p, *see* Mallows' C_p *see* Process capability index
C_{pk}, *see* Process capability index
Cell mean, 425
Central limit theorem, 161–168
  for binomial distribution, 164
  for Poisson distribution, 167
  for sample mean, 161
  for sum, 161
  sample size needed for validity, 162
Chance cause, 477
Chi-square distribution, 240
  degrees of freedom for, 240, 243
  special case of gamma distribution, 153
Chi-square statistic, 240
Chi-square test
  for homogeneity, 241–244
  for independence, 244
  for specified probabilities, 239–241
Coefficient of determination

and proportion of variance explained
  by regression, 60
 in multiple regression, 348
 in simple linear regression, 60
Column effect, 423
Column factor, 421
Column mean, 425
Column sum of squares, 426
Common cause, 477
Complete design, 421, 442
Completely randomized experiment, 398
Conditional probability, 76
Confidence bound, 186, 191, 201
Confidence interval
 comparision with prediction interval, 206
 confidence level, *see* Confidence level
 determining sample size, 184, 192
 difference between means, 269, 285
 difference between proportions, 277, 278
 for coefficients in multiple regression, 352
 for mean, 180, 198
 for mean response, 323
 for proportion, 190, 193
 for slope and intercept, 320
 one-sided, 186, 191, 201
 paired data, 296
 relationship to hypothesis tests, 225–226
 small sample, 198, 285
 Student's t distribution and, 198, 285
Confidence level, 178
 and probability, 182–184
 interpretation of, 182
Confounding, 43
 and controlled experiments, 47
 and observational studies, 47
Contingency table, 241
Continuity correction, 164–167
 accuracy of, 166
 for Poisson distribution, 167
Continuous random variable, 86, 95–102
 cumulative distribution function of, 98
 mean of, 100
 probability density function of, 96
 standard deviation of, 101
 variance of, 100, 101

Control chart
 c chart, *see* c chart
 CUSUM chart, *see* CUSUM chart
 for attribute data, *see* p chart
 for binary data, *see* p chart
 for count data, *see* c chart
 for variables data, *see* $\overline{X}$ chart
 p chart, *see* p chart
 R chart, *see* R chart
 S chart, *see* S chart
 $\overline{X}$ chart, *see* $\overline{X}$ chart
Control limits
 for c chart, 502
 for p chart, 500
 for S chart, 492
 for R chart, 483
 for $\overline{X}$ chart, 485, 493
Controlled experiment, 9
 reduces the risk of confounding, 47
Correlated, 40
Correlation, 37–47
 is not causation, 43
 population, 328
 sample, 328
Correlation coefficient, 39
 and outliers, 43
 and proportion of variance explained
  by regression, 60
 how it works, 40
 measures goodness-of-fit, 60
 measures linear association, 41
Critical point, 249
Critical value, 178
Cumulative distribution function
 continuous, 98
 discrete, 90
Cumulative sum, 505
Cumulative sum chart, *see* CUSUM chart
CUSUM chart, 505

D

Data
 categorical, 9
 numerical, 9

Data (*cont.*)
  qualitative, 9
  quantitative, 9
Dependent variable, 49, 313, 397
Descriptive statistics, 2
Discrete random variable, 87–95
  cumulative distribution function of, 90
  mean of, 91
  probability mass function of, 88, 90
  standard deviation of, 93
  variance of, 93
Dotplot, 21

E

Effect sum of squares
  in 2^3 factorial experiment, 451
  in 2^p factorial experiment, 455
Empirical model, 340, 341
Erlang distribution, 153
Error mean square
  in one-way analysis of variance, 402
  in two-way analysis of variance, 427
Error sum of squares
  in 2^3 factorial experiment, 451
  in 2^p factorial experiment, 455
  in multiple regression, 347
  in one-way analysis of variance, 401
  in simple linear regression, 60
  in two-way analysis of variance, 426
Errors
  in one-way analysis of variance, 408
  in simple linear regression, 313
Event(s), 67
  complement of, 68
  independent, 77
  intersection of, 68
  multiplication rule for independent events, 78
  mutually exclusive, 68
  union of, 68
Expectation, *see* Population mean
Expected value, *see* Population mean
Exponential distribution, 147–150
  cumulative distribution function, 148
  lack of memory property, 150

  mean, 148
  probability density function, 147
  relationship to Poisson process, 148
  variance, 148

F

F distribution, 304
  degrees of freedom for, 304
F test
  for equality of variance, 303–307
  in 2^3 factorial experiment, 452
  in 2^p factorial experiment, 455
  in multiple regression, 366
  in one-way analysis of variance, 403
  in two-way analysis of variance, 427
Factor, 396
Factorial experiment, 396
  2^3 design, *see* 2^3 factorial experiment
  2^p design, *see* 2^p factorial experiment
  fractional, *see* Fractional factorial experiment
Failure to detect, 487
False alarm, 487
Fitted value, 49
Fixed effects model, 398, 409
  for two-factor experiments, 435
Fixed-level testing, 248–251
Forward selection, 371
Fractional factorial experiment, 459–462
  aliasing in, 461
  half-replicate, 459
  principal fraction, 460
  quarter-replicate, 459
Frequency table, 22
Frequentist probability, 223
Full factorial design, 421

G

Gamma distribution, 152
Gamma function, 152
Gaussian distribution, *see* Normal distribution
Goodness-of-fit, 60
Gosset, William Sealy (Student), 130, 195
Grand mean
  population, 409, 422, 423
  sample, 399, 425

H

Half-replicate, 459
Hazard function, 156
Heteroscedastic, 336
Histogram, 22–25
    bimodal, 24
    class intervals, 22
    skewed, 24
    symmetric, 24
    unimodal, 24
Homoscedastic, 336
Honestly significant difference, 416
Hypothesis test
    Bonferroni method, 263
    Chi-square test, *see* Chi-square test
    choosing null hypothesis, 223–224
    critical point, 249
    F test, *see* F test
    fixed-level test, 248
    for difference between means, 272, 289
    for difference between proportions, 281
    for least-squares coefficients in multiple
        regression, 352
    for mean, 219, 237
    for proportion, 232
    for slope and intercept, 321
    in 2^3 factorial experiment, 452
    in one-way analysis of variance, 403
    in two-way analysis of variance, 427
    multiple testing problem, 262–264
    one-tailed, 218
    P-value, *see* P-value
    power, *see* Power
    rejecting null hypothesis, 213, 222, 249
    rejection region, 249
    relationship to confidence
        intervals, 225–226
    significance level, 249
    steps in performing, 215
    t test, *see* Student's t test
    two-tailed, 218
    type I error, 251
    type II error, 251
    with paired data, 298

I

i.i.d., *see* Independent and identically distributed
Independent and identically distributed, 110
Independent Bernoulli trials, 120
Independent events, 77
    multiplication rule for, 78
Independent random variables, 108
Independent variable, 49, 313
Inferential statistics, 2
Influential point, 58
Interaction
    and interpretation of main effects, 431
    in multiple regression, 346
    in two-way analysis of variance, 423
    sum of squares, 426
Interaction mean square, 427
Interaction sum of squares, 426
Intercept
    confidence interval for, 320
    hypothesis test for, 321
Interquartile range, 26
Interval estimate
    see Confidence interval, 175

L

Lack of memory property, 150
Least-squares coefficients, 49, 51, 313
    in multiple regression, 346
    normally distributed, 317
    relationship to correlation coefficient, 58
    standard deviations of, 317
    unbiased, 317
Least-squares line, 49–53, 56–61, 313
    computing, 50
    don't extrapolate, 56
    don't use when data aren't linear, 57
    goodness-of-fit of, 59
Level of a hypothesis test, *see* Significance level
Levels of a factor, 397
Linear combination of random variables, 107
    mean of, 107
    variance of, 109
Linear model, 313

Lognormal distribution, 143–145
  mean, 145
  outliers, 145
  probability density function, 144
  relationship to normal, 143–144
  use of z table with, 145
  variance, 145

M

Main effect, 423
  interpretation of, 431
Mallows' C_p, 371
Margin of error, 178
Mean
  cell, 425
  column, 425
  grand, *see* Grand mean
  population, *see* Population mean
  row, 425
  sample, *see* Sample mean
Mean response, 323
  confidence interval for, 323
Mean square
  for error, *see* Error mean square
  for interaction, *see* Interaction mean square
  for treatment, *see* Treatment mean square
Median
  sample, 14
  population, 134
Mixed model, 435
Mode
  of a histogram, 24
Model selection, 362–375
  art not science, 375
  Occam's razor, 363
  principle of parsimony, 363
Multinomial trial, 239
Multiple comparisons
  simultaneous confidence intervals, 416
  simultaneous hypothesis tests, 416
  Tukey-Kramer method, 416
Multiple regression
  analysis of variance table, 350, 351
  assumptions in, 347

F test, 366
least squares coefficients, 346
model selection, *see* Model selection
multiple regression model, 346
sums of squares, 347
Multiplication rule for probabilities, 78
Mutually exclusive events, 68

N

Normal approximation
  to binomial distribution, 164
  to Poisson, 167
Normal distribution, 134–142
  mean, 134
  median, 137
  outliers, 142
  probability density function, 134
  standard deviation, 134
  standard normal population, 135
  standard units, 135
  variance, 134
  z-score, 135
Null hypothesis, 213
  choosing, 223–224
  in one-way analysis of variance, 399
  in two-way analysis of variance, 424–425
  put on trial, 213
  rejecting, 213, 222, 249

O

Observational study, 9
  and confounding, 47
Observed significance level, 215
Occam's razor, 363
One-factor experiment, 397
One-way analysis of variance
  analysis of variance identity, 408
  analysis of variance table, 404
  assumptions in, 402
  error sum of squares, 401
  F test, 403
  fixed effects model, 409
  hypothesis test, 403

null hypothesis, 399
    random effects model, 409
    total sum of squares, 408
    treatment sum of squares, 400
Outcome variable, 397
Outlier, 14
    and simple linear regression, 57–58
    and Student's *t* distribution, 198
    and the correlation coefficient, 43
    and use of median, 15
    deletion of, 14
    extreme, 26
    in boxplots, 26
    lognormal distribution, 145
    normal distribution, 142

P

p chart, 499
    control limits for, 500
P-value, 215
    interpreting, 222, 225
    not the probability that H_0 is true, 223
Parameter, 173
Parsimony, 363
Percentile
    sample, 16
Physical law, 341
Point estimate, 173
Poisson distribution, 127–132
    approximation to binomial, 129
    mean, 129
    normal approximation to, 167
    probability mass function, 129
    variance, 129
Poisson process, 148
Polynomial regression model, 346
Pooled standard deviation, 286, 290
Population, 3
    conceptual, 6
    tangible, 5
Population correlation, 328
Population mean
    of a continuous random variable, 100
    of a discrete random variable, 91

Population proportion
    confidence interval for, *see* Confidence interval
    Hypothesis test for, *see* Hypothesis test
Population standard deviation
    of a continuous random variable, 101
    of a discrete random variable, 93
Population variance
    of a continuous random variable, 100, 101
    of a discrete random variable, 93
Power, 253, 259
    depends on alternate hypothesis, 255
    determining sample size, 256
    steps in computing, 253
Power transformation, 337
Prediction interval, 204–206
    comparison with confidence interval, 206
    in linear regression, 326
    one-sided, 206
    sensitive to departures from normality, 206
Principal fraction, 460
Principle of parsimony, 363
    exceptions to, 363
Probability
    addition rule, 73
    axioms of, 69
    conditional, 76
    frequency interpretation of, 68
    frequentist, 223
    multiplication rule, 78
    subjective, 223
    unconditional, 74
Probability density function, 96
Probability distribution, 88, 96
Probability distributions
    Binomial, *see* Binomial distribution
    Chi-square, *see* Chi-square distribution
    Exponential, *see* Exponential distribution
    F, *see* F distribution
    Gamma, *see* Gamma distribution
    Gaussian, *see* Normal distribution
    Lognormal, *see* Lognormal distribution
    Normal, *see* Normal distribution
    Poisson, *see* Poisson distribution
    t, *see* Student's *t* distribution
    Weibull, *see* Weibull distribution

Probability histogram, 94
Probability mass function, 88, 90
Probability plot, 157–160
    interpreting, 160
    to detect effects in factorial experiments, 458
Process capability, 507–511
    versus process control, 479
Process capability index
    C_p, 509
    C_{pk}, 508
    C_{pl}, 511
    C_{pu}, 511
Propagation of error formula
    multivariate, 113
    results only approximate, 112
    univariate, 112
Proportion
    confidence interval for, *see* Confidence interval
    hypothesis test for, *see* Hypothesis test

Q

QQ plot, 159
Quadratic regression model, 346
Quantile-Quantile plot, 159
Quarter-replicate, 459
Quartile, 15
    first quartile, 15
    second quartile, 15
    third quartile, 15

R

R^2, *see* Coefficient of determination
R chart
    comparison with S chart, 494
    control limits for, 483
    steps for using, 486
Random effects model, 398, 409
    for two-factor experiments, 435
Random sample, *see* Simple random sample
Random variable
    continuous, *see* Continuous random variable
    discrete, *see* Discrete random variable

independent, 108
    linear combination of, *see* Linear
        combination of random variables
    sum of, *see* Sum of random variables
Randomization within blocks, 442
Randomized complete block design, 441, 445
Rational subgroups, 478
Regression coefficients, 313
    confidence intervals for, 320
    hypothesis tests for, 320
Regression equation, 346
Regression line, 313
Regression sum of squares
    in multiple regression, 347
    in simple linear regression, 60
Rejection region, 249
Reliability analysis, 79
Residual
    in one-way analysis of variance, 401
    in simple linear regression, 49
Residual plot, 335
    in multiple regression, 353
    interpreting, 336, 338
Response variable, 397
Row effect, 423
Row factor, 421
Row mean, 425
Row sum of squares, 426

S

S chart
    comparison with R chart, 494
    control limits for, 492
Sample, 3
    cluster, 9
    of convenience, 4
    simple random, 3
    stratified random, 9
Sample mean, 11
    central limit theorem for, 161
    mean of, 111
    standard deviation of, 111
    variance of, 111
Sample median, 14

Sample space, 66
  with equally likely outcomes, 71
Sample standard deviation, 11–13
Sample variance, 12
Sampling variation, 5
Sampling with replacement, 8
Scatterplot, 38
Sign table, 449, 456
Significance level, 249
Simple linear regression
  analysis of variance table, 327, 328
  and outliers, 57–58
  assumptions in, 316, 335
  plot of residuals versus time, 339
  transformation of variables in, 337–338
Simple random sample, 3
Simultaneous confidence intervals,
    see Multiple Comparisons
Simultaneous hypothesis tests,
    see Multiple Comparisons
Six-sigma quality, 510–511
Slope
  confidence interval for, 321
  hypothesis test for, 320
Special cause, 478
Specification limit, 508
Standard deviation
  population, *see* Population standard
    deviation
  sample, *see* Sample standard deviation
Standard error, 178
Standard error of the mean, 17
Standard normal population, 135
Standard units, 135
Statistic, 173
Statistical significance, 222–223, 249
  not the same as practical significance, 224
Stem-and-leaf plot, 20–21
Stepwise regression, 371
Student's t distribution, 195–198
  and outliers, 198
  and sample mean, 196
  confidence intervals using, 198, 285
  degrees of freedom for, 195, 285, 286, 289
  in hypothesis testing, *see* Student's t test

Student's t test
  one-sample, 237
  two-sample, 289
Studentized range distribution, 416
Subjective probability, 223
Sum of random variables
  central limit theorem for, 161
  mean of, 107
  variance of, 109
Sum of squares
  for columns, *see* Column sum of squares
  for error, *see* Error sum of squares
  for interaction, *see* Interaction sum of squares
  for rows, *see* Row sum of squares
  for treatment, *see* Treatment sum of squares
  total, *see* Total sum of squares
Summary statistics, 11–18

T

t distribution, *see* Student's t distribution
t test, *see* Student's t test
Test of significance, *see* Hypothesis test
Test statistic, 215
Tests of hypotheses, *see* Hypothesis test
Tolerance interval, 207
Total sum of squares
  in multiple regression, 347
  in one-way analysis of variance, 408
  in simple linear regression, 60
  in two-way analysis of variance, 426
Transforming variables, 337–338
Treatment effect, 409
Treatment mean square, 402
Treatment sum of squares, 400
Treatments, 397
Tukey's method, 416
  in one-way ANOVA, *see* Tukey-Kramer
    method
Tukey-Kramer method, 416
Two-factor experiment
  fixed effects model, 435
  mixed model, 435
  random effects model, 435

Two-way analysis of variance
  additive model, 424
  analysis of variance table, 426
  assumptions in, 426
  balanced design, 422
  cell mean, 425
  column effect, 423
  column factor, 421
  column mean, 425
  column sum of squares, 426
  complete design, 421
  error sum of squares, 426
  F test, 427
  full factorial design, 421
  hypothesis tests, 427
  interaction, 423
  interaction sum of squares, 426
  interpretation of main effects, 431
  main effect, 423
  mean squares, 427
  null hypothesis, 424–425
  one observation per cell, 435
  row effect, 423
  row factor, 421
  row mean, 425
  row sum of squares, 426
  total sum of squares, 426
Type I error, 251
Type II error, 251

U
Unconditional probability, 74
Uncorrelated, 40

V
Variance
  population, *see* Population variance
  sample, *see* Sample variance

W
Waiting time, 147
Weibull distribution, 153–156
  cumulative distribution function, 154
  mean, 155
  probability density function, 154
  variance, 155
Western Electric rules, 491

X
$\overline{X}$ chart
  control limits for, 485, 493
  steps for using, 486

Z
z test, 215, 272